A Guide to Vertebrate Development

A Guide to Vertebrate Development

SEVENTH EDITION

ROBERTS RUGH
Professor of Radiology (Biology)—Retired
College of Physicians and Surgeons
Columbia University

MACMILLAN PUBLISHING COMPANY
New York
COLLIER MACMILLAN PUBLISHERS
London

COVER: The cover design consists of actual photographs of five vertebrate embryos or fetuses at comparable stages of development. It is designed to show how difficult it would be to distinguish between turtle, chick, mouse, pig, and human in early development, suggesting that development is basically similar in all vertebrates. For this reason extrapolation of data from one vertebrate to another, in comparable stages of development, is valid; also, for this reason, research data from one species can suggest what to expect from another, even the human. They all go through the same pattern of development but on different time scales (for example, 20 days for the mouse and 266 days for the human).

Printed in the United States of America

Macmillan Publishing Company
866 Third Avenue, New York, New York 10022

Collier Macmillan Canada, Inc.

Printing: 11 12 Year: 7 8 9 0

ISBN 0-02-404491-1

PREFACE

SEVENTH EDITION

This seventh edition of the *Guide to Vertebrate Development* has resulted from a specific request by the publishers that the manual include more of the experimental possibilities with vertebrate embryos and fetuses. The author has published an *Experimental Embryology,* which lists over 50 procedures for those students who have had a sufficient foundation in normal ontogeny and can understand and evaluate the responses of a developing embryo to imposed variations in its environment. Much of normal development has been learned through experimental procedures, and it is proper that the beginning student be introduced to the wide range of possibilities of experimenting with embryos and fetuses and the significance of the results.

In addition, many illustrations have been more fully labelled, some illustrations have been added, and enough information on human development has been included to indicate that the human is a vertebrate and a mammal that develops basically along lines of all vertebrates. This in no way suggests that an individual is not potentially a human from the very moment of fertilization, but rather that the basic processes of development are the same for all animals. Few institutions have human material to study, but it is well that the seventh presentation of this *Guide* shows one of the purposes of studying vertebrate development, namely that one can thereby understand one's own development. There is no question but that the 9 months prior to birth are more important in a person's whole life than the 70, 80, or even 90 years that may follow. For those interested, the author has written a book on human development entitled *From Conception to Birth: The Drama of Life's Beginnings,* with colored photographic illustrations.

The author wishes here to thank the thousands of students who have used this *Guide* each year for their helpful comments, many of which have been incorporated into this edition; the teachers who have found the *Guide* a satisfying teaching tool; and the many practicing physicians who received their introduction to human development through prior editions of this *Guide.*

Roberts Rugh

SIXTH EDITION

New editions of textbooks or manuals are often used as a device to stimulate lagging sales. The publisher reports that this manual is currently at an all-time high for adoptions. In spite of this the author has, in the light of some constructive suggestions that have been offered by users, felt a need for a revised edition. A questionnaire soliciting suggestions was sent out to those embryology professors using this manual, and the response has been most gratifying and helpful. This sixth edition is, therefore, the result of a deliberate effort to make the manual more useful, with the help of those who have had direct experience with it.

In addition to correcting a few typographical errors that persisted, we have made the following changes: First, we have added a complete series of photomicrographs of the 18-, 24-, 33-, 48-, 72-, and 96-hour chick embryo and the 6 mm and 10 mm pig, plus a sagittal section series of the 6 mm pig. These are excellent photographs of actual specimens generously supplied by Dr. Allan Scott of Colby College. Some of these pictures were taken by an especially talented student of his. By not reducing these pictures to their limit we are providing an atlas of development of these two forms that should help every student to find a section comparable to the one he proposes to study. No labels are applied, as this would detract from and, in some cases, deface the excellent pictures. Supplementary and specially selected sections are presented as drawings. Thus, the major new contributions are these photographs of the normal development of chick and pig. Second, all drawings that could be improved have been thus changed, and the labels made more legible. This has been made possible by a bioartist, Mrs. Rhoda Van Dyke, and has resulted in both uniformity and excellence. Third, the answers to the questionnaire indicated that the review questions at the end of each section were rarely used, and they have therefore been deleted. They were meant to be helpful and suggestive, but it seems obvious that each instructor would prefer to organize his own review questions. Fourth, while in previous editions spaces had been left for specific drawings which the author felt would be important in the teaching process, here again each instructor has his own ideas and plans, which are often different. In addition to a few such blank spaces, a blank page or two has been provided in most chapters where the instructor can ask for drawings of his choice to be made. Fifth, the mouse section has been expanded somewhat in light of the author's conviction that soon the mouse will supplant the pig in the teaching of early mammalian embryology. A full text on normal mouse embryology is now available for reference. Sixth, all through the text there have been changes in construction aimed at greater clarity, and the glossary (reported to be very useful) has been severely edited to make it more succinct and accurate. Thus, we hope that the composite of this new edition will assist the student in becoming familiar with and stimulated by the marvels of normal development.

CONTENTS

GENERAL INTRODUCTION

There are at least three possible reasons for any college student to study embryology. First, a properly balanced course will give him experience with a dynamic scientific discipline through an understanding of the mechanisms of development which apply to all higher animals; second, such a course is basic to experimental embryology and will prepare him for advanced study in the field; third, it will give him an understanding of human development, particularly as it relates to medical studies. There is also, for every human being, the possibility that he will be personally very much interested and concerned with the changes taking place in his own offspring from the fertilization of the ovum to the delivery of the child some 266 days later. Understanding life processes always engenders appreciation, and no aspect of life is more exciting to contemplate than those intricate changes that occur between fertilization of the ovum measuring 1/175 inch and the 7- or 8- pound child whose life is the result of the biological cooperation of two individuals. It is quite probable that more important biological events occur in the first 9 months after conception than in the 70, 80, or even 90 years that follow.

Since it is unethical and illegal to experiment with the human embryo or fetus, specimens that do become available are often imperfect. True, Dr. Arthur Hertig and Dr. John Rock of Harvard long ago arranged with married women scheduled for hysterectomies (removal of the uterus) for them to become pregnant before the operation and thus provide a limited number of perfect specimens from which much of our knowledge of human development has been obtained. But this guide is designed to demonstrate, by studying the development of four vertebrates (two of them mammals), that all higher forms develop along basically similar lines and the complete understanding of one will aid substantially in understanding all vertebrate development. At certain stages the embryos of all of these forms have similar characteristics and could be mistaken for one another related to evolutionary variations in ultimate structure. Only as one becomes familiar with the detailed development of a particular species will he appreciate the variations that do occur.

The maturation of germ cells, the mechanics of fertilization, and the processes of cleavage are treated here in a rather general manner, followed by detailed studies of the frog, chick, pig, and mouse. A section on Amphioxus is included because it is a prechordate and yet shows developmental similarities to higher vertebrates. A short section on fish is also offered because, thanks to its transparent form, one can observe internal development without sacrificing for histological study. These sections may be omitted by the instructor, but some students will be interested in comparative embryology and these accessory studies will be pertinent. Any student who completes this course should be able to appreciate and understand any human specimen or section of human embryo or fetus without further experience, should the opportunity arise. A short summary of human development is included, with illustrations, to demonstrate its basic similarity to all vertibrate development.

The study of embryology trains the student to reconstruct, in his mind, the three-dimensional embryo or fetus from dissections and serial sections mounted on slides. The microtome knife is a marvelous invention which serializes all the parts of the embryo or fetus, from one end to the other or from one side to the other, so that the student can visualize by direct examination all of the inner structures as they develop. Any embryo, or three-dimensional object, can be sliced (sectioned) in any of three planes—frontal, sagittal, or transverse—so that any and all organs can be studied from three points of view. In this way, exact dimensions can be determined. Usually the sections are of uniform thickness (6,10, or 15 microns) and are serially mounted on a slide or series of slides. This fact can be utilized in determining the three-dimensional size of any organ simply by counting the number of sections of known thickness in which it appears and measuring in two directions the

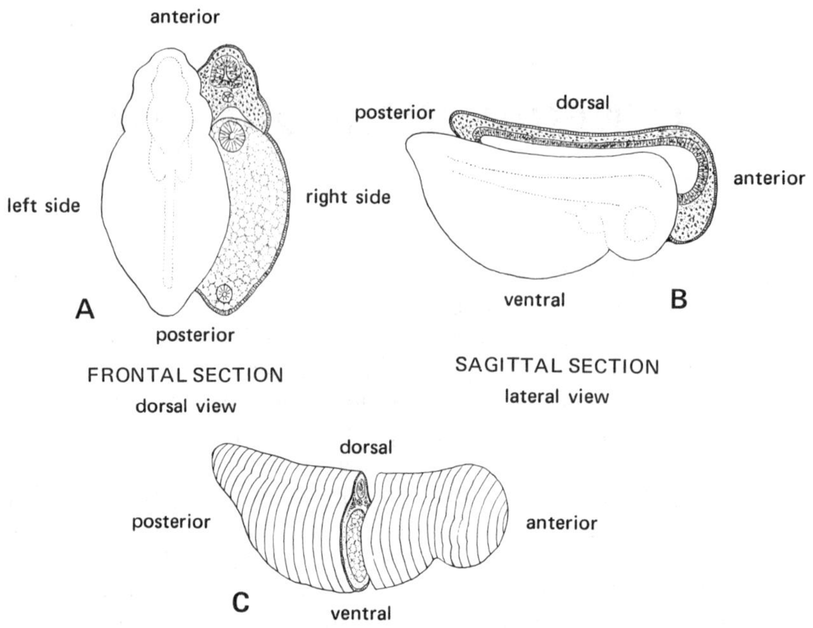

TRANSVERSE SECTION

LATERAL VIEW

largest single transverse section. Usually the most anterior or most dorsal section is mounted on the upper left-hand corner of the slide, and the series is mounted as if they were words on a printed page. If they are mounted upside down, they will appear properly oriented through the microscope, which inverts and reverses all images. Each student will quickly learn to move from left to right, examining each serial section and mentally recording changes in shape and size of the various organs as they appear and disappear. The two-dimensional sections on the slide are reconstructed as the three-dimensional embryo in the student's mind. It should be remembered that perfect sections are technically difficult to produce, so that the illustrations in this manual should be considered as probably more ideal than many of the sections which will actually be studied by the student. While the microtome dismembers, the student must put the parts together in his mind, always remembering when and where each vital organ, tissue, or cell type is appearing in the mosaic of development. It is a rare student who cannot do this, and the great majority gain a new kind of satisfaction from acquiring an indelible memory of all the minute inner parts as they form and integrate into a properly functioning and most complex organism.

To accomplish this end we have provided many drawings and photomicrographs of the various embryos to be studied. Some of them are labelled, and all are designed to help the student in this mental exercise of reconstruction. In addition, the instructor is encouraged to include living embryos or pregnant animals, which each student will be able to examine and study directly, to visualize the study of the prepared slides.

While this Guide is designed as an introduction to vertebrate embryology, most students will find that the characteristic events of development, such as fertilization, cleavage, gastrulation, cephalization, and organogenesis, never lose their fascination as they are viewed in any living embryo. To this author there remain few wonders in life as exciting to watch as the first cleavage of the recently fertilized ovum in the knowledge that it is on its way to becoming a free-living organism of great complexity; or the birth of a child after 9 months in a watery chamber when it gasps its first breath, expanding its lungs, and crying out to let the mother, doctor, and nurses know that all is well. The embryologist appreciates these events because he knows their significance. But, probably most of all, he knows what it means to develop from the single-celled zygote and better appreciates his good fortune in being alive and surviving to his present state, so as to be able to study the marvelous integrations of development.

LABORATORY EQUIPMENT

Organizing and programming vary from department to department so that only very general suggestions relating to basic equipment and procedures will be made here.

MICROSCOPES

Usually the embryology student has been schooled in the use of the microscope in prerequisite or preliminary courses. But, since the microscopes used in embryology are generally of two kinds and are often of a finer quality than those used in the preliminary courses, a few reminders may be offered.

The dissecting microscope has a low-power range of magnification, so it is useful for examining whole and living embryos, for certain experimental procedures with the living material, and for orienting the student in the arrangement of serial sections. Such a microscope is necessary in experimental embryology. The standard compound microscope, either monocular or binocular, will be used frequently by each student and is usually equipped with 10X ocular(s) and both 10X and 44X objectives. Students should be cautioned that the focal distance of the 44X objective is such that, in focusing, it should be lowered cautiously to a position very close to the slide before looking through the ocular, particularly with those slides on which there is a mount of considerable thickness. It should NOT be assumed that the objectives are parfocal, meaning that they can be rotated and will each be in focus without adjustment. It is standard practice to lower the higher objective to a position close to the slide by observing it from the side and then, while looking through the tube of the microscope, to RAISE the tube until the object comes into focus. If focusing while observing through the microscope is always upward, damage to the slides, particularly to the thicker ones, can be avoided.

Most microscopes come with a reversible reflecting mirror beneath the stage, one side with a flat and the other with a concave mirror. The best source of light is northern skylight, but there are many varieties of miscroscope lamps with adjustable diaphragms that are quite satisfactory, and some microscopes come with built-in lamps. In general the concave surface will be used with artificial light, but either side of the mirror can be so adjusted as to get a dark-field effect of illumination. Obviously, most examinations will be made under the lower objective magnification (10X), and slide after slide may be examined with very little adjustment of the microscope lens. With a 10X ocular and a 10X objective the magnification is 100 diameters, and with the 44X objective it is 440 diameters. There are, of course, much higher magnifications (up to the electron microscope), but these lower magnifications are adequate for embryological studies. There is usually a diaphragm beneath the stage of the microscope by means of which the aperture for the reflected light may be altered. In using the low-power objective it will often be found that the best light is not necessarily the most. With the higher magnification more light will be needed.

If binoculars are available on the microscope, the student will have to see that both oculars are adjusted so so to fit his eyes and that each is focused correctly before he begins his study. Once adjusted, they should not be altered. With the monocular microscope, the student must learn to observe through the microscope with one (his best) eye, leaving the other open at all times but "without looking" or focusing on anything within his field of sight. This may sound facetious, but it is a necessary trick to learn, else there will be headaches due to the squinting of the unused eye. At the beginning, the student could cover his unused eye with a card, hand, patch, or piece of cloth, but leave it open. It is possible to have the eyes relaxed and use the microscope for hours at a time without discomfort. If drawings are required, the student can learn to look with one and draw by using his other eye, achieving visual magnification in exact proportions from slide to drawing. Also, one can learn to manipulate the fine adjustment of the microscope with one hand and the slide on the stage with the other. These little tricks will help, particularly when serial sections of embryos are being studied. At the close of each laboratory, the tube is raised a bit and the objectives of each microscope should be placed in a neutral position to avoid damage to them. It would be considerate for each student to leave the lenses, both ocular and objective, as clean as he found them for use by the next student, using soft lens paper for this purpose.

SLIDES

Good slides of embryonic material are particularly difficult to obtain. Embryos must be secured in the healthy living condition, killed and fixed at precise moments in development, and properly sectioned and stained. They are generally sectioned in such a way as to show the student all structures from one end to the other, one side to the other, or from front to back. To prepare such a slide without losing a single section is an art whose teacher is experience. Small embryos can often be sectioned and mounted entirely on a single slide, even though there may be 30 to 40 sections. Large embryos, or late fetuses (of mice, for example), may require hundreds of slides for mounting, and each slide must be completely identified as to the species, age, direction of sectioning, stain, and number of the slide in the series. Thus, if a single slide of a series is damaged, it may nullify the usefulness of the entire series. For this reason, students are cautioned to handle the slides only at the uncovered end and never to put fingers or any pressure on the coverslip. Ideally, a complete set of slides should be issued for each student, or at least for each laboratory desk, to be used by the students occupying each such desk. It is suggested that at the beginning of the semester each student examine every slide assigned to him and inform the instructor should he find any damaged or missing slides. The instructor will expect each student to return the collection of slides at the end of the semester in the same condition in which he received them at the beginning of his study.

GENERAL TEXTBOOKS OF EMBRYOLOGY

AREY, L. B. 1965. *Developmental anatomy.* 4th ed. Philadelphia: W. B. Saunders Company.

BALINSKY, B. I. 1970. *An introduction to embryology.* 3rd ed. Philadelphia: W. B. Saunders Company.

BALLARD, W. W. 1964. *Comparative anatomy and embryology.* New York: The Ronald Press Company.

BARTH, L. G. 1953. *Embryology.* Rev. ed. New York: Holt, Rinehart & Winston.

BODEMER, C. W. 1968. *Modern embryology.* New York: Holt, Rinehart & Winston.

DALCQ, A. M. 1957. *Introduction to general embryology.* London: Oxford University Press.

HAMILTON, H. L. 1952. *Lillie's development of the chick.* New York: Holt, Rinehart & Winston.

HAMILTON, W. J., BOYD, J. D., and MOSSMAN, H. W. 1962. *Human embryology.* Baltimore: The Williams & Wilkins Company.

HUETTNER, A. F. 1953. *Fundamentals of comparative embryology of the vertebrates.* New York: McGraw-Hill Book Company.

McEWEN, R. S. 1957. *Vertebrate embryology.* 4th ed. New York: Holt, Rinehart & Winston.

NELSEN, O. E. 1953. *Comparative embryology of the vertebrates.* New York: McGraw-Hill Book Company.

PATTEN, B. M. 1948. *Embryology of the pig.* Philadelphia: The P. Blakiston Company.

———. 1957. *Early embryology of the chick.* Philadelphia and New York: The P. Blakiston Company and McGraw-Hill Book Company.

———. 1964. *Foundations of embryology.* New York McGraw-Hill Book Company.

———. 1968. *Human embryology.* Philadelphia and New York: The P. Blakiston Company and McGraw-Hill Book Company.

RUGH, R. 1951. *The frog: Its reproduction and development.* Philadelphia and New York: The P. Blakiston Company and McGraw-Hill Book Company.

———.1962. Experimental *embryology: Techniques and procedures.* Minneapolis: Burgess Publishing Company.

________. 1964. *Vertebrate embryology: The dynamics of development.* New York: Harcourt, Brace & World.
________. 1968. *The mouse: Its reproduction and development.* Minneapolis: Burgess Publishing Company.
SHUMWAY, W. 1946. *Introduction to vertebrate embryology.* New York: John Wiley & Sons.
TORREY, T. W. 1967. *Morphogenesis of the vertebrates.* New York: John Wiley & Sons.
WADDINGTON, C. H. 1956. *Principles of embryology.* New York: The Macmillan Company.
WILLIER, B. H., WEISS, P. A., and HAMBURGER, V., eds. *Analysis of development,* Philadelphia: W. B. Saunders Company.
WITSCHI, E. 1956. *Development of vertebrates.* Philadelphia: W. B. Saunders Company.

COLLATERAL READINGS

ALLAN, F. D. 1969. *Essentials of human embryology.* New York: Oxford University Press.
AUSTIN, C. R. 1961. *The mammalian egg.* Oxford: Blackwell Scientific Publications Ltd.
BELLAIRS, R. 1964. "Biological aspects of the yolk of the hen's egg." Advan. Morph. 4:21-72 (M. Abercrombie and J. Brachet, eds. New York: Academic Press)
BELLAIRS, R., HARKNESS, M., and HARKNESS, R. D. 1963. "The vitelline membrane of the hen's egg; a chemical and electron microscopial study." J. Ultrastructure Res. 8:339-59.
BROMAN, I. 1911. *Normale and abnorme Entwicklung des Menschen.* Wiesbaden: Bergmann.
CLAWSON, R. C., and DOMM, L. V. 1969. "Origin and early migration of primordial germ cells in chick embryo; a study of the stages of the definitive primitive streak through eight somites," Am. Jour. Anat. 125:87-111.
CORNER, G. 1945. *Ourselves unborn.* New Haven: Yale University Press.
CORNING, H. 1921. *Lehrbuch de Entwicklungsgeschichte des Menschen,* Munich: Bergmann.
DeHAAN, R. L., and URSPRUNG, H., eds. 1965. *Organogenesis.* New York: Holt, Rinehart & Winston.
DAVIDSON, E. H. 1968. *Gene activity in early development.* New York: Academic Press.
DUBOIS, R. 1969. "Le mecanisme d'entre'e des cellules primordiales dans le reseau vasculaire chex l' embryon de Poulet." J. Emb. Exp. Morph. 21:225-70.
FRAPS, R. M. 1955. "Ovulation in the domestic fowl," pp 133-62 in *Control of ovulation.* (C.A. Villee, ed.) New York: Pergamon Press.
GALLERA, J. 1971. "Primary induction in birds." Advan. Morph. 9:149-80.
GILBERT, A. B. 1967. "Formation of the egg in the domestic chicken." Advan. in Reproductive Physiology 2:111-80. (Anne McLaren, ed.) New York: Academic Press.
GILBERT, M. G. 1963. *Biography of the unborn.* New York: Hafner Publishing Company.
GREEN, E. L. 1966. *Biology of the laboratory mouse.* New York: McGraw-Hill Book Company.
HAMBURGER, V. 1960. *A manual of experimental embryology.* Rev. ed. Chicago: University of Chicago Press.
HERTWIG, O. 1906. *Handbuch der vergleichenden und experimentellen Entwicklungslehre der Wirbeltiere.* Jena: Fisher.
________. 1910. *Lehrbuch der Entwicklungsgeschichte des Menschen und der Wirbeltiere.* Jena: Fisher.
HILLMAN, N. W., and HILLMAN, R. 1965. "Chick cephalogenesis. The normal development of the cephalic regions of stages 3 through 11 chick embryos." Jour. Morph. 116:357-69.
JENKINSON, J. W. 1925. *Vertebrate embryology.* London: Oxford University Press.
JORDAN, H. E. and KINDRED, J. E. 1948. *Embryology.* New York: Appleton-Century-Cofts.
KEIBEL, F., and MALL, F. B. 1910-12. *Human embryology.* Philadelphia: J. B. Lippincott Company.
KELLICOTT, W. E. 1913. *Textbook of general embryology.* New York: Henry Holt & Company.
KOLLMAN, J. 1907. *Handatlas der Entwicklungsgeschichte der Menschen.* Jena: Fisher.
LILLIE, F. R. 1919a. *Problems of fertilization.* Chicago: University of Chicago Press.
________. 1919b. *The development of the chick: An introduction to embryology.* Revised 1952 by H. L. Hamilton. New York: Henry Holt & Company.

LORENZ, F. W. 1969. "Reproduction in domestic fowl," pp 569-608. in *Reproduction in domestic animals.* 2nd ed. (H. H. Cole and P. T. Cupps, eds.) New York: Academic Press.
MacBRIDE. E. W. 1914. *Textbook of embryology.* New York: The Macmillan Company.
McMURRICH, J. P. 1907. *The development of the human body.* Philadelphia: The P. Blakiston Company.
MEYER, D. B. 1964. "The migration of primordial germ cells in the chick embryo." Develop. Biol. 10:154-90.
NALBANDOV, A. V. 1961. "Mechanisms controlling ovulation of avian and mammalian follicles," pp. 122-32. In *Control of ovulation* (C. A. Villee, ed.) New York: Pergamon Press.
NEEDHAM, A. E. 1931. *Chemical embryology.* (3 vols.) Cambridge: Cambridge University Press.
———. 1959 *History of embryology.* Cambridge: Cambridge University Press.
NICOLET, G. 1971. "Avian gastrulation." Advan. Morph. 9:231-62. (M. Abercrombie, J. Brachet, and T. King, eds.) New York: Academic Press.
OLSEN, M. W., and FRAPS, R. 1950. "Maturation changes in the hen's ovum." J. Exp. Zool. 114:475-89.
ROSENQUIST, G. C. 1972. "Endoderm movements in the chick embryo between the early short streak and the head process stages." Jour. Exp. Zool. 180:95-104.
RUGH, R. 1962. *Experimental embryology: Techniques and procedures."* Minneapolis: Burgess Publishing Company.
SHETTLES, L. B. 1971. *From conception to birth: The drama of life's beginnings.* New York: Harper & Row, Publishers.
SHUMWAY, W., and ADAMSTONE, F. B. 1954. *Introduction to vertebrate embryology.* New York: John Wiley & Sons.
SNELL, G. D., ed. 1956. *Biology of the laboratory mouse.* New York: Dover Publications.
SPRATT, N. T. 1952. "Localization of the prospective neural plate in the early chick blastoderm." Jour. Exp. Zool. 120:109-30.
THOMAS, J. B. 1968. *Introduction to human embryology.* Philadelphia: Lea & Febiger.
WATTERSON, R. L., and SWEENEY, R. M. 1973. *Laboratory studies of chick, pig, and frog embryos.* Minneapolis: Burgess Publishing Company.
WILSON, E. B. 1900. *The cell in development and inheritance.* New York: The Macmillan Company.

In 1828, Karl Ernst von Baer wrote, "I have two small embryos, preserved in alcohol, that I forgot to label. At present I am unable to determine the genus to which they belong. They may be lizards, small birds, or even young mammals."

> ***"The human mind is first confronted with the effect and not with the cause. It is only after the effect is defined that reasons for (or, more correctly, events preceding) the effect are sought. The relationship between the two is then clarified through an analysis of the intermediate steps, usually referred to as the 'mechanism of action,' or, for disease, 'Pathogenesis.' There are few, if any, simple or single causes in biology. There are, instead, complex situations and environments in which the probability of certain events is increased."***
>
> **M. B. Shimkin, *Jour. Chron. Dis.* 8:38, 1958.**

LABORATORY PROCEDURE

DRAWINGS

The drawing of embryological material has two functions.

1. It provides a record, compiled by the student, of the material studied. This record should be of permanent value.
2. The time and concentration required to produce an accurate drawing will prove helpful in the learning process.

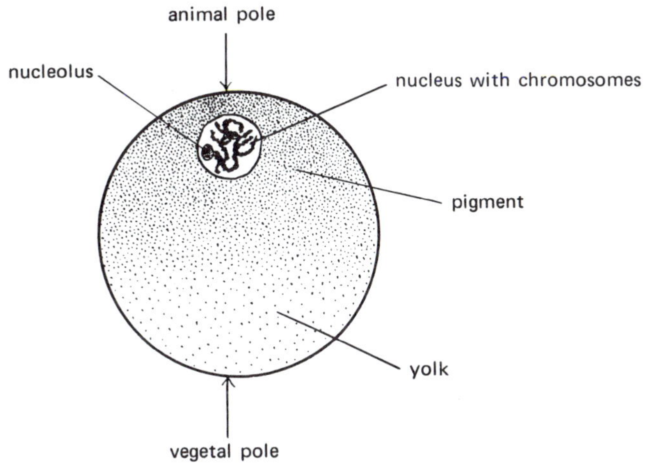

POLARITY OF THE VERTEBRATE EGG

SAMPLE DRAWING

All drawings are to be made in the laboratory, in spaced provided in this *Guide*. It is unwise to make a scratch drawing in the laboratory, when the material is at hand, and then to expect to improve on it at home where there is only the memory upon which to rely for details and accuracy. Accuracy rather than art is to be encouraged.

It is suggested that the student procure both 3H and 4H pencils and several soft gum erasers. *No drawings are to be made in ink.* Each student should have a set of good-quality colored pencils which will be used at certain times to designate germ layers or germ-layer derivatives. The same set of pencils may be used for lecture notes. The customary order of color associations is as follows.

BLUE for ectoderm and its derivatives.
YELLOW for endoderm and its derivatives.
RED for mesoderm and its derivatives.
Black ~~GREEN~~ for notochord.
ORANGE for yolk. (if it is ever shown) (later stages becomes just a yolk plug)

Every identifiable structure in every drawing should be labelled (in pencil) to indicate to the instructor that the student understands the identity of the object. By repetition, this helps the student to remember the mass of details studied. Certainly an object worth spending time to draw is worth labelling. When colored pencils are used, as suggested above, it will not be necessary to label the germ layers. Some of the general rules regarding drawing techniques are as follows:

1. Proportions should be correct in every detail, and drawings should be made sufficiently large to show detail. Proportion is more important than size.
2. Drawings should not be schematized. If the student draws what he actually sees, as if he photographed it, he cannot be wrong. The embryo is seldom wrong!

3. Shading is discouraged except when the student has an abundance of time. The method is always by stippling. In many cases, shading obscures rather than clarifies a drawing.
4. All structures observed should be included, but labels should be checked against all italicized structures in the descriptive text.
5. All labels should be printed in pencil and placed horizontally at the ends of guidelines. If the labels are incorrect they can then be changed. The ruled guidelines should never cross each other.

GENERAL PROCEDURE

The laboratory sessions will begin promptly after some oral directions by the instructor in charge. It is, therefore, of utmost importance that all students be at their desks at the beginning of each session. The oral directions supplement this *Guide* and are essential to the proper understanding of the work to be done.

The instructor and assistants may orally quiz students during any or all of the laboratory sessions. The student should understand exactly what he is trying to observe and should be prepared at all times to discuss the work with the instructor. While the student is not discouraged from asking questions, it is the general experience that the better students discover the answers to their questions by a little individual effort of their own.

Written quizzes on laboratory material are usually announced in advance. Practical (spot) quizzes may be given periodically. In this type of quiz, the student is asked to locate a number of specific structures and have their identification checked by the instructor.

GLOSSARY

One of the foundations of any science consists of an accepted list of definitions. We have, therefore, included a glossary of embryological terms in this *Guide*. The terms defined have been chosen because of their usage in vertebrate embryology. It is recommended that students be advised to make frequent use of this glossary. Its perusal has been found to be a good method of reviewing the subject as a whole.

ACKNOWLEDGMENTS

The author wishes to acknowledge the help of his teaching assistants in compiling this *Guide*, a process which has taken a number of years and a great deal of "trial and error" experience in the laboratory. Fortunately, the equipment, research, and teaching facilities at four different colleges and universities (Lawrence, Hunter, New York University, and Columbia), where the author has been privileged to teach, have been most adequate for this undertaking. This seventh edition contains the improvements suggested by some of the more than 200,000 students who have used the previous editions.

"Embryology is an ancient manuscript with many of the sheets lost, others displaced and with spurious passages interpolated by a later hand."

Cambridge Natural History, V:79

SUGGESTED SCHEDULE FOR LABORATORY WORK*

LABORATORY SESSION	EXERCISE NUMBER	SUBJECT MATTER	
1	1	Spermatogenesis } Maturation	
2	2	Oogenesis } Maturation	
3	3	Fertilization and early development	
4	4 & 5	Amphioxus and Fish (can be omitted)	
5	6	FROG:	Gonads
6	7 & 8	”	Fertilization, cleavage, and early development
7	9 & 10	”	Blastula, gastrula, neurula
8	11 & 12	”	5 mm embryo
9	13	”	7 mm embryo
10	14 & 15	”	11 mm embryo and bullfrog tadpole
11	16	CHICK:	Gonads
12	17	”	Primitive streak stage
13	18	”	24-hour stage
14	19	”	33-hour stage (can be omitted)
15	20	”	48-hour stage
16	21	”	72-hour stage
17	22	”	96-hour stage
18	23	”	Sagittal 96-hour stage
		”	Central nervous system - 5 days
		”	Eye - 7-8 days
19	23	”	Sections of 8-day stage, through level of gonads
20	24	PIG:	Entire embryos, 6 and 10 mm stages
21	24	”	Sections of 6 mm pig
22 & 23	24	”	Sections of 10 mm pig
24	25	MOUSE:	Early Embryology
25	26 & 27	Human Development and/or Experimental Embryology Techniques	

*This schedule is based on two 2-hour laboratory sessions per week, for a semester of 16 weeks, allowing about six (6) laboratory sessions for laboratory quizzes or expansion, or both, as individual instructors may wish. Exercise 20 to 23 can well be combined and 24 omitted if laboratory material is not available. However, such material can be easily prepared in the histology course and will prove to be very instructive. Or the 33-hour and 48-hour chick stages could be omitted entirely to allow more time on the 72-and 96-hour series, and hopefully, on the mouse section.

Exercise 1

THE MATURATION PROCESS:

Spermatogenesis

Mitosis in animals is the process of nuclear division which involves chromosomes and spindle apparatus and is characteristic of all living cells, both somatic and germ cells. The daughter cells resulting from such nuclear division are invariably alike in chromosomal constitution. The term *maturation process* refers to the changes requisite for the production of functional gametes (i.e., mature ova and spermatozoa) from somatic-type cells known as primordial germ cells. Maturation, therefore, includes many mitotic divisions and a qualitative division of the chromosomes known as *meiosis*. Since all somatic cells and, therefore, all primordial or primitive germ cells carry allemorphic pairs of chromosomes, the maturation process separates members of pairs of chromosomes (reductional division) so that each mature gamete is haploid with respect to chromosome number. The allelomorphic pair of chromosomes, after synaptic association is replicated, form *tetrads*. Crossing over may further complicate the picture so that reductional (meiotic) division may occur at one end of a pair of chromosomes, while equational (mitotic) division may take place at the other end of the same pair of chromosomes. Meiosis results in the *reduction* of the number of chromosomes to half, and the process of maturation includes both meiosis and mitosis. Whether *reductional* or *equational* division occurs first in a particular pair of chromosomes or whether they occur simultaneously varies with the species. The end result is the same in all cases; the ovum or the spermatozoon is haploid with more or less segregation of allelic genes.

SPERMATOGENESIS

The *testis* of the Florida lubber grasshopper, Rhomaleum, quite diagrammatically illustrates the maturational processes which change the somatic-type primordial or primitive germ cell into the functional and mature spermatozoon. The testis consists of many *lobes*, the pointed ends of which open into the *vas deferens*, or sperm ducts. Each lobe consists of numerous compartments (cysts) which are separated from each other by connective tissue partitions (*septa*). At the blunt (*apical*) end of the testicular lobe there may be seen numerous small primordial germ cells, each known as a spermatogonium, and all undergoing mitosis. With this continual increase in cell number, without concurrent reduction in cell size, there is a pinching off of groups of these presumptive germ cells into newly forming cysts. Each group, therefore, is composed of cells in exactly the same stage of development. As soon as these cells are separated from the original stock at the apical end of the testicular lobe they begin to enlarge (i.e., grow in volume without nuclear division) and are then known as primary spermatocytes. This is the first real indication that these particular cells are to become germ cells.

In examining a longitudinal section of a single lobe of the Rhomaleum testis, note that each cyst contains cells at the same stage of development and that the cysts from the blunt to the pointed end of the testis show progressive stages of maturation to the level where many mature spermatozoa are seen at the region where the testicular lobe is known to join the vas deferens. If a large testicular lobe is cut longitudinally through the center, practically all of the maturational steps should be visible in a single section. It must be remembered that the testicular lobe of Rhomaleum may not always be cut along its longitudinal axis, in which case certain stages may appear to be omitted or the heads (or tails) of the mature spermatozoa may be cut off. These false impressions must be corrected by studying ideal longitudinal sections or several sections in series. Examine many lobes before selecting the best for detailed study. A composite drawing may be necessary. The cytoplasm may not be stained, but the relative amount of cytoplasm may be determined by the proximity of the nuclei of adjacent cells. Since the maturational stages involve principally the nuclear changes, the student should attempt to locate the following.

1. SPERMATOGONIUM: These somatic-type cells are very small, crowded, and will be found toward the apical end of the testis. The chromatin threads of the nucleus may be divided into chromomeres and division is always mitotic.

2. PRIMARY SPERMATOCYTES: This term applies to all presumptive gametes (in the testis) which have grown to be larger than the spermatogonia and are found in cysts nearest those containing the spermatogonia. These cells do not undergo division for some time, but the amount of cytoplasm is increased and the nuclei enlarge and undergo a series of typical changes in anticipation of the two critical maturational divisions which follow. The changes during the primary spermatocyte prophase are:

   a. *Leptotene*—cell enlarged, nucleus possesses a spireme of chromatin threads plus a nucleolus. These are most abundant near the spermatogonia at the apical end. The sex chromosome is condensed, and the chromosomes are diploid.
   b. *Synaptene*—chromatin thread is broken up into definite chromosomes and allelomorphic chromosomes are pairing (synapsis). The chromosomes seem to converge toward the side of the nucleus nearest the centrosome.
   c. *Pachytene*—complete lateral fusion of chromosomes so that the number seems to be halved (i.e., seems haploid) and the chromosomes appear to be thicker. The fusion is known as parasynapsis. Replication results in tetrads.
   d. *Diplotene-Diakinesis*—chromosomes again separated into pairs (diplotene) which show curious (heterotypic) shapes, such as rings, coils, bars, etc. (diakinesis). Tetrads (split pairs of chromosomes forming four entities) may be seen in some species at this stage. The nucleus is now prepared for the two critical maturation divisions, reductional or equational.

3. SECONDARY SPERMATOCYTE: This term applies to all cells after the first and before the second maturational division, whether the first is equational or reductional. Generally, the number of chromosomes or the amount of chromatic material will be obviously reduced (haploid), and the cells will appear to be smaller than the primary spermatocytes.

4. SPERMATID: These are perfectly round cells with eccentric nuclei which result from the second maturation division. A dark condensation of mitochondria in the cytoplasm may indicate the nebenkern or possibly the centrosphere (idiozome or spermatosphere). A short filamentous tail will be seen protruding from the spermatid even in the earliest stages of metamorphosis. This filament represents the axial filament of the tail of the future spermatozoon.

   The earliest stage in spermatid-spermatozoon metamorphosis is seen as a typical resting cell nucleus with chromatin threads. In the next stage the cytoplasm becomes elongated to form a spindle-shaped head. Finally the chromatin material will be condensed so that the nucleus will take a uniformly dark stain. A granule (centriole) is now very distinct at the base of the tail.

5. SPERMATOZOON: These mature gametes or germ cells have long, thin, and heavily stained heads and grayish filamentous tails which may be seven to eight times as long as the heads. The chromosomes cannot be identified in the nucleus by any staining procedure of the spermatozoon.

REFERENCES

Arey: 29-34
Balinsky: 44-59
Bodemer: 14-18
Huettner: 1-22
McEwen: 1-74
Rugh: *Vert. Emb.* 2-14
Shumway: 1-43

RHOMALEUM TESTIS

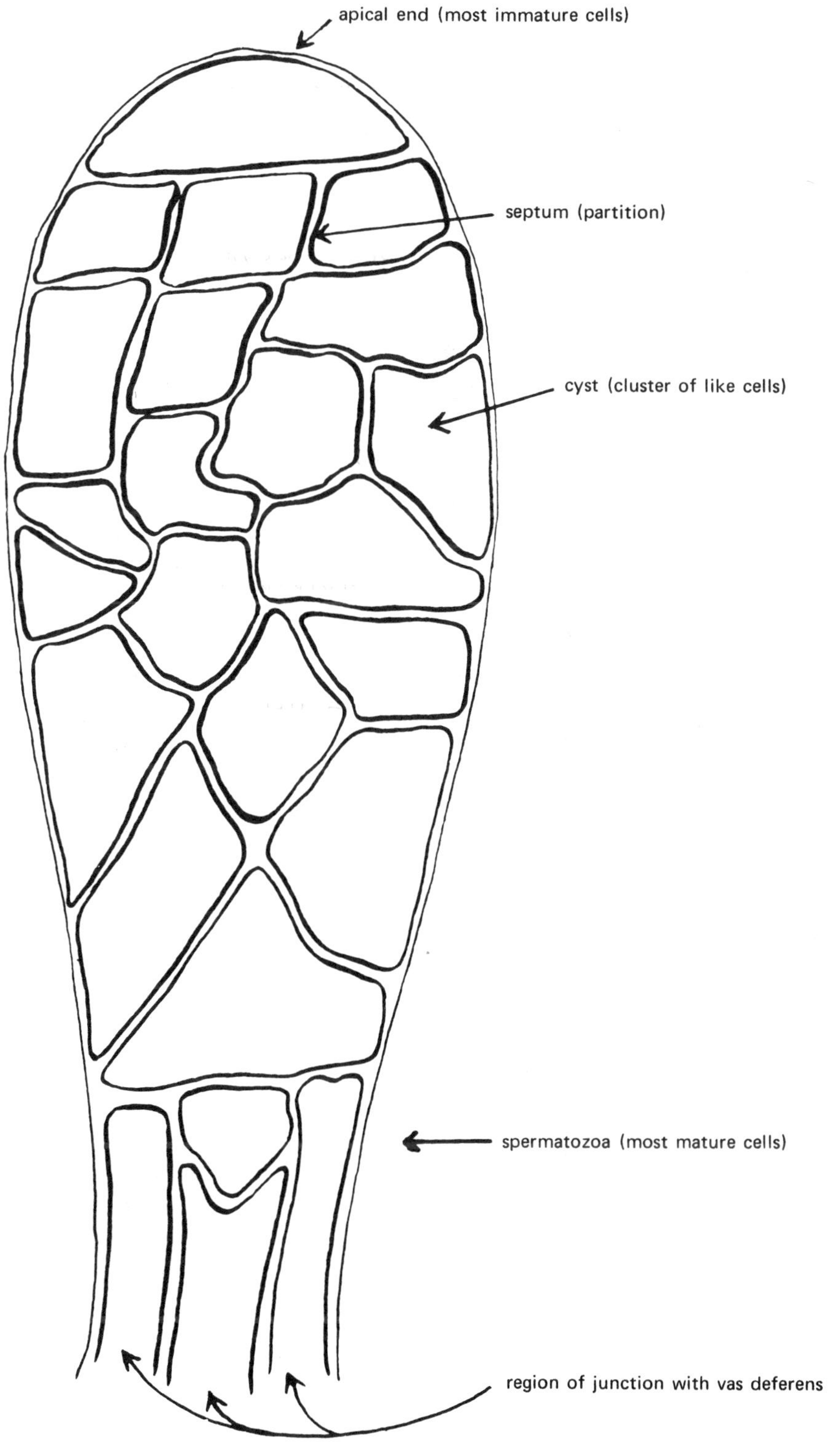

OUTLINE OF TESTICULAR LOBE OF
RHOMALEUM: GRASSHOPPER

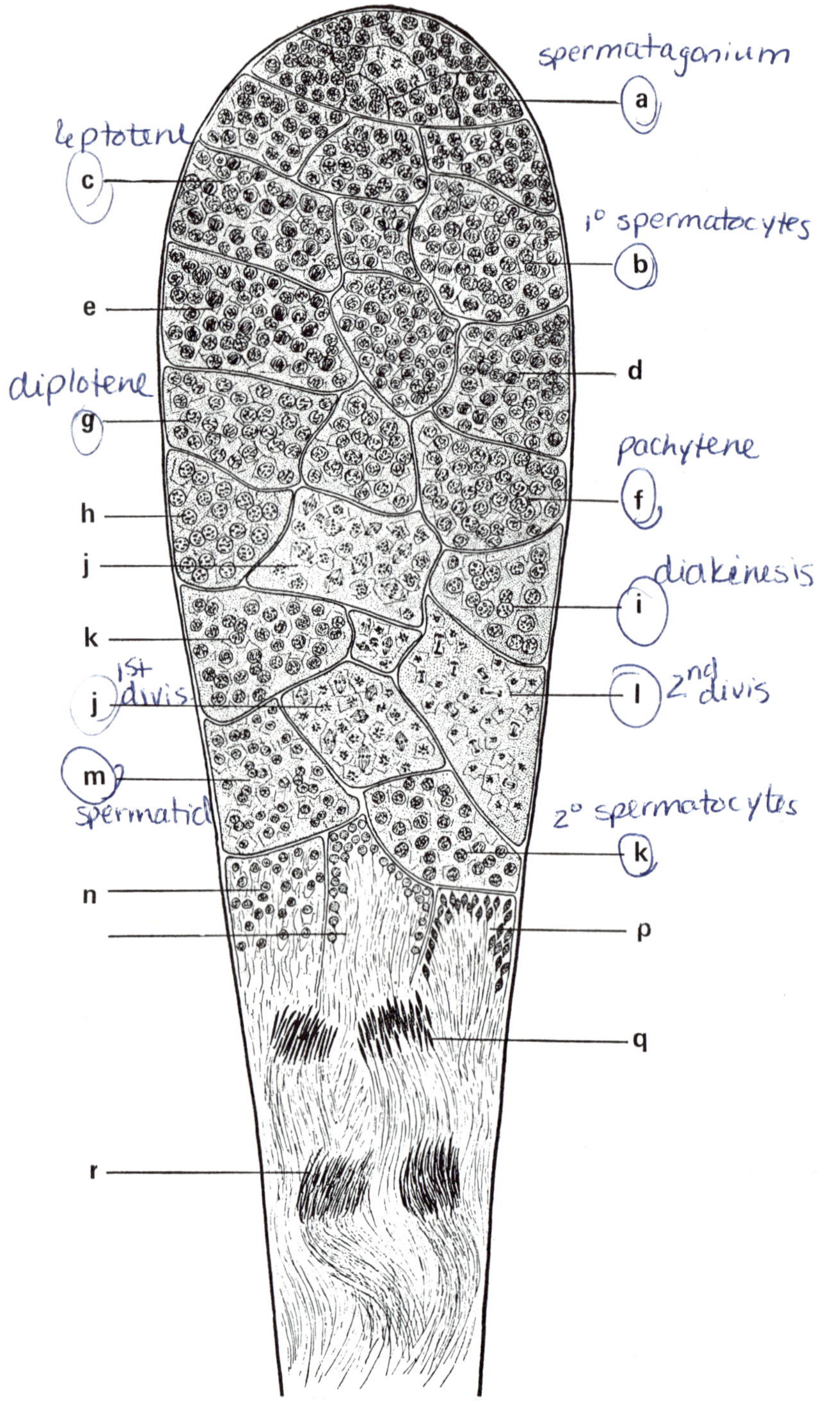

SPERMATOGENESIS IN THE LOCUST

RHOMALEUM TRICOPTERUM

Spermatogenesis in the locust *Rhomaleum tricopterum.* A median longitudinal section of one lobe of the testis is represented at the left. The nuclei contained in the various cysts are shown in greater detail on the next page. The letters attached to the cysts refer to identical muturation stages in both figures. Fig. a, spermatogonium; b, primary spermatocyte unravelling chromosomes; c, leptotene; d, amphitene; e, synaptotene; f, pachytene; g, diplotene; h, strepsinema; i, diakinesis; j, first meiosis; k, secondary spermatocyte; l, second meiosis; m, spermatid; n-r, transformation of spermatid into spermatozoon. Observe the X-chromosome form b-i. In the diplotene stage (g) when the other chromosomes are double, the X-chromosome is single. It splits in strepsinema or diakinesis (h, i) when all the other chromosomes are in the tetrad stage. Figs. j and l show the separation of the tetrads and dyads, respectively. Observe the attachment of the spindle fibers on the uppermost and lowermost tetrad in Fig. j. The manner in which the tetrads are pulled apart into dyads depends on the spindle fiber attachment and is indicated in the neighboring tetrad. The two centrally located tetrads on this meiotic figure have separated completely into dyads, and though they are both V-shaped, they differ in the arrangement of the two chromosomes composing the V-dyad. Similarly the dyads are separated into chromatids in the second meiosis (l). The organization of the tetrad chromosome is illustrated in the upper V-shaped chromosome in Fig. f. In Fig. g it appears double, in Fig. h in which it is located exactly in the upper part of the nucleus, it has divided into four components (tetrad). In Fig. i it shows its tetrad organization to the best advantage, and in Fig. j it has condensed into the typical compressed loop-shaped tetrad chromosome of the first meiosis. The paranucleus of the spermatid (m) can be traced back as far as the primary spermatocyte.

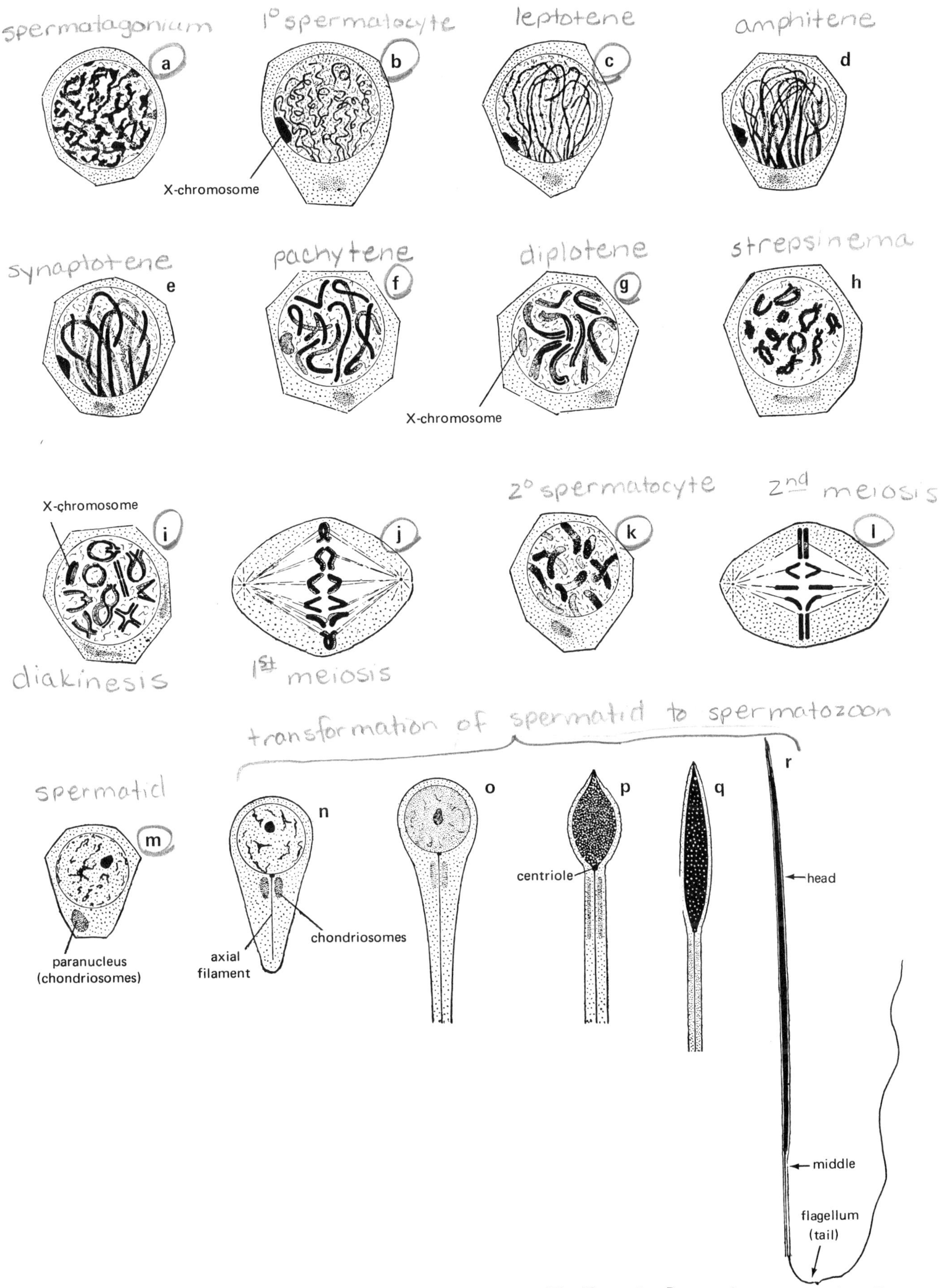
spermatagonium
a
1° spermatocyte
b
X-chromosome
leptotene
c
amphitene
d
synaptotene
e
pachytene
f
diplotene
g
X-chromosome
strepsinema
h
X-chromosome
i
diakinesis
j
1st meiosis
2° spermatocyte
k
2nd meiosis
l
transformation of spermatid to spermatozoon
spermatid
m
paranucleus (chondriosomes)
n
axial filament
chondriosomes
o
p
centriole
q
r
head
middle
flagellum (tail)

Maturation Stages

Locate the five major stages of spermatogenesis described and indicate where each is found in the sketch of Rhomaleum testis. Make an enlarged drawing of each of these cell types, in proper relative size and complete in details. Label all parts, remembering that the nuclear changes are of utmost importance.

STAGES IN SPERMATOGENESIS

Exercise 2

THE MATURATION PROCESS:

Oögenesis*

The maturation of the vertebrate egg is difficult to demonstrate, so the student will be given an opportunity to study the oögenesis of some transparent invertebrate eggs in which the processes can be recognized and are significant because they are as they might be seen in any vertebrate egg.

ASCARIS

Ascaris is the roundworm nematode parasite found in the intestine of some mammals. While the digestive and nervous systems are primitive, the reproductive system is very well developed, and the sexes are separate. The female has a Y-shaped reproductive tract. Each branch of the Y begins with the much-coiled, threadlike ovary which gradually enlarges (posteriorly) to become the thick-walled uterus. The uteri of the two branches then unite into a short muscular vagina, which opens to the exterior through a genital pore or vulva. This opens to the outside midventrally, one-third of the distance from the front end. Following copulation and insemination by the male, the eggs in the genital tubules of the female are fertilized. A single female may possess more than 25 million eggs, shedding as many as 200,000 a day. In Ascaris, the male cells are not typical spermatozoa, but small ameboid cells without tails but with large nuclei. These enter the egg before oögenesis is completed. The advantage in studying maturation in Ascaris is related to the fact that the diploid number of some species is only four, the haploid two. From a single female, it is possible to secure sections of the genital tract which will show every step in maturation, fertilization, and early cleavage. The student will be provided with prepared slides on which are mounted representative sections of the ovary and uterus of Ascaris, each containing stages to be studied. Locate the following.

1. SPERM PENETRATION: In a section through the heavy-walled uterus find many *primary oöcytes*, which are characterized by having vacuolated cytoplasm, rather inconspicuous nuclei, and thin cell membranes. Between these oöcytes find scattered heads of *spermatozoa* which will appear as small, triangular, dark bodies, each possessing a centriole at the base of the triangle. Occasionally spermatozoa may be found within the oöcyte cytoplasm, in which case the nucleus of the spermatozoon will be of a different shape and consistency. Such primary oöcytes, though immature, will be covered by a distinct fertilization membrane. The diploid number of chomosomes of some species of Ascaris is four.

2. FIRST POLAR BODY FORMATION: Further along among the sections of the uterus find primary oöcytes in which the chromosome number has doubled as a result of longitudinal splitting. Since the members of each pair of chromosomes have thus split, two groups of four beadlike chromosomes will be seen in each complete cell, where the diploid number is four. These groups of four chromosomes are known as *tetrads* and consist of a single pair of allelomorphic (homologous) chromosomes longitudinally split (replicated) in anticipation of the first maturation division. Generally the *tetrads* will be found near the periphery of the primary oöcyte. Locate such an oöcyte in which half of each tetrad is being extruded from the cell surface into the first or *primary polar body*. This division is generally considered as meiotic or

*Alternative studies are offered here, one with Ascaris and the other with a combination of Asterias, Crepidula, and Arbacia, if such can be procured. Either study alone is entirely satisfactory or both could be omitted. The embryos of Ascaris are so hardy they cannot be killed readily, even with strong fixatives.

ASCARIS

SPERM PENETRATION

FIRST POLAR BODY FORMATION

SECOND POLAR BODY FORMATION

FUSION OF PRONUCLEI – FERTILIZATION

EARLY CLEAVAGE – MITOSIS

reductional, the resulting haploid cell being designated as a *secondary oöcyte.* Discarded polar bodies containing the haploid number of dyads may be seen.

The entrance of the spermatozoon into the primary oöcyte initiates the process of oögenesis, but the spermatozoon (male pronucleus) will always be found near the center of the cell, awaiting the completion of the maturation process before fusing with the female pronucleus. Note the thicker fertilization membrane (not to be confused with the very thick egg shell) and the perivitelline space between the membrane and the oöcyte. This space is filled with a fluid, probably derived from the oöcyte as it shrinks.

3. SECOND POLAR BODY FORMATION: Still further along among the sections of the uterus locate a secondary oocyte in which the haploid chromosomes (dyads) are being arranged on a spindle, which will extrude one half of each longitudinally split chromosome into the *second polar body.* When this has occurred, the process of maturation is completed and an ovum has been formed. This mature ovum is, of course, haploid with respect to the material chromosome.

4. FERTILIZATION: In the remaining sections of the uterus find ova in which the male and *female pronuclei* are adjacent to each other and near the center of the cell. What is the condition of the chromatin material as compared with the nucleus of the sperm head before penetration? The fusion of these pronuclei is the final process in the fertilization reaction, producing the diploid zygote. This *zygote* is immediately ready to initiate the series of mitotic divisions which will carry it through the 2-, 4-, 8-, and 16-division stages successively to many cells of the Ascaris embryo.

THE STARFISH–ASTERIAS (OR ARBACIA, THE SEA URCHIN)

The unfertilized starfish egg possesses a large prematuration nucleus called the *germinal vesicle,* within which is located the distinct *nucleolus.* A *vitelline membrane* is present, but it is difficult to see until it becomes separated from the egg after fertilization. These are entire eggs (i.e., not sections of eggs) lightly stained. A few eggs will appear to lack germinal vesicles, but this is due to the fact that they have already been fertilized and the egg nucleus (germinal vesicle) has broken down in preparation for the maturation processes. Saltwater aquaria and living sea urchins are available, and, in certain seasons, ova and sperm can be procured from them for observations of fertilization and early development.

THE STARFISH EGG, SHOWING GERMINAL VESICLE

(See page 15.)

CREPIDULA

The eggs of this mollusc are quite large and have been mounted beneath elevated coverslips to avoid crushing. The student is, therefore, cautioned to use only the low-power objective at first and to shift to the highpower objective with great care. These are the most satisfactory eggs available for the demonstration of *polar bodies* and their formation. The polar bodies (three) may be seen *between* the vitelline membrane and the egg remaining through many cleavages, lodged in the *cleavage furrows* at the animal pole. Note differences in animal and vegetal poles. The *vitelline membrane,* encloses both the egg and the polar bodies.

In the diagrams below, and on the slides of Crepidula eggs, note the following.

1. Granules of chromatic material within the polar bodies (discarded nuclei).
2. The number of polar bodies at the different stages of maturation, finally three.
3. The relative sizes of the male and female pronuclei and the positions as they approach each other. Is there any uniformity in the position of the female pronucleus?
4. Appearance of micromeres during the third cleavage, i.e., at the 8-cell stage. It is important that the student distinguish between the large *macromeres* (light colored, alveolar) and the small *micromeres* (dark and granular) and their respective nuclei. Also, identify the polar bodies at all stages. The *micromere nuclei* may be confused either with the micromeres themselves or with the polar bodies (see Glossary).

On each slide there may be some later stages of development of the Crepidula egg, some of which exemplify *epiboly* or the growth of micromeres down over the *macromeres,* and others which show clearly the peculiar *spiral* type of *cleavage,* in which the micromeres tend to rotate just enough to lie within the furrows between the macromeres. (See page 14.)

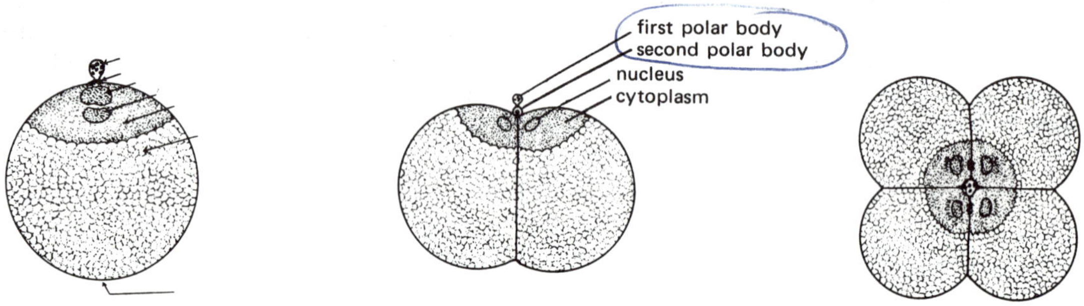

ONE CELL TWO CELL FOUR CELL: POLAR VIEW

CREPIDULA

MATURATION AND EARLY CLEAVAGE OF CREPIDULA

It is important for the student to remember that there is no general rule about the stage of maturation of the egg at which activation takes place, but that the stage is constant for any given species. In some eggs (e.g., the starfish egg), maturation does not begin until the sperm enters the egg cortex; in others (e.g., the frog or Amphioxus), the egg achieves one maturational division before it is fertilizable; and in a few instances (e.g., the sea urchin egg) both maturational divisions must be complete before normal fertilization can occur.

Note: See page 29 for oögenesis of whitefish eggs.

REFERENCES

Arey (1965): Chapters III, IV, and V
Balinsky: 59-98
Bodemer: 21-32
Huettner: 23-26
MacBride: Vol. I
McEwen: 2-74
Nelsen: 3-177
Patten (1964): 47-67
Rugh: *Vert. Emb.* 2-14
Shumway: 53-71

". . . criticism, an open and honest look at the world and at ourselves, must be desired actively and courted assiduously. It must become the mistress of the scientist. The scientific critic must be a creative artist, wise, knowing, fair, with taste and talent, who also remembers with Petrarch that love, hate, and envy are the three poisons which kill sound criticism."

W. B. Bean, *Perspectives in Biology and Medicine* 7:224, 1958

FERTILIZATION AND EARLY DEVELOPMENT

Exercise 3

FERTILIZATION*

Arbacia

The sea urchin egg, like most eggs, possesses a *vitelline membrane.* This membrane is so closely adherent to the egg cell membrane that it cannot generally be seen until after the egg is fertilized, when the egg shrinks or the membrane appears elevated. It is thenceforth known as the *fertilization membrane.* This is such a dependable reaction that embryologists consider the elevation of this membrane as sufficient evidence of successful fertilization. The space between the egg and the vitelline (or fertilization) membrane is known as the *perivitelline space.*

Since the egg of Arbacia must complete maturation before it is fertilizable, very few of these eggs will show germinal vesicles. Such eggs should be recognized as immature since they possess the prematuration nucleus. The absence of a germinal vesicle does not imply, however, any lack in the nuclear components, but rather that they are not in a stainable state. The fertilization reaction is almost instantaneous, so that by the time these eggs are preserved many of them will have developed elevated fertilization membranes, which helped to resist all excess spermatozoa immediately after the penetration of the first spermatozoon. On each slide there will be a few eggs around which the student will find innumerable small dots (stained *sperm heads*), one of which is in the process of invading just within the egg cortex and all other (excess) spermatozoa free from the egg and rejected by it. Note the relative sizes of the two types of germ cells under high magnification, and sketch below.

ARBACIA EGG AT MOMENT OF FERTILIZATION, SHOWING SPERMATOZOA

FERTILIZED EGG WITH ELEVATED FERTILIZATION MEMBRANE

*This section on fertilization may be omitted by those who studied Ascaris for oögenesis.

CLEAVAGE*

Asterias

The starfish egg exemplifies the *holoblastic type* of cleavage in a typical *homolecithal egg*. On the slide identify the 1-, 2-, 4-, 8-, and 16-cell and the morula stages. Compare this (*radial*) *cleavage* with the early *(spiral) cleavages* of Crepidula, previously studied. (Check with Glossary.)

EARLY CLEAVAGES OF ASTERIAS, THE STARFISH

BLASTULA

The starfish egg develops a coeloblastula or a hollow sphere containing a fluid in a cavity known as the *blastocoel.* The single layer of peripheral cells is not of uniform thickness. Locate the side where the cells are thicker, the side that will subsequently invaginate in the process of gastrulation. This is the *vegetal pole,* and the opposite side is the *animal pole,* which will give rise largely to ectoderm and ectodermal derivatives.

**BLASTULA OF ASTERIAS,
THE STARFISH**

*Compare with photographs of starfish (Asterias) development on page 15.

GASTRULA

Asterias

The starfish gastrula exemplifies the simplest method of didermic development, i.e., the formation of a two-layered embryo having both *ectoderm* and *endoderm*. Gastrulation begins toward the end of the first day of development, and the coeloblastula loses its spherical shape and becomes flattened on one side (see Exercise 4 on Amphioxus development for comparison.) This flattened side represents the future posterior region of the embryo, the center of which invaginates (pushes in) to form a new cavity, the *archenteron* ("primitive gut"). The opening into this archenteron (also called the gastrocoel) is really misnamed the *blastopore*. It is neither a blastula stage nor does the "pore" open into the blastocoel. This opening becomes smaller and smaller and generally disappears and in most animals gives rise to the region of the anus. It must be emphasized that the blastopore opens into the gastrocoel, not the blastocoel. The original blastocoel remains quite large as the invagination progresses toward the animal pole.

Locate a gastrula in lateral view and note that within the blastocoel there are large stellate cells budding off from invaginating mass. There are wandering, amoeboid, *mesenchyme cells*, which will give rise to the bulk of the mesodermal derivatives, so that the third germ layer (i.e., the *mesoderm*) begins to form before the completion of gastrulation. In late gastrulae these mesenchyme cells may be strung together into a network of strands. The most anterior (apical) portion of the archenteron bulges to form diverticula known as vesicles. The entire process is gastrulation by invagination. (See photographs on the next page.)

LATERAL VIEW OF STARFISH GASTRULA

BLASTOPORAL VIEW OF STARFISH GASTRULA

GASTRULA OF STARFISH

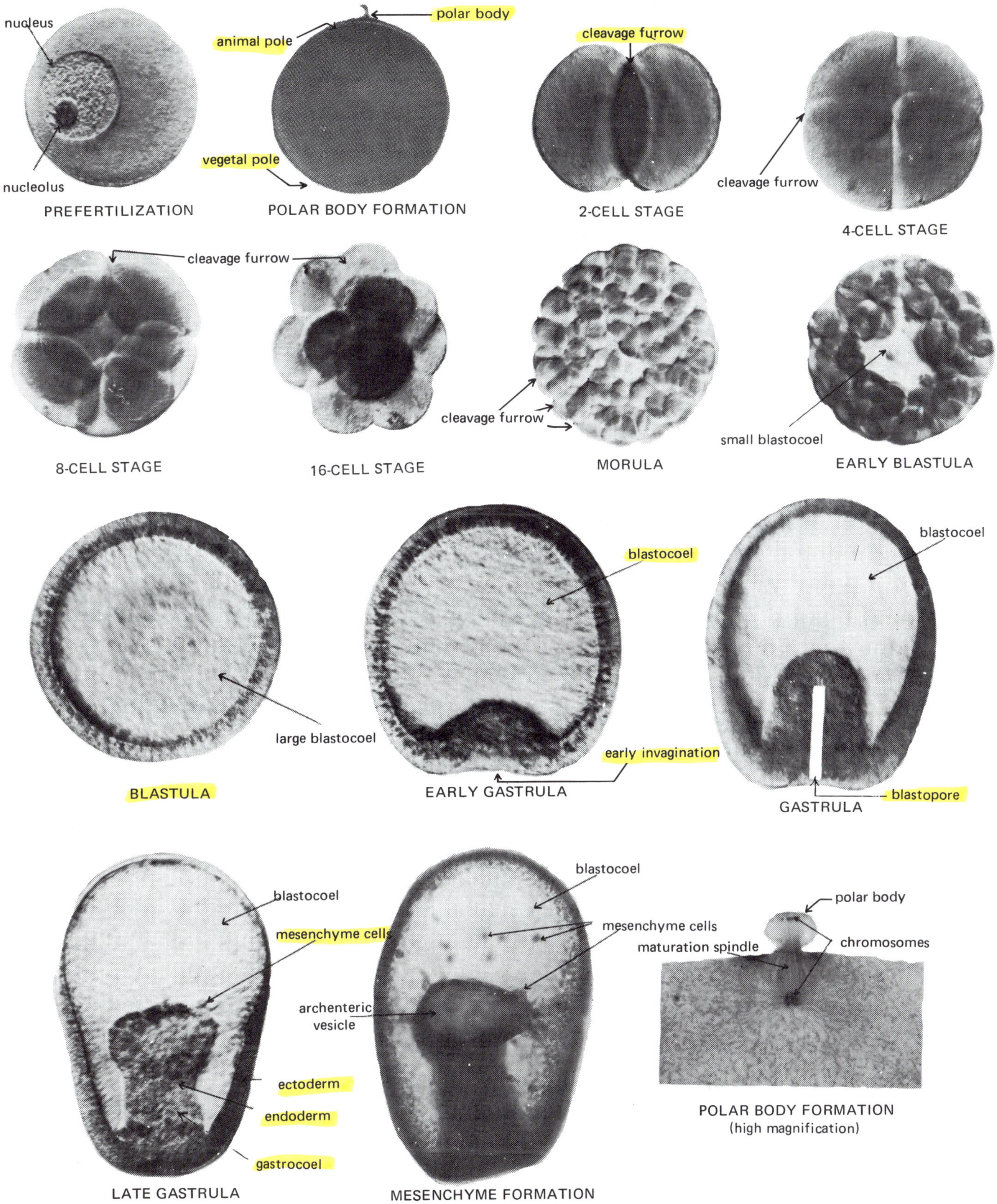

DEVELOPMENT OF STARFISH

HOLOBLASTIC CLEAVAGE

Reprinted by permission of Turtox, Chicago.

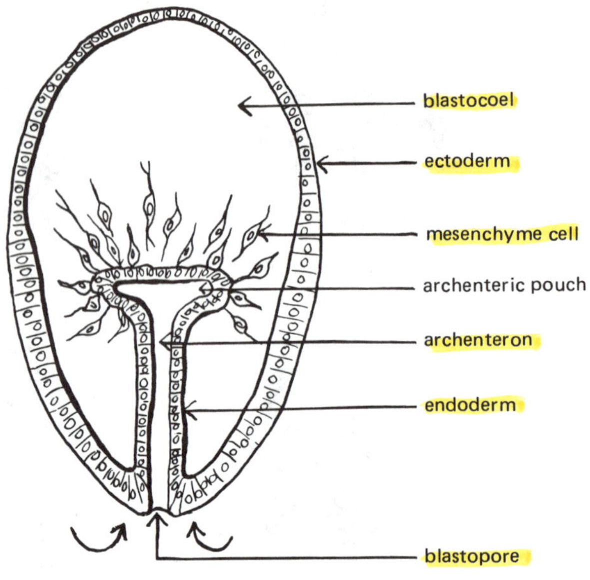

GASTRULA OF STARFISH

(Add all labels)

Crepidula

Blastula and gastrula formation in the mollusc Crepidula is rather complicated, but it is suggested here that the student examine the later stages in development in order to understand what is meant by *spiral cleavage* and *epiboly*. Spiral cleavage will be most evident in the 8-, 16-, and 32- cell stages, while epiboly (overgrowing of the macromeres by the micromeres) may be seen in later stages.

CREPIDULA: 8-CELL STAGE

CREPIDULA: EPIBOLY

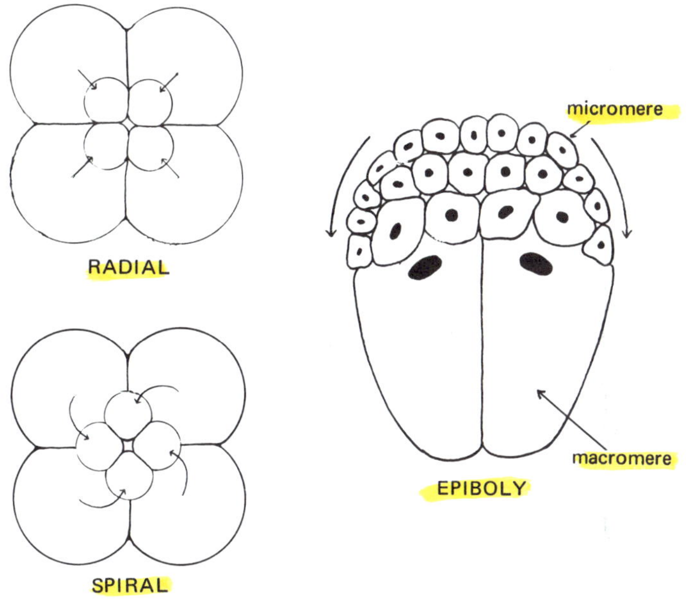

HOLOBLASTIC, UNEQUAL CLEAVAGE

REFERENCES

Arey: 43-48
Balinsky: 107-87
Bodemer: 54-68
Huettner: 40-70
Rugh: *Vert. Emb.* 14-19
Shumway: 43-52, 72-86

"Every child starts as an invisible unit with a weight of only 5/1,000 of a milligram and gains during the first weeks of life more than a million percent in weight. Which industry, whatever direction or planning boards there may be, can claim such an increase in output?"

G. M. H. Veeneklaas, 1957
Ann. Paediat. Renniae 3:718

"Living things are analyzed into organs, tissues, cells, chromosomes, genes, and their functions into tropisms, reflexes, and forced movements, while the synthesis of all these elements into the broader aspects of the organism and its relation to environment are too much neglected."

E. G. Conklin, 1944

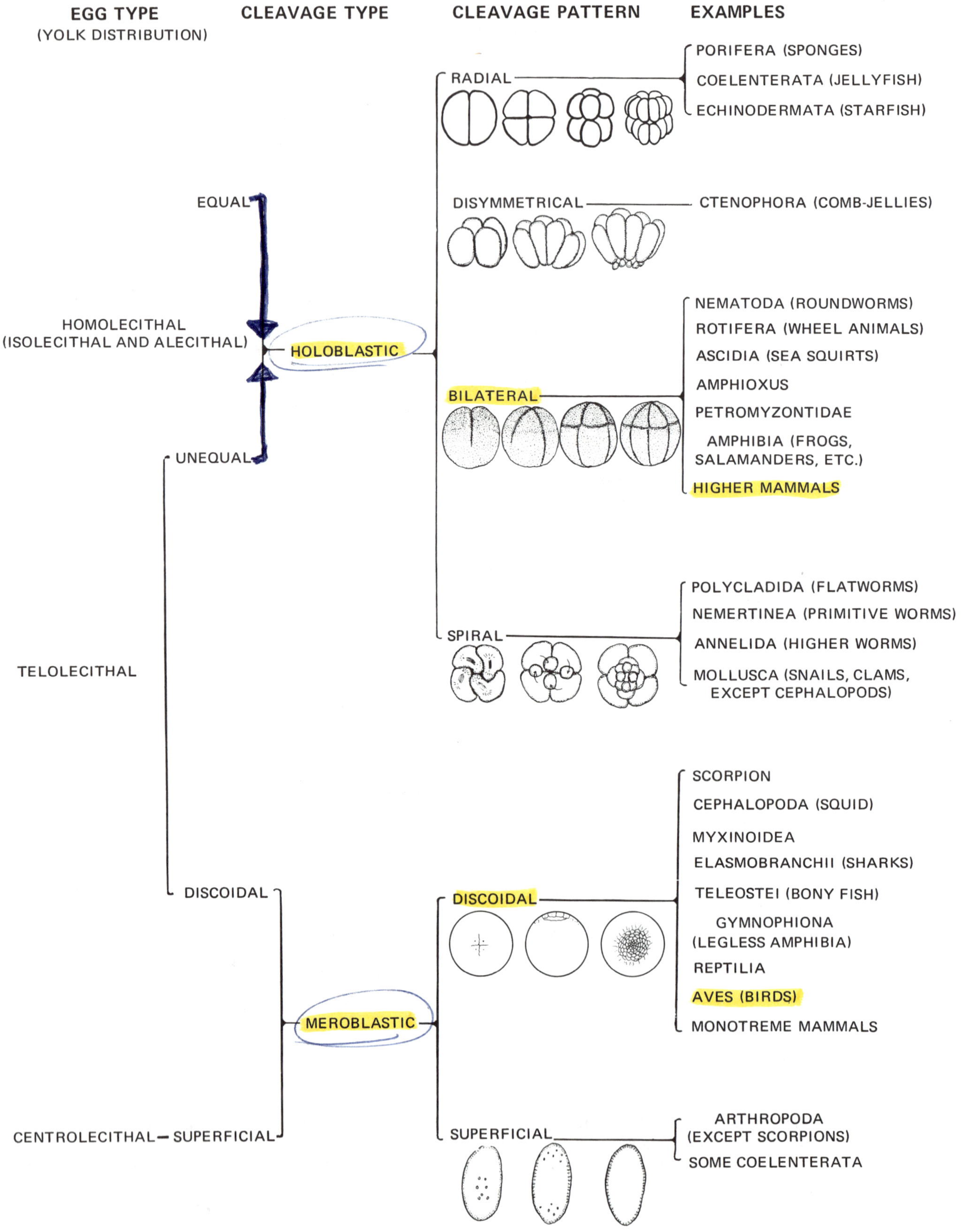

CLASSIFICATION OF EGG AND CLEAVAGE TYPES

Exercise 4 AMPHIOXUS

AMPHIOXUS DEVELOPMENT

Ovarian eggs of Amphioxus are always attached to the ovarian membrane at the animal pole, contrary to the situation found in most invertebrate eggs. The polar bodies are formed about 35° from the anterior end of the future larva, as in ascidians and amphibia. The spermatozoon enters the egg near the vegetal pole, after which a peripheral layer of cytoplasm accumulates over the lower hemisphere. Syngamy occurs between the nuclei at a point above the equator and posterior to the center. The first cleavage amphiaster appears at right angles to the egg axis, and concurrently the peripheral cytoplasm is spread out as a crescent on the posterior surface parallel to the amphiaster. This is known as the *mesodermal crescent.* On the anterior side, opposite the mesodermal crescent, appears another crescent from which the chorda and neural plate develop. Most of the animal hemisphere gives rise to ectoderm.

Examine the series of drawings of Amphioxus on pages 23 and 24 as you read the following description. The first cleavage establishes the bilateral symmetry of the embryo, dividing the egg into right and left halves. The second cleavage is at right angles to the first and divides the egg into anterodorsal and posteroventral quarters. The third cleavage is horizontal and a little above center, providing four upper micromeres and four lower macromeres. The mesodermal crescent is contained in the smaller posteroventral cells. The fourth cleavage is bilaterally meridional, the fifth bilaterally latitudinal but irregular. Synchrony in cleavage fails by the 128-cell stage when the smaller ectoderm cells divide more rapidly than do the large endoderm cells.

The blastocoelic jelly appears between the cells as early as the first cleavage, and the swelling of this jelly enlarges both the blastocoel and the blastula as a whole. In the spherical blastula the smaller, rapidly dividing mesodermal (crescent) cells surround the yolk-filled endodermal plate on its posterior and lateral borders. The presumptive ectodermal cells are intermediate in size between those large ones that will become endoderm and those smaller ones that will become mesoderm and occupy most of the animal hemisphere.

The blastula soon flattens dorsoventrally at the region of the mesodermal crescent. The endodermal plate, which is roughly triangular in outline, sinks into the blastocoel first at the posterior side. At the anterior side the invagination is sharpest, and here the dorsal lip of the blastopore forms. It is composed of both chorda and neural plate cells. The mesodermal crescent is then inflected as two lateral lips, which are rounded and contain the remnants of the blastocoel. These form the right and left mesodermal pouches. For a considerable time no ventral lip forms.

The gastrula elongates and brown pigment appears, which marks the outer end of the ectoderm cells of the ventral lip. The chorda plate elongates and becomes narrower to form the notochordal groove, and this indicates the animal axis. The chorda cells are quite indistinguishable from endoderm until after closure of the blastopore.

The neural plate follows the pattern of the chorda cells but is never quite so long. The neural plate rolls up from its sides to form the neural groove and eventually the neural tube.

The lumen of the tube is limited largely to the anterior end, which opens to the exterior by way of the permanent neuropore.

The mesodermal grooves, on either side of the chorda, become constricted into a series of pouches with the largest the most anterior. These separate from the gut to give rise to the somites. Two diverticula arise, one on either side and in front of the first somite. The right diverticulum becomes the head cavity in the rostrum.

The right wall of the anterior gut becomes thickened, and its cells elongate in a horseshoe-shaped region known as the endostyle. The mouth is formed first as a plate in the thickened left wall, ultimately to break through.

A club-shaped gland forms just posterior to the endostyle and posterior to the gland is the branchial primordium. This is essentially a groove which shortly acquires gill slits whose lining cells are fimbriated.

The gut is a small, round tube posterior to the branchial region, ending in an anal diverticulum to the left about one-third the distance from the posterior end of the body.

Conklin (Jour. Morph. 54:69, 1932), who made the study from which the above description was abstracted, stated that Amphioxus exhibits a localization pattern or fate map similar to that of amphibia and ascidians.

Slides of Amphioxus development are difficult to obtain. Three slides are offered which represent three stages of development, such as early cleavage, neurulation, and larva. The student should compare each specimen with the Conklin diagrams, on pages 23 and 24, and draw three stages in Amphioxus development indicating, in each case, the egg and the embryonic axes.

The student may fill in the following cross-section diagrams of Amphioxus with appropriate colors. (Ectoderm = blue; endoderm = yellow; mesoderm = red.)

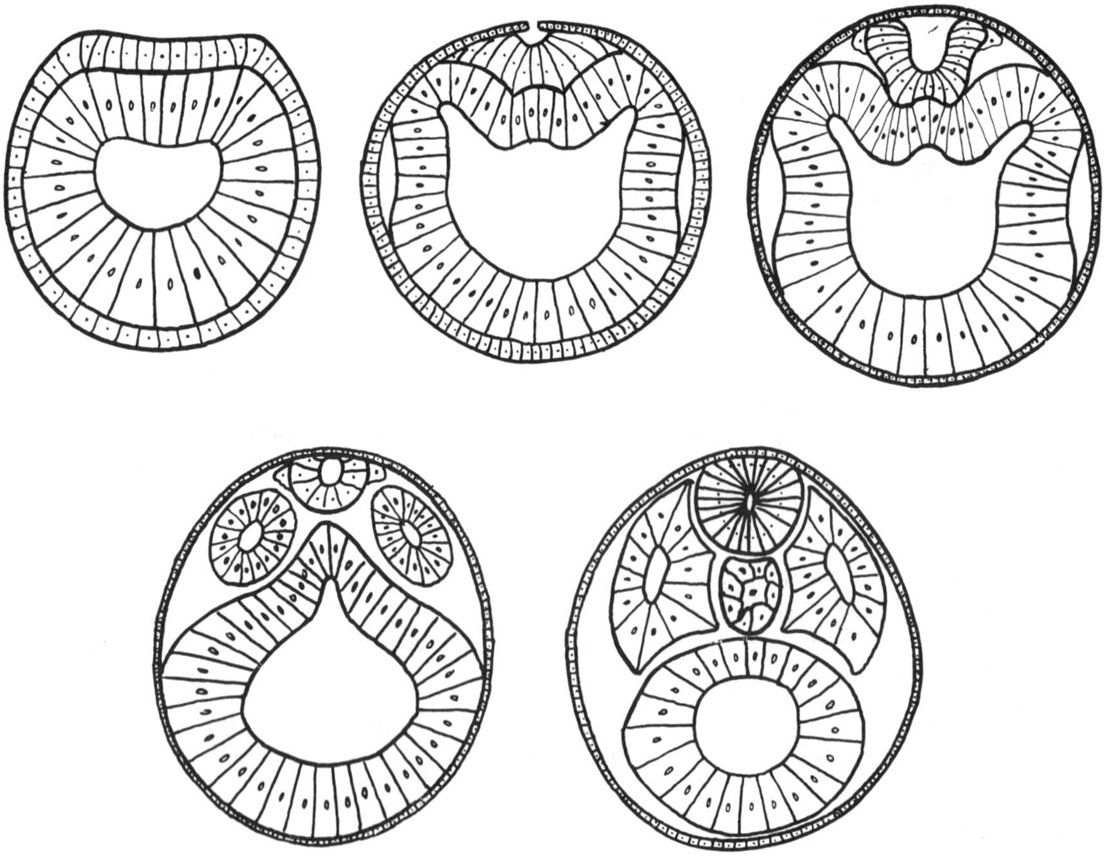

KEY TO THE LABELS FOR AMPHIOXUS DRAWINGS (pages 23 & 24)

A – anterior
bc – blastocoel
ba – branchial primordium
bv – brain vesicle
cg – club-shaped gland
cgd – duct of club-shaped gland
ch – chorda
cn – chorda-neural crescent
D – dorsal
dl – dorsal lip of blastopore
es – endostyle
L – left
ld – left diverticulum of gut (preoral pit)
m – mouth
mg – mesodermal groove
mp – mesodermal pouch
ms – mesodermal crescent
nc – neurenteric canal
ng – neural groove
np – neural plate
P – posterior
pg – pigment in posterior ectoderm
pp – preoral pit
ps – pigment spot
R – right
rd – right diverticulum of gut (head cavity)
V – ventral
vl – ventral lip of blastopore

EXPLANATION OF FOLLOWING AMPHIOXUS FIGURES

Normal cleavage stages 1 to 4 hours after fertilization.

1. Unsegmented egg 1 hour after fertilization, viewed from posterior side, as shown by position of egg and sperm nuclei nearer that side (upper surface of figure); the second polar body marks the animal pole; peripheral layer of more deeply stained substance (stippled in drawing) is thicker at vegetal pole than elsewhere.
2. Telophase of first cleavage 1¼ hours after fertilization, viewed from posterior side; the asters, represented by dotted circles, lie slightly nearer the posterior surface than the anterior.
3. Metaphase of second cleavage 2 hours after fertilization, viewed from the vegetal side (opposite polar body); the anterior side is here marked by a blob of exuded material (blastocoel jelly) in the furrow between the blastomeres, the posterior side by a thicker zone of deeply staining substance (mesodermal crescent); the two spindles diverge anteriorly from the median plane; centrosomes are granular and elongated transversely.
4. Late anaphase of second cleavage, 2 hours after fertilization, oriented as in preceding figure; chromosomal vesicles at edge of astral areas; posterior pole marked by thicker zone of dark granules; polar body seen through the egg on the farther side; cleavage cavity appearing between blastomeres.
5. Four-cell stage, 2 hours after fertilization, seen from vegetal pole, anterior cells larger than posterior ones. Latter contain most of granular substance (mesodermal crescent). The polar body is seen through the egg slightly posterior to the cleavage cavity.
6. Eight-cell stage, 2½ hours after fertilization, seen from right side; posterior cells (below at the right) contain most of the granular substance (mesodermal crescent); each nucleus a mass of chromosomal vesicles. The arrow marks the approximate anteroposterior axis of the future larva. The polar body is posterior to the animal pole (i.e., crossing of first and second cleavage furrows).
7. Eight-cell stage, 2½ hours after fertilization, seen from animal pole; polar body posterior to animal pole; anterior cells at top of figure, posterior ones at bottom; former slightly larger than latter; all cells in metaphase of fourth (meridional) cleavage.
8. Late anaphase of fourth cleavage, 2½ hours after fertilization; orientation as in Fig. 7; cross furrow turns to left in first cleavage when seen at animal pole, to right at vegetal pole.
9. Sixteen-cell stage, 2½ hours after fertilization, seen from vegetal side; polar body posterior to animal pole on opposite side; cleavage widely open at vegetal pole, opening narrower at opposite pole; nuclei dividing in fifth (latitudinal) cleavage.

10. Side view of 32-cell stage (four zones of eight cells each, somewhat schematized), 2½ hours after fertilization; every nucleus in metaphase or anaphase of sixth cleavage; cleavage cavity open at vegetal pole.
11. Left-side view of 64-cell stage, 3¼ hours after fertilization; egg axis marked by broken line, anteroposterior axis of future larva by the arrow; cells at animal (ectoderm) pole smaller than those at the vegetal (endoderm) pole, and those at the posterior (mesoderm) pole smaller than those at the anterior-dorsal (chorda-neural) pole; cleavage cavity large and its polar openings to exterior closed.
12. Anterior view of 64-cell stage, 3¼ hours after fertilization; animal (ectoderm) pole above, vegetal (endoderm) pole below; cleavage cavity closed, all nuclei in resting stage.
16. Anterior view of blastula (dorsal-anterior side below, ventral-posterior above) of approximately 256 cells, 4 hours after fertilization; endoderm cells at lower margin in resting stage (nuclei striped), ectoderm cells dividing; dorsal lip of blastopore (dl) will form at lower (posterior) border of ectoderm cells along heavily shaded line.
25. Surface view of blastula, rotated slightly toward the left, so that it is seen from the right-ventral side and the egg axis appears to pass through the mesodermal crescent, but the stippled outlines of the endoderm cells below the mesodermal crescent show that the vegetal pole lies at the posterior border of the endoderm cells. The blastula begins to be pointed at the posterior pole.
26. Optical section of a blastula 6 hours after fertilization, showing the left inside. The egg axis is shown by the broken line, the anteroposterior axis by the arrow; the large cells to the left of the broken line at the lower pole are endoderm; the small dividing cells to the right are mesoderm.
28. Surface view of right side of blastula 5½ hours after fertilization, showing small mesoderm cells dividing and large endoderm cells and small ectoderm cells in resting stage.
32. Optical section. Mesoderm cells dividing, endoderm shown by coarse stipples.
35. Optical section, showing the left inside of a blastula. The small cells of the mesodermal crescent are dividing, the large endoderm cells are shown with coarse stipples. The whole blastula is distinctly pear shaped, the pointed end being at the mesodermal pole.
36. Optical section of a blastula of the same stage as Fig. 35, but with fewer cells and much more flattened over the endodermal area; the endoderm and mesoderm cells are dividing. The approximate anteroposterior axis of the larva is shown by the arrow.
57. Optical sections in plane of bilateral symmetry of gastrulae, viewed from the right. The mesodermal pouch (mp) of the right side is above the median plane in Fig. 57. Polar body is posterior-ventral to the middle of the gastrular arch.
58, 59. Optical sections of 9½ hours gastrula, nearly in the plane of bilateral symmetry, showing the left inside of the gastrocoel in Fig. 58 and 59. The ventral (vl) and lateral lips contain most of the remnant of the blastocoel, and, in the narrowing of the blastopore in Fig. 60, the dividing cells of the mesodermal crescent (ms) extend all the way from the ventral to the dorsal lip. The polar body is ventral to the apical pole, and the ventral side of the gastrula is lengthening more rapidly than the dorsal side. In Fig. 59, the ventral endoderm is thrown into folds which will still further lengthen the ventral side when they straighten out. The cells of the mesodermal pouches adjoin the ventral lip and extend around to the dorsal side.
62. Nearly median section, showing the left inside. The polar body is distinctly ventral to the apex of the gastrula, and the dorsal side of the gastrula is less corved than the ventral. The ventral lip is bending in, thus closing the blastopore more rapidly on the ventral side.
68. View of a gastrula, showing nuclei of the blastopore lips and mesodermal ring and groove (mg) dividing. The dorsal side is flat, and the ventral lip is turned inward in a sharp curve.
73. Showing the left inside of the gastrocoel. Cells are dividing in both dorsal and ventral lips of the blastopore and at the anterior end of the gastrocoel.
74. Right inside of gastrula of 15 hours showing the ectoderm overgrowing the neural plate as a single layer of cells, beneath which is the neurenteric canal. This overgrowth is V-shaped when seen from the dorsal surface. Cells at the posterior end of the neural plate and notochord are dividing. Pigment granules are found in the ectoderm cells of the posterior end. Cells lining the gastrocoel differ in shape in different regions.

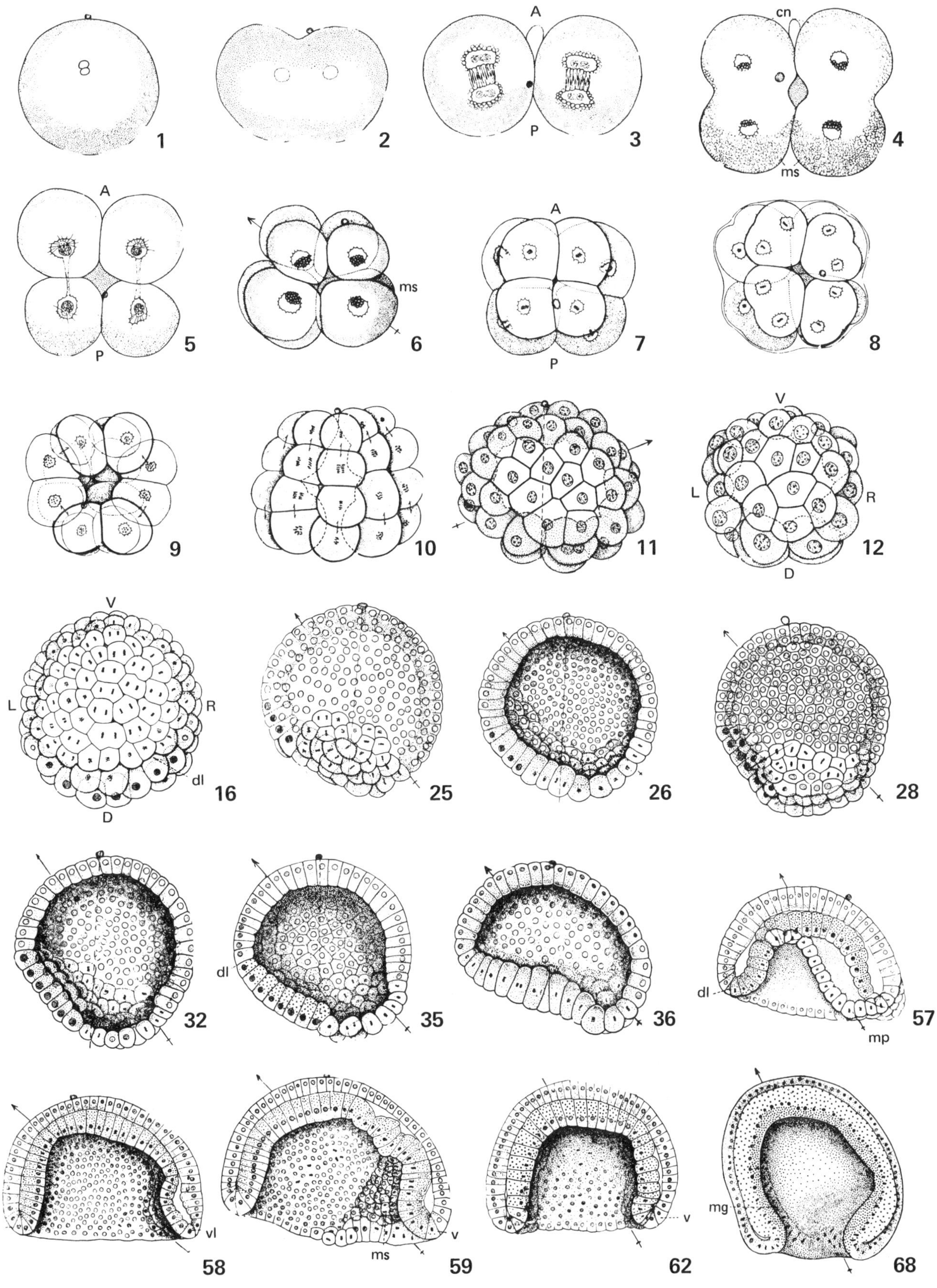

(Identification numbers are from the original papers.)

Reprinted by permission of Dr. E. G. Conklin from *Jour. Morph.* 54:69 (1932).

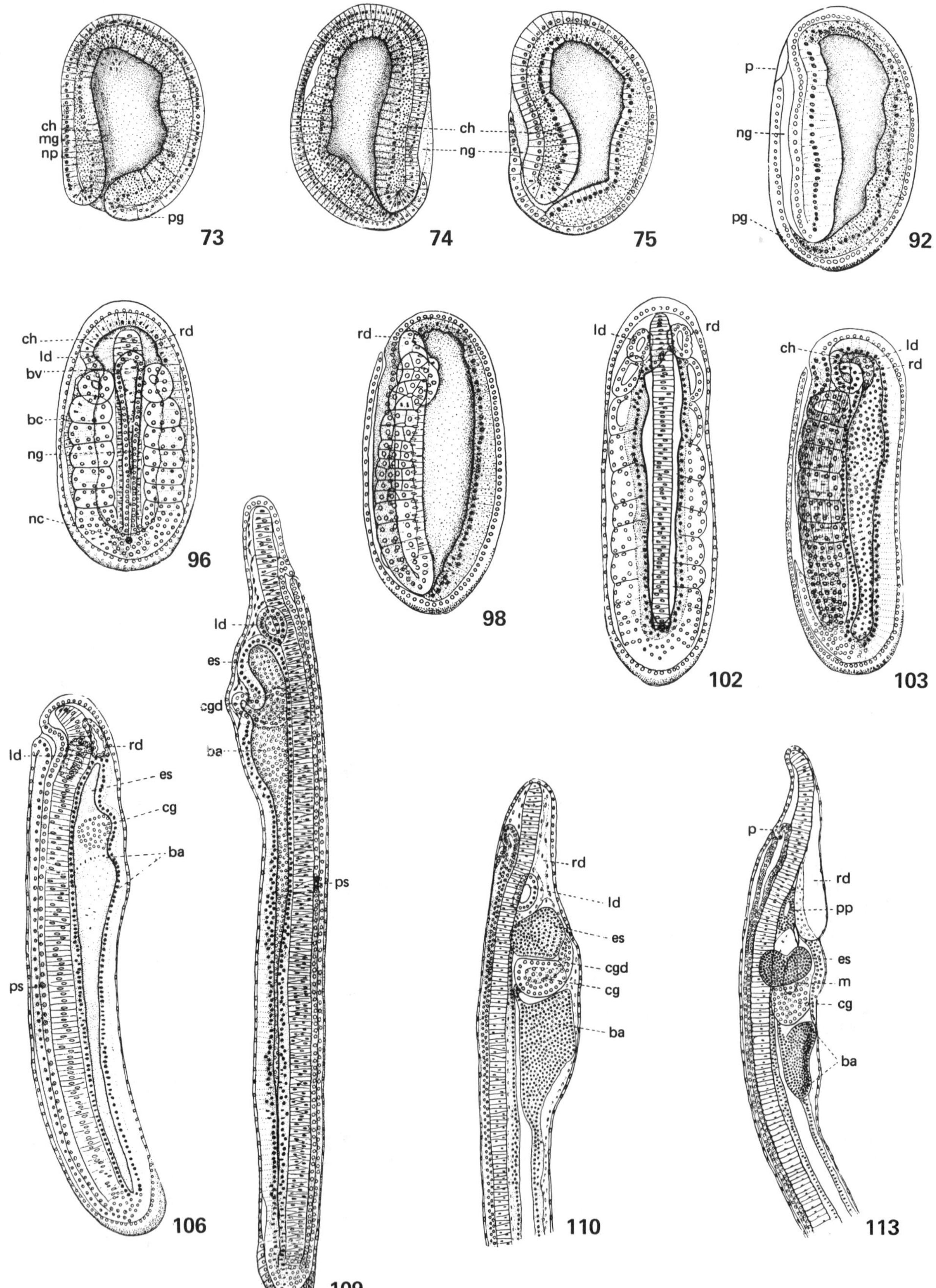

Reprinted by permission of Dr. E. G. Conklin from *Jour. Morph.* 54:69 (1932).

92. The endoderm in the floor of the gastrocoel shows transverse folds.

96. Embryo of 18 hours, viewed from the dorsal side, showing at the highest focus the neural groove (ng) closed in its posterior third and open in its anterior two-thirds, with the primordium of the brain vesicle (bv). At the posterior end of the neural tube is the neurenteric canal (nc); beneath the tube is the notochord with cells and nuclei (ch) flattened in the anteroposterior axis. Lateral to the neural tube and notochord are the somites, the anterior pair containing cavities, and beneath these is the outline of the gastrocoel. The right (rd) and left (ld) dorsal diverticula of the gastrocoel are shown in front of the first pair of somites.

98. Embryo of 20 hours, viewed from right side, showing nine somites on that side. The anterior end of the gastrocoel extends up on the side of the notochord as a pocket (right diverticulum, rd).

102. Dorsal view of embryo of 24 hours, showing the right (rd) and left (ld) diverticula in front of the first pair of somites. The notochord and somites are shown at the highest focus, and the gut, in heavy shaded outline, at the lowest focus.

103. Right view of larva of 23 hours. The neural tube is closed only at its posterior end. Nine or ten somites are visible on the right side. The dorsal diverticulum of the left side (ld) is shown at a low focus, that of the right side (rd) at a high focus. Longitudinal lines through the somites represent muscle fibrillae. In the right wall of the gut where the club-shaped gland will form, nuclei are shown with a central dot.

106. Right view of larva of 26 hours, showing the right dorsal diverticulum (rd, head cavity) at the anterior end of the gut and the left diverticulum (ld) below the notochord; the primordium of the club-shaped gland (cg) is shown at a high focus and as a depression on the ventral wall of the gut; the branchial primordium (ba) lies behind this. The pigment spot (ps) is beginning to form in the floor of the neural tube near its middle.

109. Left view of larva of 48 hours showing in median optical section all the organs described in the previous figure. The duct of the club-shaped gland (cgd) is on the left-ventral side below the wall of the gut. Branched mesoderm cells lie in the space under the anterior end of the notochord. Left diverticulum (ld), endostyle (es), branchial primordium (ba), and pigment spot (ps) as in preceding figure.

REFERENCES

Arey: 64, 65, 71, 72
Balinsky: 178-86, 467, 192-200, 505
Conklin: *Jour. Morph.* 54:69 (1932)
Huettner: Chapter V
McEwen: Chapter III
Shumway: 10, 27, 32, 49, 76-78, 87-89, 103-05

"Vitalistic hypotheses assume some sort of intelligence, or will, or psychic principle in organisms themselves that act as guiding or directing causes of adaptation – a sort of deus in machina. Among these hypotheses are the 'perfecting principle' of Aristotle and Nageli, the 'indwelling soul' of Plato and Bruno, the 'active teleological principle' of Kant, the 'will' of Schopenhauer, the 'desire, need, and appetency' of Erasmus, Darwin and Lamarck, the 'unconscious purpose' of Hartmann, the 'vital force' of Bunge, Wolff, and Virchow, the 'élan vital' of Bergson, the 'entelechy' of Driesch, and the 'psychism' of Pauly, Boveri, and Spemann. The chief objection to these hypotheses is that they are mere names, ghosts without substance, not open to experiment or analysis. Therefore they tend to prevent further inquiry and are a hindrance to research, unless they are recognized as signposts to the unexplored."

Conklin, 1944

EARLY DEVELOPMENT OF THE FISH

Exercise 5

It is very desirable that the student of embryology become acquainted with the early stages of normal fish development. Fish eggs and embryos are unique because they are largely transparent, and the internal development of vertebrate structures can be observed in the living state. This may be accomplished by using the living eggs of the fresh water Medaka, *Oryzias latipes.* Other fish such as the Siamese fighting fish (Beta) or the jewelfish (Hemichromis) may be used, but their breeding habits are not as dependable as those of the Medaka. The zebra fish, *Danio rerio,* is an excellent egg producer. Even the salt-water Fundulus can be used in an aerated marine aquarium.

The Japanese Medaka may be kept in colonies of 50 in a single 15-gallon fish tank; they may be fed on alternate days with Enchytrea (white worms) or. Tubifex (red worms) and mixed dry food. If light, temperature, and food are properly regulated, a good portion of the females in such a tank will produce eggs each morning shortly after dawn, and, by 8:00 to 9:00 A.M., many of the eggs will be in early cleavage stages. The eggs (1 to 80) remain attached to the female for many hours unless they are eaten off by other fish. They may be picked off of the female with forceps and placed in Syracuse dishes with some aquarium water, where they will develop normally through hatching (about 6 days). Following is a condensed schedule of early development of this fish egg. If material is available, the student should study it, classify it, and attempt to compare the major development features with those of embryos previously studied.*

ORYZIAS LATIPES, THE JAPANESE MEDAKA

1. Egg becomes translucent 10 minutes after fertilization.
2. First cleavage within 1 hour. Cleavage meridional; nucleoli visible. Blastodisc.
3. Second cleavage at 1½ hours, at right angles to first. Meridional cleavage with equal blastomeres.
4. Third cleavage at 2 hours, parallel to first. Blastodisc rectangular.
5. Fourth cleavage at 3 hours, parallel to the second. Central blastomeres smaller.
6. Fifth cleavage at 3½ hours. Unlike most teleost eggs this does not give rise to layers of cells.
7. Sixth cleavage at 4 hours. Cleavages no longer synchronous; blastodisc.
8. Blastula at 6 to 8 hours. Blastodisc elevated to form subgerminal cavity and marginal syncytium; beginning of epiboly.
9. Gastrulation begins when the blastodisc has grown over about one-third of the egg, accomplished by invagination and accompanied by delamination and epiboly. The blastopore is the uncovered yolk area.
10. Embryonic development:

First day	–	development of embryonic shield by migration of cells, covering about three-eighths of yolk by blastodisc; optic vesicles appear; somites number about 10.
Second day	–	blastodisc covers five-eighths of the yolk; heart beats 50-60 times per minute; blood is colorless; Kupffer's vesicle appears: 18 somites.
Third day	–	embryo covers three-fourths of the yolk; heart beats as in adult 150 to 170 per minute; blood slightly pink; pigment in eyes; 20 to 28 somites.
Fourth day	–	embryonic movement begins; blood is red; 30 to 38 somites.
Sixth day	–	numerous chromatophores; hatching time. (Maturity in 1 to 1½ months; life span about 1 to 2 years.)

*Experimental studies with eggs of various fish, marine and fresh-water, are described in the author's *Experimental Embryology,* Burgess Publishing Company.

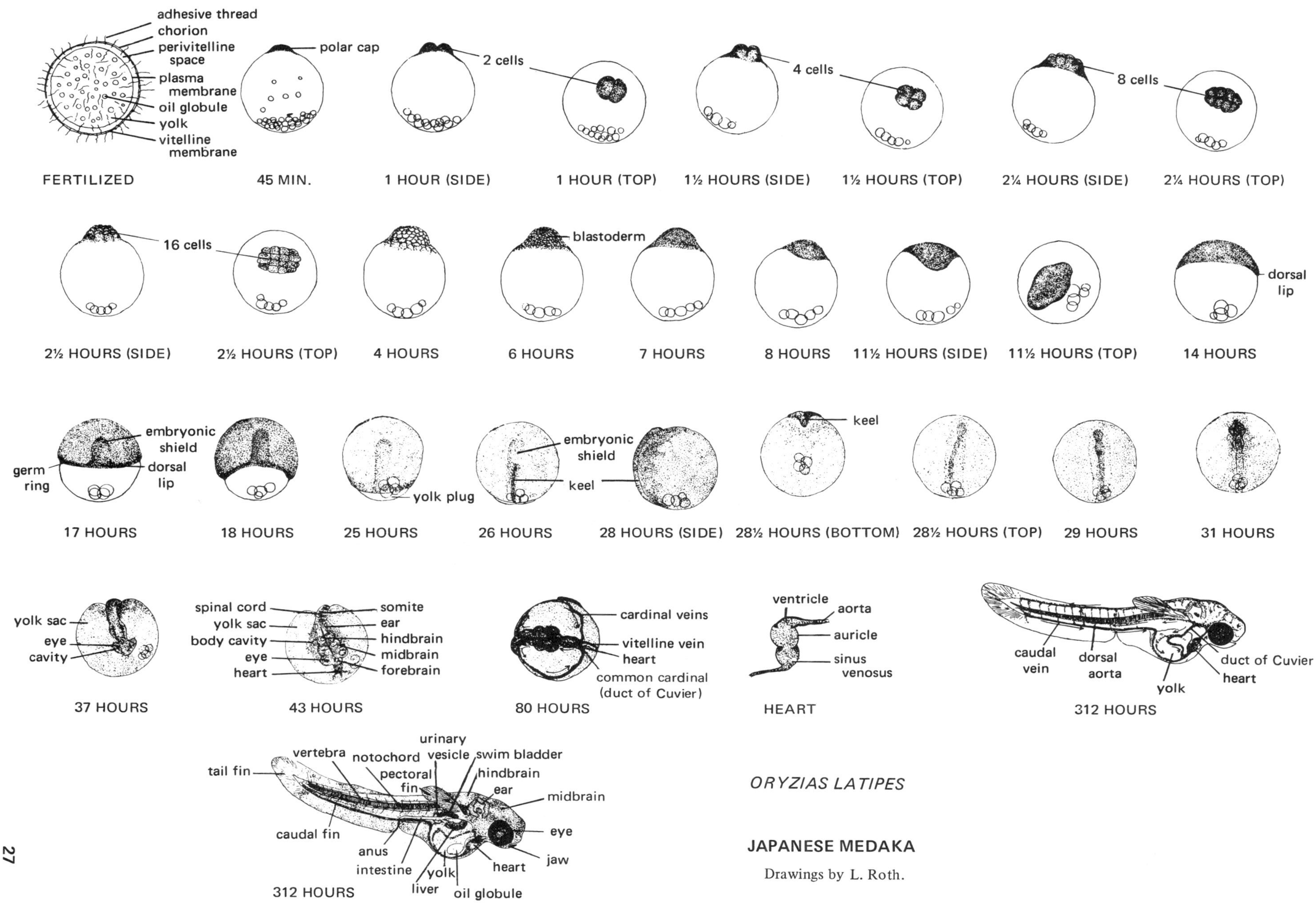

ORYZIAS LATIPES

JAPANESE MEDAKA

Drawings by L. Roth.

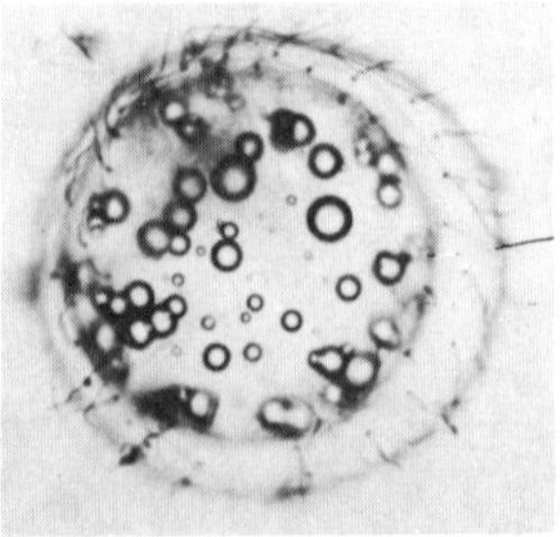

JUST FERTILIZED

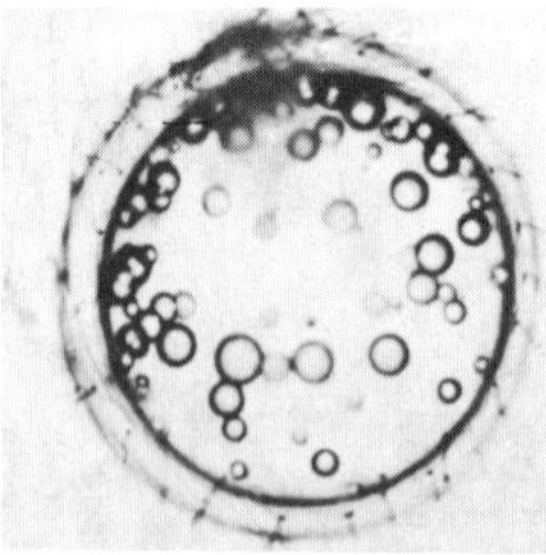

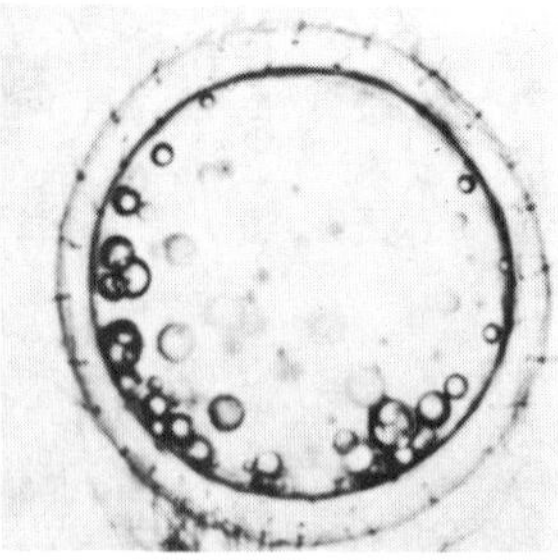

CONCENTRATION OF PROTOPLASM AT POLAR CAP

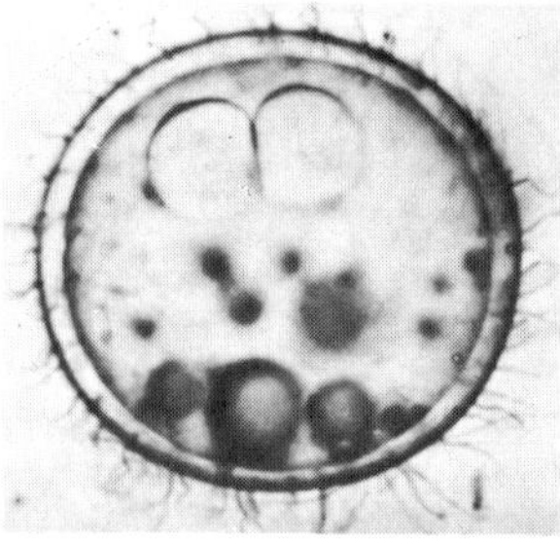

2 CELL – 1 HOUR

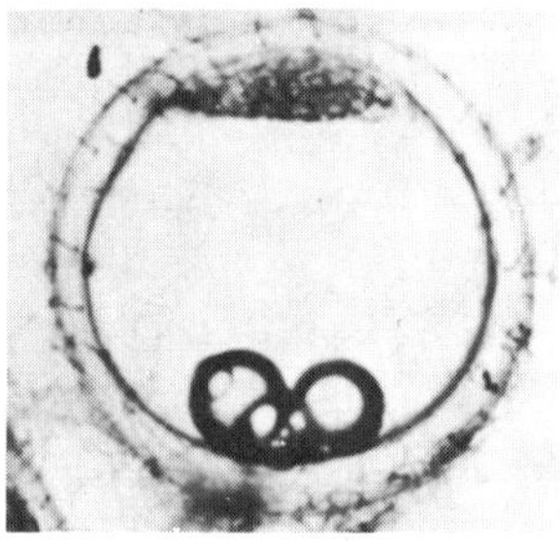

BLASTULA – 3 HOURS

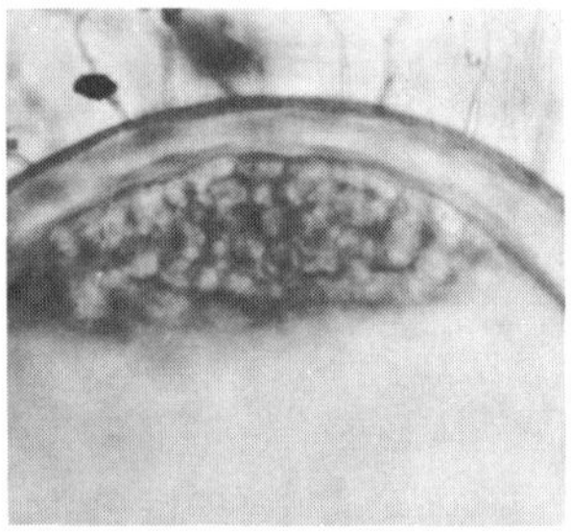

3 HOURS – H.P.

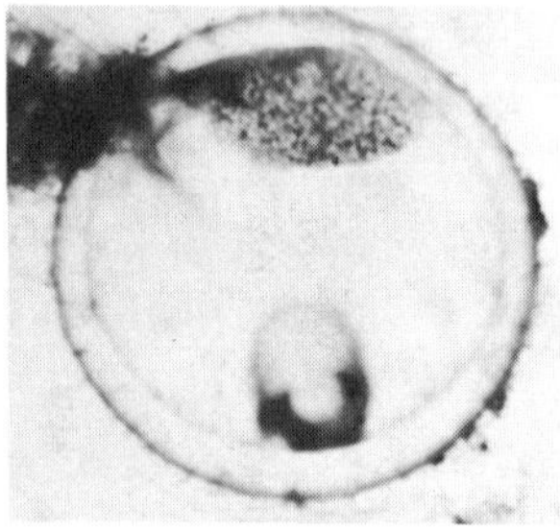

10 HOURS

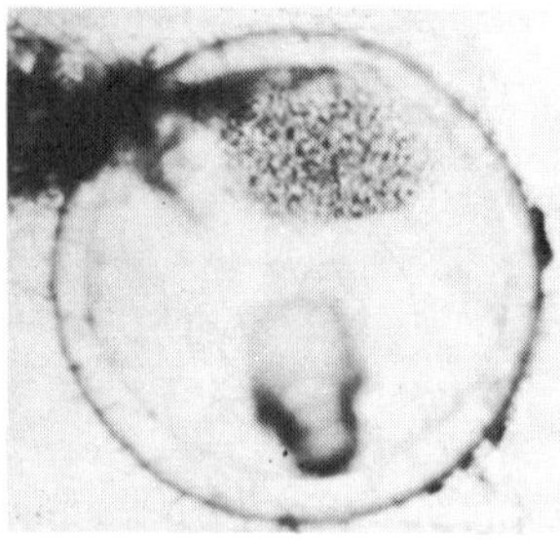

12 HOURS

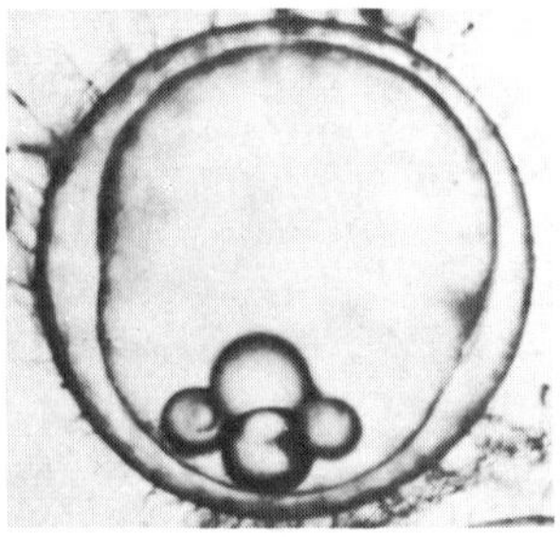

15 HOURS

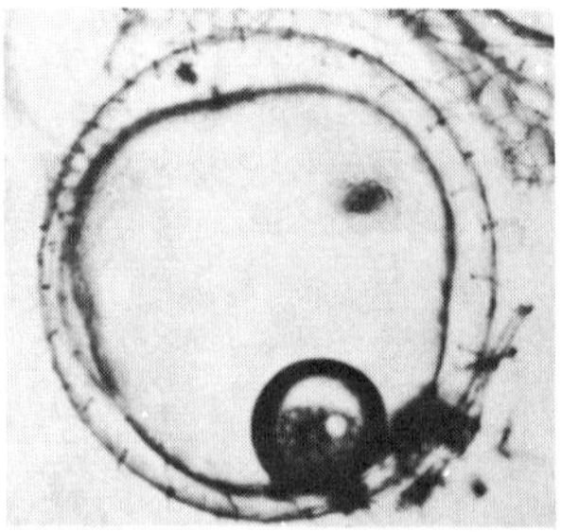

18 HOURS

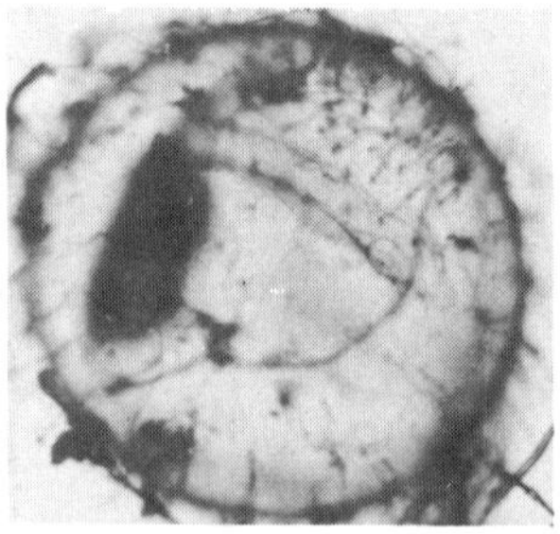

35 HOURS

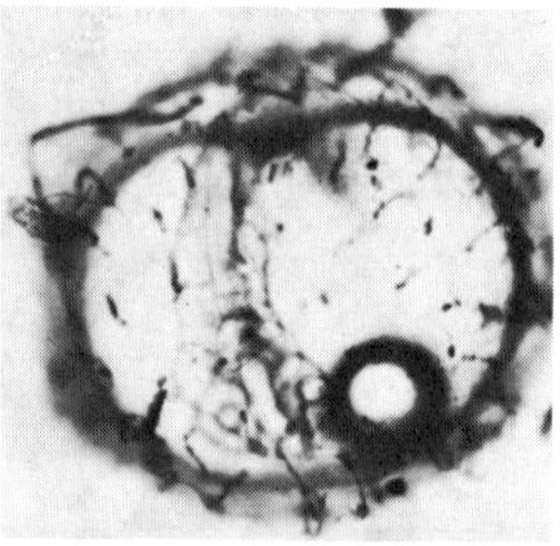

40 HOURS

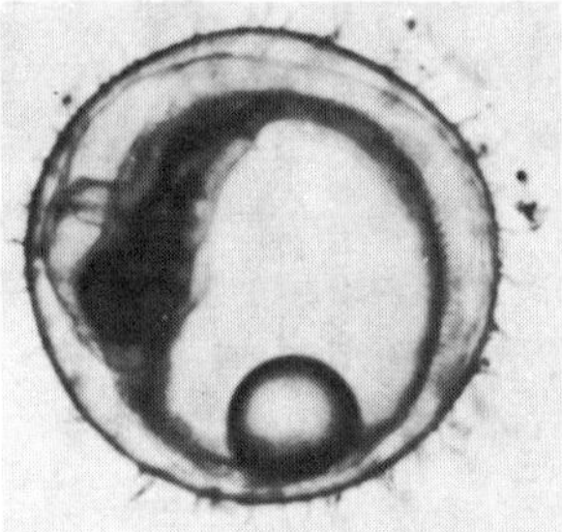

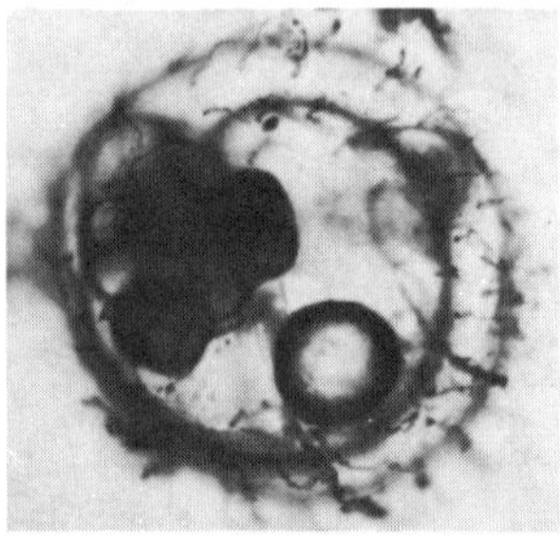

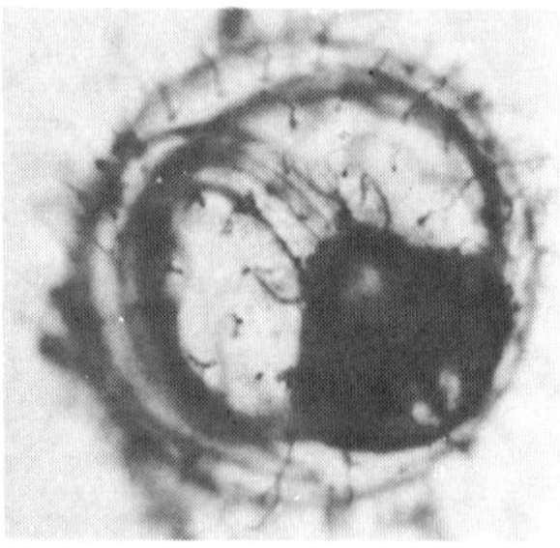

2 DAYS TO HATCHING

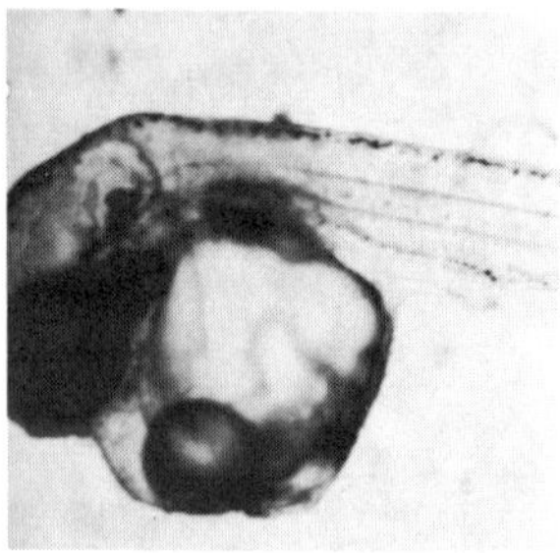

YOLK SAC – HATCHED

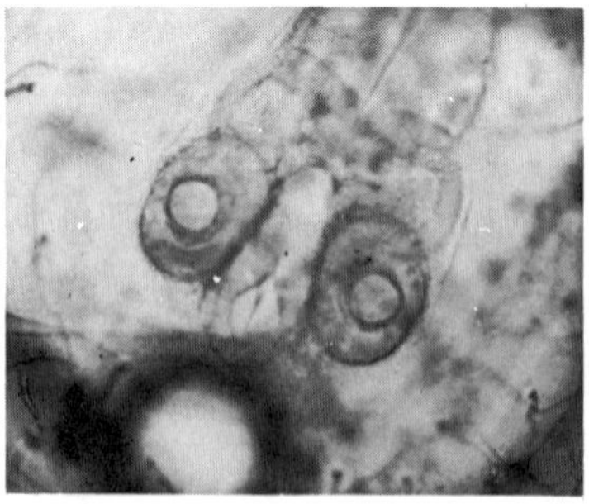

OPTIC VESICLES

EARLY DEVELOPMENT
OF
ORYZIAS LATIPES

THE JAPANESE MEDAKA

Photographs by L. Roth.

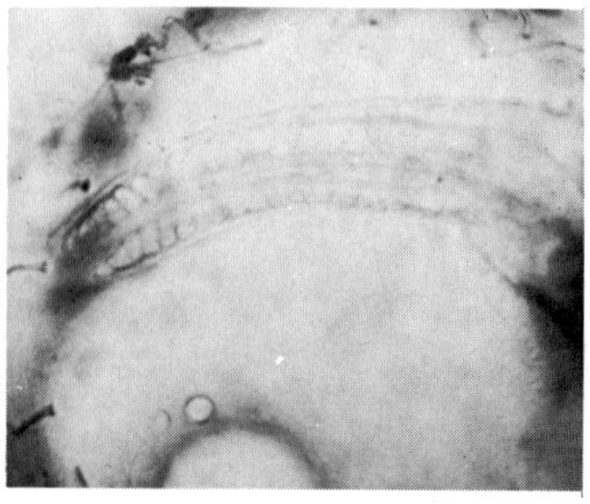

TAIL SOMITES

DRAWINGS: FISH EMBRYOS

NORMAL DEVELOPMENT OF FUNDULUS HETEROCLITUS AT 25°C.

Age in hours	Oppenheimer stage	Developmental stage
0	1	Unfertilized egg.
1	2	1 cell.
1½	3	2 cell.
2	4	4 cell.
2½	5	8 cell.
3	6	16 cell.
3½	7	32 cell.
4	8	64 cell.
4½		128 cell.
5		256 cell.
5½-6		Early high blastula.
7-9	9	Late high blastula.
10-12	10	Flat blastula.
13-15	11	Expanding blastula. Blastula enlarges.
16	12	Gastrulation begins. Early gastrula.
18		Blastoderm about one-third over surface yolk.
19½	13	Blastoderm about one-half over surface of yolk. Middle gastrula.
21		Blastoderm about two-thirds over surface of yolk.
22		Blastoderm about three-fourths over surface of yolk.
23		Embryonic shield condenses to form keel.
24	14	Optic vesicle first visible as an expansion of the forebrain. Large yolk plug.
25	15	Small yolk plug.
26	16	Blastopore closes.
27		First somites formed.
28		Four somites.
31	17	Optocoele develops.
33	18	Auditory placode forms. Optocoele connects across brain.
34	19	Optic cup forms, and lens develops. Neurocoele develops. About 10 somites.
38	20	Expansion of the midbrain to form the optic lobes.
40		Melanophores first appear on yolk.
42		Melanophores appear on embryo.
44	21	Heart pulsates. No circulation. Hindbrain enlarges.
46	22	Circulation begins, through dorsal aorta and vitelline vessels.
48		Circulation through ducts of Cuvier.
60	23	Otoliths develop.
72		35 somites developed.
78	24	Pectoral fin bud appears.
84	25	Retinal pigmentation begins. Urinary vesicle formed. Caudal fin begins to develop.
90	26	Liver develops. Cartilage begins to differentiate.
102	27	Pectoral fin round.
108		Lens of eye just obscured by retinal pigmentation.
114	28	Pigmentation of peritoneal wall.
120	29	Circulation in pectoral fin.
126	30	Fin rays in caudal fin visible.
144	31	Air bladder develops.
168		Neural and hemal arches in vertebrae in tail are developed.
192		Head flexure begins to straighten out.
216		Head flexure nearly straightened out.
240		Mouth opens.
264	32	Hatching.
	33	Pigmentation of air bladder.

From Solberg, 1938: *Jour. Exp. Zool.* 78:445.

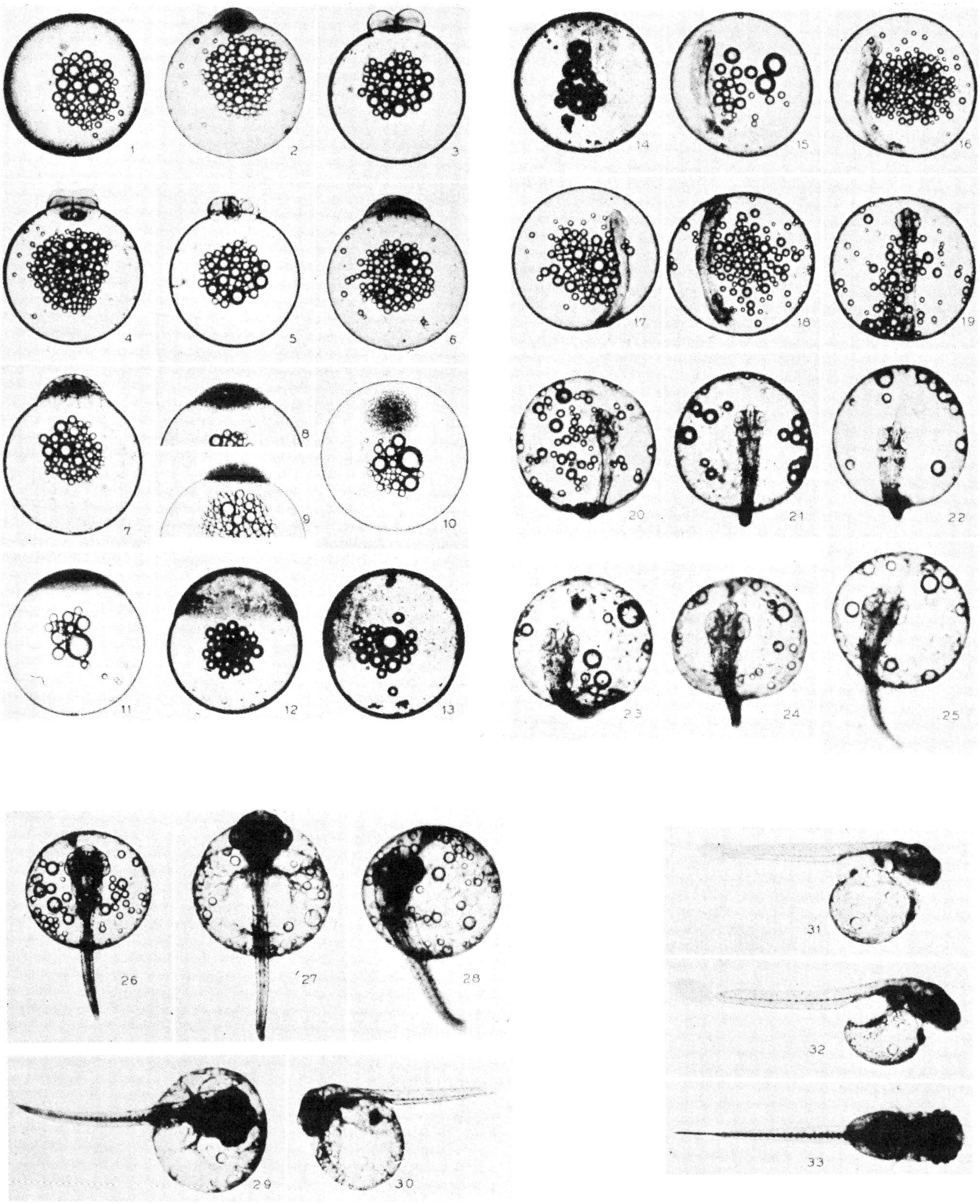

NORMAL STAGES OF FUNDULUS

Photographs by Jane M. Oppenheimer

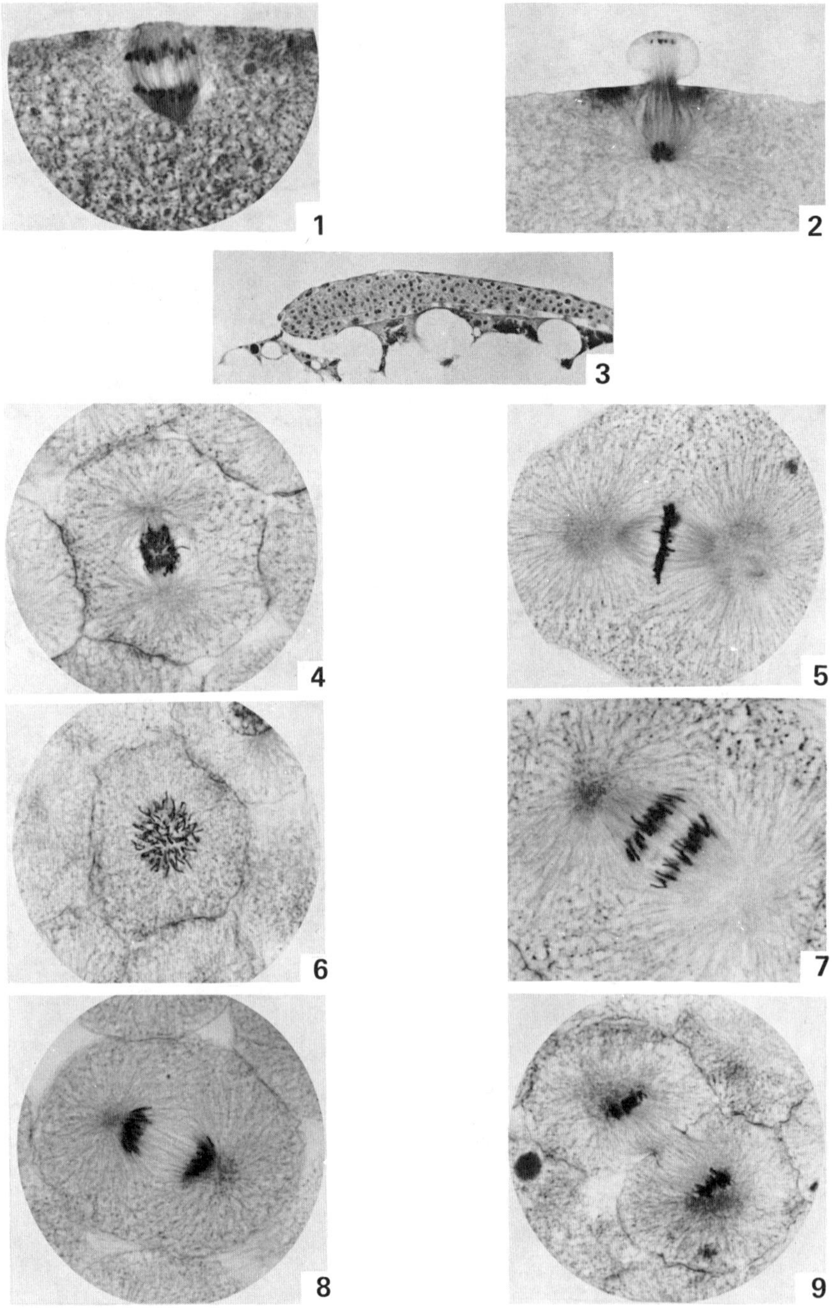

WHITEFISH DEVELOPMENT

Fig. 1 – First polar body formation in anaphase.
Fig. 2 – First polar body formation in telophase. The polar body about to be pinched off.
Fig. 3 – Blastula formation. Note distinct cells above yolk.
Fig. 4-9 – Mitosis in whitefish blastula cells.

Fig. 4 – Late prophase.
Fig. 5 – Metaphase, lateral view.
Fig. 6 – Metaphase, polar view.
Fig. 7 – Anaphase, lateral view.
Fig. 8 – Late anaphase, lateral view.
Fig. 9 – Telophase, lateral view.

Exercise 6

THE FROG:

Gonads and Maturation

METHODS OF REPRODUCTION IN THE FROG

The Female

The female leopard frog *(Rana pipiens)* is identified by the lack of the thumb pad peculiar to the male and the possession of a flabby rather than muscular abdomen. The paired *ovaries* are hollow and lobed. There may be from seven to twelve distinct lobes, all connected by an internal cavity and held to the dorsal body wall by means of mesentery, the *mesovarium.* Between May and August the oöcytes undergo the growth phase of the maturation process. From August until the breeding season in spring, the ovary will be distended with eggs ready to be released (i.e., ovulated).

The eggs are located in the periphery of the ovary, each within its individual sac or *follicle.* The eggs to be ovulated possess their full complement of yolk and pigment, while the younger *oöcytes,* representing immature eggs which are to mature the following year, appear as small, white bodies interspersed among the mature ova (i.e., full-sized *primary oöcytes*). The young oöcytes possess no yolk or pigment.

As the eggs are released from the ovary, they pass into the body cavity, which is lined almost entirely with cilia. They are carried by these cilia to the *ostia* (openings into the oviducts). These ostia are located anteriorly in the body cavity at the junction of the liver, lungs, and oviducts. Each ostium is highly ciliated and elastic. *The eggs are forced into the openings by currents caused by body cavity and ostial cilia and then are spiralled through the oviduct by oviduccal cilia.*

Within the oviducts the eggs receive their three gelatinous coats of *albumen.* The first polar body begins to form at the time of ovulation and is given off in the upper part of the oviduct. The second maturation division progresses to the metaphase stage by the time the egg reaches the *uterus.* Maturation is stopped at this point until the egg is fertilized. Normally frog's eggs are fertilized externally after oviposition during amplexus. (See page 45.)

It is possible to induce breeding (ovulation and amplexus) by the injection of the anterior pituitary hormone. An actively ovulating female can be dissected for demonstration purposes in the laboratory.

The Male

The male frog possesses large black thumb pads, a muscular abdomen, and can produce a respectable croaking sound. The *testes* are internal in these cold-blooded (*poikilothermic*) forms and are held to the kidneys by means of a fold of mesentery known as the *mesorchium.* Through this double-layered mesorchium numerous small tubes (vasa efferentia) pass from the testis into the substance of the kidney where the ducts join the malphigian corpuscles. The sperm are, therefore, carried into the uriniferous tubules and across the kidney into the *Wolffian* (mesonephric) duct. This duct may be regarded as a functional urinogenital duct during the breeding season. The sperm are collected in a saccule (*seminal vesicle*) at the lower end of the Wolffian duct, where they remain until, during amplexus, they are emitted into the water over the eggs.

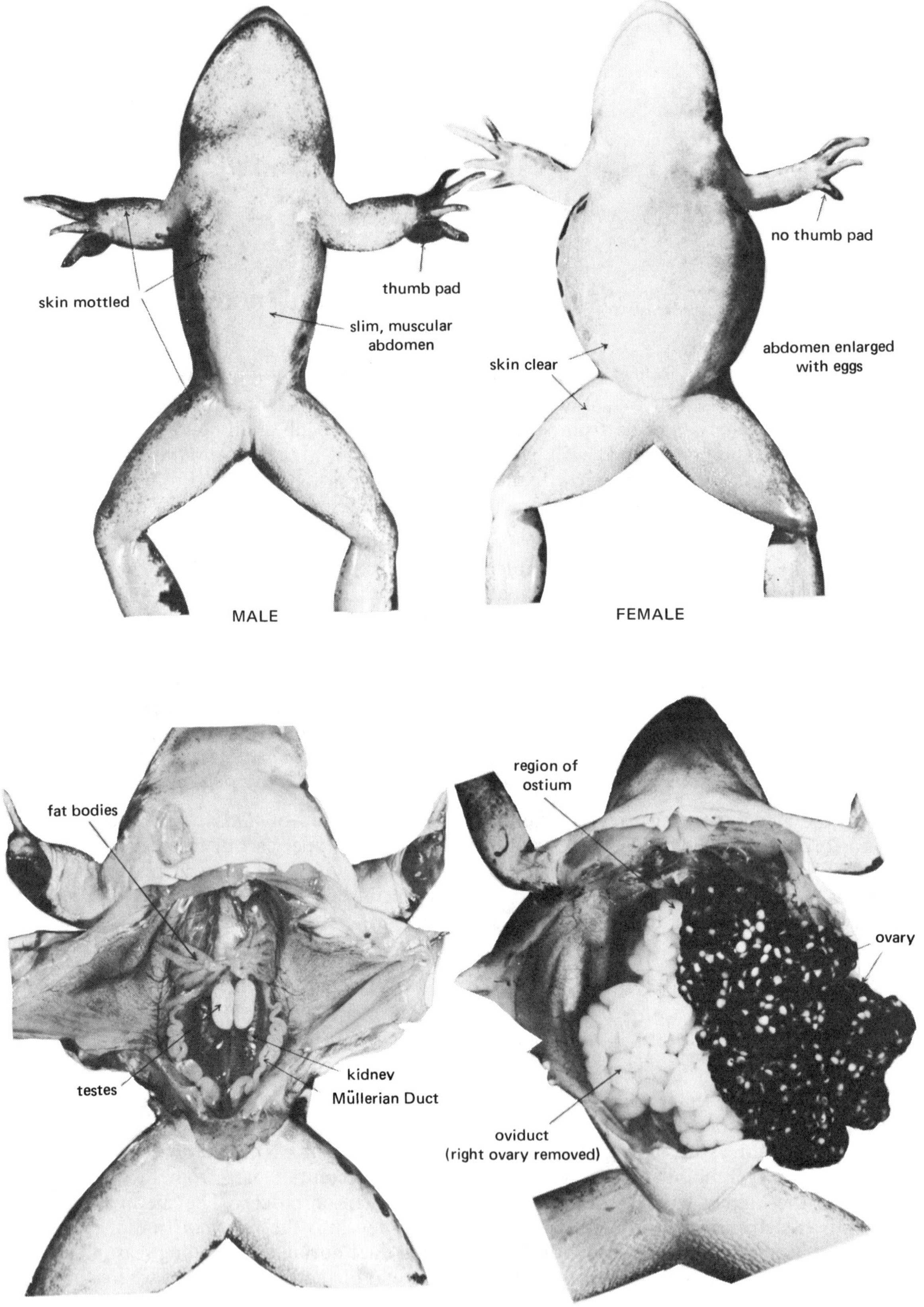

THE LEOPARD FROG: *RANA PIPIENS*

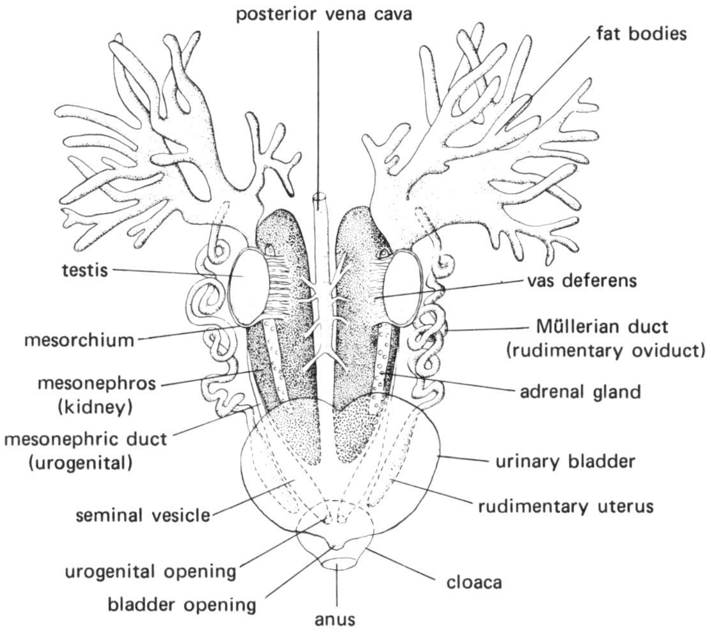

UROGENITAL SYSTEM OF ADULT MALE FROG

From R. Rugh. *Vertebrate Embryology.* 1964. Harcourt, Brace & World, New York.

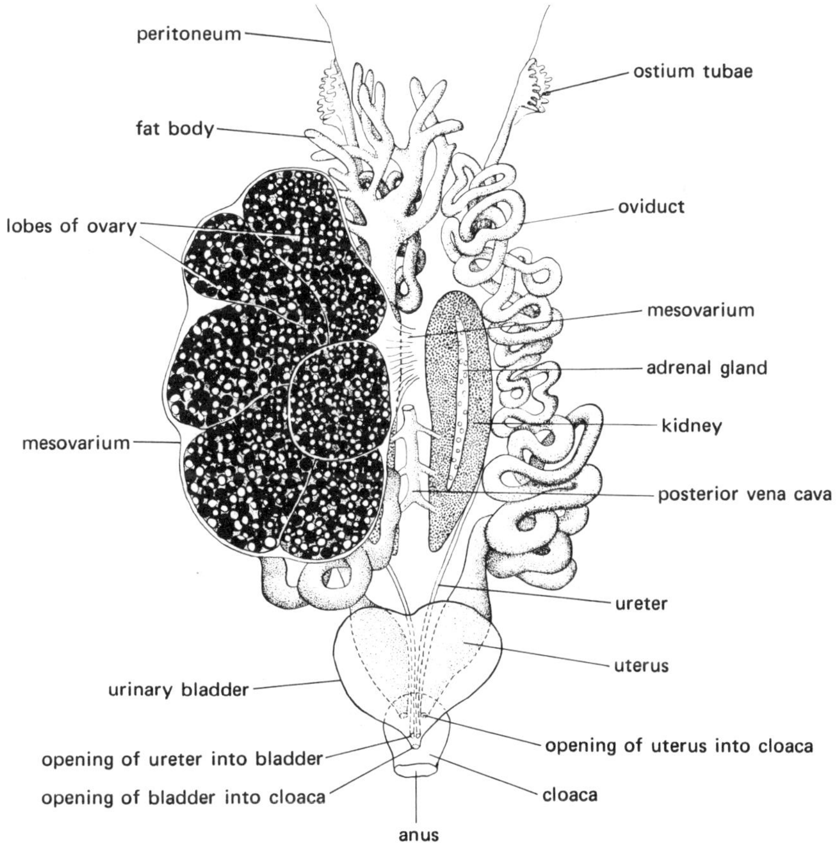

UROGENITAL SYSTEM OF THE ADULT FEMALE FROG

From R. Rugh. *Vertebrate Embryology.* 1964. Harcourt, Brace & World, New York.

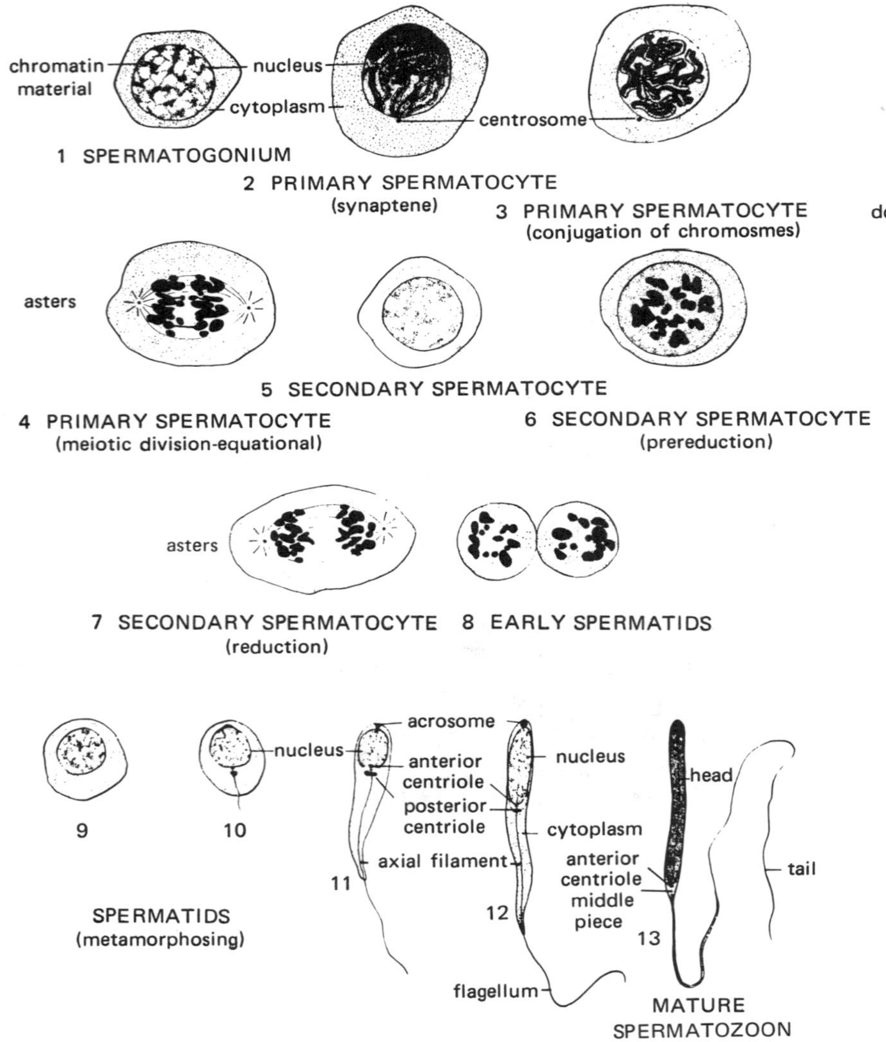

SPERMATOGENESIS IN THE FROG

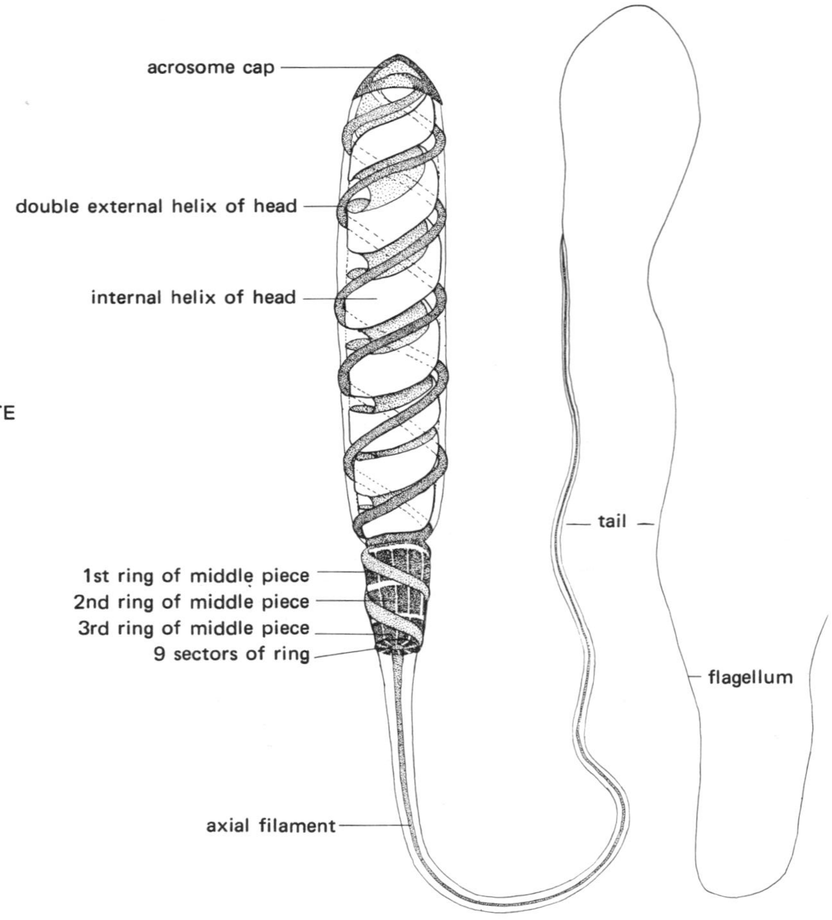

ULTRA-STRUCTURE OF RANA SPERMATOZOAN

After Serra and Vicente, 1960. *Rev. Port. Zool.* 2:223.

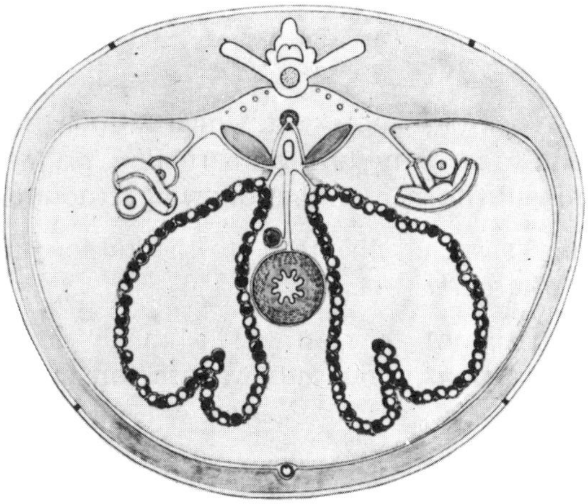

Egg transplanted to nonovulating female but acquiring normal a-mount of jelly from inactive oviduct.

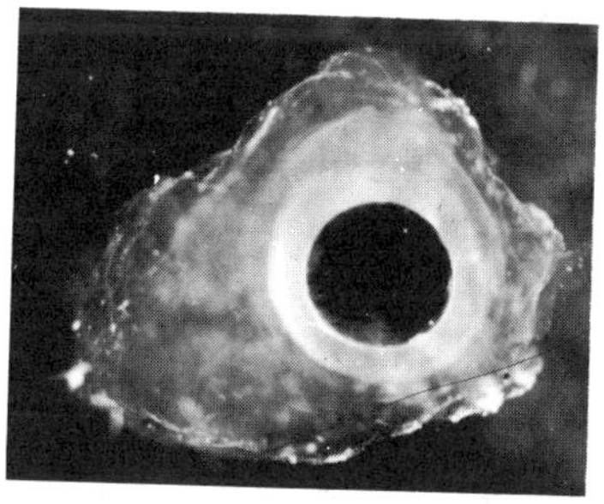

Midbody schematized section of female frog, showing suspended and hollow ovary (see page 47).

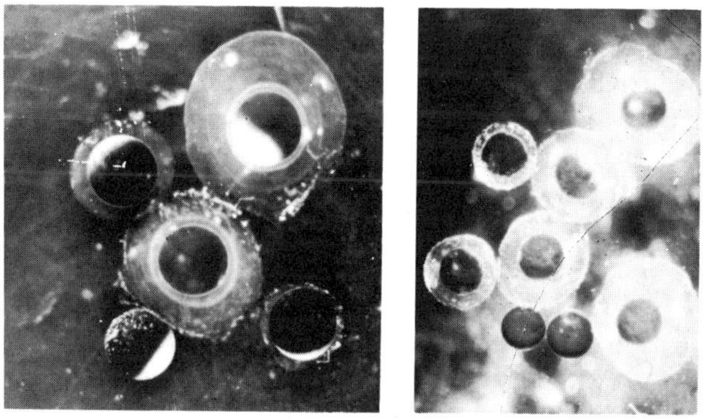

Egg taken from the body cavity and various levels of the oviduct to show increasing deposition of jelly.

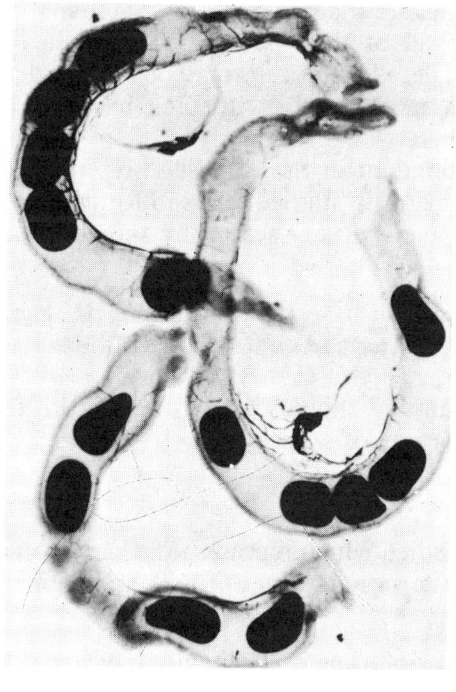

Oviducts made transparent to show eggs being propelled toward the uterus. Note distortion of the eggs, which occurs without injury to them, and deposition of jelly.

MICROSCOPIC STUDY OF THE FROG GONADS

The Testes

The slides of the frogs testes are made at two different seasons: in August, which is the period of high spermatogenetic activity, and in February, a period of inactivity during hibernation, prior to breeding activity in April. Study carefully both slides, noting differences in the following structures.

1. MESORCHIM – The portion of the peritoneum which suspends the testis to the kidney and between the layers of which pass the tubes of the *vasa efferentia.*

2. TUNICA ALBUGINEA – outer connective tissue capsule of the testis. This tough sheath extends into the substance of the testis as *septula* or partitions, which subdivide the organ into many compartments known as seminiferous tubules.

3. SEMINIFEROUS TUBULES – the many subdivisions of the testis, connected with the vasa efferentia, by way of the collecting tubules. Contain cells in various stages of spermatogenesis.

4. COLLECTING TUBULES – these are few in number and are lined with nonciliated cuboidal epithelium. They generally contain some loose spermatozoa and are continued into the vasa efferentia.

5. INTERSTITIAL CELLS – these constitute the interstitial tissue between the seminiferous tubules and will show great variation when one compares the February and the August testis. In higher forms this tissue is known to have endocrine function, as it probably does in the frog.

6. CELLS WITHIN THE SEMINIFEROUS TUBULE (high magnification):

   a. The numerous small peripheral cells which possess granular nuclei are the *spermatogonia,* which are to give rise to spermatozoa for future seasons.

   b. The most centrally located cells with long black heads and longer but gray or yellowish tails projecting into the central cavity (lumen) of the tubule, are the *spermatozoa.* These mature gametes are clustered into groups of 20 to 40 with all of their heads embedded in the cytoplasm of a large nutritive or nurse (Sertoli) cell. These Sertoli cells may be obscured by the many sperm heads, but they are elongated columnar type cells with their bases at the periphery of the seminiferous tubule.

   c. *Maturation stages* will also be found and are most abundant in the August testis. Identify such stages as you can find by comparison with the previously studied series in Rhomaleum. (The mature spermatozoa may be liberated from the testis at any season by the action of the anterior pituitary hormone.)

Fat bodies may be found associated with the testis. These will have the appearance of finger-lobed, yellowish bodies consisting of vacuolated cells and should not be confused with the mesorchium.

If time permits, the student should study living frog sperm and should compare them with the prepared slides of salamander and human spermatozoa. Note variety of shapes and activity.

The Ovaries

Several stages in the development of the ovary are to be studied which represent the early gonad of a very young frog and the maturing ovary containing eggs in various stages of yolk and pigment accumulation.

1. THE IMMATURE OVARY – examine under low magnification. The outer membranous coat of the ovary is made up of a thin peritoneal covering which holds the ovary to the body wall (the *mesovarium)* and an underlying thin layer of tissue called the *theca externa.* This external theca is quite vascular except for a small area over each egg, the region where the matured egg will rupture into the body cavity. Within the external theca is a thin theca interna which extends around each follicle as the *cyst wall,* made up principally of smooth muscle fibers. This *internal*

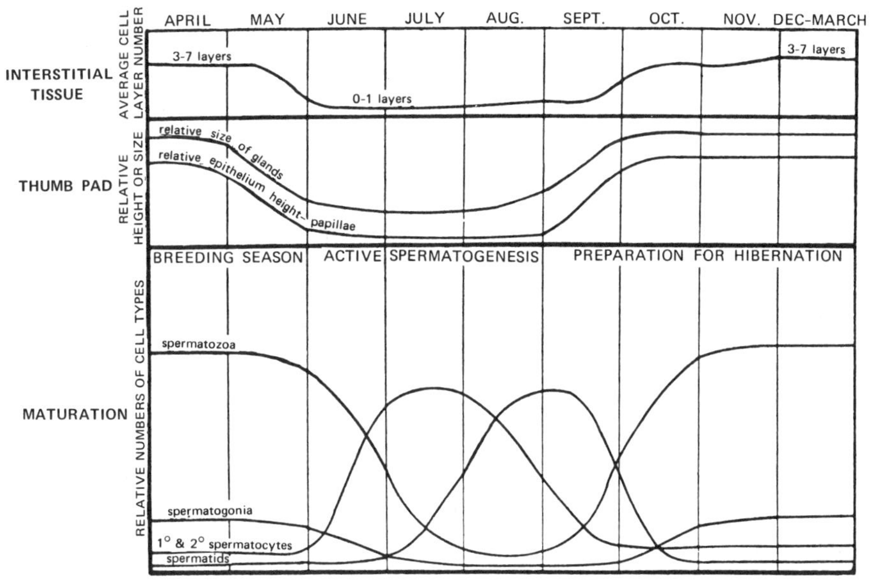

NORMAL CYCLIC CHANGES IN PRIMARY AND SECONDARY SEXUAL CHARACTERS IN THE FROG, *RANA PIPIENS*

From Glass and Rugh, 1944: *Jour. Morph.* 74:409.

OUTLINE OF FROG TESTIS (LOW MAGNIFICATION)

CELL TYPES IN SEMINIFEROUS (TUBULE) AUGUST TESTIS (HIGH MAGNIFICATION)

THE FROG TESTIS

theca does not cover the egg in the region of future rupture, and at the time of ovulation it helps to force the egg out of the follicle. Within the tissue of the ovary locate some of the masses of *yellow pigment,* which are the sources of supply of pigment to be accumulated in the maturing eggs. Locate also the blood vessels containing oval and nucleated corpuscles. (Remember that these are sections of whole ovaries and that some of the eggs may be so cut as to appear to be within the lumen of the ovary, an artefact.)

Under low magnification select an egg which appears to be complete, showing a distinct nucleus and follicle wall, and carefully switch to high magnification to identify:

a. Cyst wall – the internal theca is the thin membrane of smooth muscle fibers (containing spindle-shaped nuclei) which surrounds the egg except where the follicle is exposed to the body cavity.
b. Follicle cells – the granular, ovoid cells lying between the cyst wall and the egg cell membrane. Note that there is no vitelline membrane around these early eggs.
c. Oöcyte nucleus – the egg cell is regarded as an oöcyte as soon as it begins to grow in size, to distinguish it from the more primitive *oögonium.* The nucleus (*germinal vesicle*) has a serrated wall or membrane and contains many nucleoli. The peripheral nucleoli are large and often vacuolated, while the small and more central nucleoli are solid and darker. The larger nucleoli probably migrate into the cytoplasm to function as *yolk nuclei* (not true nuclei, however) which help as centers of organization and distribution of accumulating yolk in the growing oocyte. Very little yolk or pigment will be seen in these youngest oōcytes.

DRAWINGS OF THE FROG OVARY

2. THE MATURE OVARY – the fundamental tissues of the mature ovary are identical with those of the immature gonad, but the distinction between the ova present will be clearly evident. During the growth phase of maturation the ovum increases in size with the accumulation of yolk and pigment, the latter being more or less limited to the periphery of the egg over one hemisphere. The original cyst wall and follicle cells are so stretched by this growth of the egg cell that they will now appear as thin lines closely adherent to the egg. But there has also appeared another structure, a noncellular *vitelline membrane* between the *follicle cells* and the *egg cell membrane.* This membrane gets its name from the fact that it encloses the egg yolk or vitellus. It is a secretion formed by the cooperation of the follicle cells and the egg itself.

Each fully grown ovarian egg with yolk and pigment is known as a *primary oöcyte* as long as the germinal vesicle is intact. The bulk of the egg consists of *yolk*, which under high magnification, appears to be made up of oval granules similar to those found in milk emulsions. Between these more mature eggs may be seen ŏogonia and young oŏcytes similar to those previously studied, cells which are to be developed in a subsequent season.

The nucleus of the primary oöcyte undergoes no maturational divisions while the egg is still within the (frog's) ovary, so that the germinal vesicle observed represents the diploid nucleus of

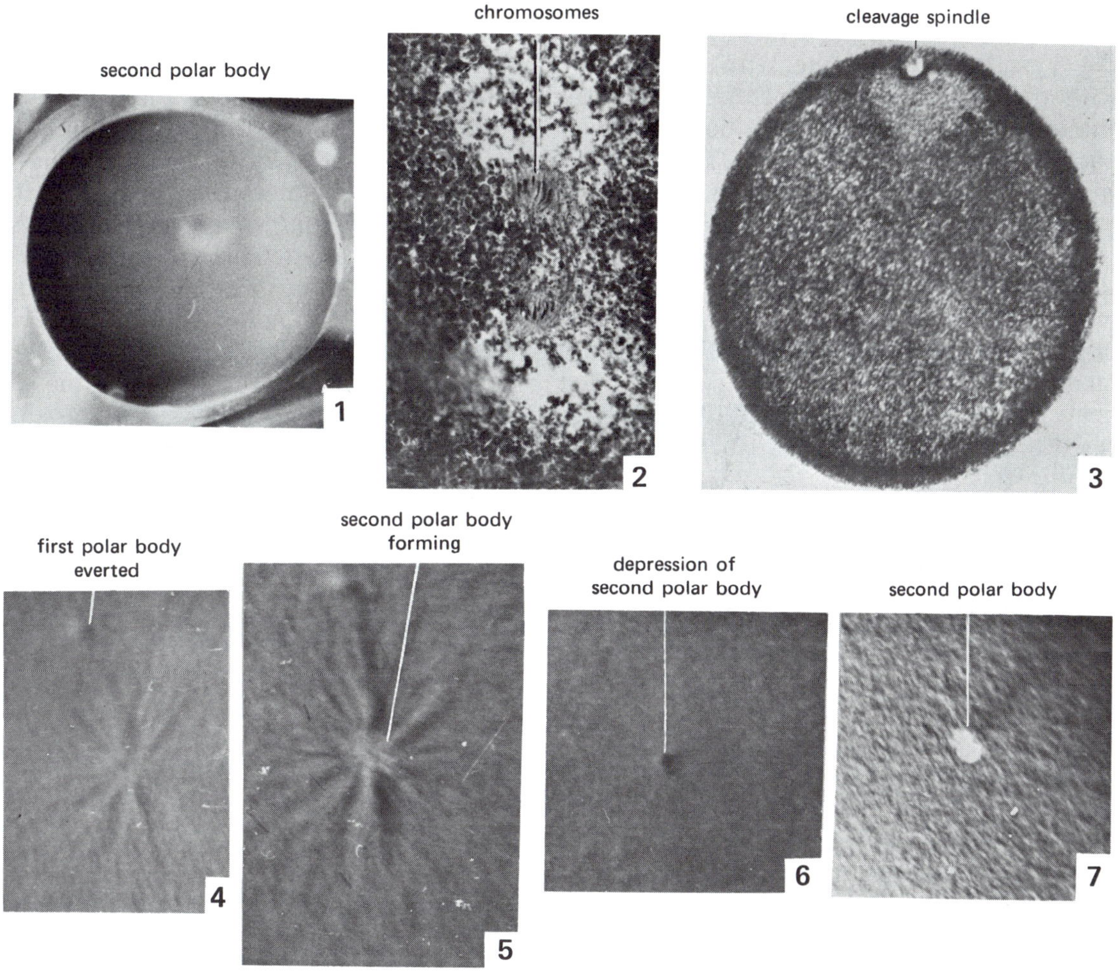

SECOND POLAR BODY FORMATION IN FROG

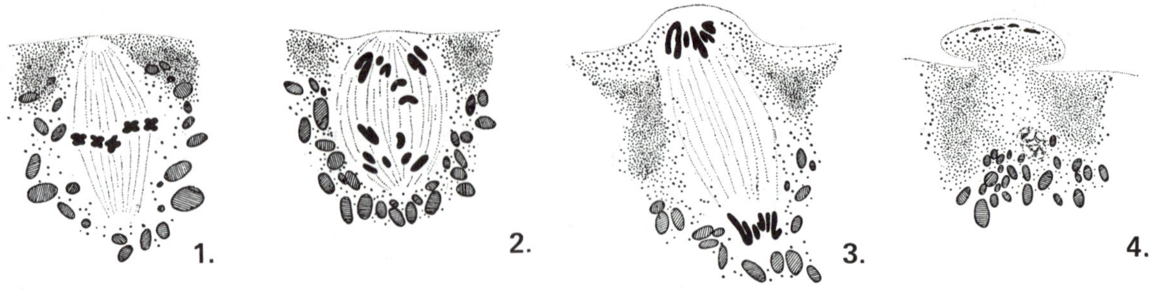

POLAR BODY FORMATION: ***RANA PIPIENS***

From Porter, 1939: *Biol. Bull.* 77:233.

Figs. 1- 4. Semidiagrammatic representations of four stages in second polar body formation of *R. pipiens* eggs. Drawings were made with camera lucida and given exact distribution of pigment granules, yolk platelets, and chromosomes, only part of which are shown. Selected from considerable material sectioned at 10 μ. (Eggs inseminated and kept at 12°C.)
Fig. 1. Division spindle as in egg at time of insemination.
Fig. 2. Anaphase of maturation division. Stage at which spindle can be seen from exterior of egg as small black dot. Egg fixed 35 minutes after insemination.
Fig. 3. Early telophase. Egg fixed 50 minutes after insemination.

the typical primary oöcyte. With appropriate staining, some feathery chromosomes may be seen within this nucleus, or the entire germinal vesicle may be dissected out of the living egg and studied in isolation. (This may be done by the instructor as a demonstration.) The number of nucleoli (in the stained material) will be somewhat reduced, and they will appear to be more vacuolated than the nucleoli seen in the preceding study of living material. Due to the affinity for certain dyes, the heavily stained bodies within the nucleus are known as *chromatin nucleoli.*

In most mature eggs, it will be clear that the pigment is distributed around the periphery of about one-half of the egg (the animal pole or hemisphere), but it must be remembered that sections may not be cut exactly through the egg axis and that this polarity may not be evident in all eggs. The nonpigmented half is known as the *vegetal hemisphere.* (Examine whole eggs, models, and the photographic series presented in this *Guide.)* The nucleus lies within the animal hemisphere so that the *egg axis* is an imaginary line which runs through the egg nucleus and the center of the egg, consequently passing through the center of both the animal and the vegetal hemispheres.

REFERENCES

Arey: 62-66
Bodemer: 69-84
Huettner: 92-98
McEwen: 104-13
Nelsen: 3-176
Patten (1964): 30-67
Rugh: *Frog.* Chapter I through III
Rugh: *Vert. Emb.* 20-26
Shumway: 32, 44-45

TECHNIQUE FOR PROVIDING AMPHIBIAN EGGS AND EMBRYOS*

Since 1929 when independent investigators in Brazil and in Wisconsin published descriptions of a procedure for inducing ovulation in frogs and toads, it has been possible to provide students with living amphibian eggs and embryos from early September until the normal breeding season in the spring. It is suggested that the instructor demonstrate the following procedure and provide each student with living embryos for study concurrently with the preserved material. Departments which have the facilities may also give the students the opportunity to do some simple experiments with living embryos.*

*A detailed description of the method for inducing ovulation, fertilizing eggs, and raising embryos, as well as numerous experiments in the field of embryology, will be found in the author's *Experimental Embryology,* 1962, Burgess Publishing Company, Minneapolis.

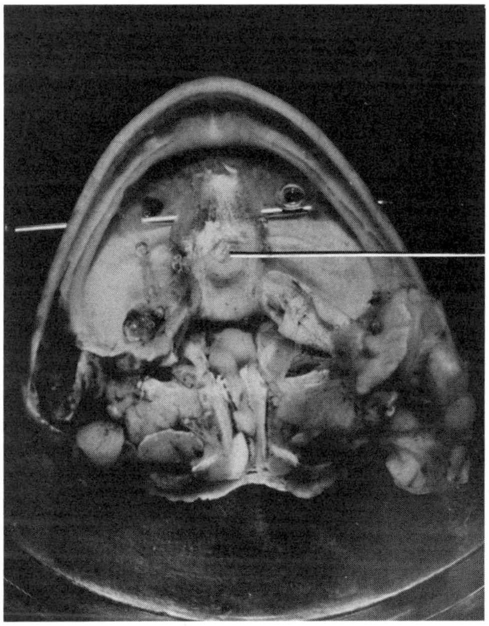

REMOVAL OF PITUITARY GLAND TO INDUCE OVULATION

The Induction of Ovulation

All species of frogs that have been tested will respond to the treatment described, but the leopard frog. *Rana pipiens,* has been found to be the most uniformly satisfactory. Each adult female will give about 2,000 eggs, but to obtain these it is necessary to secure from collectors seven large females and two large males, freshly caught. Laboratory-kept animals deteriorate quickly, due both to high temperature of most laboratories and to starvation. If kept moist and at 10°C, frogs may be used for about 3 weeks.

Select the largest female frog for the purpose of ovulation and put it aside in a covered battery jar with a small amount of tap water. Remove the pituitary glands from each of the other six females. This is done by inserting scissors in the angle of the jaw, cutting posteromedially behind the eyes, then across the head through the occipital region, and finally to the other jaw, thereby removing the cranium. Invert the skull and locate the large cross formed by the bones in the floor of the cranium. The pituitary gland lies just posterior to the optic chiasma beneath the point where the bones cross each other. Insert the sharp scissors into the posterior brain cavity, but lateral to the brain tissue, and cut the bone anteriorly through the floor of the cranium on either side of the brain stem. Avoid injury to the brain tissue. The parallel cuts should extend as far as the orbits. With forceps, deflect forward the floor of the cranium and locate the pinkish, kidney-shaped pituitary gland. It will be attached either to this cranial floor or to the optic chiasma and will usually be surrounded by white endolymphatic tissue which has no known endocrine function.

The pituitary gland should be lifted out with forceps and placed in less than 2 cc of water. Glands from the other five females are to be placed in the same water, and injection should follow within a few minutes. The glands tend to lose their potency at room temperatures after a few hours. Six whole female pituitary glands will be adequate to induce complete ovulation in a female between September and January, after which time the dosage may be reduced, because of the approach of the normal breeding season and the greater sensitivity of the gonads. Three female glands will be sufficient in March and April. The anterior pituitary glands from female frogs are about twice as potent in sex-stimulating capacity as the similar glands from male frogs.

The whole glands are sucked up into the barrel of a 2-cc hypodermic syringe, and then a large bore (#18) hypodermic needle is added. The female to be injected is held firmly by the legs, and the injection is made into the posterolateral abdominal cavity. The frog should be held in such a position that the injection is downward since the glands are heavier than water and will sink into the exit of the syringe. Avoid injury to the large cutaneous veins, the ventral abdominal vein, and to the vital organs. The injected female is then placed in a battery jar containing water to a depth of about 1 inch. The males need no special treatment.

At the usual laboratory temperatures of 23°-25°C, eggs will first appear in the uteri of the frog in about 24-36 hours. If the ovulating female is kept at 10°C, the time lapse between injection and appearance of the uterine eggs is about 5 days, so that ovulation time can be controlled by controlling the temperature at which the injected female is kept. It has been found that the first few eggs to reach the uteri are not generally fertilizable or that fertilization percentages are best when most or all of the eggs are allowed to accumulate in the uteri before stripping. An interval of about 3 days at 25°C is, therefore, considered best.

The female may be tested within 24 hours by gentle stripping. This is done by bending the body forward at the pelvis so that the legs are at right angles to the body. Pressure is then applied by gentle squeezing from anterior to posterior of the abdomen, within the closing hand. Pressure on the body at the level of the lungs will force the eggs from the uteri, if they are already there. Avoid damage to the eggs and hemorrhage. If the eggs emerge very easily and in great numbers, they are ready for insemination.

Fertilization of the Eggs

After the female has been tested to indicate the presence of uterine eggs, remove the paired testes from the two male frogs and cut them up with biologically clean scissors into 20 cc of filtered pond water. Spring water or even tap water may be used if pretested to show that it is nontoxic. Let the sperm suspension stand for about 20 minutes at laboratory temperature and then divide it into two clean, flat-bottom fingerbowls.

Remove the ovulating female from its container; gently squeeze a few eggs out of the uteri and wipe them away with a clean paper towel; and proceed to strip the remaining eggs into each of the two fingerbowls of sperm suspension, distributing them evenly. Rotate the fingerbowls so that the suspension comes into contact with every egg, and let them stand for 3 minutes. Flood the eggs with the same water used to make up the sperm suspension and let stand for 2 hours, undisturbed.

After the eggs have rotated so that the animal poles are uppermost (about 15-20 minutes or less), they should be gently separated from the bottom of the fingerbowls by means of a biologically clean, flexible section lifter. If this rotation of most of the eggs does not occur, it generally means that they were not fertilized. The entire mass of eggs should be allowed to float freely for about 15 minutes and then may be cut up with scissors into clusters of less than 15 eggs each. In order for development to be normal there should be no more than about 30 eggs per fingerbowl of 50 cc of water.

There are two methods of providing a series of developmental stages. *First,* successive batches of eggs from the same female may be inseminated at planned intervals. It is possible to remove several hundred eggs and leave the others for a later stripping and insemination, with fresh sperm suspension, providing the total time at room temperature is less than 3 days. *Second,* within the range of about 13°C to 28°C, *Rana pipiens* eggs will develop normally. The above simultaneously inseminated eggs may be separated into a series of fingerbowls kept at different temperatures within this range. At any one time quite a series of stages will thus be available, simply by using the fact that lower temperatures retard development.

SCHEDULE OF NORMAL DEVELOPMENT OF THE FROG'S EGG

The rate of development of the egg and embryo will depend upon the temperature at which they are kept. The approximate schedule of development at two different temperatures is given below.

Those who wish to carry the tadpoles through to later development and even through metamorphosis into frogs must begin to feed them at about the time that the external gills appear. The food consists of small bits of green lettuce leaves, washed thoroughly and wilted in warm water. The water in the fingerbowls (or larger tanks) must be cleaned frequently to remove debris and faecal matter and to prevent bacterial growth. If the tadpoles are not crowded, they will grow faster. Water surface is important for aeration, so, if the volume is increased, also increase the surface exposed. After about 10 days the number should be reduced to about five tadpoles per fingerbowl of 50 cc of water. After metamorphosis, the young frogs must be fed small living worms (Enchytrea) or force fed small pieces of fresh liver.

TIME SCALE FOR DEVELOPMENT OF THE FROG EMBRYO AND TADPOLE

STAGE	AT 18°C	AT 25°C
Fertilization	0	0
Gray crescent	1	½-1
Rotation	1½	1
Two cells	3½	2½
Four cells	4½	3½
Eight cells	5½	4½
Blastula	18	12
Gastrula	34	20
Yolk plug	42	32
Neural plate	50	40
Neural folds	62	48
Ciliary movement	67	52
Neural tube	72	56
Tailbud	84	66
Muscular movement	96	76
Heartbeat	5 days	4 days
Gill circulation	6 days	5 days
Tail fin circulation	8 days	6½ days
Internal gills, operculum	9 days	7½ days
Operculum complete	12 days	10 days
Metamorphosis	3 months	2½ months

"The development of an egg into a finished embryo is essentially due to factors residing in the egg itself."

Wilhelm Roux

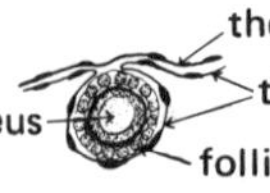

FIGURE 1

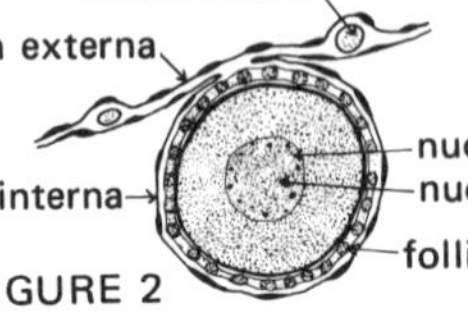

FIGURE 2

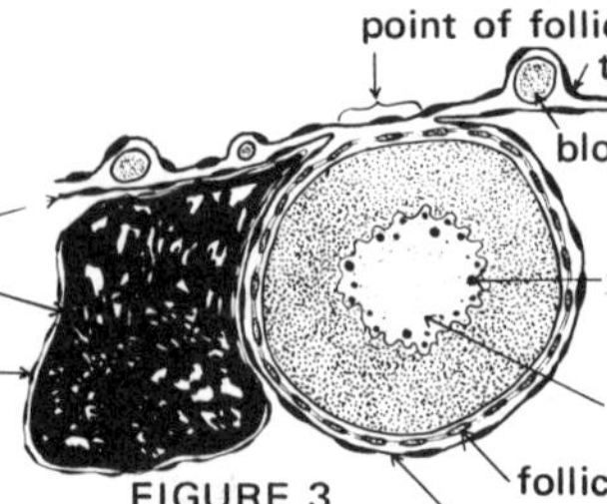

FIGURE 3

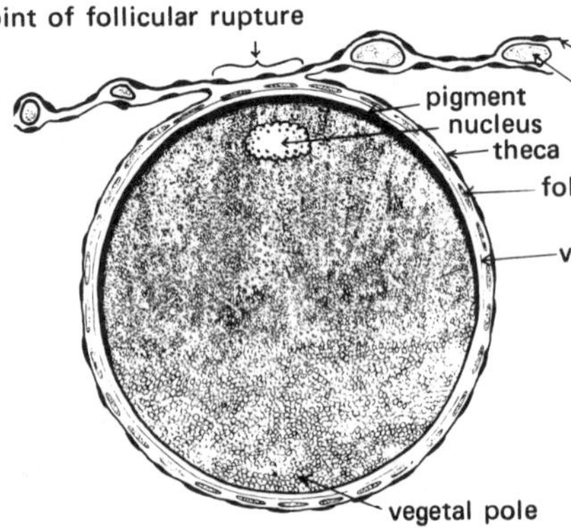

FIGURE 4

GROWING OÖCYTES OF THE FROG

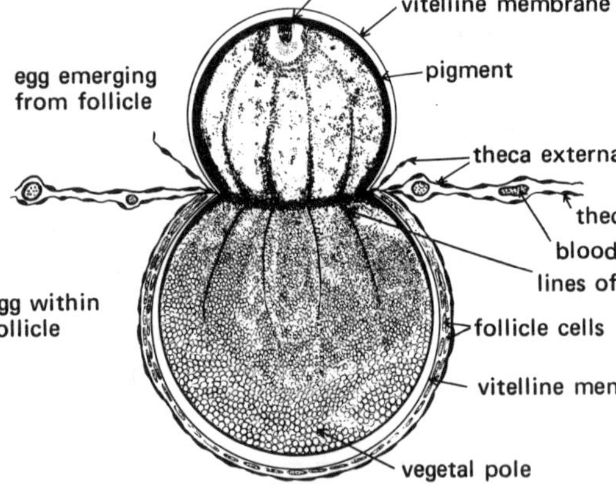

FIGURE 5

FOLLICULAR RUPTURE (OVULATION) IN THE FROG

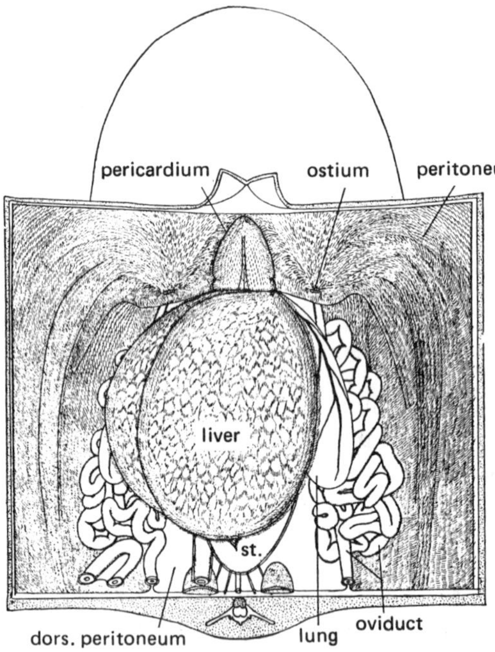

Diagrammatic representation of the open body cavity of an adult female *Rana pipiens*, showing the distribution of coelomic cilia and their convergence toward the two ostiae. The phrenum has been cut, and the heart has been deflected forward to expose underlying organs, and the left lobe of the liver has been displaced to show lung and oviduct on the left side of the frog.

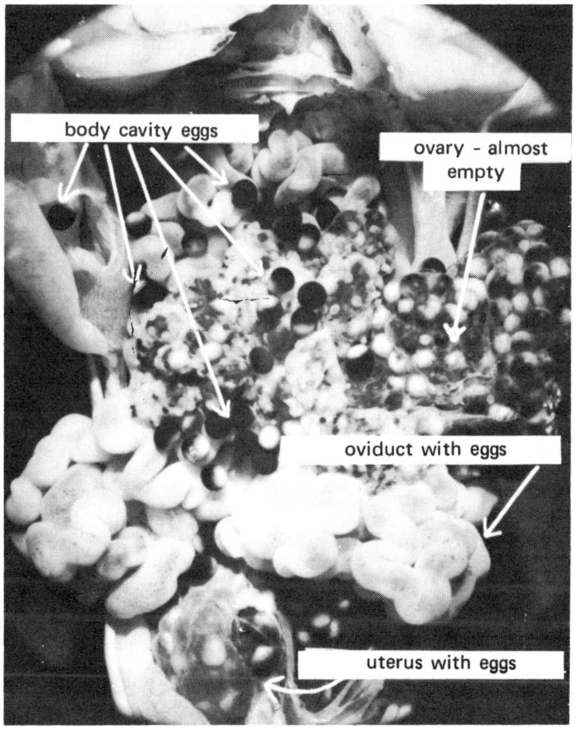

OVULATING FEMALE FROG
16 HOURS AFTER PITUITARY INJECTION

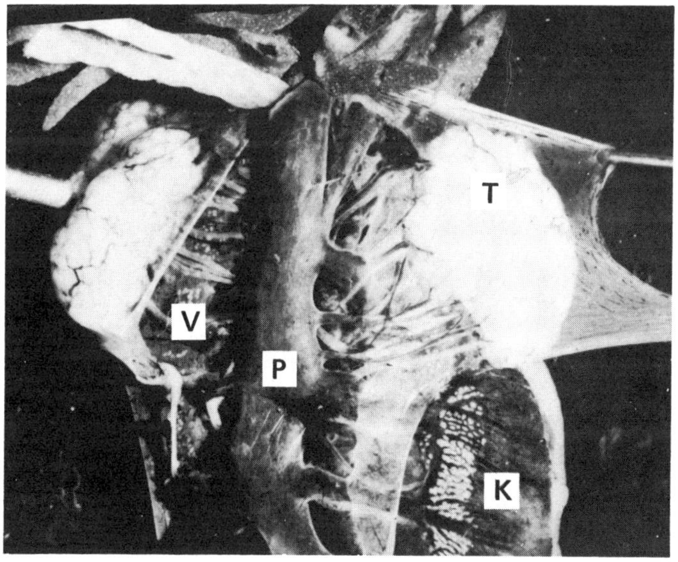

K – KIDNEY
P – POSTERIOR VENA CAVA
T – TESTIS
V – VAS EFFERENS
MALE

RANA PIPIENS

RANA PIPIENS
AMPLEXUS
(BUFO FOWLERI)

DRAWINGS: EARLY FROG EMBRYO

EARLY DEVELOPMENT OF THE FROG

Exercise 7

THE NORMAL DEVELOPMENT OF THE FROG'S EGG: MACROSCOPIC

Each student will be provided with a series of preserved specimens which illustrate the normal development of the frog's egg from fertilization to the tadpole. It is of utmost importance that the student become acquainted with the macroscopic changes in early development so that the microscopic sections will be interpreted properly. The series will include the following and should be compared with the accompanying photographic series (pages 52, 53).

1. Eggs from the body cavity provided with vitelline membrane alone, unfertilized.
2. Eggs from the uterus, provided with jelly, still unfertilized.
3. Fertilized egg in the 2-cell stage. Look for gray crescent, often split by first cleavage.
4. Four-cell stage. Examine both animal and vegetal poles.
5. Eight-cell stage. Which plane represents the third cleavage?
6. Blastula stage. Note germ ring; animal and vegetal poles; variations in cell sizes.
7. Gastrula stage. Prominent yolk plug at blastopore.
8. Neurula stage. Note open neural groove of slightly elongated embryo.
9. Tailbud stage. This is the period of extensive organogeny.
10. Tadpole stage. Note definite tail fin and external gills.
11. Tadpole stage. Note loss of external gills and appearance of opercular (spiracular) pore on left side of head. This opening leads from the opercular cavity containing the internal gills.

The instructor will attempt to provide living material for direct observation in the laboratory. The gray crescent cannot be identified in sections of eggs, but from 15-45 minutes after the egg is fertilized it can generally be seen on the living egg surface as a gray area between the animal and vegetal poles.

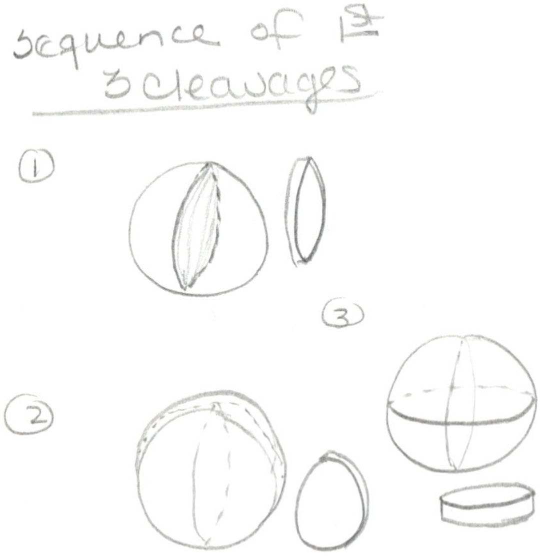

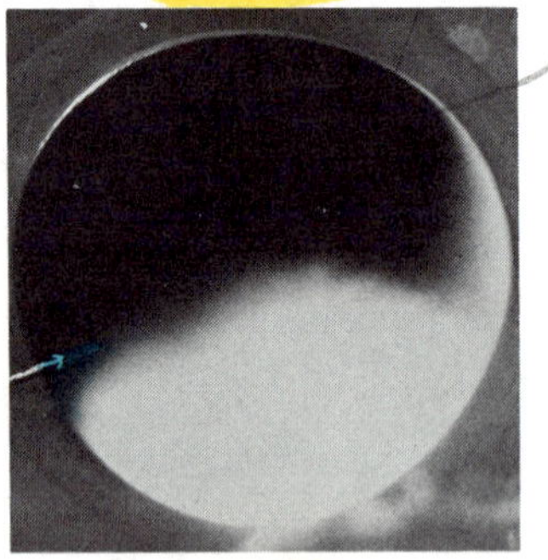

NEWLY FERTILIZED FROG'S EGG

REFERENCES

Rugh: *Frog*, Chapters I through III

KEY TO FROG DEVELOPMENT PHOTOGRAPHS
(Following pages)

1. Unfertilized egg plus full complement of jelly.
2. Fertilized egg with jelly and vitelline membrane removed, lateral view to show animal and vegetal poles.
3. Polar view showing spreading of pigment at region of second maturation spindle.
4. Second polar body near area where it was eliminated.
5. Lines of tension in the process of cleavage.
6. First cleavage, 1- to 2-cell stage.
7. Second cleavage, 2- to 4-cell stage.
8. Third cleavage, 4- to 8-cell stage.
9. Animal pole view, 8-cell stage.
10. Lateral view, 8- to 16-cell stage, near animal pole.
11. Lateral view, 8- to 16-cell stage, near vegetal pole.
12. Early blastula, 16- to 32-cell stage.
13. Early blastula, 32-cell stage.
14. Blastula, 32- to 64-cell stage.
15. Blastulae, (stages 15 to 21).

22. Germ ring.
23. Beginning of pigment migration to form dorsal lip of blastopore.
24. Dorsal lip of blastopore as pigmented crescent.
25. Gastrulation, early yolk plug stage.
26. Medium-sized yolk plug stage, active gastrulation.

27, 28. Small-sized yolk plug stage.

29, 30. Late yolk plug and beginning of medullary plate.

31. Medullary plate and neural folds.

31, 36. Neurula stage.

32. Neural folds.
33. Neural folds almost closed, early neurula.

37, 42. Tailbud stages (between closed neural tube and earliest tadpole).

43. Early tadpole, about 4-5 mm size.

"If there be a better way of securing understanding than by the method of science, human history and human experience have not yet revealed it."

A. J. Carlson

"The very egg itself must have specific characters although they may be invisible, and the eggs of different animals are really distinct from one another as are their adults, the distinction becoming more and more visible as development proceeds."

O. Hertwig

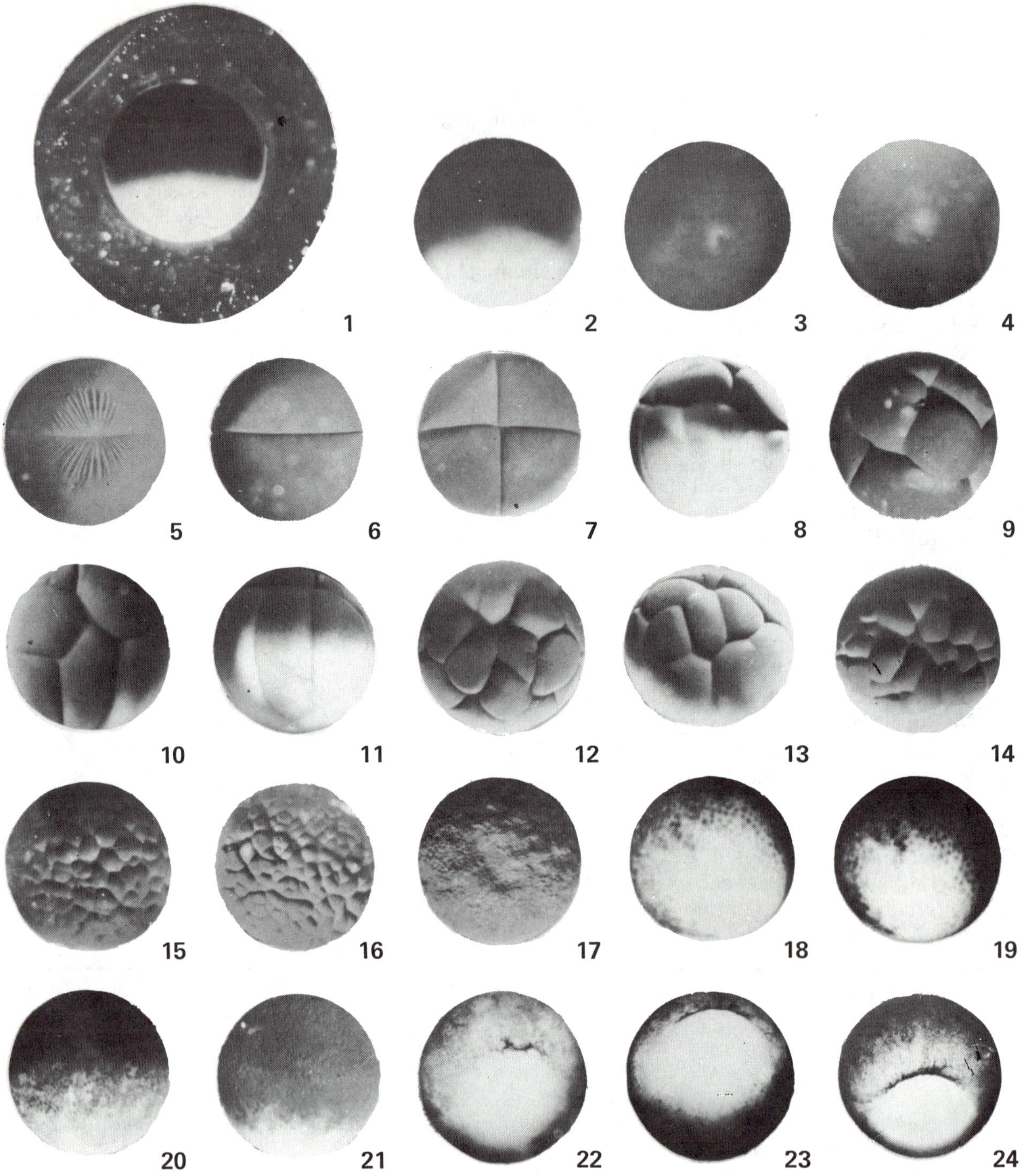

The above developmental stages of the frog have been given stage numbers by Dr. Waldo Shumway (*The Anatomical Record,* vol. 78, 1940) as follows:

Photograph Number	Shumway Stage	Developmental Stage	Photograph Number	Shumway Stage	Developmental Stage
1	1	Unfertilized egg	8-11	5	8- to 16-cell stage
2	2	Gray crescent	12-13	6	Early blastula
4		Polar body extrusion	14	7	64+ cells
5		Cleavage tension lines	15-16	8	Pregastrulation
6	3	2-cell stage	17-21	9	Epiboly of pigment
7	4	4-cell stage	22-24	10	Involution of gastrulation

EARLY DEVELOPMENT OF THE FROG EMBRYO

RANA PIPIENS

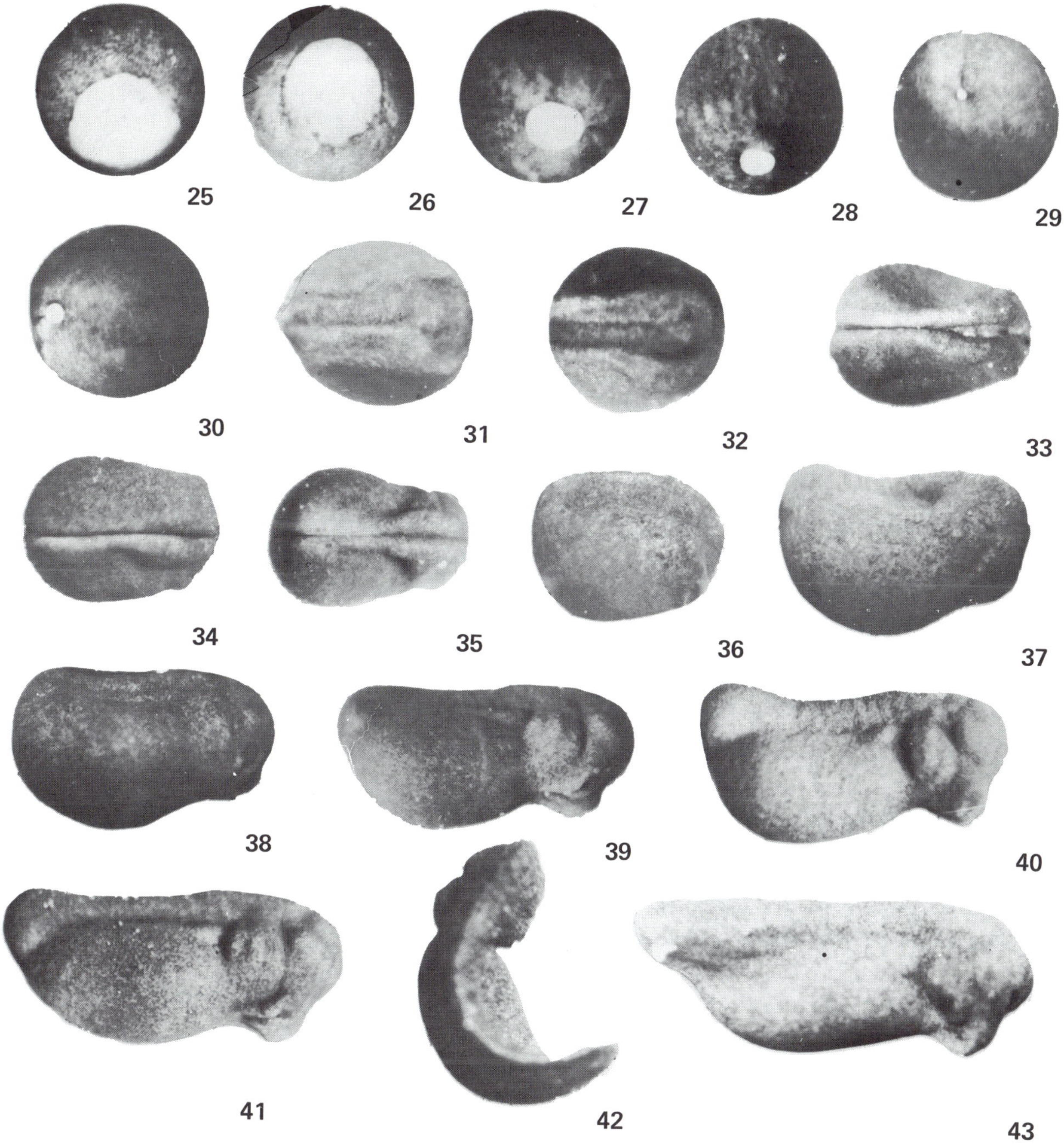

Photograph Number	Shumway Stage	Developmental Stage
25-26	11	Early yolk plug
27-28	12	Disappearing yolk plug
29-30	13	Neurenteric canal stage
31-32	14	Early neural plate
33-36	15	Neural folds
37-38	16	Neurula, closed tube
39-40	17	Body elongation
41-42	18	First body movements
43	19	Tailbud stage

LATE DEVELOPMENT OF THE FROG EMBRYO

DRAWINGS: FROG EMBRYOS

THE FROG'S EGG:

Fertilization and Cleavage

Exercise 8

MICROSCOPIC STUDIES

The Unsegmented Egg

The egg of the frog *Rana pipiens* is normally spherical and has a diameter of about 1.75 mm. The pigment of this egg is concentrated on the surface at the *animal hemisphere*, while the yolk is found throughout but is more concentrated at the *vegetal pole*. There is very little cytoplasm, the bulk of the egg consisting of oval-shaped fat globules. Surrounding the egg is the noncellular and nonliving *vitelline membrane* and triple-layered jelly coat. This (uterine) egg will not show any clear nuclear structure, because the chromatin material is in the metaphase of the second maturation division and the mitotic figure is so small that it is difficult to find. In most cases the jelly covering has been removed to allow better preservation of the egg substance. The section of this egg will appear much like that of the egg within the mature ovary.

The Penetration and/or Copulation Paths

About 20 to 30 minutes after the frog's egg has been inseminated, it completes the second maturation division. During this time the spermatozoon has penetrated the cortex of the egg in the animal hemisphere and has made some progress through the egg substance.

As the sperm passes into the egg, it carries with it a *trail* of peripheral *pigment* so that even if the sperm itself cannot be seen in these sections, the much broader trail of the spermatozoon is clearly marked. The initial pigment path is known as the *penetration path*, and, if this is seen to veer in the direction of the egg nucleus, the new course is known as the *copulation path*. This path leads to the fusion of the sperm and the egg pronuclei, the fusion itself known as *syngamy*. Eggs which are killed 45 to 60 minutes after insemination will show these pigment paths, but it is a rare section that is cut parallel to the paths.

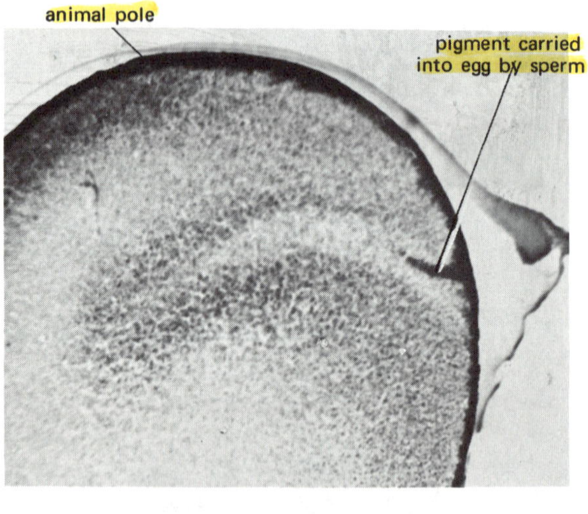

PENETRATION PATH OF SPERM

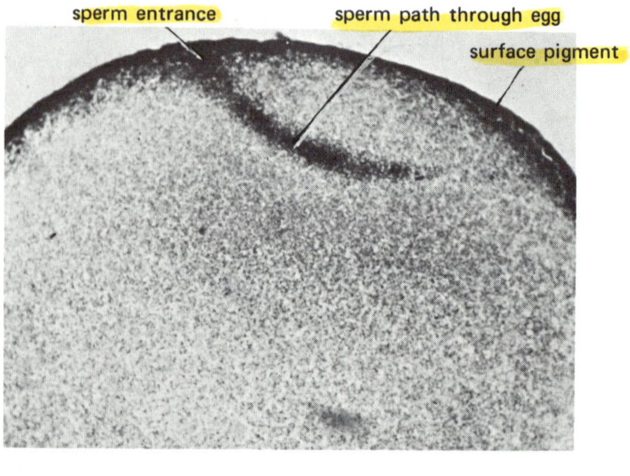

COPULATION PATH OF SPERM

The First Cleavage

When the 2-cell stage is sectioned through the egg axis and at right angles to the *cleavage furrow,* you will note that the furrow does not cut through the vegetal half of the egg but is carried only about half the distance from the animal toward the vegetal pole. The cleavage furrow on the surface will eventually encircle the egg, but it never quite separates the egg into two blastomeres. This is explained on the basis of yolk resistance. (Compare with starfish egg p. 15.) Nevertheless you will note that the furrow is clearly marked by pigment and that the animal hemisphere of the egg is definitely separated into two parts by the pigmented furrow. At normal laboratory temperatures

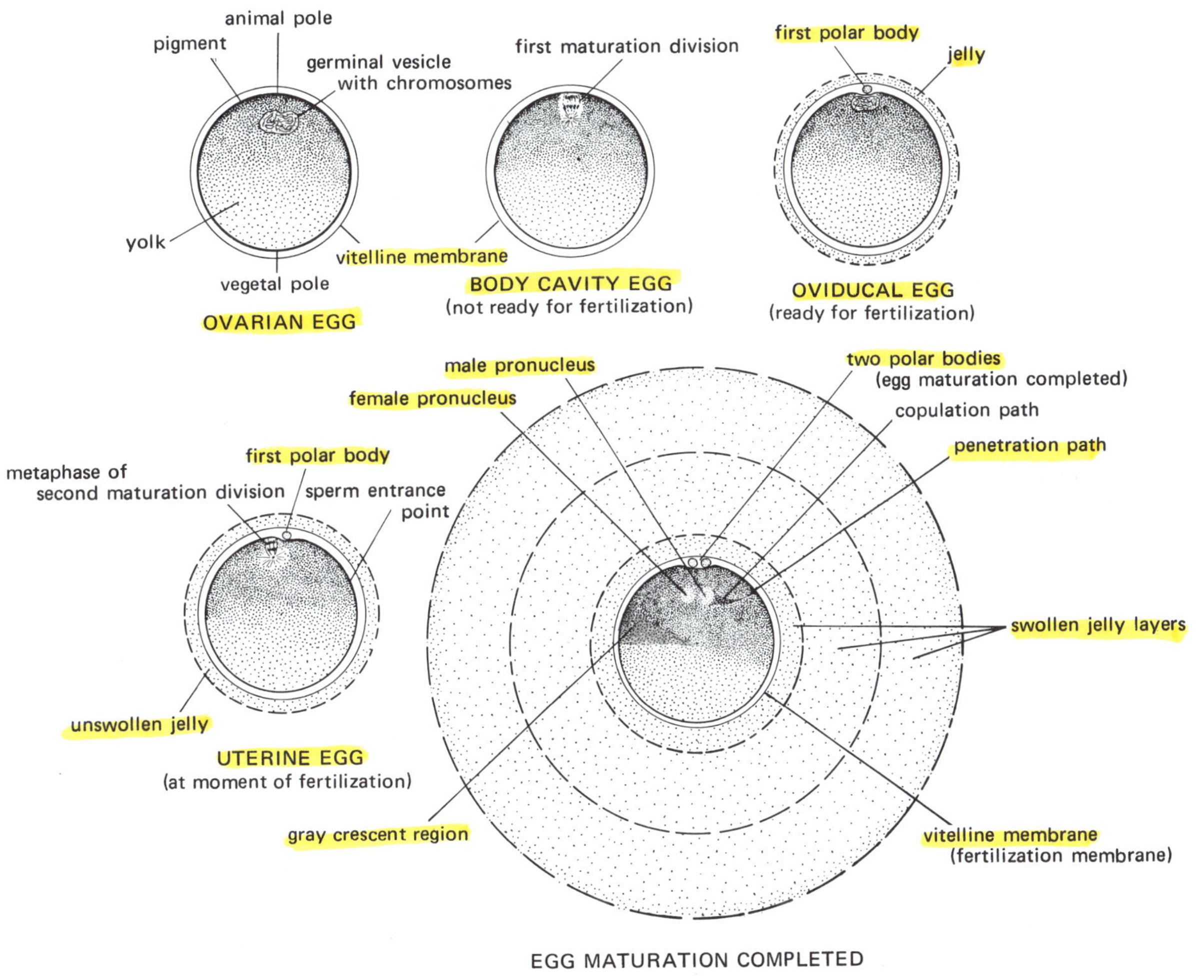

MATURATION AND FERTILIZATION OF THE FROG'S EGG

RANA PIPIENS

Growth period of egg and development of vitelline membrane–May to September.
Hibernation period, no changes–October to March.
Breakdown of germinal vesicle–at time of ovulation { Natural in March & April; Induced September to March
Elimination of first polar body-during transit through oviduct (jelly applied from oviduct).
Second maturation to metaphase stage–uterus and until fertilization or death.
Elimination of second polar body–15-30 minutes after insemination.
Jelly swells rapidly when egg is placed in water.
Fusion of pronuclei–1½-2 hours after insemination.
First cleavage–2½ hours after insemination.

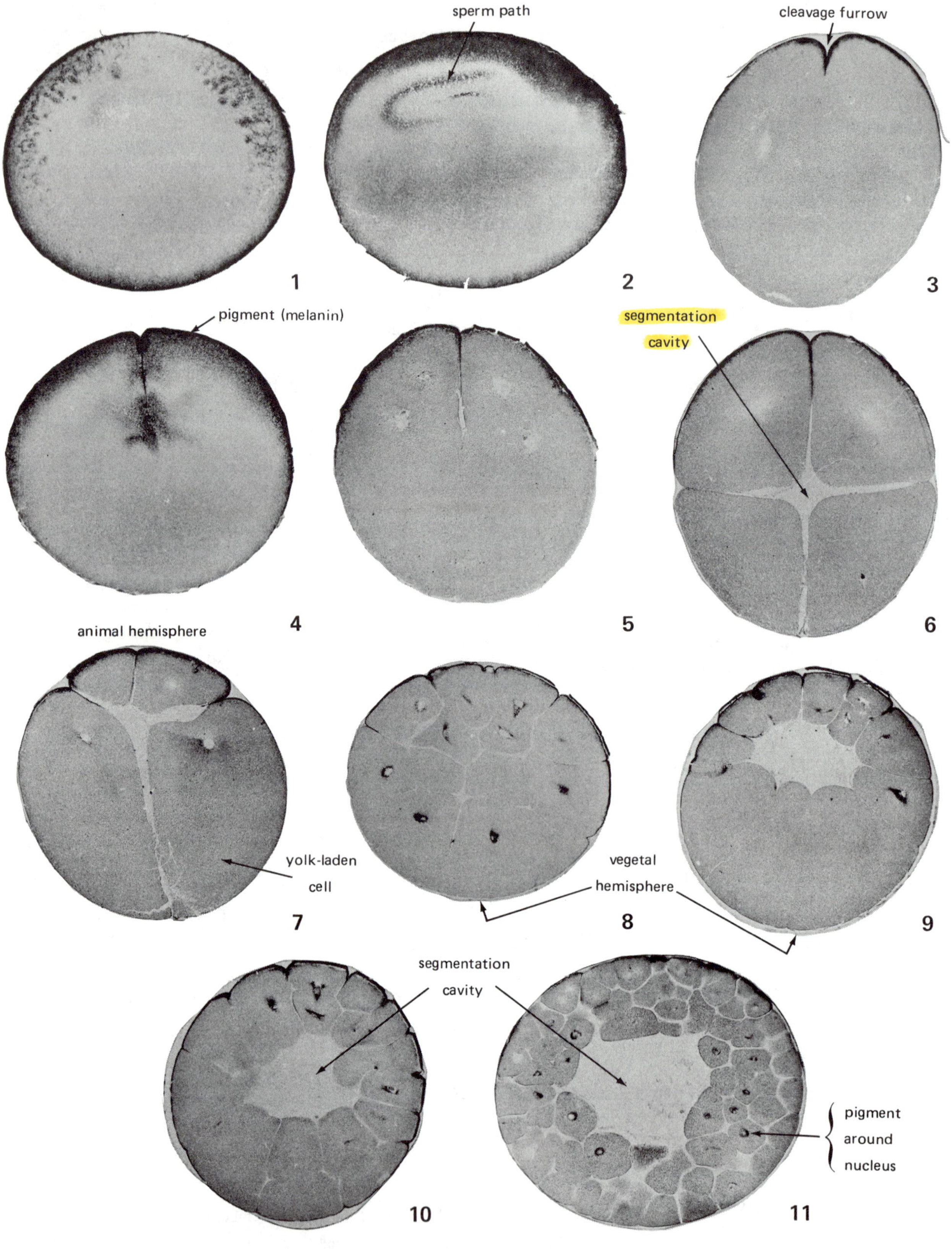

FERTILIZATION AND EARLY CLEAVAGE OF THE FROG

Courtesy H. Wurmback.

this first furrow occurs within 2½ hours after fertilization. In living material and in the photographs on page 52 note the superficial lines of tension which appear at the beginning of cleavage, making it appear as though a drawstring were being tightened around the egg from the animal to the vegetal pole. The asters of the forming blastomeres may be seen slightly off center and toward the animal pole in sectioned eggs. This eccentric position of the nuclei is also related to the uneven distribution of yolk.

The Second and Third Cleavages

These sections will show clearly that the cleavage furrows do not penetrate the vegetal hemisphere within the egg and that the furrows are rather superficial. The fully formed and separate blastomeres of the frog's egg are found only in the animal hemisphere at this stage of development. The second cleavage normally occurs at right angles to the first, within 1 hour at laboratory temperatures, and begins at the animal hemisphere, progressing down and around the egg. The fully formed blastomeres are heavily outlined with pigment which was carried into the egg by the advancing cleavage furrows. The first two cleavages are vertical (meridional), while the third is horizontal (latitudinal). Attempt to determine how each section studied must have been cut.

Later Cleavages

These sectioned eggs may be from 8- to 32-cell stages and the sections will be in various planes. It is important to determine the plane of each section by means of the pigment distribution. Obviously, no single section will include all the cells, but it will be possible to estimate the approximate stage by comparing the cell sizes with previous sections. Note that the better formed blastomeres are always in the animal hemisphere, that cleavage furrows now appear in the vegetal hemisphere, and that there is the beginning of an inner segmentation cavity *(blastocoel)* represented by the intercellular space slightly above the center of the egg. Why is there a disparity in cell size which cannot be explained as due to the angle of section?

REFERENCES

Balinsky: 107-72
Bodemer: 95:118
Huettner: 98-05
McEwen: 104-26
Nelsen: 279-339
Patten (1958): 67-69
Rugh: *Frog*, Chapters IV and V
Rugh: *Vert. Emb.* 20-36
Shumway: 50, 79-82

"A graduate from one of our larger universities, when asked why he changed from biology to philosophy said, 'Well, I found that there was so much to be learned in biology that I had no time to think, so I took up philosophy where there is nothing to be learned and I had all my time to think'."

E. M. East, Harvard

DRAWINGS: FROG EMBRYOS

"Anyone considering the human body is repeatedly tempted to exclaim: 'This is the most wonderful of all organs.' However, one soon recalls that what was previously learned concerning other organs: the fantastic structure of the bones with the bone marrow, which produces several hundred million cells every second; the beating heart, which for seventy years sucks in and expels one sixth of a quart of blood every second; the liver, which within a thimbleful of its substance carries on as many chemical processes as a six-story factory; the brain with its millions of telephone connections and mysterious functions of consciousness–truly it is impossible to give preference to any region of the body."

F. Kahn, 1944

THE FROG EMBRYO:

Blastula and Gastrula

Exercise 9

THE BLASTULA

The living blastula of the frog is spherical, being divided progressively into smaller cells, and has an eccentric *blastocoel* (segmentation cavity). Since the animal pole cells were, from the first, smaller than the vegetal pole cells, they divide more rapidly and gradually carry their pigment downward toward the vegetal pole of the blastula. The advancing margin of rapidly dividing cells, thus marked, is known as the *germ ring**, which is best identified either on whole eggs or in photographs. However, in sections of blastulae the germ ring will be seen to include cells lateral to the segmentation cavity as well as the peripheral pigmented cells. It is, therefore, not limited to the surface cells only. This germ ring will continue to converge toward the vegetal pole until all but a small area, the yolk plug, is covered.

In vertical sections of the blastula note that the roof of the blastocoel is made up of several layers of cells, the outermost one being the most highly pigmented. The single outer layer is known as the *epithelial layer*, while those layers below it, within the roof, represent the *nervous layer* of ectoderm. The blastocoel is filled with a noncellular fluid, which, upon fixation, becomes reticular in appearance. This fluid is probably derived from the cells immediately surrounding the blastocoel. Because of this cavity, some sections may show a collapsed roof, that should be regarded as an artefact. Many scattered nuclei, or associated pigment, will be seen throughout, but it will be quite clear that there are more and smaller cells within the animal than in the vegetal hemisphere.

THE GASTRULA

The margin of the germ ring at the level of the lower limit of the original gray crescent becomes separated from the underlying cells and begins to turn in and under the previously mentioned blastocoelic roof. The balance of the germ ring continues to expand down around the vegetal pole, forming an ever-decreasing ring around the exposed yolk cells. This involution of cells is the very beginning of gastrulation, i.e., the formation of the two-layered or didermic embryo having *ectoderm* and *endoderm*. The ectoderm includes most of the prospective mesoderm. (Study the series of photographs and drawings on pages 65 to 69 to understand these changes in every detail.)

The sections to be studied were made from embryos representing *early* and *late gastrulae*. The earliest involution at the region of the *dorsal lip* of the future blastopore can be identified by the superficial pigment, which is carried inwardly along with these cells. The line of demarcation between *epiblast* and *endoderm* changes constantly, but it is arbitrarily regarded as the point of involution known as the dorsal lip of the blastopore. The cells of the original gray crescent region, which are the first to turn in, will become either endoderm, notochord, or mesoderm, depending in part on exact positional relation. It is the final position of cells which may be related to their fate. The terms ectoderm, mesoderm, and endoderm are positional and do not carry any other implication. (See diagrams on the next page.)

*The term "germ ring" is more universally used than the term "marginal zone" in the Amphibia.

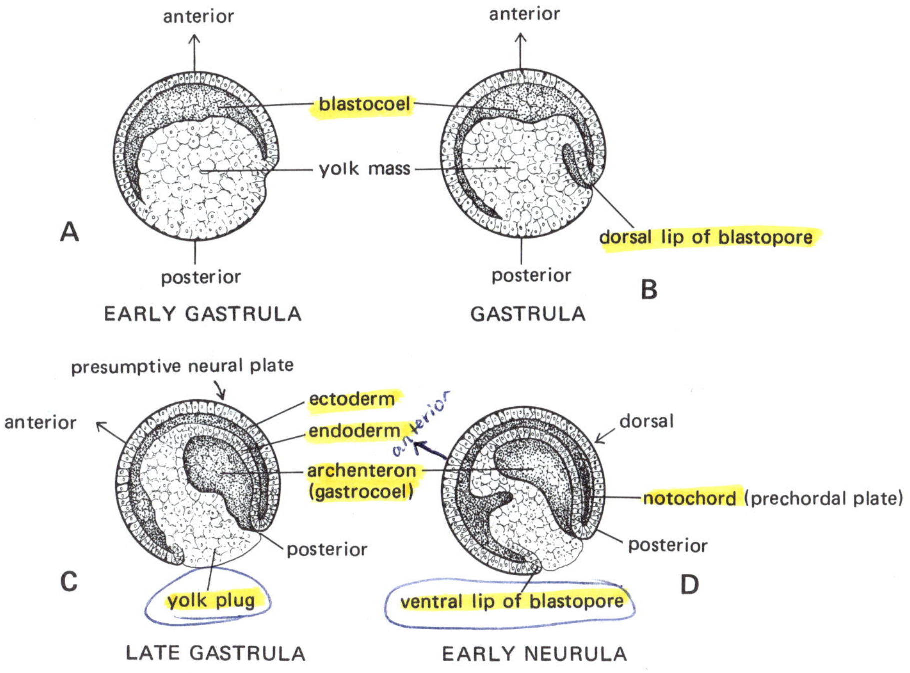

GASTRULATION IN THE AMPHIBIA
MEDIAN SAGITTAL SECTIONS
(arrows indicate change in embryonic axis)

In sections, you will note that the inturned cells are clearly separated from the yolk mass of the vegetal hemisphere, leaving a slitlike space between these cells and the overlying ectoderm. This space is continuous with the blastocoel but is known as the *gastrular slit* (an unfortunate name). The inturned cells form the roof of a new cavity, the *gastrocoel* or *archenteron* (so named because it is the forerunner of the embryonic gut), and the foremost of these advancing endoderm cells will shortly turn under themselves and begin to form the floor of the same cavity. The most anterior endoderm cells cannot be easily identified except by the presence of some of the original surface coat pigment and the relatively greater amount of contained yolk. While these gastrular movements are regular and predictable, the forces that initiate and control them, leading to the three-layered embryo, are still not well understood.

The epibolic expansion, which has been carrying the germ ring down over the yolk mass, is directed inward at the point of dorsal involution, but the balance of the germ ring continues to encircle the vegetal pole cells until only a small circular area (the yolk plug) is visible in the whole egg. The cells at the dorsal lip, in the later gastrula, will be seen to have progressed far enough inward to form both a roof and a floor of the new cavity, the *archenteron,* and the floor is continuous with the blastopore. Thus, some original outer ectoderm turns inward to become endoderm. In any section the roof of the archenteron may always be distinguished by the smaller, pigmented, and relatively yolk-free cells, while the floor is made up of large, almost nonpigmented, and highly yolk-laden cells. Obviously this newly formed cavity will progressively dwarf the original blastocoel until the latter is completely obliterated. The blastocoel may be identified by its contained jellylike reticular coagulum. In some sections a single layer of floor cells may be seen to separate the archenteron (gastrocoel) from the blastocoel. These comprise the *completion bridge,* which has no known function. It is an incidental consequence of the three-dimensional pattern of gastrulation movements.

While the archenteron is forming, two other changes occur. The ventral margin of the germ ring (i.e., that opposite the formed dorsal lip) becomes clearly marked by an involution of its cells similar to that at the dorsal lip. Actually, the process began at the dorsal lip and spread laterally as a crescent. Finally, the distal ends of the crescent-shaped margin of inturned cells join to form a complete circle around the exposed yolk plug. The involuted cells of the *ventral lip* immediately become *peristomial mesoderm.* The second notable change is the thickening of the ectoderm, which will give rise to the *medullary plate,* over the archenteric roof. This is in response to an inductive influence of the underlying archenteric roof.

The process of gastrulation seems to differ considerably in urodeles and anura, so that it must be emphasized that the description here applies to the embryo under study, i.e., that of the frog. Apparently during and after involution at the dorsal lip some cells are proliferated off into the gastrular slit from between the ectoderm and the endoderm of the newly forming archenteric roof. The most dorsal and central cells are destined to become *notochord,* and the lateral sheets of cells to become *mesoderm.* This source of mesoderm is known as *gastral,* distinguishing it from ventral or *peristomial mesoderm*, although the two are continuous and similar. (See page 63, Fig. D, for remains of blastocoel, which is now continuous with the gastrular slit. This latter space is formed by the ectoderm and endoderm moving in opposite directions and will be filled, by proliferation, with presumptive mesoderm and notochord cells.)

In all drawings of these stages the student must indicate the egg axis (*animal-vegetal pole*) and the newly developing embryonic axis (*anteroposterior axis*). This latter axis may be determined in part by the position of the archenteron and by the fact that the blastopore occurs near the position of the future anus.

REFERENCES

Arey: 73-76
Balinsky: 163-88, 201-45
Bodemer: 119-68
Huettner: 105-15
McEwen: 126-46
Nelsen: 340-453
Patten (1964): Chapter 5
Rugh: *Frog,* Chapters VI through IX
Rugh: *Vert. Emb.* 27-47
Shumway: 89-93

"The cells of the cerebral cortex are the most unnecessary of all the cells of the body. A creature can live without a cerebral cortex, but in these cells that are unnecessary for life the creature becomes conscious of its own self and the world; with them it recognizes and enjoys, it gathers memories and experiences, thinks and feels, speaks and writes, makes music and paints, dreams and loves–and suffers. They are life, knowledge, feeling, and enjoyment; they are the I, the personality. We is the sum total of the cortical cells of our brain; our I is the giant concert which is the greatest of all radio stations, this station of microscopic tubes, antennas, coils, condensers, and transformers, broadcasts as thought and feeling to the microcosmos of the cell body, and as word and deed to the wide world."

Fritz Kahn, 1943 in *Man in Structure and Function*, A. Knopf.

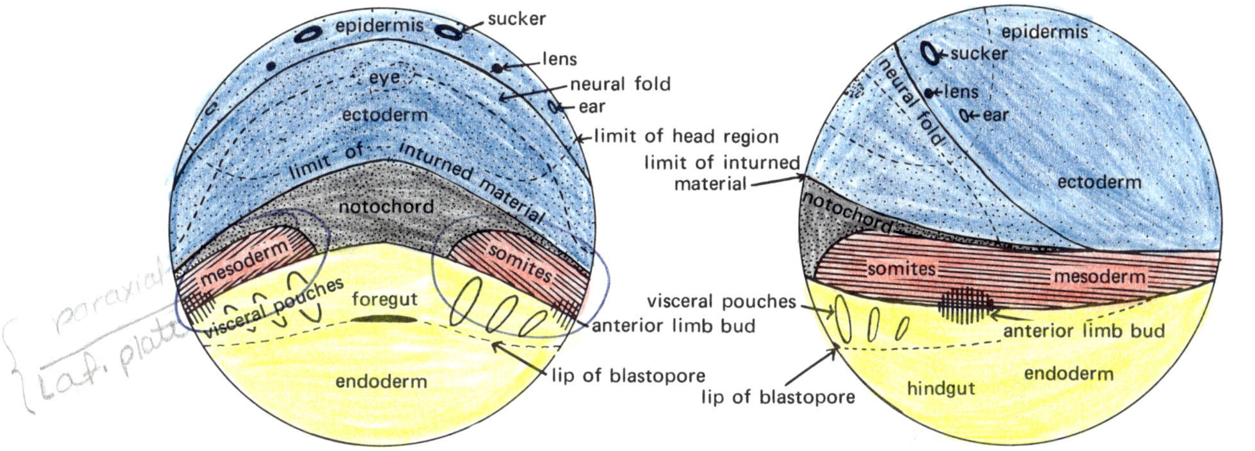

POSTERIOR-DORSAL VIEW

RIGHT SIDE VIEW

PRESUMPTIVE REGIONS OF THE LATE ANURAN BLASTULA

Adapted from Vogt: 1929.

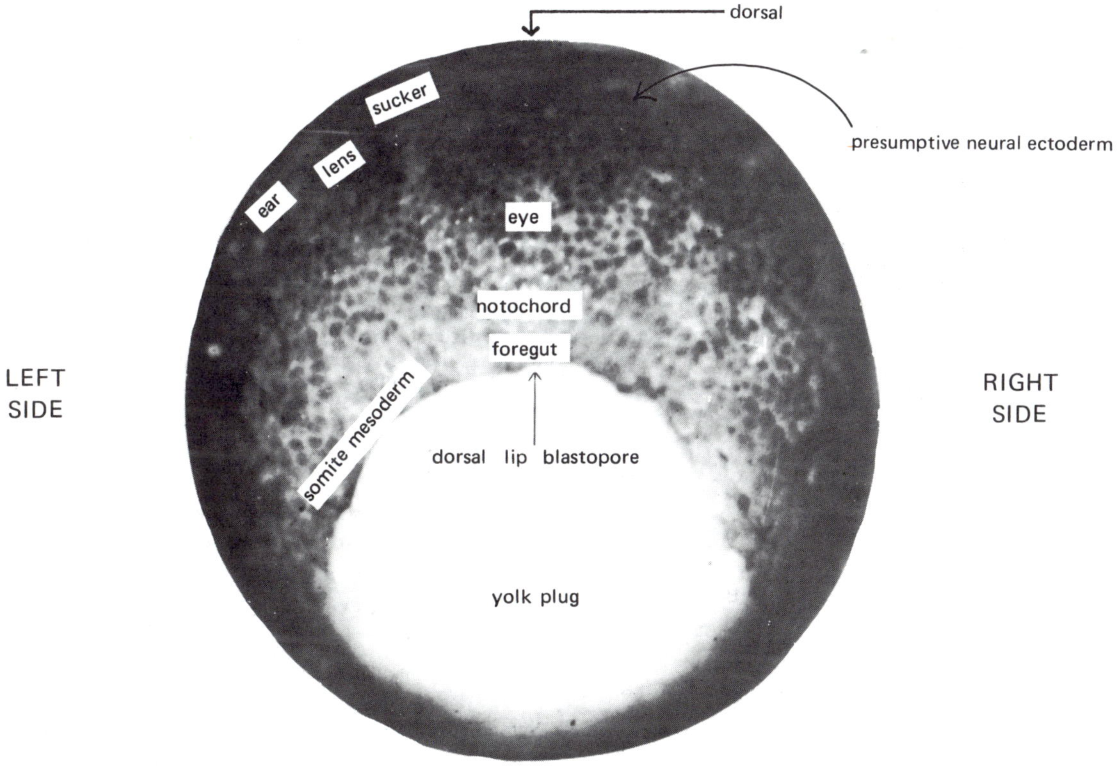

GASTRULATION IN THE FROG: PRESUMPTIVE ORGAN PRIMORDIA LABELED, POSTERIOR VIEW

Modified from Vogt: 1929.

Note: These organs are not yet formed but, based upon experience, this prediction of the position of their various primordia is correct.

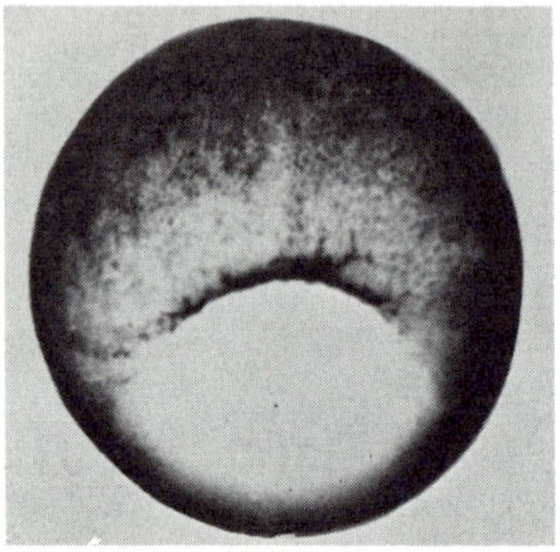

EARLY

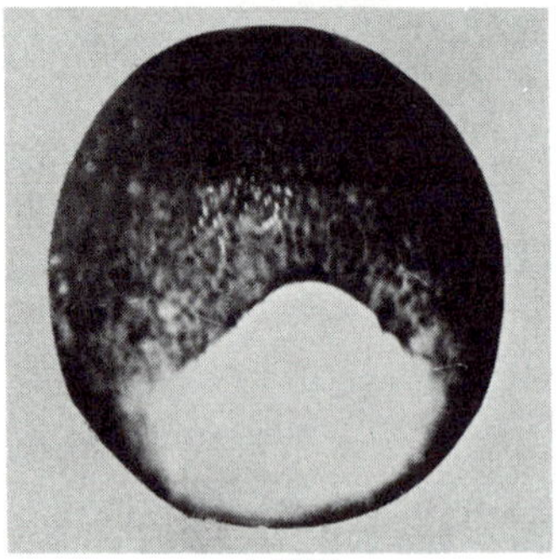

LATE

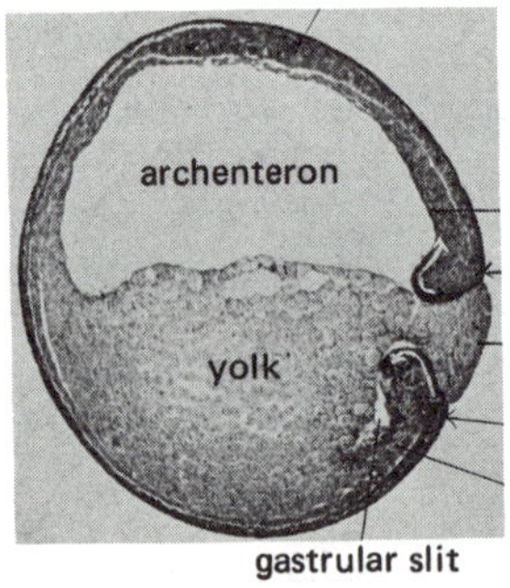

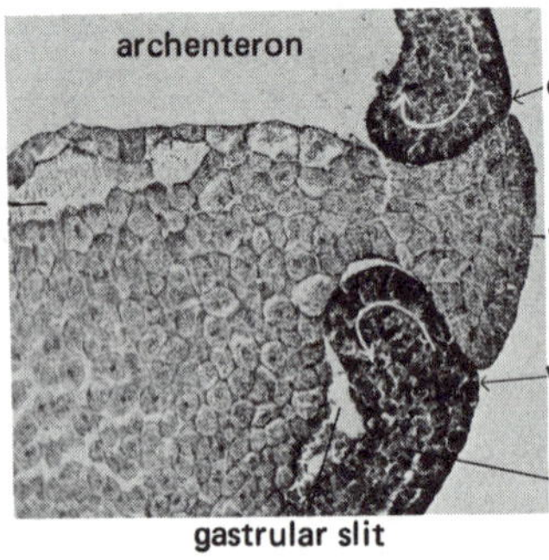

FROG GASTRULATION

SHOWING INVOLUTION OF DORSAL LIP OF BLASTOPORE

GASTRULATION IN THE FROG

SAGITTAL SECTIONS

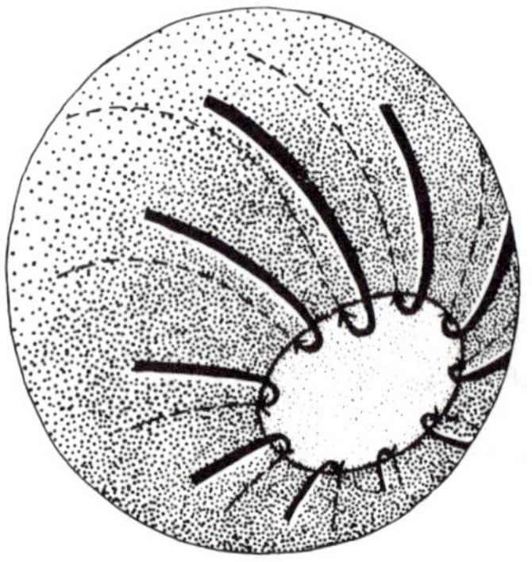

MORPHOGENETIC MOVEMENTS OF THE LATE GASTRULA

This drawing depicts the movement of cells over the dorsal lip first and most actively, with inturning of cells to form the lateral and finally the ventral lips of the blastopore. Simultaneously with the inturning, there is an epibolic over-rolling of the lips so that the exposed yolk (plug) appears smaller and smaller and finally disappears as neurulation begins.

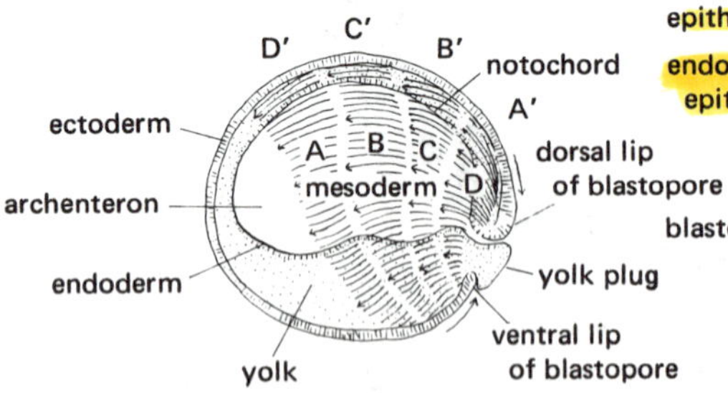

MESODERM FORMATION IN FROG

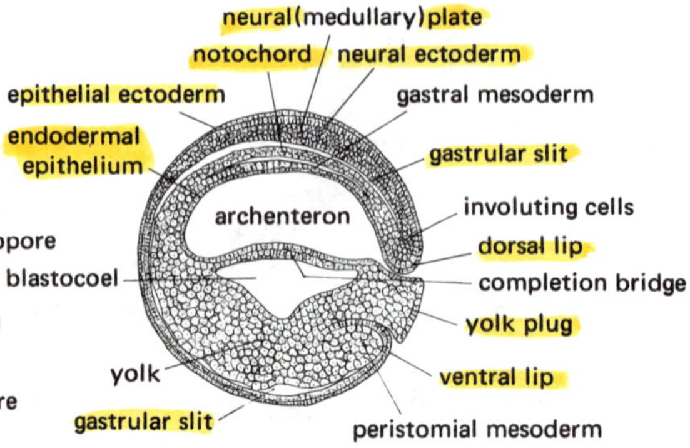

LATE GASTRULA IN FROG

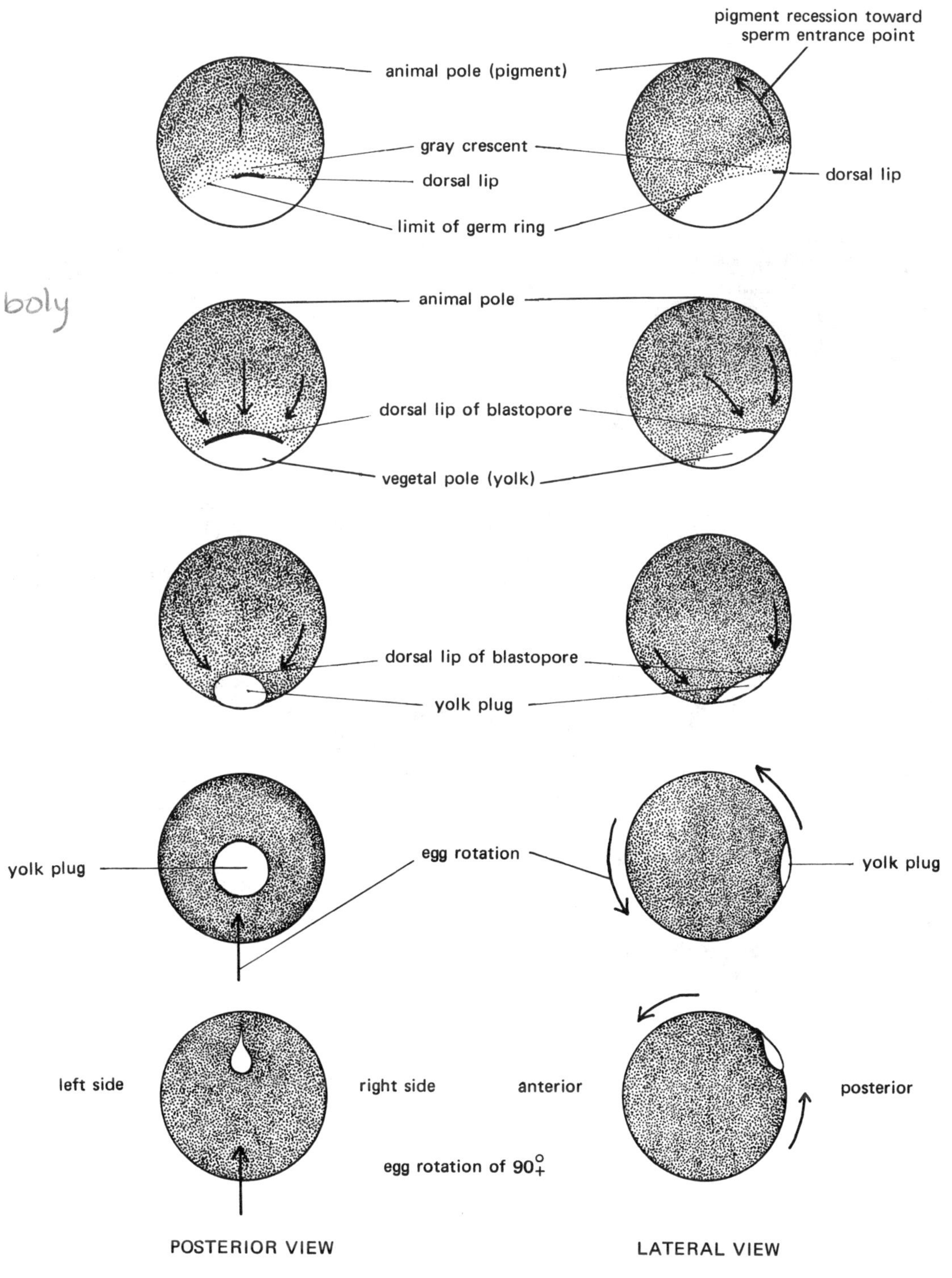

GASTRULATION AND EGG ROTATION

Arrows indicate surface movement during gastrulation, without overgrowth of the yolk-endoderm.

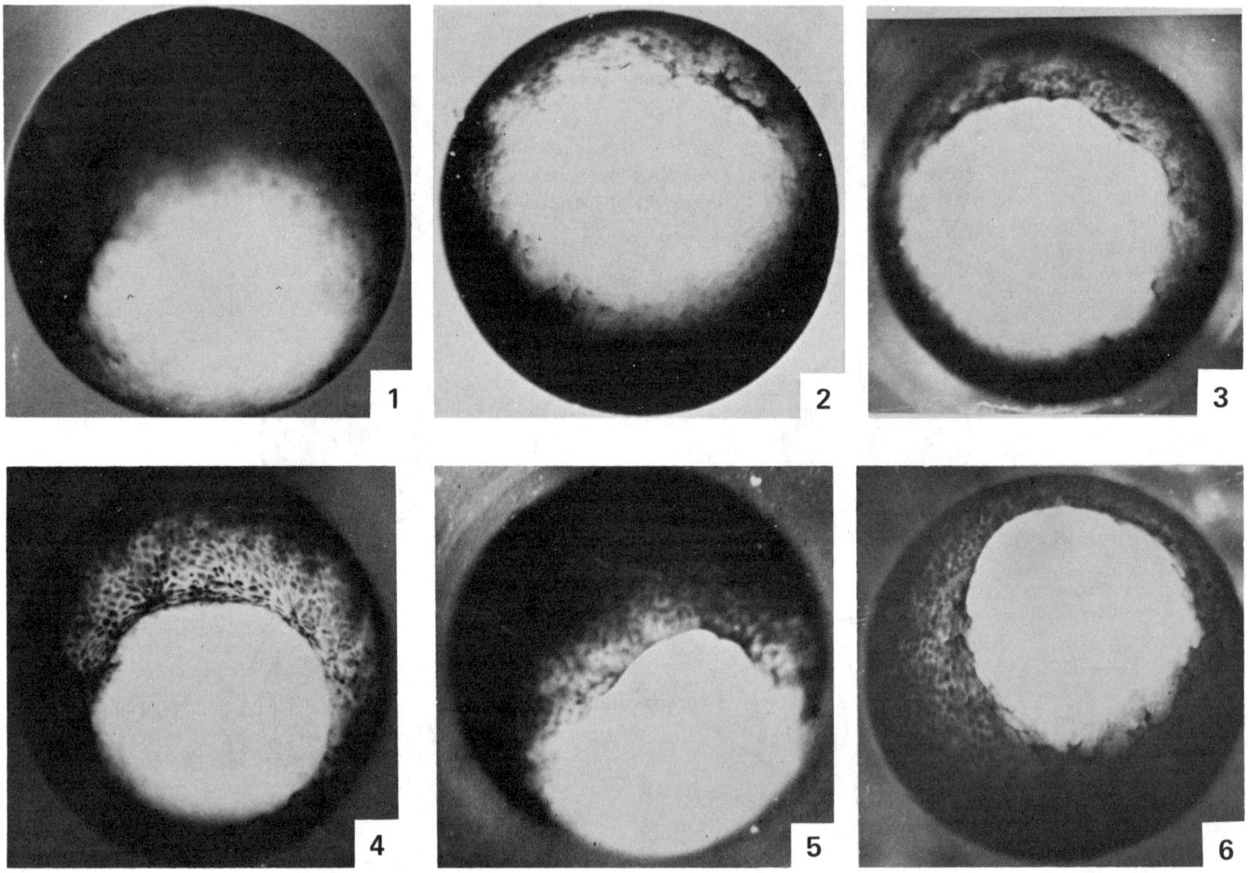

GASTRULATION IN THE LIVING FROG'S EGG

1 – Epiboly of the animal hemisphere cells.
2 – Earliest indication of involution at dorsal lip.
3 – Involution becoming circumferential.
4 – Involution active, forming a bulging lip.
5 – Conspicuous dorsal lip, notochord formation within.
6 – Yolk plug stage or completion of active gastrulation.

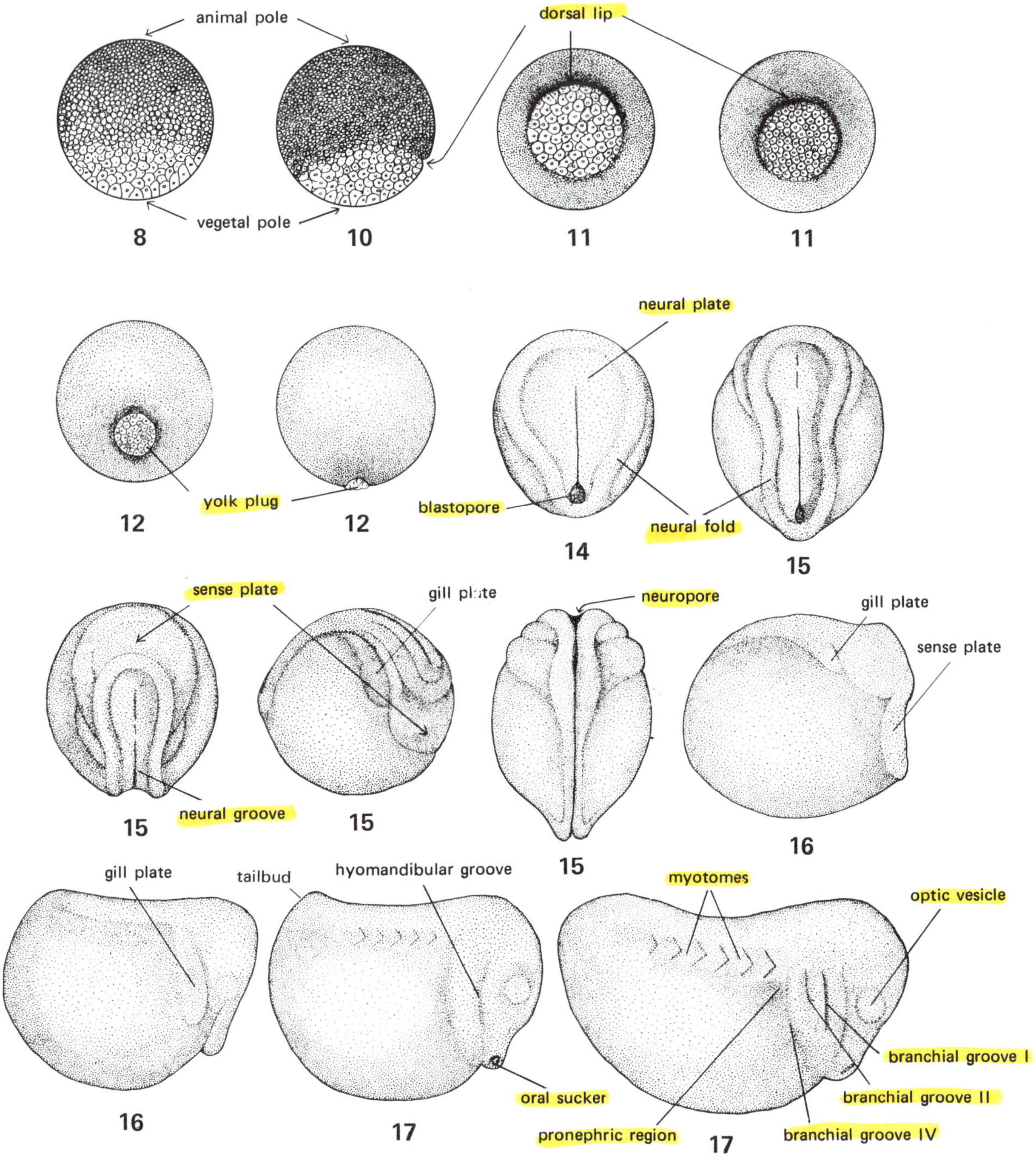

NEURULATION

BLASTULA TO TAILBUD STAGES IN
FROG (*Rana pipiens*)

Numbered according to Shumway.

DRAWINGS: FROG EMBRYOS

DRAWINGS: FROG EMBRYOS

THE FROG EMBRYO:

Neurulation

Exercise 10

THE EARLY NEURULA – TRANSVERSE SECTIONS

Several stages in the early development of the neural tube may be mounted on the same slide and are to be drawn in sequence. The earliest stage shows a flat but thickened *neural ectoderm,* known as the *medullary plate,* in the mid-dorsal region of the animal hemisphere. Directly beneath the medullary plate find the oval group of cells which comprise the *notochord,* possibly derived from the dorsal region of the hypoblast (or archenteric roof). Beneath the notochord, find the thin, pigmented roof of the *archenteron.* Lateral to the notochord and dorsal to the archenteric roof are sheets of *mesoderm,* coming also from the chorda-mesodermal region of the dorsal lip. Follow the mesoderm around the archenteron and note that it merges rather imperceptibly into the ventral yolk cells.

The next stage shows the formation of the *lateral neural folds,* elevations on either side of the thickened medullary plate (neural ectoderm) formed by the convergence of cells from the two sides. Note the slight depression in the midline, lined with highly pigmented *(epithelial ectoderm)* cells. This *neural groove* will give rise to the neural canal, or *neurocoel,* which can always be identified by the pigmented and ciliated lining. It is correct, therefore, to say that the lining of the neurocoel was at one time continuous with the outer integument of the embryo. Frequently in this stage it is possible to identify a lingering group of cells between the notochord and the archenteric roof, known as the *hypochordal* or subnotochordal *rod.* The stages in neurulation are neural or medullary plate, neural fold, and neural tube.

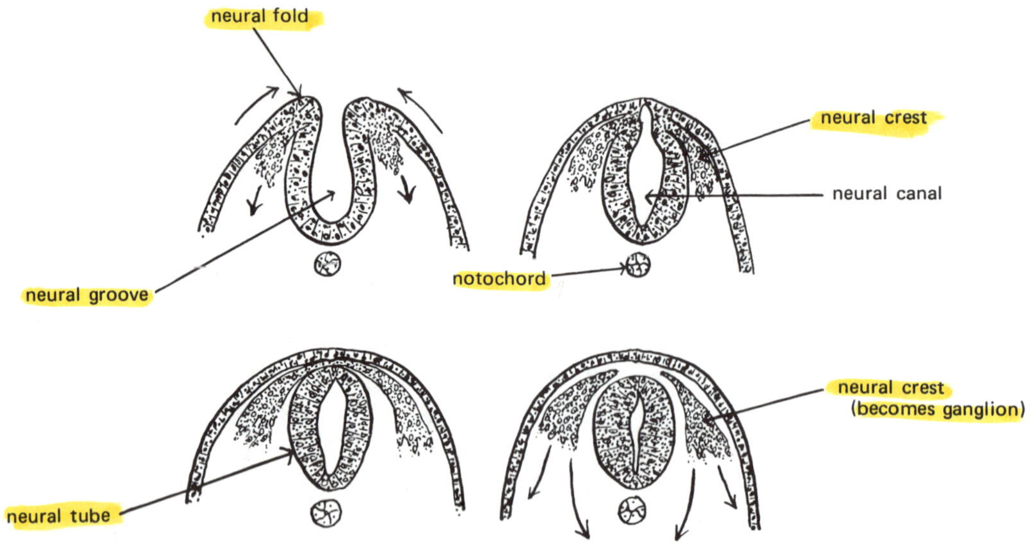

NEURAL TUBE AND CREST FORMATION

Observe in the successive sections the progressive expansion of the mesoderm. Lateral to the notochord are the large masses of *epimeric mesoderm*, which will become segmented from anterior to posterior into metameric somites. Later these somites will differentiate into *sclerotome, myotome*, and *dermatome*. In a more ventrolateral position locate the rather solid and oval mass of *intermediate* or mesomeric *mesoderm* from which will develop the bulk of the excretory system. More ventral is the uniformly thin hypomeric or *lateral-plate mesoderm*, which encircles the yolk cells close beneath the outer belly ectoderm. Subsequently these sheets of hypomeric mesoderm will split to give rise to an intermediate *coelomic cavity*. The outer layer of mesoderm is then known as the *somatic* layer and the inner one as the *splanchnic* layer, the latter giving rise to the bulk of the smooth muscle of the gut. These three regions of mesoderm are continuous in the earlier stages and in cross section would have a horseshoe shape open dorsally at the region of the notochord and nerve cord.

THE LATE NEURULA – TRANSVERSE SECTIONS

In these sections the neural tube is closed, showing the neural canal (neurocoel) lined with ciliated and pigmented epithelium. Ventrolaterally to the neural tube the somite material may also show some pigmented cells, indicating the probable common origin of the archenteric roof and this mesoderm from the region of the dorsal lip. The *notochord* and *hypochordal* rod are now clearly distinguishable from all other tissues, and the three major regions of the mesoderm are distinct. Due to the upward pressure on the yolk mass from the ventral mesodermal sheets, the *midarchenteron* will appear to be reduced almost to a horizontal slit, but the roof and floor may still be identified by their cellular and pigmentary differences. The closed neural tube now provides a vertically projecting ridge along the mid-dorsal line of the entire embryo. With the final closure of the neural folds, there are left behind, on either side in the space between the neural tissue and the ectoderm, some loose cells which form a continuous column from anterior to posterior, known as the *neural crests*. These will later become intermetameric (between the somites) and give rise to the spinal ganglia and sympathetic system.

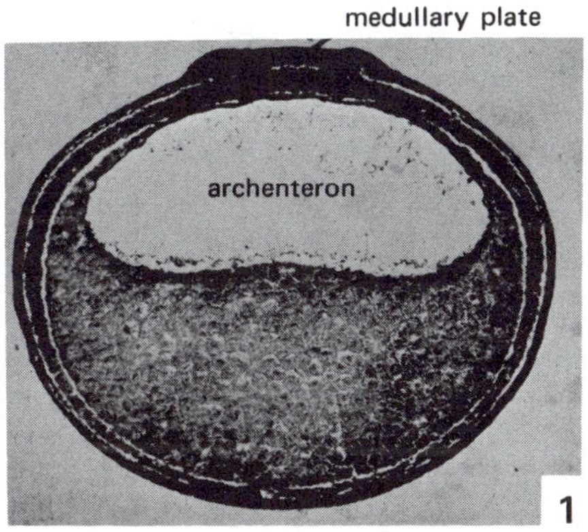

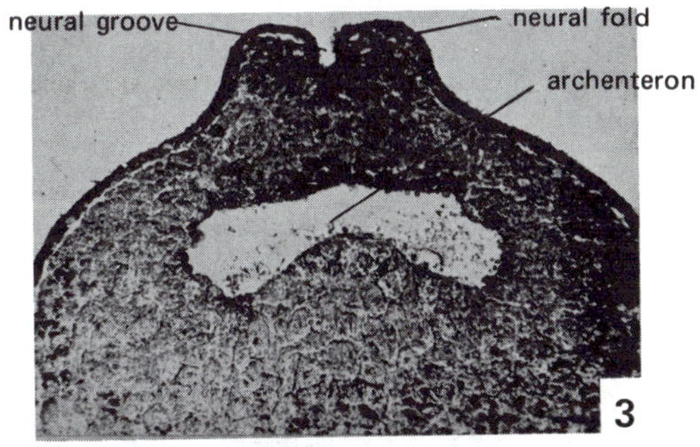

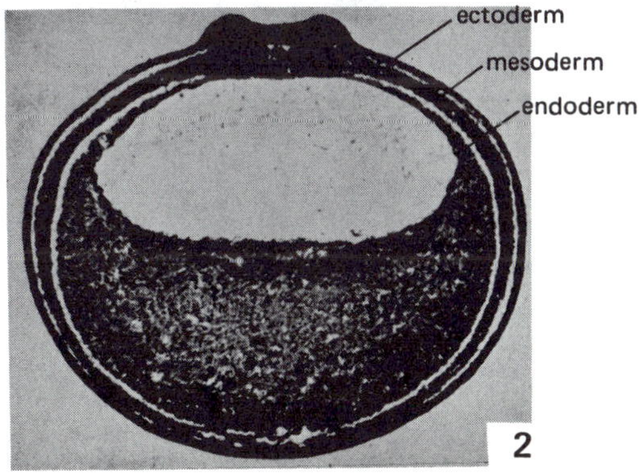

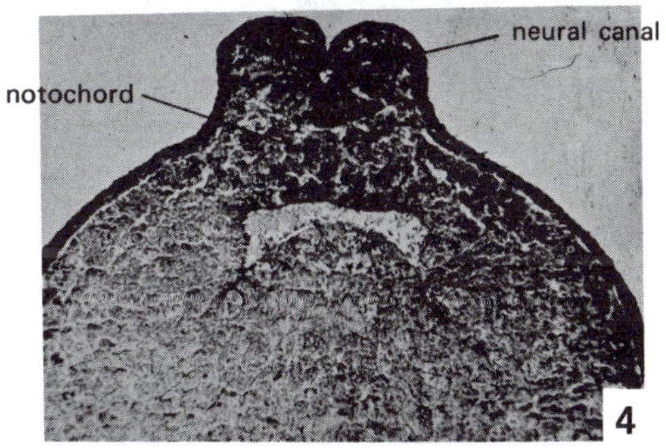

NEURULATION IN THE FROG

Fig. 1 – Medullary plate
Fig. 2 – Early neural folds
Fig 3 – Closing of neural folds
Fig. 4 – Neural canal stage

archenteron
- forms gut
- anterior - pharynx
- liver diverticulum

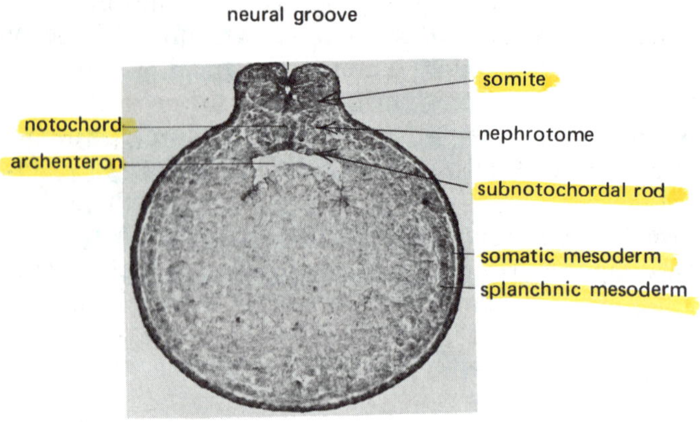

FROG NEURULA

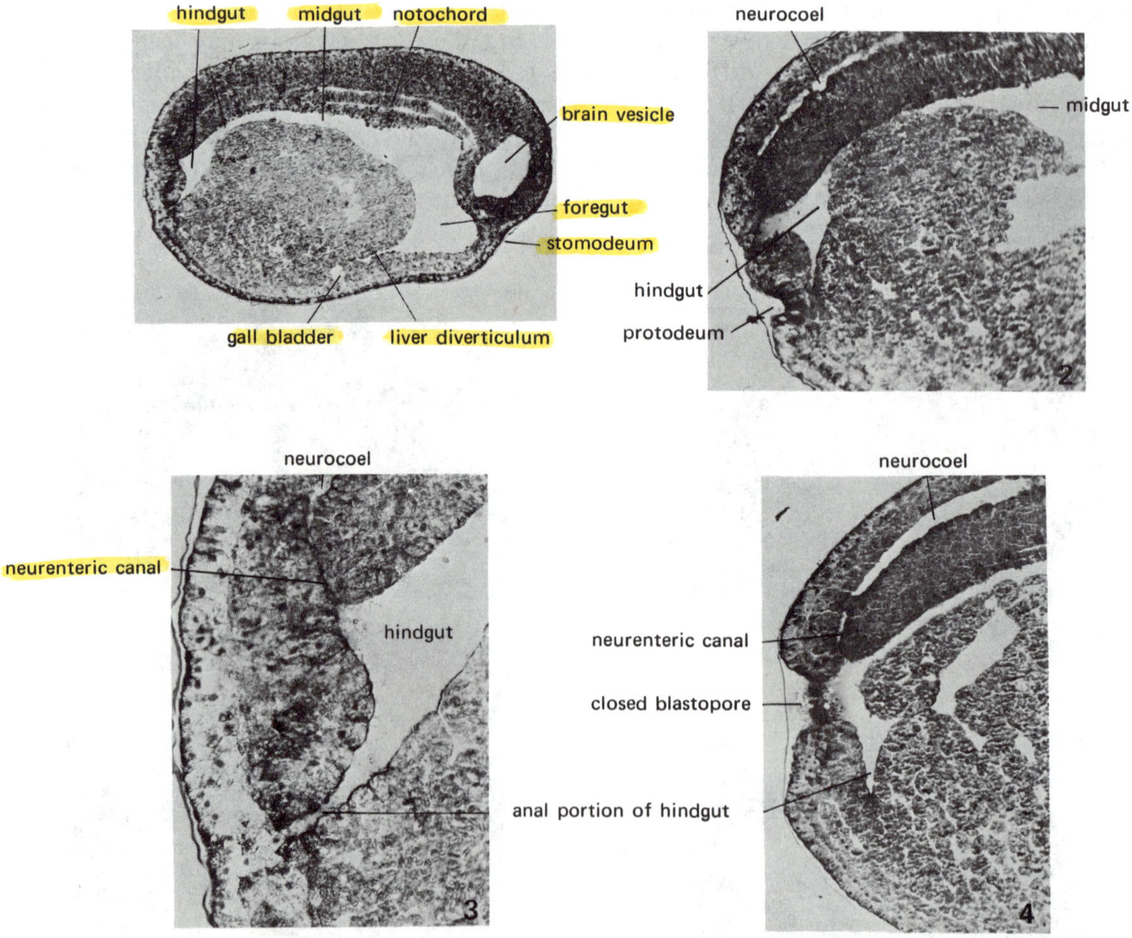

TEMPORARY NEURENTERIC CANAL OF FROG

SAGITTAL SECTIONS

THE EARLY TAILBUD STAGE – SAGITTAL SECTIONS

Simultaneously with the closure of the blastopore, the anus (ectodermal proctodeum) of the frog embryo opens slightly ventral to the position of the original yolk plug. Just within the body of such an embryo one can discern a temporary connection between the posterior end of the neurocoel and the archenteron via the residual blastopore, the canal being called appropriately a *neurenteric canal.* Since this connection is very temporary and small, the student is provided with serial sagittal sections among which he is to find evidence of this neurenteric canal. It may appear closed, but it is possible to find a line of pigmented cells which connect the neurocoel and archenteron, representing the canal cells.

Select for drawing a median sagittal section which will show the major part of the central nervous system and the alimentary canal. Note that the anterior end of the neurocoel is flexed ventrally and that it is quite bulbous, forming the *mesencephalon* (midbrain) at the region of flexure and the *prosencephalon* (forebrain), the most anterioventral portion. This embryonic brain is actually bent (ventrally) around the tip end of the notochord, the floor of the brain at this point being the *tuberculum posterius.* Since it is extremely difficult to section embryos of this stage in exactly a median plane, a continuous neurocoel may not be evident, but all portions of it must be identified. The notochord is marked by vertical striations of its large, vacuolated cells, and lateral to the notochord may be seen the *metameric somites*. (See schematic diagrams on pages 76 and 77.)

Beneath the forebrain, but anterior to the cavernous *foregut* (anterior archenteron), locate a group of pigmented cells which are ectodermal and seem to be growing inwardly between the brain cavity and the gut. These will become the *hypophysis* and, in conjunction with a portion of the forebrain (*infundibulum*), will give rise to the pituitary gland. Locate next the ventroanterior outpocketing of the foregut, which will give rise to the mouth when it breaks through to the stomodeal ectoderm. Ectodermal thickenings ventrolateral to the oral region are the *suckers.* Beneath the floor of the foregut and anterior to the yolk mass locate the loose mesodermal cells, which will become organized into the *heart*. A median posteroventrally directed diverticulum of the foregut may be identified as the forerunner of the liver. The bulk of the ventral body of the tailbud stage is filled with *yolk* cells (endoderm), whose yolk reserve is continually absorbed, making growth and differentiation possible. (See figure below.) Small yolk platelets are still present in all embryonic cells until circulation begins.

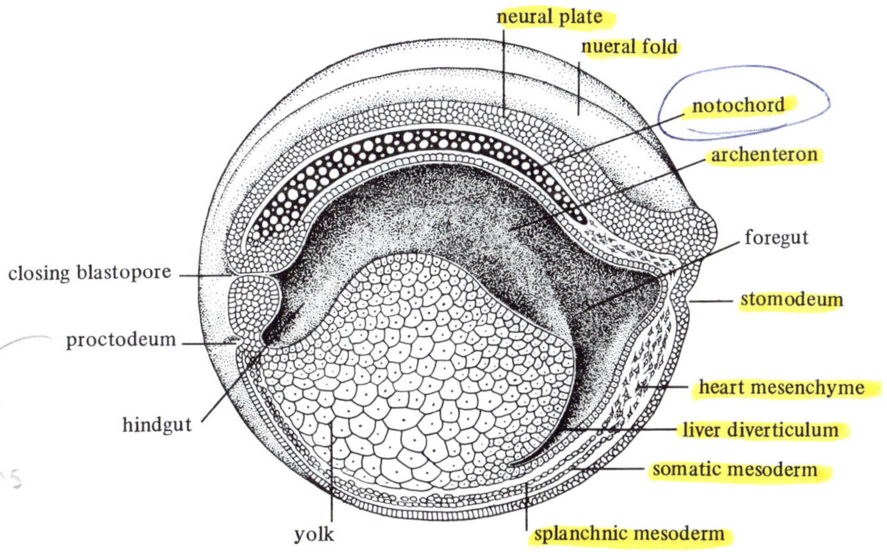

SAGITTAL SECTION OF OPEN NEURAL FOLD STAGE IN FROG

From R. Rugh. *The Frog*. 1951. McGraw-Hill, New York. Adapted from Huettner.

brain
3 regions
fore -
mid -
hind -

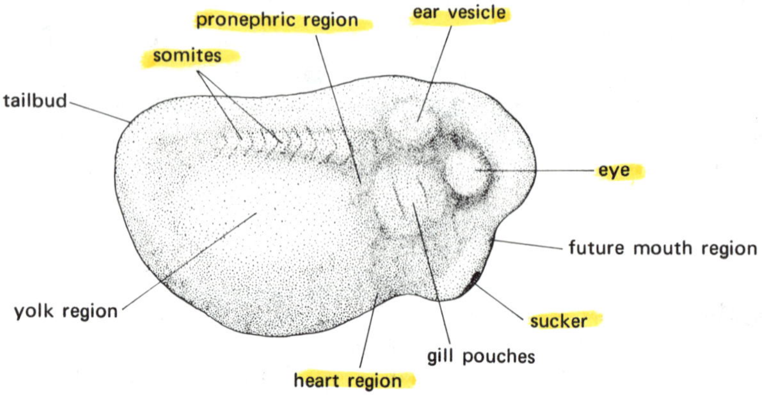

FROG EMBRYO IN EARLY TAILBUD STAGE
EXTERNAL VIEW OF RIGHT SIDE

brain
diencephalon (optic vesicles forming)
prosencephalon
mesencephalon
rhombencephalon

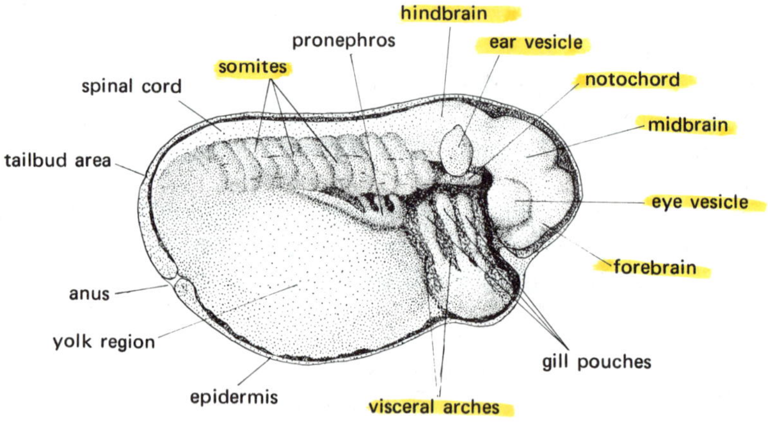

FROG EMBRYO IN EARLY TAILBUD STAGE
SKIN REMOVED

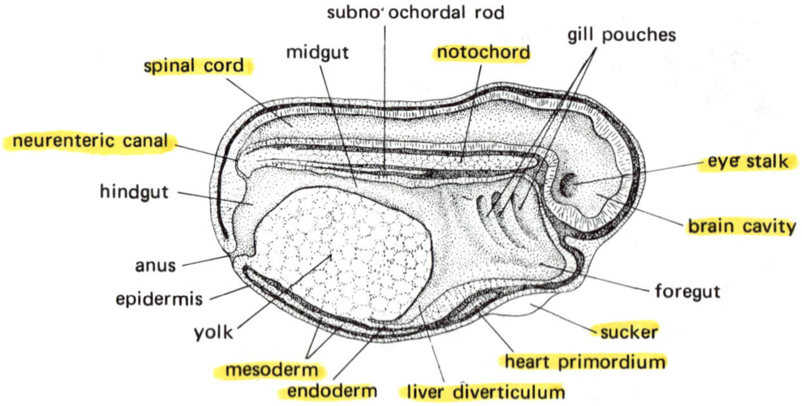

FROG EMBRYO IN EARLY TAILBUD STAGE
MEDIAL SAGITTAL SECTION

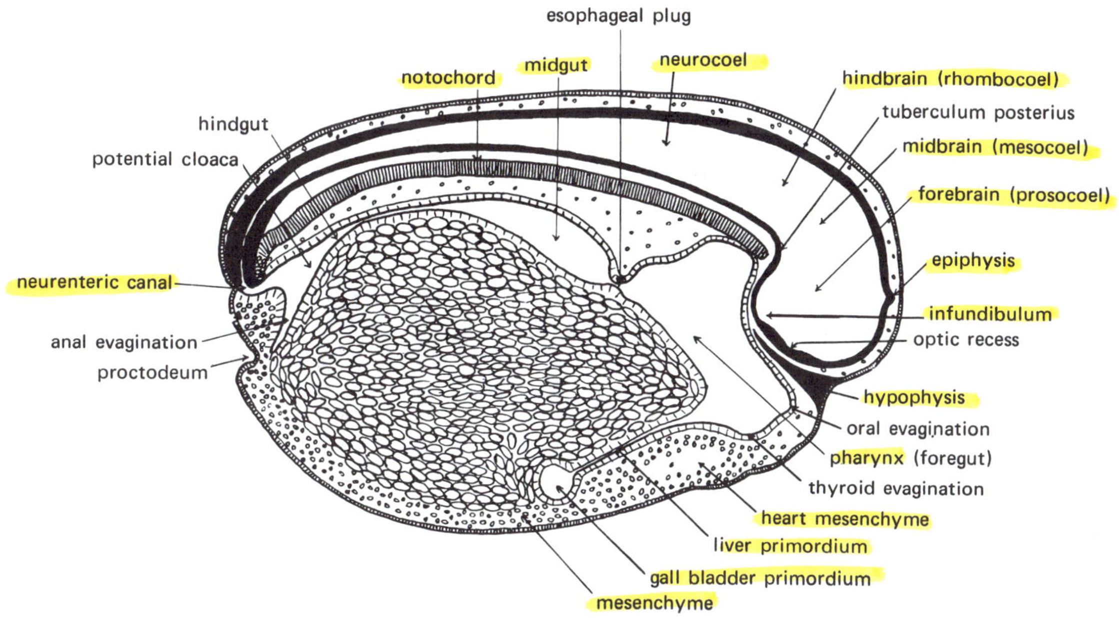

THE 3-mm FROG TADPOLE

MIDSAGITTAL RECONSTRUCTION

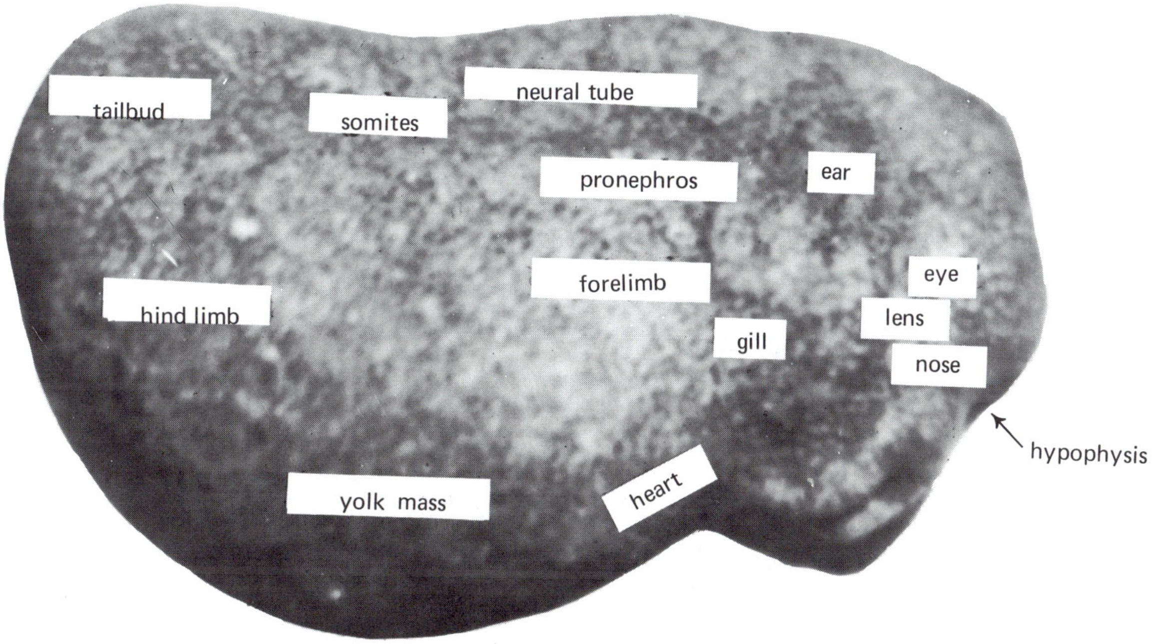

RANA PIPIENS

REGIONS OF ORGAN PRIMORDIA

The organs named are not developed, but the regions indicated, under normal conditions, will give rise to the specific organs listed, by a process of differentiation (see page 65).

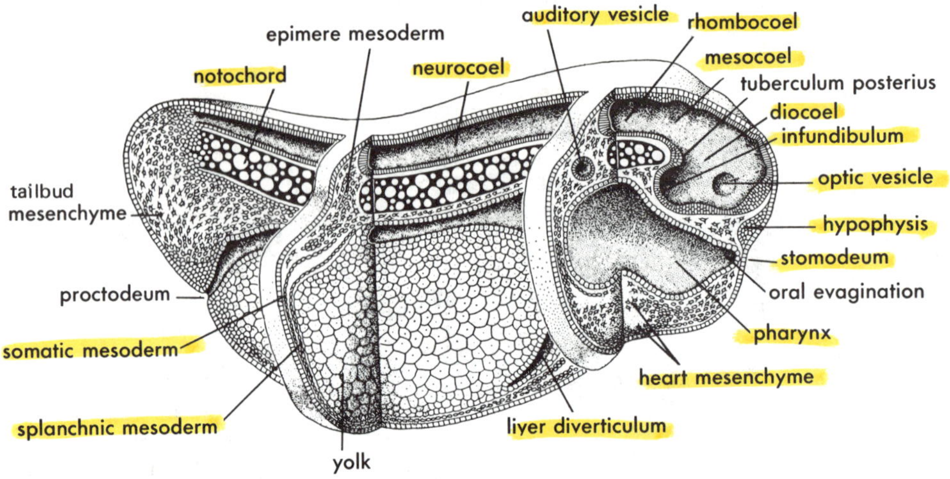

THREE–DIMENSIONAL DIAGRAM OF TAILBUD (4-mm) STAGE IN FROG

From R. Rugh. *The Frog.* 1951. McGraw-Hill, New York. Adapted from Huettner.

REFERENCES

Arey: 73-76
Balinsky: 367-423
Bodemer: 104-09
Huettner: 115-19
Nelsen: 454-554
McEwen: 147-61
Rugh: *Frog*, Chapter X
Rugh: *Vert. Emb.* 36-48
Shumway: 101-03, 105-07, 271

"The search for Truth is in one way hard and in another easy. For it is evident that no one can master it fully nor miss it wholly. But each adds a little to our knowledge of Nature, and from all the facts assembled there arises a certain grandeur."

Aristotle

"When it is recalled that the 9,200,000,000 cells in the human cerebral cortex are the nervous elements of this organ and that they collectively constitute rather less than a cubic inch of protoplasm, it seems almost incredible that they should serve us as they do. They are the materials whose activities represent all human states, sensations, memories, volitions, emotions, affections, the highest flights of poetry, the most profound thoughts of philosophy, the most far-reaching theories of science, and, when their action goes astray, the ravings of insanity. It is this small amount of protoplasm in each of us that our whole educational system is concerned with training and that serves us through a lifetime in the growth of personality."

G. H. Parker

PHARYNGEAL DERIVATIVES		
Ectoderm	Mesoderm	Endoderm
	visceral and aortic arches I (maxillary and mandibular)	
visceral groove I	visceral cleft I (hyomandibular)	visceral pouch I (auditory tube)
	visceral and aortic arches II (hyoid; operculum)	
visceral groove II	visceral cleft II (first branchial)	visceral pouch II
	visceral and aortic arches III (carotid and gill I)	
visceral groove III	visceral cleft III (second branchial)	visceral pouch III
	visceral and aortic arches IV (systemic and gill II)	
visceral groove IV	visceral cleft IV (third branchial)	visceral pouch IV
	visceral and aortic arches V (gill III)	
visceral groove V	visceral cleft V (fourth branchial)	visceral pouch V
	visceral and aortic arches VI (vestigial) (pulmocutaneous and rudimentary gill IV)	
visceral groove VI (vestigial)	visceral cleft VI (vestigial) (fifth branchial)	visceral pouch VI

"I have no theory to propose to explain the origin of organic fitness and human purpose other than the endowment of living beings with the differential sensitivity and reactivity tropisms, organic memory, trial and error behaviour, leading in higher animals to intelligence and purpose. I do not doubt the reality of the mechanistic principles of physics and chemistry and natural selection in the evolution of organisms with all their adaptations. But with crude mechanism that finds everything the result of accident and chance I have no more sympathy than with transcendental vitalism. Those who say there are no ends, values, purposes in the living world may understand atoms, molecules, and even genes, but they do not understand organisms. They see the elements of which life is composed, but they fail to see the pattern and beauty of the entire tapestry of life.

E. G. Conklin, 1944

DRAWINGS: FROG EMBRYOS

THE FROG TADPOLE:

Five-Millimeter Stage

Exercise 11

This is the first serial section study, of which the student will have many before the conclusion of the course. It is necessary to realize that these sections are uniform slices of embryo and that (mental) reconstructions of organ systems are possible only by studying consecutive sections. Depending upon the size of the embryo, the thickness of the sections, and the proximity of the mountings, a single embryo may be spread over one or many slides. Such a series of slides is generally labelled in sequence, so that the most anterior slices appear at the upper left-hand corner of the first slide, while the most posterior slices are found toward the lower right-hand corner of the last slide of the series. When wax reconstructions are made, each section is reproduced in minutest detail and in accurate proportions, and then the wax models are brought together, making it possible to determine exactly the relative size of any organ or organ system. In a somewhat similar manner, the student must mentally reconstruct the whole from the parts mounted on these slides. Several points must be emphasized, however:

1. Ideal transverse (cross) sections of bilaterally symmetrical embryos should show bilateral organs. (See pages 84 and 96 for illustrations.) It is a rare technician that can produce exact transverse sections on such a microscopic scale. Therefore, in some sections the student may see, for instance, but a single eye while a few sections further along the other eye will appear. In making drawings the student should represent exactly what he sees without any improvisation, even though bilaterality may not be obvious. *The embryo is less apt to be wrong than is the student.*
2. Occasionally, a technician mounts the entire series backwards so that the first slide in a series may have the last sections on its upper left-hand corner. In a few rare cases the sections may even be mounted from right to left. The first thing to do, therefore, is to become acquainted with the sequence of the series on your particular slides. This can generally be done with the naked eye.
3. While a particular set of slides might be labeled "5-mm tadpole," it is quite possible that certain embryos might have been 4½ or 5½ mm in length. These differences are small in terms of our usual measuring devices but may be very great in terms of embryonic development. There are probably no diagrams in any textbook which exactly represent the slides the student is to study, although many diagrams will aid in the identification of structures.
4. The staining of sections brings out the differences between the nucleus and the cytoplasm and occasionally the differences between cells. But it must be emphasized that *the stain in no sense corresponds to the germ layer colors used in the lecture drawings.*
5. The drawings called for are supposed to represent the various levels, and these should be chosen to show the greatest number and best developed organ systems.

THE 5-mm TADPOLE – TRANSVERSE SECTIONS

Locate the most anterior sections in the series and identify the following structures as you examine the series progressively toward the posterior.

1. An oval group of *head ectoderm* cells which enclose another group of *brain ectoderm* cells, with an innermost cavity known as the *prosocoel* (cavity of the forebrain or prosencephalon). Ventrolaterally, locate two pigmented invaginations of head ectoderm on either side of the prosocoel, representing the very beginnings of the *olfactory* (nasal) *pits*. In the midventral position locate the dorsally directed invagination of this head ectoderm, which is the beginning of the *stomodeal* cleft, an ectodermal invagination which will give rise to the lining of the

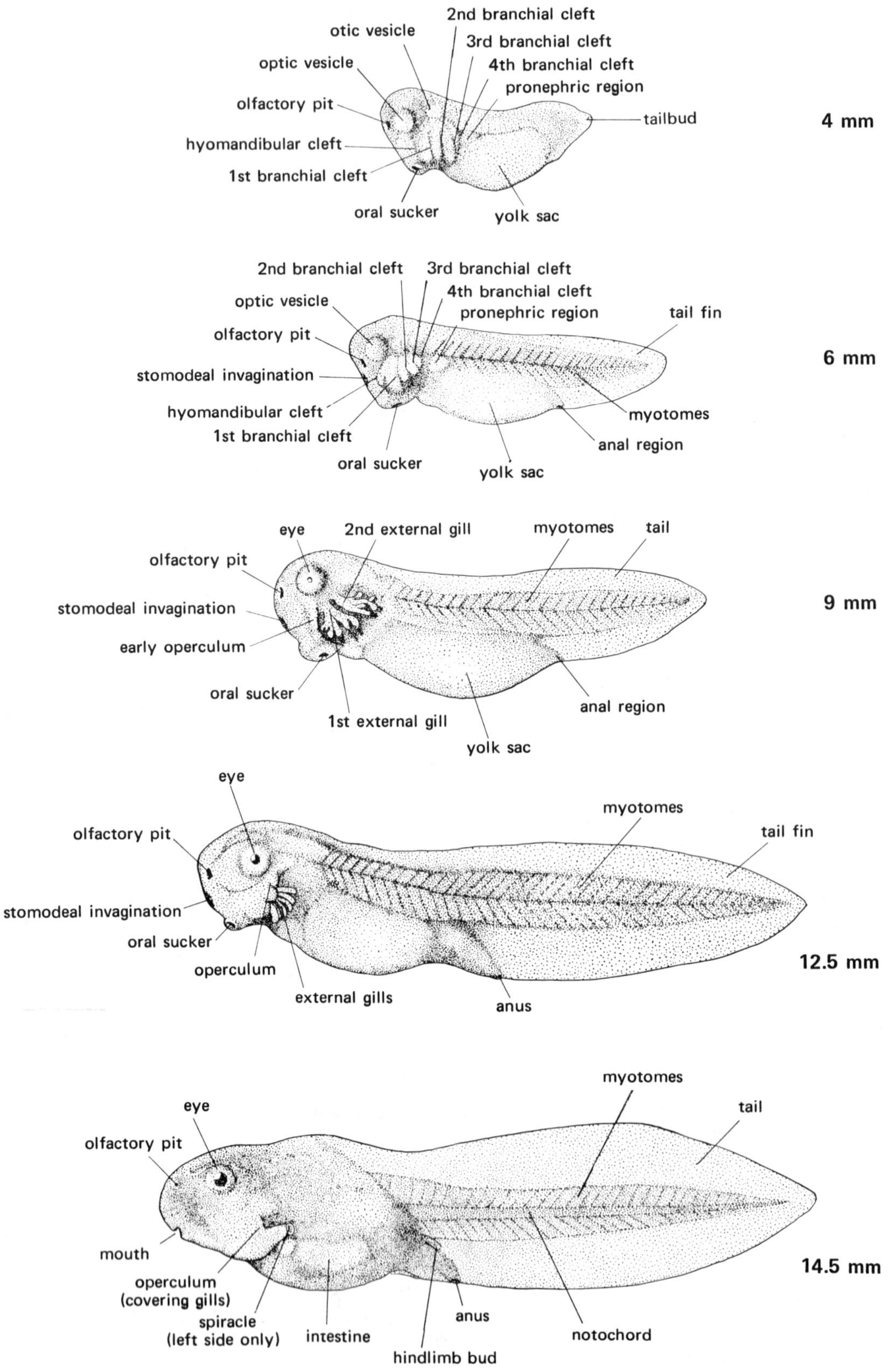

DEVELOPMENT OF GILLS IN THE FROG

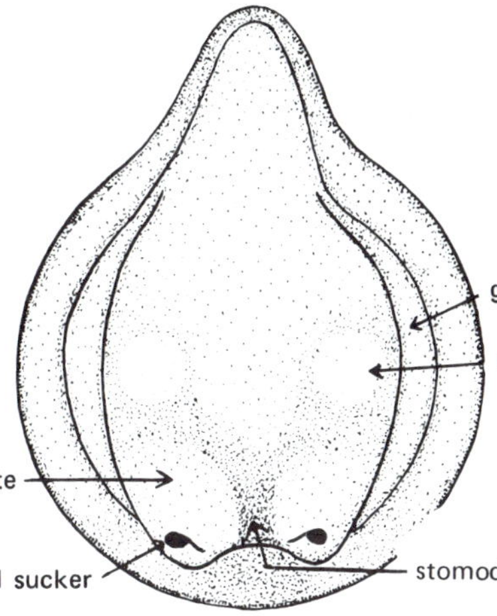

FACE VIEW

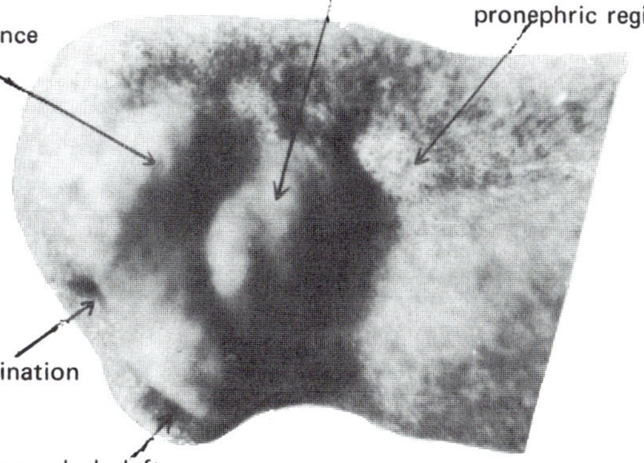

THE 5-mm FROG TADPOLE

mouth. Between the brain and the head ectoderm identify the scattered embryonic mesoderm cells known as *head mesenchyme.* In this vicinity locate also a slight mid-dorsal evagination of the brain vesicle known as the *epiphysis,* a structure which becomes the pineal body in the adult of many vertebrate forms.

2. Several sections more posteriorly note that the nasal pits have disappeared entirely, as have also the stomodeal cleft and epiphysis. This will convey the impression of the relative sizes of these structures. The brain cavity at this level is now much larger and is known as the mesocoel or cavity of the thick-roofed midbrain *(mesencephalon).* Near the position of the original stomodeum there appears a triangular space surrounded by large but unpigmented cells. This is the pharynx, which later will join the invaginating ectodermally lined *stomodeum* to form the mouth. Directly dorsal to the pharynx find a nodular aggregation of cells lying between the pharyngeal and brain cavities. This is the ectodermal *hypophysis.* The larger, more anteroventral of the brain cavities is the prosocoel, to be distinguished from the mesocoel by position and the thinness of its walls. In this vicinity, there appear two lateral evaginations of the prosocoel, known as the *opticoels* because they will give rise to portions of the optico-ocular apparatus. The lateral walls of these opticoels are much thicker than the adjacent walls of the brain. In later stages these lateral optic evaginations will begin to invaginate (i.e., fold in again toward the brain cavity) to form the *optic cups.* The most lateral and thick wall of the original opticoel will

become the nervous or *retinal portion* of the eye. As the dorsal margin of this optic cup comes into contact with the superficial head ectoderm, it induces (causes) the development of the *lens placode,* which thickens and will later invaginate. Therefore, both brain and superficial ectoderm are utilized in the formation of the eye. The most ventral portion of this optic cup does not come into contact with the outer ectoderm but remains to form the inverted groove known as the *choroid fissure.* This is very difficult to identify in transverse sections but may be seen in sagittal sections. Along this groove there will later develop the optic nerve and the blood vessels which supply the retina. The connection of the optic cup with the brain is the *optic stalk.* The surrounding loose *head mesenchyme* is easily identified in these sections, the cells of which will give rise to the connective tissue and skeletal structures in the region of the eye. In slightly older sections, the lateral *branchial pouch* evaginations of the pharynx may be seen.

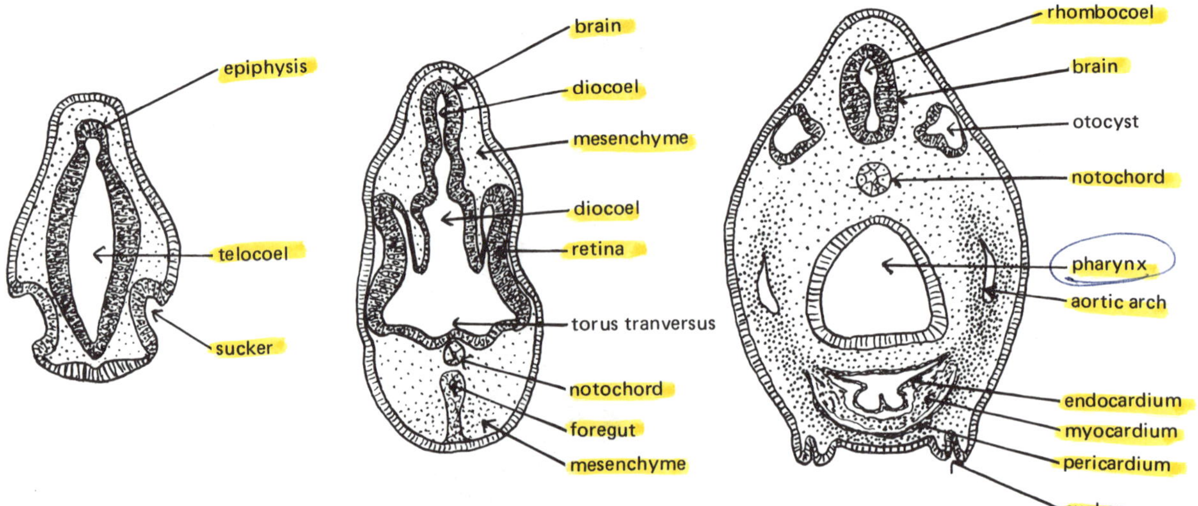

THE 5 mm FROG TADPOLE
TRANSVERSE SECTION

3. In more posterior sections note that the dorsal brain cavity is much reduced at the level of the hindbrain, or *rhombencephalon.* The cavity of this hindbrain is the rhombocoel and may be distinguished by its thin roof. Lateral and slightly ventral to the brain locate paired invaginations of the superficial ectoderm known as the *auditory vesicles.* These may already be closed off from the surface. The ears, unlike the eyes, do not develop from any portion of the brain itself but begin at the surface as placodes and invaginate. Midway between these auditory placodes or vesicles is the well-formed *notochord* with its large and vacuolated cells. Loose *mesenchyme,* some of which will give rise to auditory capsule, is abundant.

Lateral to the enlarged foregut locate the slitlike openings which are the beginnings of the *aortic arches,* not yet supplied with blood corpuscles. These arches will connect the *dorsal* and *ventral aortai.* Beneath this level of the foregut the mesoderm is beginning to form the *embryonic heart.* Heart mesoderm migrates from the two sides to this midventral position dorsal to a large semicircular cavity which will become the *pericardial cavity.* The surrounding mesodermal membrane will become the *pericardium.* Within the pericardium find the much thicker mass of mesoderm known as the *myocardium*, the precursor of the heart muscle. Scattered within the myocardial cavity are loose mesenchymal cells which will soon become organized into a continuous endothelial lining of the heart known as *endocardium.* At about this same level locate the two pigmented ventrolateral ectodermal invaginations known as the *oral suckers.*

4. Examine serial sections more posteriorly until the foregut is much reduced dorsally but has a slitlike connection with a more ventral cavity, the *liver primordium.* It should be clear that the foregut above and the liver below are lined with the same type of yolk-laden endodermal cells.

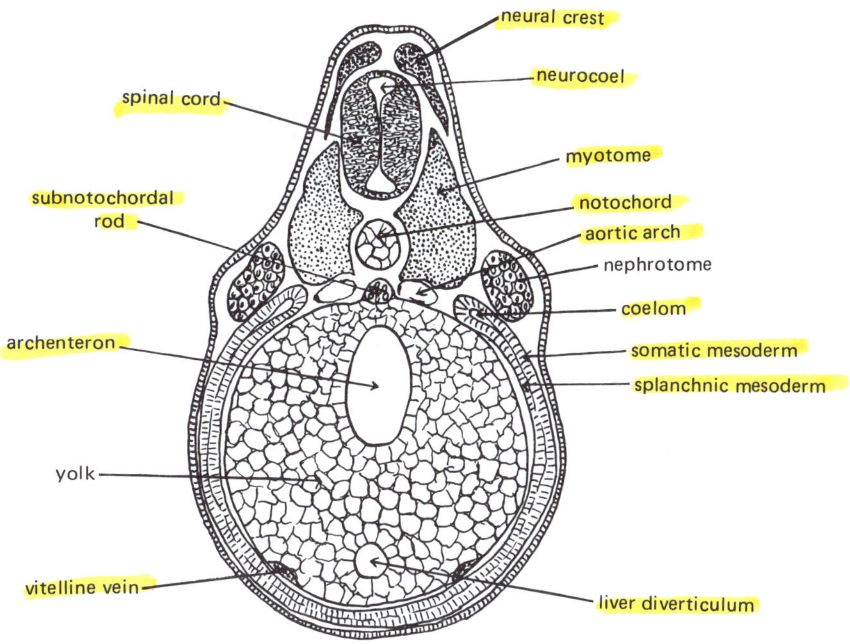

THE 5-mm FROG TADPOLE
SCHEMATIC TRANSVERSE SECTION

Surrounding this entire area is a mass of undigested *yolk.* Ventrolaterally to the liver, at the margins of the yolk and splanchnic mesoderm, locate the paired vitelline veins. Actually, these veins are within the splanchnic layer of mesoderm. In later stages numerous *vitelline veins* will appear. The *hypomeric mesoderm,* consisting of this *splanchnic layer* plus the outer *somatic layer,* now completely surrounds the yolk mass except for the mid-dorsal region. One can now see that the outer somatic mesoderm is in close association with the body ectoderm, the two being designated as the *somatopleure.* The coelomic split is more extensive than previously noted, and, when the inner splanchnic mesoderm becomes associated with the gut endoderm, the two layers will be known as the *splanchnopleure.*

Directly above the gut is the subnotochordal *(hypochordal)* rod and dorsal to that the large vaculated cells of the *notochord.* On either side of the hypochordal rod find the paired dorsal aortae. Lateral and ventral to the aortae, the *intermediate* (mesomeric) *mesoderm* is now organized into circular groups of cells which will become the pronephric tubules. The presence of this mass may be detected as a bulge on the side of the 5-mm tadpole just below the third to fifth somite, posterior to the small gill primordium (see page 83). These pronephric tubules later fuse to form the segmental duct, the whole being considered as the primordium of the ancestral excretory system which, in the frog, functions only in the larval stage.

Just lateral to the notochord find the *epimeric mesoderm,* which is segmented into *somites.* At about this level locate the ectodermal cells, which bulge inwardly but remain as part of the ectoderm known as the *lateral line system.* These are the posterior extensions of the tenth (vagus) *cranial ganglia.*

The brain cavity at this level is still the *rhombocoel,* but the roof is obviously much thinner than the floor. From the junction of the closed neural folds and the outer ectoderm there will appear aggregates of cells, lying dorsolaterally to the developing central nervous system. These are the *neural crests,* the forerunners of the spinal ganglia, chromatophores, and the sympathetic system.

5. Several serial sections further posteriorly, note that the liver *cavity* is separated from the foregut; the *pronephric tubules* are better formed, and the loose *sclerotomal* cells have broken away from the somites to cluster about the neural tube and notochord. These cells will give rise to the axial skeleton. The major part of the somite will remain as the rather solid *myotome* (muscle), while an outer layer of cells will become closely associated with the outer ectoderm as the *dermatome,* to form the connective tissue and blood vessels of the skin.
6. Sections through the midbody level will show much the same structures as previously identified, except that the size relations will be different. The liver, heart, and foregut structures will no longer be visible.

REFERENCES

Balinsky: 367-536
Bodemer: 107-15
Huettner: 120-33
McEwen: 147-68
Nelsen: 454-55
Rugh: *Frog*, Chapter XI
Rugh: *Vert. Emb.* 48-103
Shumway: 270-79

DRAWINGS: FROG EMBRYOS

DRAWINGS: FROG EMBRYOS

FROG TADPOLE: 5 MM

Exercise 12

THE FROG TADPOLE:

Five-Millimeter Stage

SAGITTAL SECTIONS

The sagittal sections are to be studied after full acquaintance with the transverse sections of the same stage. The student should locate the most mesial (central) sagittal section, i.e., the one which shows the maximum of the brain cavity and the neurocoel. The most important "landmarks" in this section will be the *tuberculum posterius* and the *dorsal thickening*. Note that the forebrain vesicle is flexed (ventrally) around the tip end of the *notochord* at the region of the tuberculum posterius and that the most anterior projection of the forebrain is the *epiphysis*. In the floor of the *prosencephalon* locate the two thickenings, the posterior *optic chiasma* and the anterior *torus transversus*, with the intervening depression called the *optic recess*. At this time, it is possible to draw an artificial line between the anterior *telocoel* and the *diocoel* (subdivisions of the prosencephalon) by locating the posterior limit of the *choroid plexus* and the anterior limit of the optic recess and joining the two. Posterior to the optic chiasma, the diocoel bulges beneath the notochord as the *infundibular cavity*, the ectodermal (brain) portion of which will aid in the formation of the *hypophysis*. Posterior to the dorsal thickening and dorsal to the notochord locate the hindbrain which merges with the neurocoel.

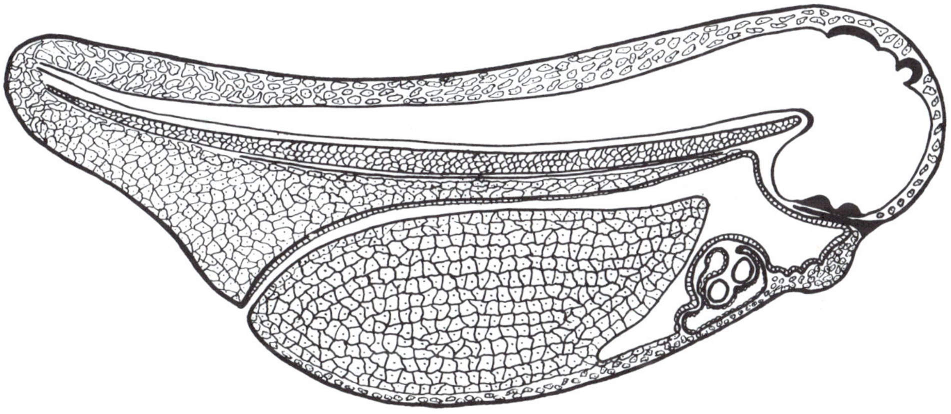

RECONSTRUCTION OF 5-mm TADPOLE IN SAGITTAL SECTION

Locate each of the following structures:

anus
auricle
diocoel
ectoderm
endoderm
epiphysis
foregut primordium
hindgut primordium
hypophysis
infundibulum
lamina terminalis
liver primordium
mesencephalon
mesenchyme
mesoderm
midgut
neurocoel
notochord
optic chiasma
optic recess
oral plate
pericardial cavity
pharynx
prosencephalon
rhombencephalon
subnotochordal rod
telocoel
thyroid primordium
torus transversus
tuberculum posterius
ventricle
yolk

In these sections, it will be possible to determine the extent of the notochord. Its most anterior tip is between the tuberculum posterius and the infundibulum, and the most posterior limit is in the developing tail. Notochordal cells are always easily distinguished, because they are large and vacuolated. The *subnotochordal rod* may be difficult to locate, largely because it is so small.

The relationship of the stomodeal hypophysis to the infundibulum (floor of the diocoel) and the roof of the foregut *(pharynx)* will be clearly indicated, since the stomodeal hypophyseal cells are pigmented and grow in between these regions. The *heart,* it must be remembered, forms as bilateral tubes, then becomes a single tube, and finally develops as a bent tube, so that in these sagittal sections parts of the heart may appear as separate from each other, although all will be enclosed within the *pericardial cavity and membrane.* The *thyroid* is a single evagination from the floor of the pharynx just anterior to the heart, and the *liver* is a prominent diverticulum of the foregut which projects posteriorly and ventrally behind the heart.

At the posterior end, the anal opening should be located as a newly formed ectodermal invagination (*proctodeum*) directly ventral to the position of the original blastopore. Both the proctodeum and the remnant of the *neurenteric canal* may be identified by pigmented cells. (Compare with the stomodeum.)

In the more lateral of the sagittal sections endeavor to locate the *pronephric ducts;* the *auditory vesicles;* the *optic stalk* and, still more laterally, the *choroid fissure;* the segmental *somites;* and, in the region of the pharynx, the lateral evaginations of endoderm, which, together with the opposing ectodermal invaginations, will go to form the *visceral clefts.* The most anterior pouch, later to form a cleft, is the *hyomandibular.* This is followed by the first to the fifth branchial or the first to the fifth gill pouches, which later break through as clefts.

FRONTAL SECTIONS

These frontal sections are mounted from the most dorsal to the most ventral so that the first longitudinal cavity to be seen will be a part of the central nervous system, in all probability the *midbrain* and *hindbrain* region with posterior *neurocoel.* The first organs to appear will be the *epiphysis* and the lateral *otic* (auditory) *vesicles.* Just anterior to these vesicles locate aggregations of neuroblasts, representing the large *fifth* and the more posterior *seventh* and *eighth cranial ganglia.* The seventh and eighth arise together, later to be separated as the facial and auditory. The auditory is closely adherent to the otic (ear) vesicle and the more posterior of the two. These sections are particularly good to show fundamental vertebrate bilaterality and metamerism.

In more ventral sections, at about the level of the *notochord,* locate the *optic cups* with *lenses,* just anterior to the region of the fifth cranial ganglia. At this level the small *ninth cranial ganglia* and the *tenth,* with their posterior extensions, the *lateral line nerves,* may come into view. The only portion of the brain visible at this level is the most anterior *prosencephalon.* Just posterior to the *prosocoel,* where the notochord is disappearing, locate the most dorsal portion of the pharynx (*foregut*).

Locate next the *olfactory* (nasal) *placodes* lined with pigmented ectodermal cells; the *optic stalks* and *vesicles;* the *infundibulum;* and the *midgut and hindgut.* On some slides it may be possible to see the metamerically arranged *spinal ganglia* on either side of the *notochord* or *neurocoel.* Within the *visceral arches* note the large empty blood vessels (*aortic arches*), some of which will later grow out into the developing external gills as branchial vessels. The *hypophysis* will be seen anterior to the pharyngeal cavity as a median column of pigmented cells just ventral to the disappearing *infundibulum* as the more ventral sections are reached.

More ventrally, the midgut disappears. Just posterior to the pharynx the *heart* structures will appear, consisting of the paired *vitelline veins* converging from the posterior, the *endocardium, myocardium,* and *pericardium.* These are particularly good sections in which to determine the changes in shape of the embryonic heart. Between the heart and the pharynx is the rather large mass of darkly staining cells which forms a pocket and represents the *thyroid gland,* later to become

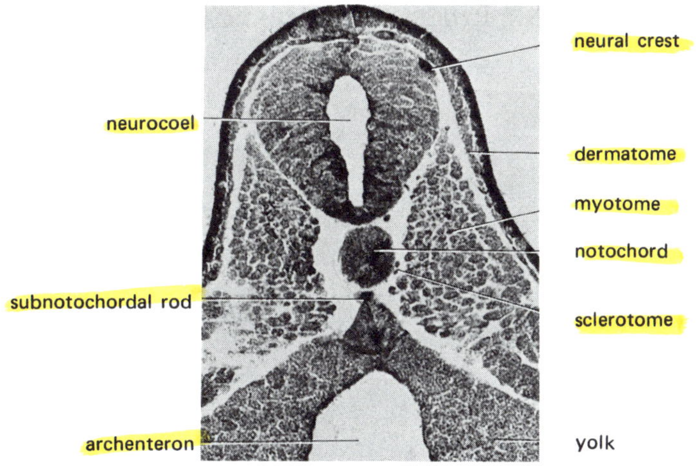

FROG TADPOLE X-SECTION
5-mm STAGE

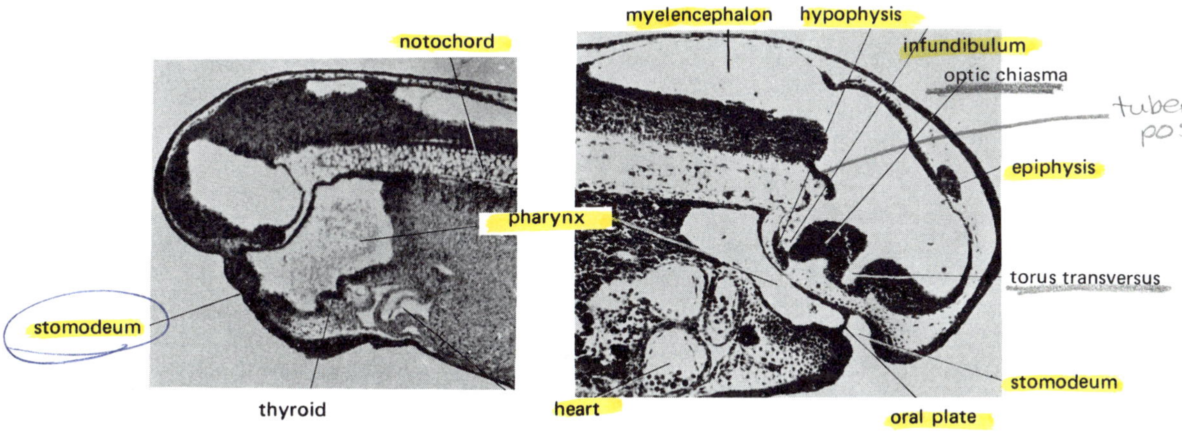

5-mm **7-mm**

FROG EMBRYO
SAGITTAL SECTIONS

bilobed and to migrate posteriorly. Posterior to the heart, and apparently embedded in the yolk, locate the rudiment of the *bile duct* leading into the more ventral *liver diverticulum.* At the posterior end of these sections note the pigmented *proctodeum.* Still more ventrally, at the anterior end, find the *suckers.*

REFERENCES

Balinsky: 367-536
Bodemer: 109-18
Huettner: 120-33
McEwen: 147-68
Rugh: *Frog,* Chapter XI
Rugh: *Vert. Emb.*, Chapter 3, 20-103
Shumway: Chapters 7-11, frog sections only

DRAWINGS: FROG EMBRYOS

Exercise 13

THE FROG TADPOLE:

Seven-Millimeter Stage

TRANSVERSE SECTIONS

The 7-mm frog tadpole has well-formed external gills and a functional heart. The oral plate is very thin and ready to break through to form the mouth. The student is provided with serial transverse sections from the anterior end of the tadpole through the level of the liver, and the following structures should be located in a study of the sections beginning at the anterior end.

1. The *stomodeum* now appears as a deep invagination of the pigmented, midventral ectoderm at the anterior end. It is an inverted longitudinal groove more extensive than the ultimate mouth.
2. The original nasal placodes have separated from the ectoderm and have acquired large pigmented *nasal pits*, possessing so many cells that the cavity itself may seem to be obliterated. It may be possible to identify some cells growing in the direction of the stomodeum, ventromesially, later to become the tubular connections with the internal nares (choanae).
3. The *epiphysis* is now separated from the brain (*diencephalon*) as a circular knob of cells, later to become the pineal body.
4. The *oral plate* is from one to two cells in thickness, between the invaginating *stomodeum* and the evaginating floor of the *pharynx*. This membrane breaks through at about the 7-mm stage to form the mouth, the margins (lips) of which are actually the *mandibular ridges* (outer edges of mandibular arches). The student should be able to distinguish by cell structure and pigmentation between the stomodeum, pharynx, and forebrain.
5. Lateral to the pharynx there will appear condensations of head mesenchyme known as the *mandibular arches*.

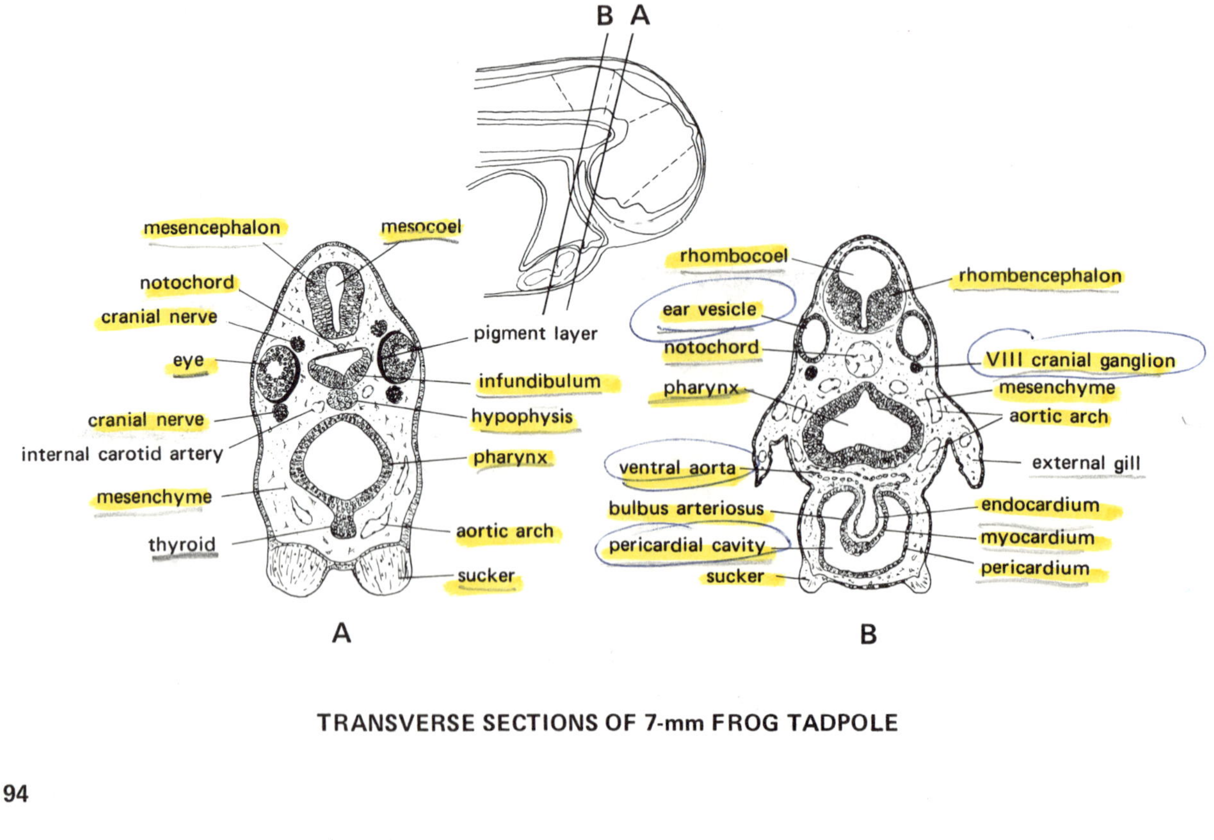

TRANSVERSE SECTIONS OF 7-mm FROG TADPOLE

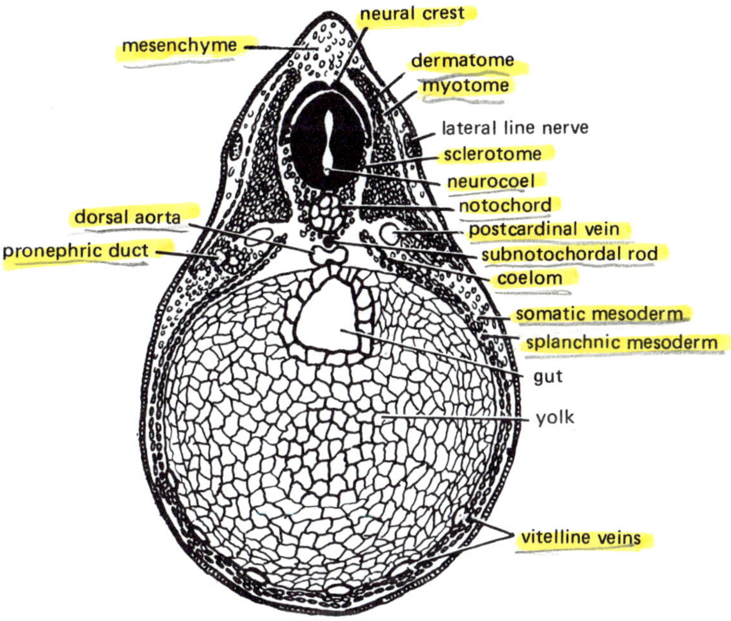

THE 7-mm FROG TADPOLE

TRANSVERSE SECTION THROUGH THE MIDBODY LEVEL

6. Portions of the eye will come into view, and it will be simple to distinguish between the outer *pigmented* and the inner *retinal portions* of the original *optic cup*.
7. The *optic stalk* is now much smaller and obviously tubular, connecting with the ventral portion of the eyecup itself. The *choroid fissure* (inverted groove in the optic stalk) is difficult to identify in transverse sections, since these sections run parallel with the fissure itself. It is within this fissure that small nerves and blood vessels run into the eye chamber. Note the well-formed lens, with evidences of the original *lens* vesicle.
8. The *pharynx* at this level is large and more rounded. Directly above the pharynx locate the pigmented group of *hypophyseal stomodeal cells*, which persist through a number of sections. The presence of these cells helps in the identification of the nearby *infundibulum*. At this same level, but ventral to the pharynx, will be seen an evagination of endodermal cells of the pharyngeal floor, forming the *thyroid gland*.
9. The second pair of visceral (*hyoid*) arches may be identified as those nearest the thyroid primordium. The beginnings of the arches appear as condensations of mesenchyme around the pharynx. These arches become rather rigid supports between the future *visceral clefts* (slits), each arch having blood vessels (*aortic arches*) and nerves.
10. At the level of the most posterior section of the hypophyseal cells note that the infundibular portion of the diencephalon has been cut off from all other parts of the brain in this section. These brain ectoderm cells will join those of the stomodeal hypophysis to form the anterior pituitary gland. The most dorsal cavity at this level is probably the *rhombocoel* (cavity of rhombencephalon). The uncertainty is due to various possible angles to transverse sections, but the rhombencephalon is safely identified by the very thin roof and the nearby *otic vesicles*. The *mesencephalon* (midbrain) has the *dorsal thickening* as its roof.
11. As the infundibulum closes off (posteriorly), the anterior tip of the notochord will take its place, in progressively posterior sections. That portion of the brain floor just anterior to the tip of the notochord has already been identified as the *tuberculum posterius*.
12. Just posterior to the level of the suckers in these transverse sections locate the anterior parts of the heart. It may be possible to find the *aortic arches* which carry blood from the heart through the *visceral arches* (around the pharynx) to the developing *dorsal aortae*. This anterior portion of the heart is the *bulbus* or *truncus arteriosus*. The layers of the heart wall can be identified as the lining *endocardium*, the thick-walled *myocardium*, and the thin outer *pericardium*.

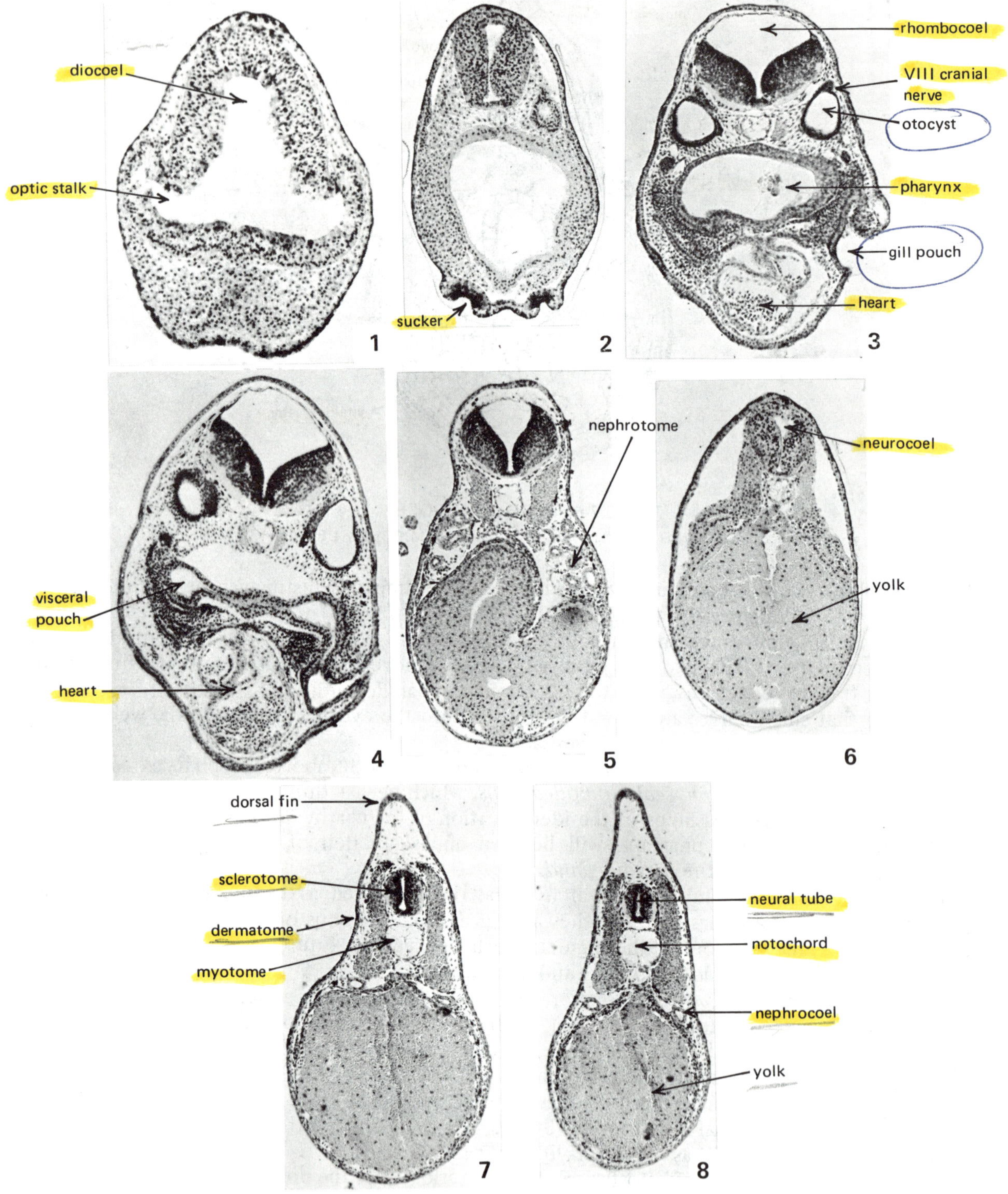

7-mm TADPOLE TRANSVERSE SECTIONS

Courtesy H. Wurmbach

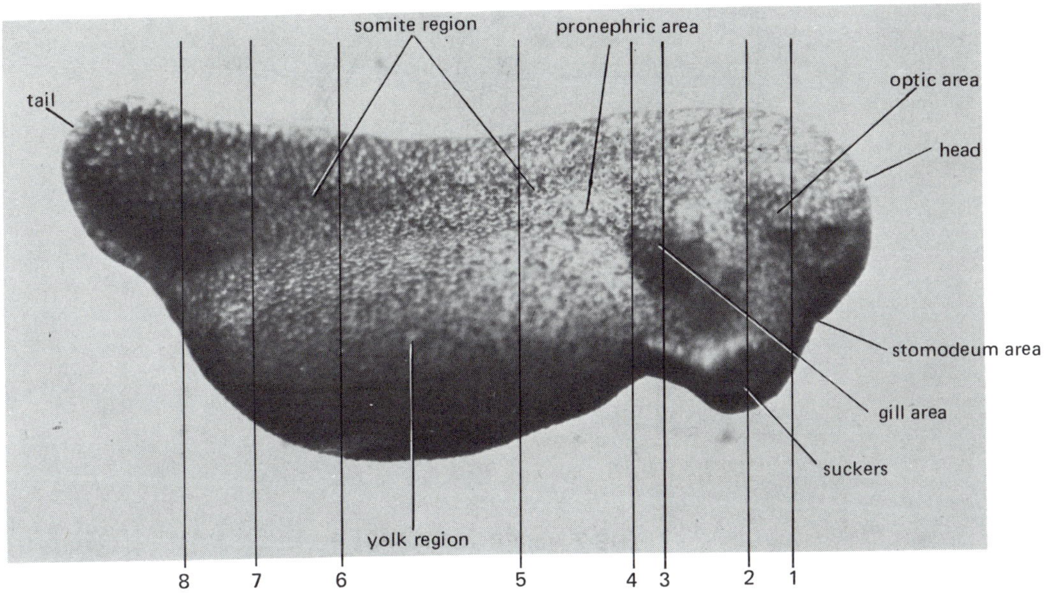

LIVING 7-mm TADPOLE
LOCATION OF SECTIONS FROM OPPOSITE PAGE.

13. At about this level, neither the optic nor the auditory organs will be seen, but there will appear paired condensations of ectodermal cells in the general vicinity of these sense organs. These will be the primordia of the *fifth, seventh,* and *eighth cranial ganglia.* The fifth ganglia are always the largest, while, in more posterior sections (as the auditory vesicle is approached), the seventh and eighth cranial ganglia may be identified together close to the vesicle. Note particularly the differences in cell thickness in the walls of the *otic* (auditory) *vesicle.*

14. The brain cavity above should still be the thin-roofed *rhombocoel,* but it should be noted that its ventrolateral walls are now very thick. This portion of the brain (hindbrain) is associated with all cranial ganglia from the fifth through the tenth in the frog, as in all vertebrates.

15. Lateral or ventrolateral to these head sections, there will appear loose sections of tissue which are pieces of the *external gills.* These gills are extremely thin and contain loops of blood vessels used for respiration.

16. Within a few sections of this level, it will be noticed that the foregut cavity becomes occluded and disappears entirely. The type of cells which normally line the gut (endodermal) are still discernible, but the cavity itself has been blocked off. This blocking mass is known as the *esophageal plug,* later to be broken through to reestablish the gut as a continuous tube.

17. The paired *dorsal aortae* may now be seen just beneath and slightly lateral to the *notochord.*

18. As the gut reopens and forms the ventrally directed *liver diverticulum* note in the same sections the lateral well-formed *pronephric tubules.* These embryonic excretory units come from the intermediate mesoderm (*mesomere*), located between the *epimere* and the *hypomere* (or lateral plate mesoderm).

19. The epimere (dorsal mesoderm) now begins to give off loose cells which encapsulate the nerve cord and notochord, comprising the *sclerotome.* The outermost loose cells of this epimere become the *dermatome* and give rise to connective tissue and blood vessels of the skin, while the larger (central) mass of this epimeric mesoderm becomes the *segmental myotomes,* or muscle bundles.

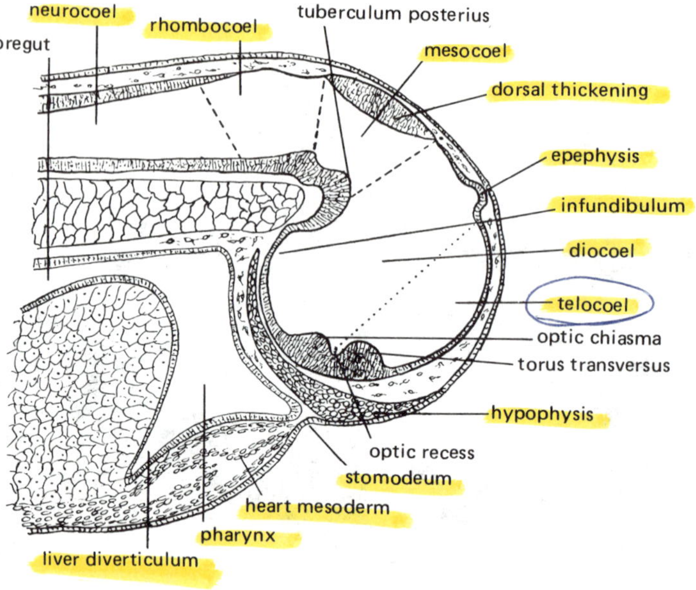

THE 7-mm FROG TADPOLE

SAGITTAL SECTION: ANTERIOR END

From R. Rugh. *The Frog*. 1951. McGraw-Hill, New York.

FRONTAL SECTION

These sections are cut and mounted from the most dorsal to the most ventral, so that the first sections on the slide (as in the 5-mm stage) will show the anterior *brain vesicles* and the *neurocoel* of the spinal cord as continuous cavities. At about the level of the *epiphysis* also locate portions of the eyes (*optic cups*) and ears (*otic vesicles*). Just posterior to the eyes find the large mass of neurons comprising the *fifth cranial ganglia*. The *seventh* pair of cranial ganglia are slightly posterior and mesial to the *fifth*, while the *eighth* pair are close to the auditory vesicle. Posterior to the ear locate the *ninth* and finally the *tenth* pair of cranial ganglia. At this age the related lateral line nerves should be seen growing posteriorly from the tenth (vagus) ganglia to a position just beneath the body skin.

A few sections lower the *nasal placodes* come into view. The neurocoel disappears and in its place appear the highly vacuolated cells of the *notochord*. On either side of the notochord are the bilaterally paired clumps of neurons which constitute the beginnings of the *spinal ganglia*, derived originally from the neural crests. Lateral to these ganglia are the segmental *myotomes* (muscle bundles) which extend posteriorly into the tail. Still further laterally, just posterior to the head region, find the slightly convoluted pronephric or *segmental tubules* and their *ducts*. Anterior to these ducts, running into the head mesenchyme, are the paired *carotid arteries* which come from the heart and extend through the third pair of visceral arches. At this stage one is likely to find *erythroblasts* (erythrocyte-forming cells) within the blood vessels.

At a still lower level the convolutions of the lateral pronephric tubules will become more prominent. Note the external *gills*, apparently detached from the body. In the head notice that, in sections posterior to the diocoel, the large *pharyngeal cavity* takes it place. Identify as many of the *visceral arches* and *clefts* as can be seen. Just posterior to this pharyngeal cavity there appears a rather oval mass of tissue (largely yolk) in the center of which is a cross section of a vertically directed tube. This is the primordium of the *bile duct*. On either side of this enlarged liver diverticulum are large blood vessels, originally the *vitelline veins*, which now empty directly into the heart. The *heart* will be seen just anterior to this liver mass and now consists of a bulbous cavity

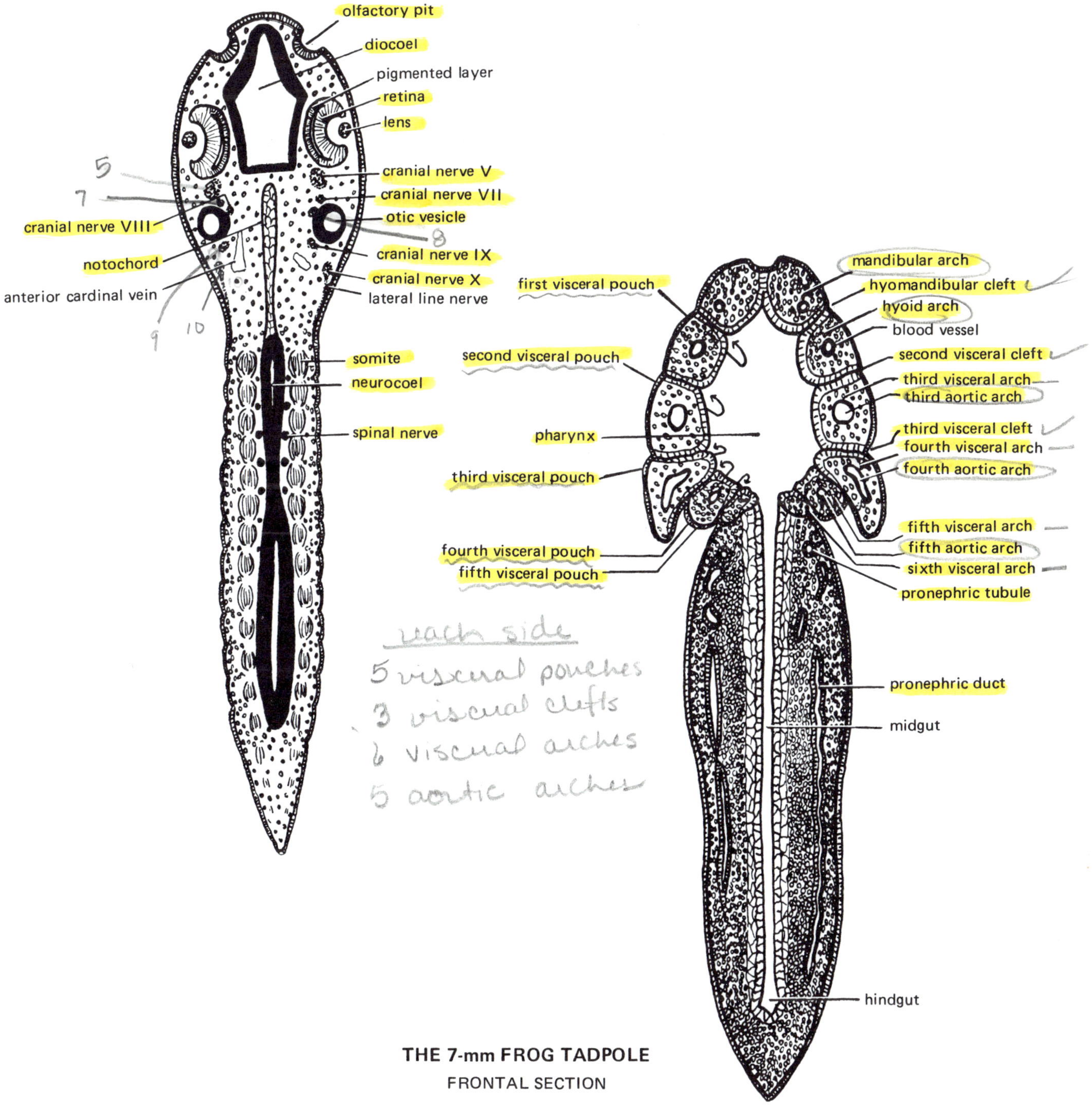

THE 7-mm FROG TADPOLE

FRONTAL SECTION

filled with blood corpuscles. Identify parts of the heart at this level and such parts of the aortic arches as may be seen.

Anterior to the heart find the enlarged *thyroid mass.* The tubular origin of this endocrine (ductless) gland should be evident at this stage. Locate the more ventral *stomodeum* and *oral plate* (which may have broken through) and the mesenchymal condensations laterally, masses which will give rise to the *mandibular arches* (muscles of the jaws). Most of the endoderm at this level is yolk, although it will be possible to identify the *liver* at the anterior end and *proctodeum* at the posterior end. Just beneath the body wall attempt to find the *somatopleure* and *splanchnopleure* with the intervening *coelomic cavity.*

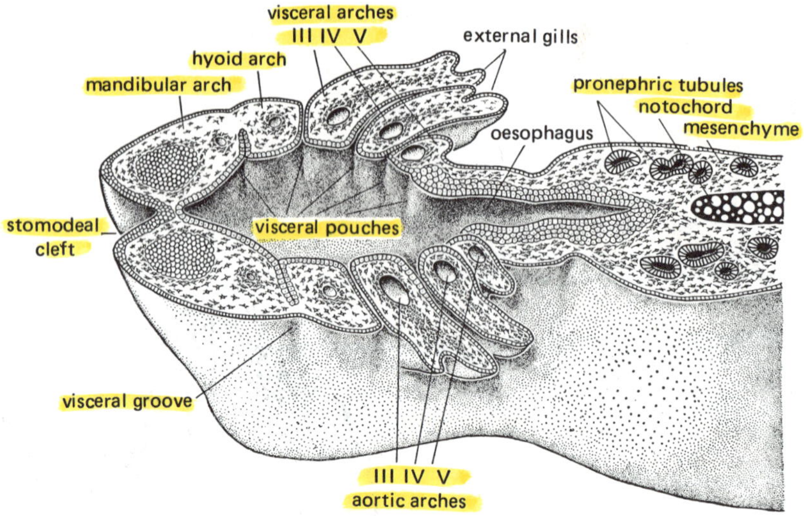

FRONTAL (HORIZONTAL) RECONSTRUCTION OF THE EXTERNAL GILL STAGE OF THE FROG LARVA

From R. Rugh. *The Frog.* 1951. McGraw-Hill, New York.
Redrawn and adapted from Huettner.

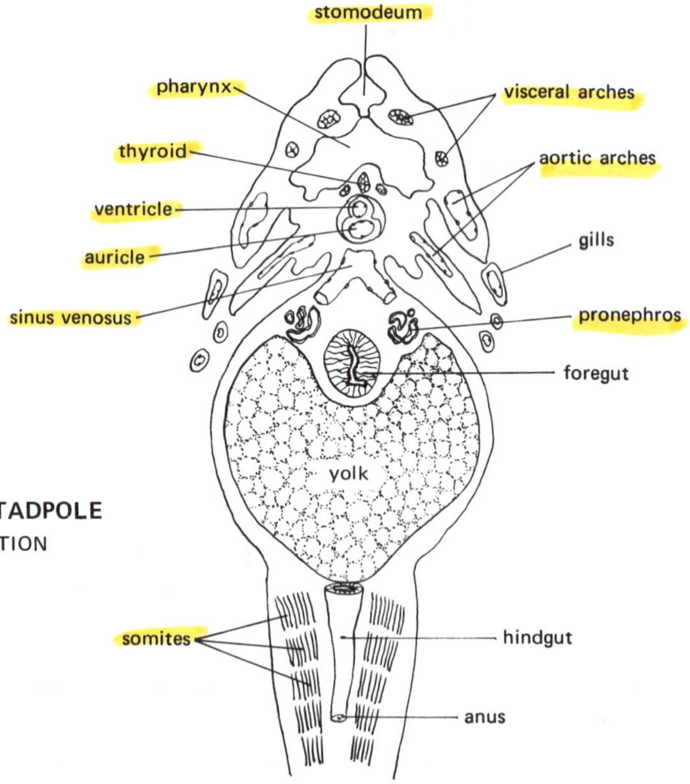

THE 7-mm FROG TADPOLE
FRONTAL SECTION

REFERENCES

Balinsky: 367-536
Bodemer: 109-18
Huettner: 134-91
McEwen: Chapter VI
Rugh: *Frog,* Chapter XII
Rugh: *Vert. Emb.* 48-103
Shumway: Chapters 12-16, section on the frog

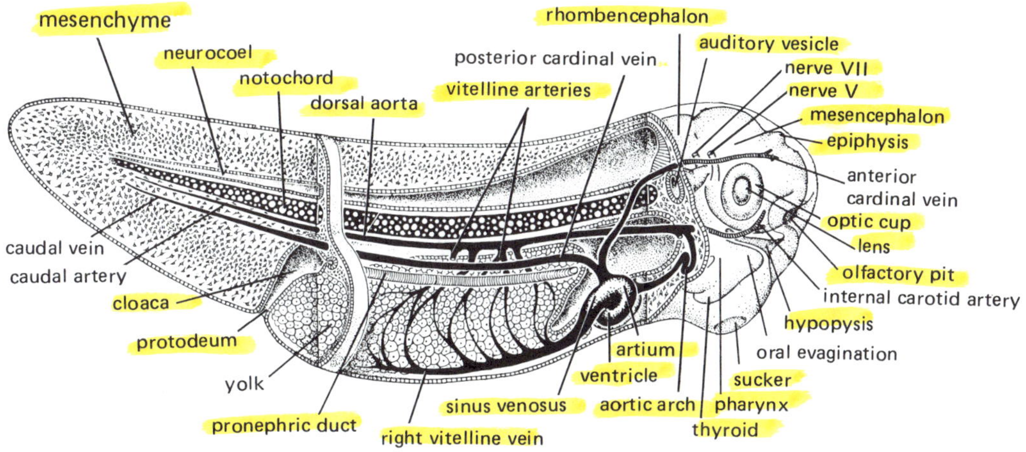

THREE-DIMENSIONAL DIAGRAM OF 7-mm TADPOLE

From R. Rugh. *The Frog.* 1951. McGraw-Hill, New York.
Redrawn and modified from Huettner.

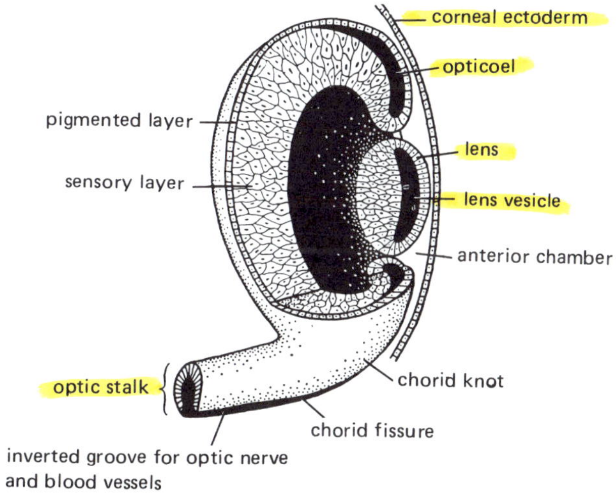

DIAGRAM OF EYE DEVELOPMENT IN FROG

From R. Rugh. *The Frog.* 1951. McGraw-Hill, New York.

DRAWINGS: FROG EMBRYOS

"The campaign must begin with the study of developmental biology, for we have hardly began to understand the forces and reactions that drive and guide the development of the fertilized ovum into a human infant. Embryology of the classical morphological type, experimental embryology, cytology, histochemistry, all have their contribution to make. The genes, and the whole sequqnce of events by which they determine the infinite detail of bodily structure, await new discoveries. There are countless unsolved questions in the physiology of reproduction, touching on the maternal environment of the embryo and its control by hormones and other chemical agents of the body. The chemistry of respiration and nutrition must be called upon to explain the marvelous homeostatic balances through which, after all, the child generally enters the world sound and healthy. Microbiology must detect for us all the various pathogenic organisms which invade and damage the fetus in utero; clinical medicine and obstetrics must teach us what illnesses of the mother may affect her offspring in the susceptible earliest days of development."

G. H. Corner, 1960, First Ind. Conf.
Congenital Malformation, Landa.

THE FROG TADPOLE:
Eleven-Millimeter Stage

Exercise 14

The tadpole by this stage (10 days at 25°C) has the familiar shape with head, trunk, and laterally compressed tail. It is regarded as the larva of the developing frog, but the term "tadpole" is used for the larvae of many forms, both vertebrate and invertebrate. It is free swimming and is considered as a physiologically independent organism. It has lost its external gills by absorption and has developed opercular folds of head skin (by posterior growth of the second visceral or hyoid arch) which cover the external gills. The newly formed gill cavity is now known as the *opercular cavity,* with a single external opening (the spiracle) at the posterior margin of the *opercular fold,* visible on the left side of the head. The mouth is now open and is surrounded by horny lips. There are large eyes, nasal openings, and a cloacal aperture. Serial transverse sections through the head to the level of the heart and pronephros of such tadpoles are provided. From anterior to posterior the student should locate the following.

OLFACTORY ORGANS – the olfactory placodes will be seen in the most anterior sections, growing inwardly so as to lie close to the telocoel. The anterior wall of the olfactory vesicle is very thin, but in more posterior sections there will appear a slightly pinched-off dorsolateral vesicle, known as the *lateral appendix,* which is rather thick with the original placode cells. The position of the *internal nares* (choanae) may be determined, although the tubular connection of these with the external nares will not be established until later. The first cranial (olfactory) nerves grow from these original placodes to the olfactory lobes (telencephalon) of the brain.

PHARYNX – in these anterior sections locate the mouth opening with partial (*stomodeal*) lining of pigmented cells and surrounded by frilled organs which constitute the *embryonic lips.* Cornified ectoderm cells within the mouth give rise to conical elevations regarded as *embryonic teeth.* The entire mouth region is practically surrounded by groups of cells which begin to take on the semblance of cartilage. These are derived from head mesenchyme and represent mandibular derivatives of the first visceral arches. From these arches come most of the jaw parts. There will be no notochordal tissue this far anteriorly, but there will be abundant chondrification of head mesenchyme beneath the divided telocoels, as well as around the oral cavity. Later in development, this cartilage will be largely replaced by bone. Most posteriorly, at about the level of the eyes, note that the midventral floor of the pharynx becomes thickened and that the cells in the vicinity take a very dark stain. This is the beginning of the formation of the *tongue,* which structure will receive some support from the second pair of visceral (*hyoid*) arches. Note the relation of the tongue to the more lateral cartilages. Two clusters of cells beneath the tongue may represent the *thyroid gland,* already divided into two lobes.

GILLS – beneath the tongue level locate a continuous cavity which runs from one *gill* (opercular) *chamber* to the one on the other side. This is the *opercular groove* through which water passes from the right opercular cavity to the left opercular cavity and out through the *spiracle,* which is always located on the left side of the head. The respiratory organs at this stage consist of the *internal gills* only, the external gills having atrophied. It is important that the student study the entire respiratory apparatus represented over a series of these serial sections, locating anteriorly the *velar plates* which are lateral growths of the floor of the pharynx that partially separate the pharynx from the gill chambers; the filamentous *gill rakers* which sift the water; and the *opercular cavities* with their contained internal gills. (See schematized drawing on page 105.)*

*It is suggested that the instructor set up a demonstration, at this point in the study, to show circulation in the living tadpole gills and tail. The tadpoles can be anesthetized by MS-222 (1/3000 concentration) or dilute chloretone, and the circulation observed under the dissecting microscope.

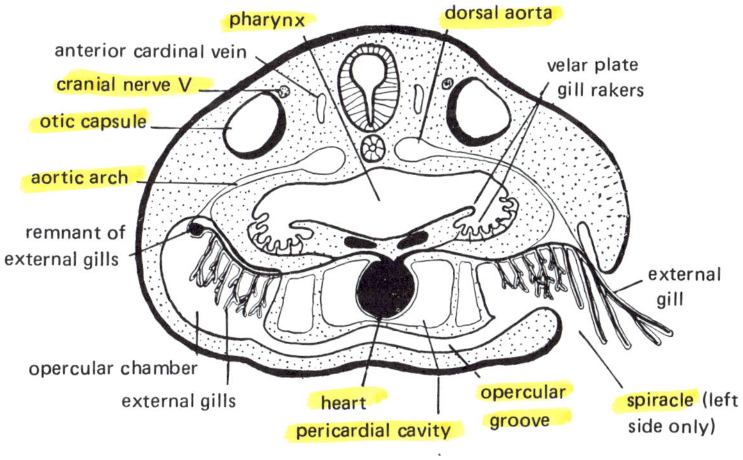

RELATION OF THE PHARYNX TO THE INTERNAL AND EXTERNAL GILLS OF THE FROG, TRANSVERSE SECTION

From R. Rugh. *The Frog.* 1951. McGraw-Hill, New York.

EYES – at this stage the tadpole eye is essentially like that of the adult. Note the fibers of the *lens;* the positional relation of the lens to the *anterior* and the *posterior chambers;* the *iris;* and the overlying *cornea.* The layers of the retina are now structurally distinct. Since the eyes develop anterior to their connections with the *diencephalon,* the optic nerves may be located slightly posterior to the eyes and will appear as small clusters of blue-staining cells, possibly as fibrous tracts.

EARS – the ears now consist of a number of related cavities, the most dorsal of which lies nearest the *rhombencephalon* and is the remnant of the *endolymphatic duct* or original tubular connection with the outside of the head. Lateral to the endolymphatic duct is the *anterior vertical semicircular canal,* below which is the large cavity, the utricle. Directly lateral the horizontal *semicircular canal* may be identified and ventrolateral the enlarging *saccule.* Note the scattered *cartilage cells* which are beginning to surround the entire auditory apparatus to form the *auditory capsule.* Locate, in close association with the utricle, the *eighth cranial nerve.*

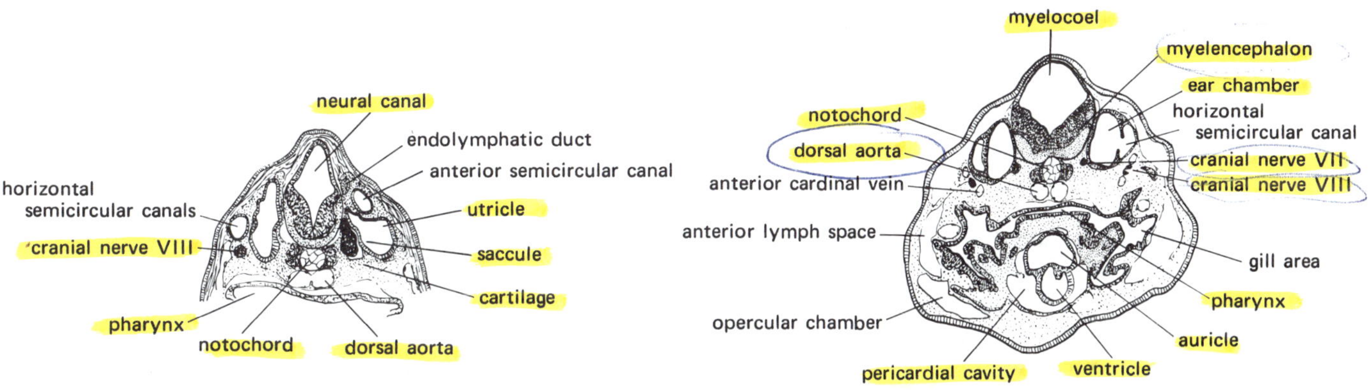

11-mm FROG–EAR LEVEL (TRANSVERSE SECTIONS)

PITUITARY GLAND – at the level of the infundibulum locate again the *stomodeal hypophyseal cells* which have clustered against the floor of the *infundibular recess* and are obviously intermingling with the infundibular cells. This combination of ectoderm from the brain and from the head integument gives rise to three components of this gland.

HEART – at about the level of the internal gills find the thin-walled and saccular auricle, often filled with blood cells. Associated with the auricle, but in a more ventral position, locate the thick-walled *ventricle.* Both of these parts of the embryonic heart are lined with an *endocardium,* but the ventricular portion is generally devoid of corpuscles. Surrounding all parts of the heart is the *pericardial space* and *membrane.* More anteriorly find the single tubular *bulbus arteriosus* which feeds the numerous paired lateral *aortic arches* passing up through the *visceral arches.* Posterior to the auricle note that the *hepatic veins* (originally the vitelline veins) bring blood into the *sinus* venosus at the level of the *ducts of Cuvier,* which convey blood from the cardinal system.

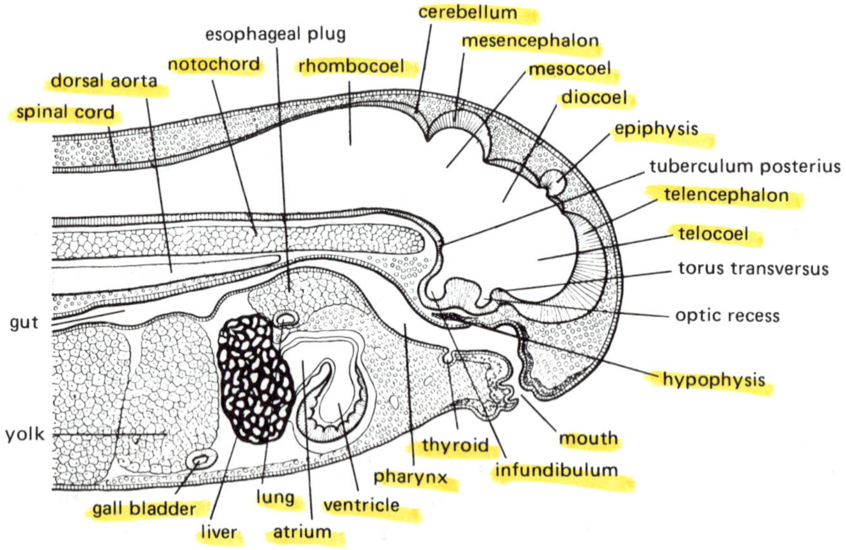

SAGITTAL SECTIONS SHOWING DEVELOPMENT OF BRAIN AND ANTERIOR STRUCTURES IN TADPOLE AT 11 mm.

From R. Rugh. *The Frog.* 1951. McGraw-Hill, New York.

LUNGS – at this stage the lungs consist of only a median ventral diverticulum of the floor of the foregut just posterior to the level of the heart. This is the primary *lung bud* and is lined with the same type of endoderm found in the gut itself. This diverticulum will later bifurcate into paired embryonic lungs, which grow posteriorly to remain unused until the time of metamorphosis (generally after a winter of hibernation), when the method of respiration shifts from gills to lungs for the young frog.

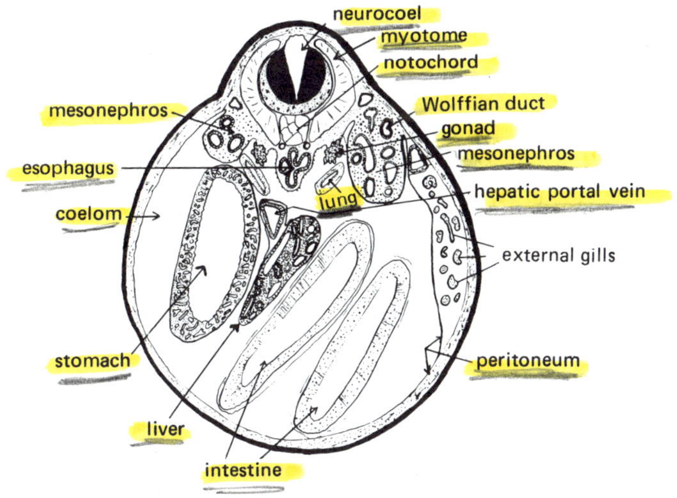

11-mm FROG–GONAD LEVEL

VISCERA – locate as many of the following as can be identified in the remaining sections.

LIVER – highly vascularized organ.

INTESTINES and STOMACH – thick-walled tubular organs having columnar-type epithelial lining.

MESONEPHROS – mass of tubules similar in structure to the pronephric tubules previously studied. Attempt to locate a nephrostomal opening from the body cavity into this embryonic kidney mass. These persist even on the adult kidney.

GONAD PRIMORDIUM – associated with the kidneys, an ovary primordium showing the beginnings of a secondary genital cavity. Find germ cell nests and large primordial germ cells, and the darker rete cord cells.

DORSAL AORTAE– generally paired at this level, ventrolateral to notochord.

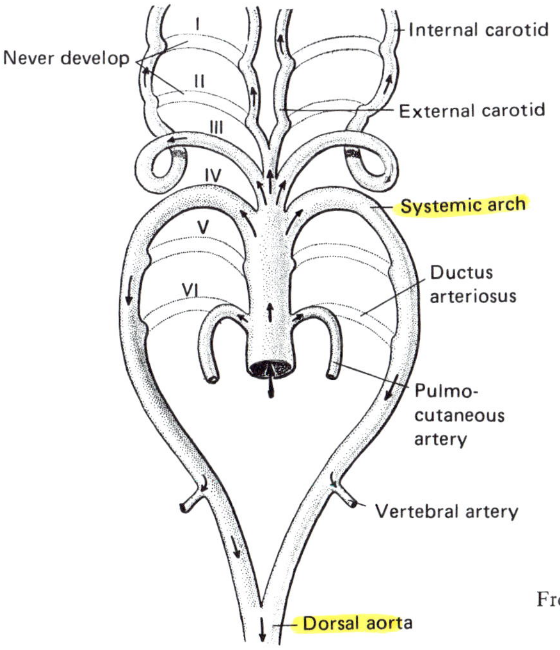

VENTRAL VIEW SHOWING FATE OF AORTIC ARCHES IN FROG

From R. Rugh. *The Frog.* 1951. McGraw-Hill, New York.

NOTOCHORD – large, vacuolated cells directly ventral to the neurocoel.

SCLEROTOME – aggregations of epimeric mesodermal cells which will give rise to the internal skeleton around the notochord and central nervous system.

SPINAL GANGLIA – now identified as the dorsal root ganglia, associated with the spinal cord derived from the original neural crest cells.

REFERENCES

Balinsky: 109-18
Bodemer: 367-536
Huettner: 155-91
McEwen: Chapter VI
Rugh: *Frog*.Chapter XIII
Rugh: *Vert. Emb.* 48-103
Shumway: Chapter 17 (Experimental Embryology)

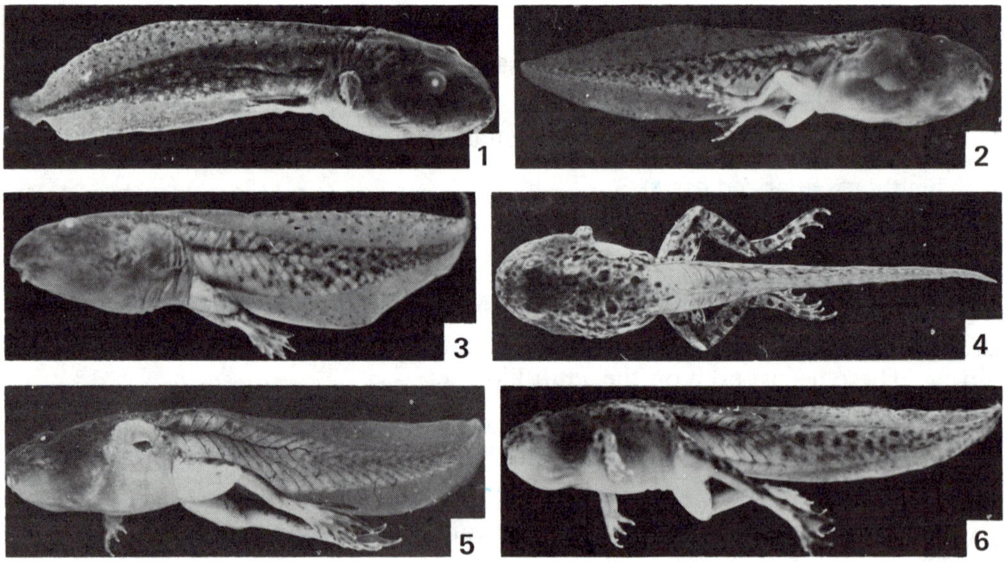

METAMORPHOSIS OF *RANA PIPIENS*

1. Tadpole stage, no appendages.
2. Hind limb stage. These emerge rapidly and together.
3. Left side showing closed tympanum.
4. Dorsal view showing right forelimb beginning to emerge.
5. Tympanic membrane rupture by growing left forelimb.
6. Left forelimb emerging through tympanum. (Tail would then be resorbed and the frog spend more time on dry land.)

SUMMARY OF EMBRYONIC DEVELOPMENT: 11-mm TADPOLE

Ectodermal Derivatives

EPIDERMIS – thickened, ciliated on tail only.

CENTRAL NERVOUS SYSTEM – differentiated into four brain vesicles and spinal cord.

PROSENCEPHALON (Forebrain)

Telencephalon – anterior to optic recess; thin-roofed.
Cerebral hemispheres develop around each of lateral ventricles.
Anterior choroid plexus as a thin median dorsal invagination into telocoel and diocel, which becomes vascularized.
Olfactory (I) nerves to floor of each olfactory lobe from the mesial sides of nasal tubes.

Ectodermal Derivatives (Continued)

NERVE	ORIGIN	DESTINATION	TYPE
I olfactory	telencephalon	olfactory organ	afferent
II optic	diencephalon	retina	afferent
III oculomotor	mesencephalon	superior rectus, inferior rectus, median rectus, and inferior oblique eye muscles	efferent
IV trochlear	mesencephalon	superior oblique eye muscle	efferent
V trigeminal	rhombencephalon and first crest segment and placode (semilunar)	opthalmic, maxillary, and mandibular regions	mixed
VI abducens	rhombencephalon (neuroblasts of medulla)	lateral rectus eye muscle	efferent
VII facial	rhombencephalon and second crest segment and placode	hyomandibular region	mixed
VIII acoustic	rhombencephalon and second crest segment and placode	inner ear	afferent
IX glossopharyngeal	rhombencephalon and third crest segment and placode	visceral arches II and III, mouth, tongue, and pharynx	mixed
X vagus	rhombencephalon and fourth crest segment and placode	lateral line and visceral arches IV and V viscera	mixed efferent

Ectodermal Derivatives (Continued)

Diencephalon – from epiphysis to dorsal thickening to tuberculum posterius.

Epiphysis as mid-dorsal evagination just posterior to choroid plexus.

Optic recess – just posterior to torus transversus or region of lamina terminalis (anterior fusion of neural folds) as median ventral evagination.

Optic chiasma – thickened floor where optic nerves cross (posterior to recess).

Optic (II) nerve – to chiasma from the eye, may be seen joining retina.

Infundibulum – bulbous evagination in floor of diencephalon between optic chiasma and tuberculum posterius, ventral to notochord.

MESENCEPHALON – bounded by dorsal thickening and tuberculum posterius.

Optic lobes from dorsal thickening, bilobed, become corpora bigemina.

Cavity – mesocoel, iter, or aqueduct of Sylvius.

Oculomotor (III) nerve – mesiolateral, from floor.

Trochlear (IV) nerve – dorsolateral, just posterior to optic lobes and difficult to find, being small.

RHOMBENCEPHALON – from dorsal thickening to spinal cord. Metencephalon and myelencephalon not distinguishable in the frog.

Cerebellum – from thickened roof, anterior.

Posterior choroid plexus – from thin, vascular posterior roof.

Otic vesicles – found at level of cerebellum.

Cranial nerves V to X are associated with this portion of the brain, all paired.

Trigeminal (V) nerve – anterior to otocyst, large, sends branches to mandibular and maxillary processes of first visceral arch and ophthalmic branch to the eye muscles.

Ectodermal Derivatives (Continued)

Abducens (VI) nerve – from floor of Rhombencephalon, anterior and ventral to origin of Trigeminal. Supplies the lateral rectus eye muscle (as do Oculomotor and Trochlear).

Facial (VII) and auditory (VIII) nerve – arise as a single ganglion, anteromesial to otocyst, supplying the facial and hyoid muscles and the saccule and utricle of the ear.

Glossopharyngeal (IX) nerve – arises posterior to otocyst, sending branches to the first branchial arch. Ganglion.

Vagus (X) nerve – arises with IX, sending branches to second, third, and fourth branchial arches, to the lateral line organs and to the viscera.

SPINAL CORD – from rhombencephalon into tail, with continuous cavity.

Spinal (central) canal – original neurocoel, constricted centrally.

Ependymal layer – elongate, ciliated cells lining the canal.

Mantle layer – gray matter, consisting of compact cell bodies, lateral.

Marginal layer – white matter, consisting of outermost axons.

Dorsal root – afferent fibers passing dorsomesially, to join dorsal root ganglion.

Dorsal root ganglion – paired thick bundle of neurons dorsolateral to spinal cord, derived from neural crests.

Ventral root – efferent fibers passing ventrolaterally from spinal cord to spinal nerve trunk.

Spinal nerve – mixed bundle of afferent and efferent fibers associated with distal organs (striated muscle, etc.).

Dorsal ramus – branch of spinal nerve to dorsal skin and muscles.

Ventral ramus – branch of spinal nerve to ventral muscles.

Communicating ramus – fibers from spinal nerve to sympathetic ganglion.

Sympathetic ganglion – nerve cells lateral to dorsal aorta, derived from neural crest.

SPECIAL SENSE ORGANS – found in the head region.

EYE – well developed in 11-mm stage, some parts mesodermal.

Nervous retinal layer – sensory portion consisting of rods and cones, inner and outer granular layers, and ganglion cell layer, comprising the retina.

Pigmented retinal layer – thinner layer outside of nervous layer.

Lens – arising as vesicle, now a solid ball lying in opening of optic cup.

*Choroid coat – thin, vascular, and pigmented layer of mesenchyme immediately around retina.

*Sclerotic coat – loose mesenchyme becomes tough connective tissue investing the choroid layer.

*Cornea – outer tissue component of sclerotic layer together with the specialized corneal epithelium; therefore both ectoderm and mesoderm.

*Eye muscles – small bundles of muscles which control the eye movements, innervated by cranial nerves.

EARS – arising from solid auditory placodes, then become the hollow otocyst.

Utricle – mesial and dorsal, giving rise to the three semicircular canals, two of which have developed at this stage, i.e., the anterior dorsoventral and the outer horizontal canals.

Saccule – Endolymphatic duct – between utricle and brain, joining saccule.

Cochlea – mesial portion of saccule, pigmented and ciliated.

NOSE – arising from olfactory placodes, which hollow out to form choanae.

External nares – glandular organ of Jacobson and olfactory sensory epithelium lining the cavity which opens to exterior.

*Part or all of the starred structures are derived from mesoderm.

Ectodermal Derivatives (Continued)

Internal nares – extensions of tubular openings from external nares into pharynx, called internal choanae.

STOMODEUM and PROCTODEUM – ectodermally lined openings into mouth and cloaca.

Endodermal Derivatives

The derivatives of the primary digestive tube, from anterior to posterior, are:

MOUTH – with the exception of that (stomodeal) part from oral aperture to point of internal nares (choanae).

PHARYNX – dorsoventrally flattened cavity into which the mouth leads.

VISCERAL CLEFTS – junction of endoderm lining of the visceral pouches with the ectodermal invaginations from the side of the head to form clefts; associated with second, third, fourth, and fifth visceral pouches. The sixth pouch never forms a cleft. By the time the clefts are open they are covered by the operculum so that they lead into the opercular chamber ventrolaterally. External gills degenerating.

THYROID GLAND – now separated from floor of pharynx, dividing posteriorly into two lobes. Found just anterior to the heart as paired pigmented masses.

THYMUS GLANDS – derived from dorsal part of first and second visceral pouches (the first degenerating) and migrating posteriorly to position near the ear.

PARATHYROID GLANDS (epithelioid bodies) – from ventral portions of third and fourth visceral pouches. Difficult to locate.

ULTIMOBRANCHIAL BODIES – small pigmented masses at level of the trachea ventral to the sixth visceral pouches. Difficult to locate.

TRACHEA and LUNGS – laryngotracheal groove from median ventral floor of pharynx near visceral clefts which leads into tubular trachea that bifurcates into two primary bronchi. These expand posteriorly into lungs.

ESOPHAGUS – gut from laryngotracheal groove to stomach, bending behind the liver.

DUODENUM – from pyloric end of stomach to coiled intestine.

LIVER – highly vascularized and enlarged organ which incorporates the vitelline veins. Original liver diverticulum becomes the gall bladder with the common bile duct (ductus choledochus) leading into the dorsal wall of the duodenum.

PANCREAS – within curvature of stomach, duct joining common bile duct.

INTESTINE – two complete spiral coils posterior to duodenum. The cloaca is as yet undifferentiated but the pronephric ducts open into its posterior end.

Mesodermal Derivatives

MESENCHYME – loose, primitive mesoderm found scattered throughout the tadpole. There are condensations in the head region where cartilage-forming centers are developing, later to give rise to the neurocranium (except the dura and pia mater which come from neural crest). Connective and blood vascular tissues are also to be derived from mesenchyme. Around the mesodermal notochord are sclerotomal (mesenchyme) cells which will form the axial skeleton.

EPIMERE – most dorsal mesodermal masses appearing as metameric somites, the most anterior of which are being transformed. About 12 myotome pairs in the trunk and 32 in the tail remain distinct.

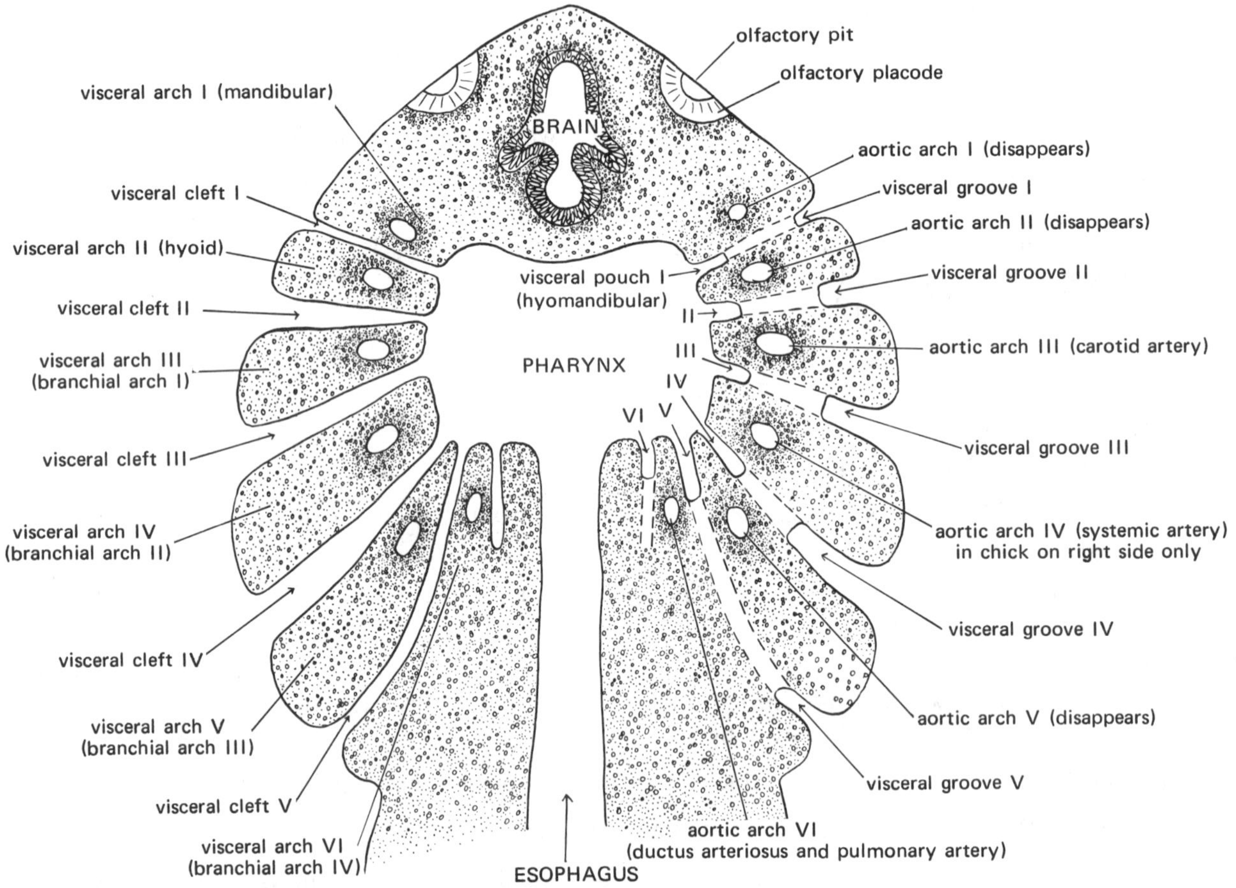

SCHEMATIC DRAWING TO SHOW PHARYNGEAL DERIVATIVES IN FROG AND CHICK

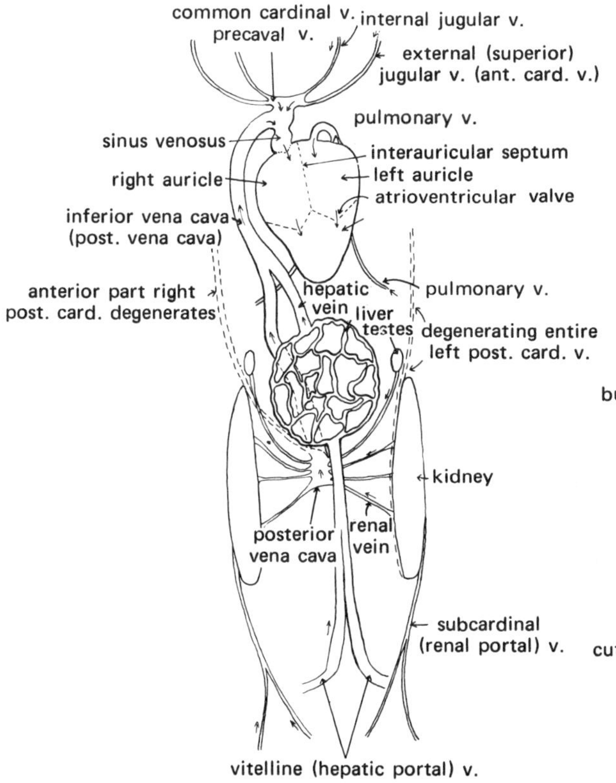

VENOUS SYSTEM FROG TADPOLE

VENTRAL VIEW

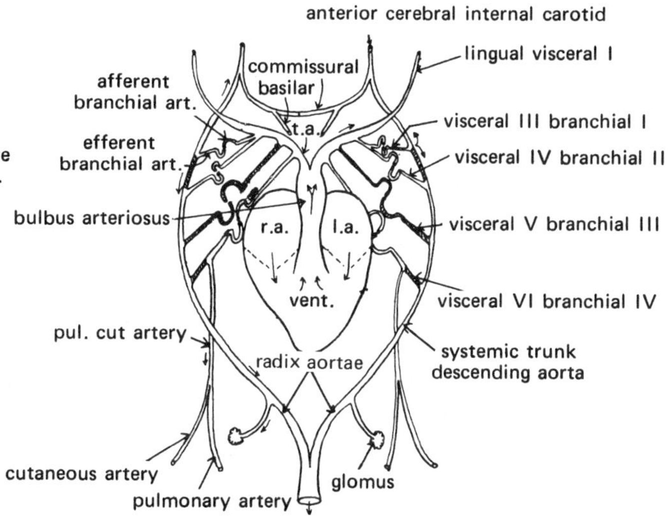

ARTERIAL SYSTEM OF THE FROG TADPOLE

VENTRAL VIEW

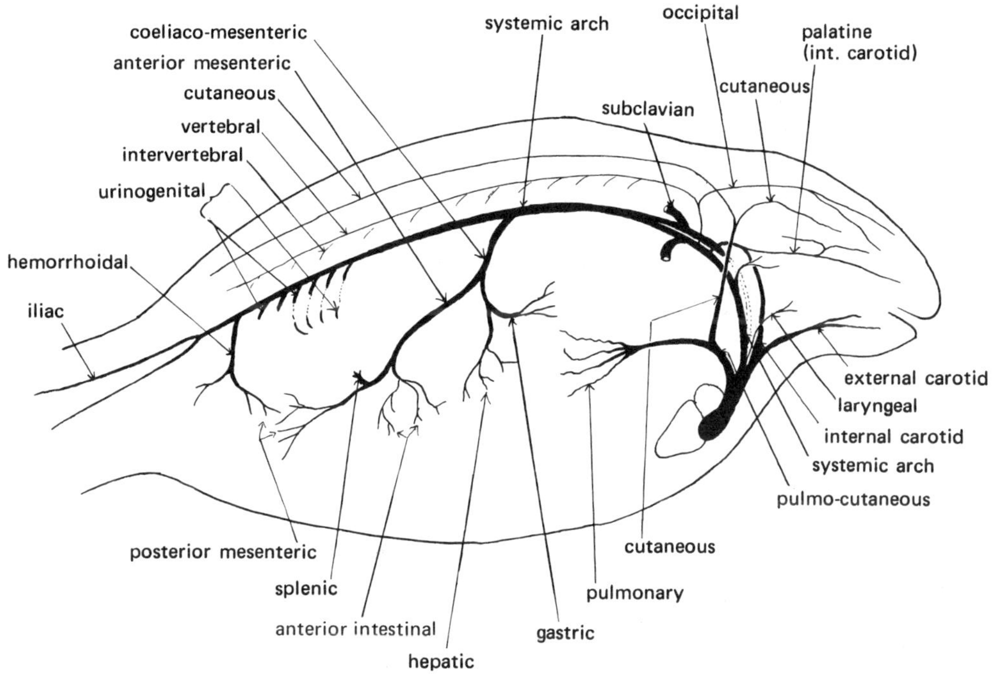

SCHEMATIC DIAGRAM OF FROG ARTERIAL SYSTEM

LATERAL VIEW

Mesodermal Derivatives (Continued)

Dermatome – distinct band of mesenchymatous cells lying dorsolaterally just beneath the ectoderm, to form the dermis (cutis) of the skin.
Myotome – central portion of somite which is being organized into muscle bundles, divided in the tail region into dorsal and ventral bundles.
Sclerotome – thin layer of mesenchyme cells surrounding the nerve cord and notochord from which will develop the axial skeleton.

MESOMERE – intermediate mesodermal mass from which the urogenital system is derived.

Pronephros – primary, embryonic kidney made up of few (three) coiled tubules, ciliated nephrostomes, and a glomus on each side.
Pronephric ducts – lateral to the dorsal aorta and dorsal to the posterior cardinals, these ducts lead from the pronephros posteriorly to join each other just before they fuse with the cloaca.
Mesonephros – large rudimentary mass of nephrogenic tubules, with nephrostomes, forming permanent frog kidney.
Gonads – cell masses hanging into the coelom from dorsal peritoneum between the mesonephros and the gut. (See diagram page 107.)

HYPOMERE – lateral plate mesoderm, ventral to the mesomere, and consisting of somatic and splanchnic layers with the intermediate coelomic cavity.

Pericardial cavity – surrounds heart, not yet separated from true peritoneal cavity.
Peritoneal cavity – body cavity surrounding viscera.
Mesenteries – double-layered dorsal mesentery supports the viscera. A remnant of the ventral mesentery (gastrohepatic omentum) is found between the stomach and the liver.
Spleen – accumulation of cells lateral to mesenteric artery in dorsal mesentery.

CIRCULATORY SYSTEM – differentiated as heart, arteries, veins, lymphatics, and the contained corpuscles.

HEART – already a three-chambered structure as in the adult.

Sinus venosus – contractile chamber known as the pacemaker; junction of postcaval and anterior cardinal veins, leading to the right atrium.
Atria – thin-walled, paired heart chambers, at least partially separated from each other by the interatrial septum. Right atrium receives the sinus venosus and the systemic veins while the left atrium receives the pulmonary vein from the lung buds.
Ventricle – thick-walled, muscular, and generally ventral to the atria. Leads into the bulbus and truncus arteriosus.
Bulbus arteriosus – tubular extension of the early embryonic heart which leads directly into the aortic arches.
Truncus arteriosus – short ventrally directed vessel leading from the bulbus arteriosus. (Syn. ventral aorta.)

ARTERIES – major arteries only.

Afferent branchial arteries –
Lingual (external carotid) – extensions of ventral aorta into the lower jaw. First branchial within third visceral arch.
Second branchial – forked portion of first branchial found within the fourth visceral arch.
Third branchial – temporarily found within fifth visceral arch; degenerates at metamorphosis.
Fourth branchial – within sixth visceral arch.
Efferent branchial arteries –
Anterior cerebral (internal carotid) – gives rise to basilar and commissural arteries; the

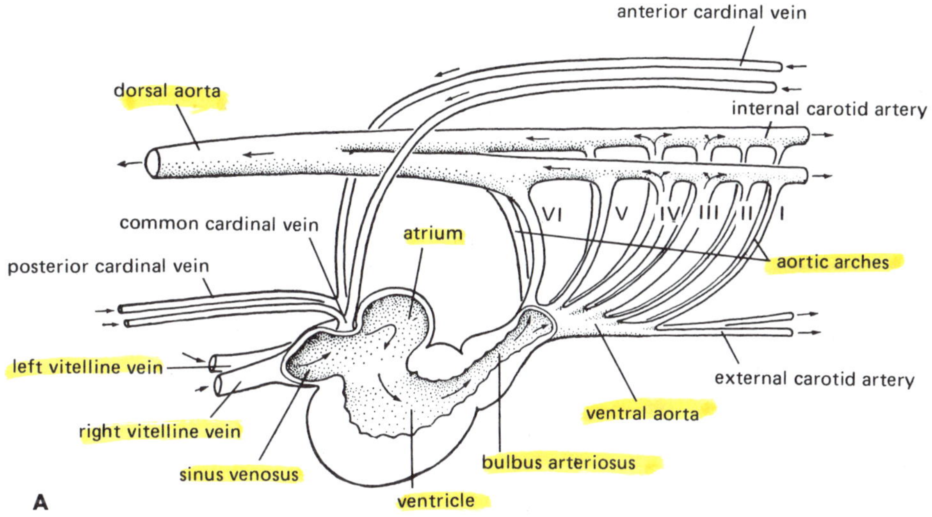

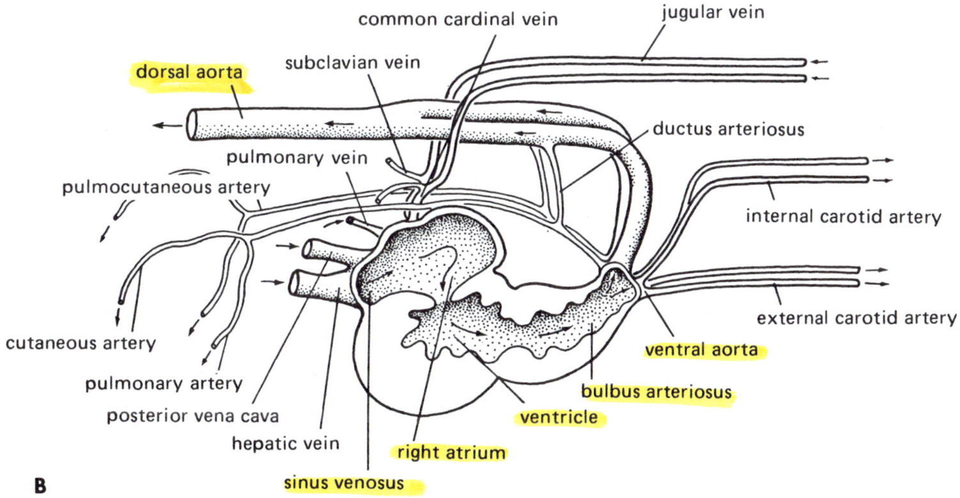

LATERAL VIEWS SHOWING DEVELOPMENT OF BLOOD VASCULAR SYSTEM, IN FROG EMBRYO, A, EARLY EMBRYO: B, LATE EMBRYO

From R. Rugh. *The Frog*. 1951. McGraw-Hill, New York.

latter pass beneath the infundibulum to the other side of the head (see diagram on page 113). Extensions of first branchial artery from within the first visceral arch.

Second branchial – within the fourth visceral arch, enlarges and grows posteriorly as the descending aorta to join the corresponding vessel of the other side to form the systemic trunk. Junction about the level of the liver.

Third branchial – within the fifth visceral arch; degenerates.

Fourth branchial – within the sixth visceral arch, loses its connection (ductus arteriosus) with the descending aorta and grows posteriorly to form the pulmonary and the cutaneous arteries. (See diagram on page 113.)

Dorsal aorta – single large artery formed by the junction of two descending aortae (radices aortae) and extending posteriorly into the tail, giving off various smaller arteries en route.

Intersegmental arteries – paired and metameric arteries which grow from the dorsal aorta dorsolaterally between the myotomes.

Glomi – short branches from the radices aortae which grow toward the pronephric chambers; really undeveloped glomeruli.

Mesenteric artery – single large vessel growing ventral from the dorsal aorta to supply the viscera. Branches into the coeliac and mesenteric artery.

Caudal artery – extension of the dorsal aorta into the tail.

VEINS – major veins only. Generally, the walls are thinner than arterial walls.

Cardinal system –

Anterior cardinals (superior jugulars) – irregular in cross section, found in head, closely associated with internal carotid arteries.

Inferior jugulars – bring blood from lower jaw; accompany the external carotid arteries and join the common cardinals.

Posterior cardinals – ventrolateral to the dorsal aorta, dorsolateral to the nephrogenic tissue. Carry blood to the common cardinals.

Subcardinals – smaller veins, mesial to mesonephric tissue. Bring blood from the caudal vein. Become renals.

Common cardinals – ducts of Cuvier which receive blood from the anterior and posterior cardinal vessels and the pronephric sinus and convey it to the sinus venosus at its posterolateral margin.

Postcaval vein – posterior vena cava or inferior vena cava. This vessel has grown posteriorly from the hepatic vein and incorporates the mesonephric level of the right posterior subcardinal vein. It therefore passes through the liver and into the sinus venosus posterior to the entrance of the right common cardinal vein. (See diagram on page 113.)

Hepatic system – derivatives of the original vitelline veins.

Hepatic vein – fused vitelline veins carrying blood from the liver sinuses to the sinus venosus. Short and difficult to distinguish from postcaval vein to which it gives rise.

Portal vein – common vessel receiving blood from the pancreas, stomach, intestine, etc., representing the original vitelline veins, carrying it to the liver.

Lymphatic system – not well developed at this stage.

CHONDROCRANIUM – cartilage centers within or around which bone will develop. May be found in the embryonic skull.

NEUROCRANIUM – skeleton around brain and primary sense organs. Anterior to posterior.

*Labial (suprarostral) cartilages – paired, short, within the upper lip and lateral to the mouth.

*May be ectodermal.

Mesodermal Derivatives (Continued)

*Cornua trabecularum (trabecular horns) – paired cartilage bars which accompany the olfactory tubes.

Ethmoid (internasal) plate – median fused trabeculae, anterior to ascending process of pterygoquadrate.

*Pterygoquadrate (palatoquadrate) – derives from maxillary portion of the first visceral arch and joins the posterior end of the trabecula by its posterior ascending process.

Basicranial fontanelle – skeletal cavity for pituitary gland.

Trabeculae – paired cartilage bars on either side of basicranial fontanelle, converging slightly anteriorly.

Basilar plate – single, fused cartilage mass immediately anterior to the notochord. Joins parachordals and trabeculae.

Parachordals – lateral to the notochord at its anterior end.

SPLANCHNOCRANIUM – derived from the visceral arches. Anterior to posterior.

Mental (infrarostral) cartilages – paired, but meet in midline of lower lip.

*Meckelian cartilages – laterally projecting flanges to which muscles will be attached, joining the pterygoquadrate.

*Ceratohyals – large cartilages arising within the hyomandibular arches, posterior to the meckelian cartilages.

*Basibranchial (copula) – single cartilage between the ceratohyals.

*Hypobranchials – paired, posterior to the ceratohyals, and derived from the third visceral arches. Articulate also with basibranchial.

*Ceratobranchials – three or possibly four slender processes within the third to the sixth visceral arches, extending from the hypobranchial on each side.

(It is recommended that the student quickly thumb through the preceding pages and locate for special attention those structures which he cannot quickly and accurately define from an embryological point of view.)*

"Life has been compared to a beautiful tapestry, woven in intricate design of many threads and colors. By means of physics, chemistry, physiology, anatomy, embryology and genetics we unravel this texture, separate its constituent threads and colors, but lose the pattern as a whole. These analytical sciences have enormously increased our knowledge of life's constituent elements and processes, but the pattern of the tapestry is usually neglected or ignored."

". . . no philosopher, scientist, or average human being can avoid asking the question WHY, or fail to feel that the end, goal, meaning or purpose of any phenomenon in nature is the most significant inquiry that can be made about it. . . . In his 'De Partibus Animalium' (Aristotle) maintained that the essence of a living animal is found not in what it is or how it acts, but why it is as it is and acts as it does."

E. G. Conklin, 1944 Trans. N.Y. Acad. Sci. 6:125

*May be ectodermal.

DRAWINGS: FROG EMBRYOS

Exercise 15

THE BULLFROG TADPOLE

Each student will be provided with a preserved bullfrog tadpole *(Rana catesbiana)* in which the external gills have been absorbed and the internal gills have been covered by the opercular folds. It is important to realize that, while this tadpole is quite large, it is structurally very much like the 11-mm tadpole just studied in serial transverse sections. The bullfrog tadpole remains in this stage of development through two summers and then rapidly undergoes metamorphosis into a terrestrial frog during the latter part of August.

Note that the eyes are well developed. Are there lids or nictitating membranes? With forceps and sharp scissors remove an eye and dissect to remove the lens. Note the color of the lining of the vitreous chamber of the eye and the position of the optic nerve.

The external nares may be visible as a pair of small openings dorsal and anterior to the mouth. With a small probe one may follow these into the mouth cavity by way of the internal nares or choanae.

The margin of the mouth is serrated by fleshy lips. There are rows of horny teeth and horny beaklike jaws in the mouth of the frog tadpole.

There are no external ears on the tadpole or frog, but, at the time of metamorphosis from one to the other, the tympanic membranes appear.

Through the thin ventral abdominal wall, note the much coiled intestine of the normally vegetarian tadpole, an intestine which may be eight or ten times the length of the body. This intestine will be quickly shortened at the time of metamorphosis until it is only slightly longer than the body of the frog. In the throat region find the bulbous heart.

One must dissect the tadpole in order to locate the respiratory organs, except for the spiracle. The spiracle is the pore on the left side of the head at the posterior margin of the opercular fold. Insert fine scissors into this spiracle and cut through the throat skin (only) until you reach a corresponding position on the right side of the head. In doing this, you will be exposing the opercular tube which connects the opercular cavities of the two sides. Locate one of these opercular cavities, and, with a blunt probe, find the passage from the mough cavity into the opercular cavity. You may be able to identify the gill rakers, the velar plate, and the numerous gill filaments of the internal gill system.

Make a median ventral cut through the skin and body wall from this branchial region to the base of the tail, and fold the flaps back to expose the viscera. The stomach is quite normal in proportions but appears small. The intestine is very long, thin walled, coiled, and has a grayish appearance. Locate the pancreas, spleen, lungs, large mesonephros, and the embryonic gonads.

With careful dissection one should be able to identify the pericardium, pericardial cavity, bulbus arteriosus, ventricle, and auricles.

EXPERIMENTAL PROCEDURES WITH FROG EMBRYOS*

A description of a method for procuring viable frog's eggs and fertilizing them at almost any

*See final exercise in this Guide, which describes basic techniques and equipment needed for experiments with embryos. The amphibian embryos are admirably suited to experimentation.

GROWTH AND METAMORPHOSIS IN THE BULLFROG
RANA CATESBIANA

time of the year will be found on pages 43, 44. The various species of frogs, toads, and salamanders shed their ova at various times (seasons) of the year, and hibernating animals may be kept in artificial hibernation in order to extend the prebreeding period. Since amphibia are widespread in their natural distribution, it is possible for any students almost anywhere to enjoy watching and experimenting with the development of these vertebrates, whenever embryology is studied.

For those who have the time, the basic background in morphological embryology, and the curiosity, some experiments with these embryos may be suggested. The author's *Experimental Embryology: A Manual of Experimental Procedures* gives details of procedure and might be found useful by the graduate student or premedical candidate who wishes to pursue the experimental method further. We will simply list here some of the experiments that anyone could carry out with a minimum of equipment.

INDUCED OVULATION (see page 43):

a. Determination of quantity of fresh or frozen male or female anterior pituitaries necessary to induce ovulation at the various seasons, and environmental temperatures.
b. Effect of aging on ova: Induce ovulation but do not remove ova from uterus until various times thereafter, always fertilizing with fresh sperm. Combine with temperature variables.
c. Effect of synthetic hormones or pituitaries from other species, even mammals, to determine whether they too would induce ovulation in the frog. Induction of amplexes by pituitary injection of male *Rana pipiens* with any other species or female.

HYBRIDIZATION:

a. Attempt to fertilize frog ova with sperm from other species of frogs, as well as toads and salamanders. Test reciprocal crosses.
b. If fertilization is achieved, determine how far development will proceed with the particular cross attempted.

PHYSIOLOGICAL CONDITIONS OF EARLY DEVELOPMENT (see table on page 45):

a. The effect of various temperatures upon the rate of development. Water volume relations must be constant.
b. The temperature tolerance of embryos at various stages, using both high (37°C) and low (4°C) temperatures. Vary times at extreme temperatures.
c. The effect of crowding on survival normality and rate of development. Fertilized ova may be placed 5, 50, 100, 200 per fingerbowl, with adequate spring water (100 cc), properly covered to reduce evaporation but kept at uniform temperature. Determine the optimum ratio of embryo to water volume.
d. Effect of changes in salinity and pH on development.

ANDROGENESIS AND GYNOGENESIS:

a. Androgenesis–it is possible to remove the nucleus of the fertilized ovum before it unites with the sperm nucleus. This is done with a glass needle under a dissecting microscope 20 minutes after fertilization.
b. Gyogenesis or parthenogenesis–development without benefit of spermatozoa. Stimulate by chemical or physical means, and determine how far development progresses.

PHYSICAL ALTERATIONS DURING DEVELOPMENT:

a. Effect of unequal pressures on cleavage planes.
b. Effect of centrifugation at various stages of development.

SURGICAL PROCEDURES (amphibian embryos are very hardy and can tolerate extensive surgical intervention with a minimum of asepsis):

a. Removal of the dorsal lip of the blastopore (organizer).
b. Transplatation of organizer to a different region of another embryo.
c. Lateral (parabiosis) or terminal (telobiosis) fusion of embryos in the later neurula stage.

d. Transplantation of organ primordium, such as eyes of anura, or eyes and limbs of urodela.
e. Tissue culture of isolated organ primordium.
f. Parabiosis—lateral fusion of neuulae to form Simaese-type twins.

LATE DEVELOPMENT:

a. Iodine and its effect on metamorphosis through the thyroid gland.
b. Growth as affected by diet: lettuce or spinach (wilted) vs. fish food or liverwurst. Avoid overfeeding and resultant bacterial contamination.
c. Cannibalism resulting from mild starvation.

If time permits, this would be an opportunity for the ambitious student to learn the essentials of research and experimentation. The prime consideration is always to plan for adequate and appropriate controls—embryos treated in a normal manner so that other and identical embryos subjected to a single imposed variable will graphically indicate the effect of that variable. A course in biostatistics would be helpful. The general outline of an experiment would be something like the following.

Purpose:	Major objective of the experiment.
Materials:	Animals or embryos, solutions, drugs, surgical and glass equipment, and any special equipment for the particular protocol.
Method:	Experimental design includes, with proper checks: a. Controls. b. Experimentals.
Results:	Data collected and analyzed, with evidence of statistical adequacy.
Conclusions:	If the quantity of data and the results allow for any clear-cut and unequivocal conclusion, this may be stated here at the conclusion of the study.
References:	If the experiment is to be written up, it should include such references as are directly pertinent to the study. If there are divergent conclusions, some explanation should be offered.

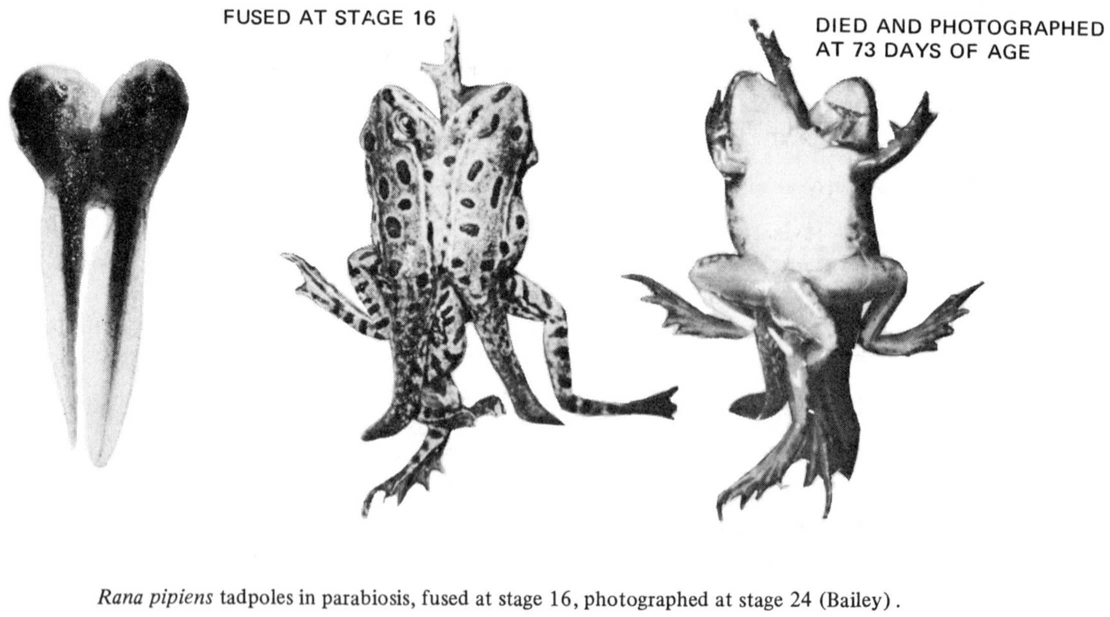

Rana pipiens tadpoles in parabiosis, fused at stage 16, photographed at stage 24 (Bailey).

PARABIOSIS IN METAMORPHOSED FROGS

RANA PIPIENS

Exercise 16

THE CHICK:

Gonads and Maturation

The chick embryo is the classical form used for centuries in the study of development, upon the basis of which many theories have been advanced. Probably, the most significant is the theory of epigenesis as opposed to that of preformation, since it is possible to watch, almost with the naked eye, development evolve step by step out of apparently no preformation. In contrast to the frog, the chick is a warm-blooded form (104°F), with sufficient yolk to carry it through 21 days of development and devices to prevent evaporation and desiccation, all contained within a porous shell until such time as the newly hatched chick is prepared to emerge and fend for itself. Within certain limits we can obtain very accurately timed stages, checked by known marks of development (e.g., somites), or we can make a window in the shell and watch the developmental process from hour to hour, day to day. The major complication is that the warm yolk, if exposed to bacteria of the air, is an ideal medium in which infection may develop. At the end of this section on the chick, we will suggest some practical experiments suitable for use in any laboratory by any student acquainted with the normal development of the chick embryo.

The Rooster Testis

In the walls of the avian *seminiferous tubule* it is possible to locate all stages in the maturation of the spermatozoon, although the cells are very small and closely packed together. The *spermatogonia* are the most peripheral, smallest, and most abundant cells within the tubule, possessing vesicular nuclei in which the chromatin is in the form of threads. The *primary spermatocytes* are the largest cells within the tubule, and their nuclei will tend to exhibit more definite chromatin, frequently in the form of chromosomes in a division phase leading to the secondary spermatocyte stage. These *secondary spermatocytes* are somewhat smaller cells and are rather difficult to distinguish from the spermatid-spermatozoon transformation stages. *Spermatids* generally have very dark chromatin bodies, which anticipate the solidly stained head of the mature spermatozoon, always found in the same vicinity. The *head* of the avian sperm is slightly bent; the *middle piece* is elongated and merges with the long and filamentous *tail.* Here, as in all forms studied, the tails of the mature spermatozoa lie free within the lumen of the seminiferous tubule, while the sperm heads are buried within the *Sertoli cells.* These latter cells are generally obscured by the sperm heads and are, therefore, difficult to identify. There is very little *interstitial tissue* in the avian testis. (See Patten, *Chick,* page 16.) A single insemination of the hen by the rooster provides her with viable sperm for all the ovulations that occur in the succeeding weeks.

The Hen's Ovary

It must be remembered that the ovary of the hen possesses no ŏogonia. This stage in gametogenesis was reached by the fourteenth day of incubation (prior to hatching) in all potential females, and no ŏogonia develop thereafter. It must also be remembered that a fully matured hen's egg measures about 40 mm in diameter (exclusive of the albumen); hence, such an egg is much too large to mount in sections on microscope slides. Therefore, the mounted sections of hen's ovary contain only intermediate stages in ŏogenesis, particularly those showing the growth and acquisition of yolk (deutoplasm) in the development of the *primary ŏocyte.* Maturation divisions appear at the time of ovulation and of fertilization.

Within the section of the ovary locate some of the small but distinct cells, each possessing a clear *germinal vesicle* and encircled by small rounded follicle cells. These are the progrowth *primary*

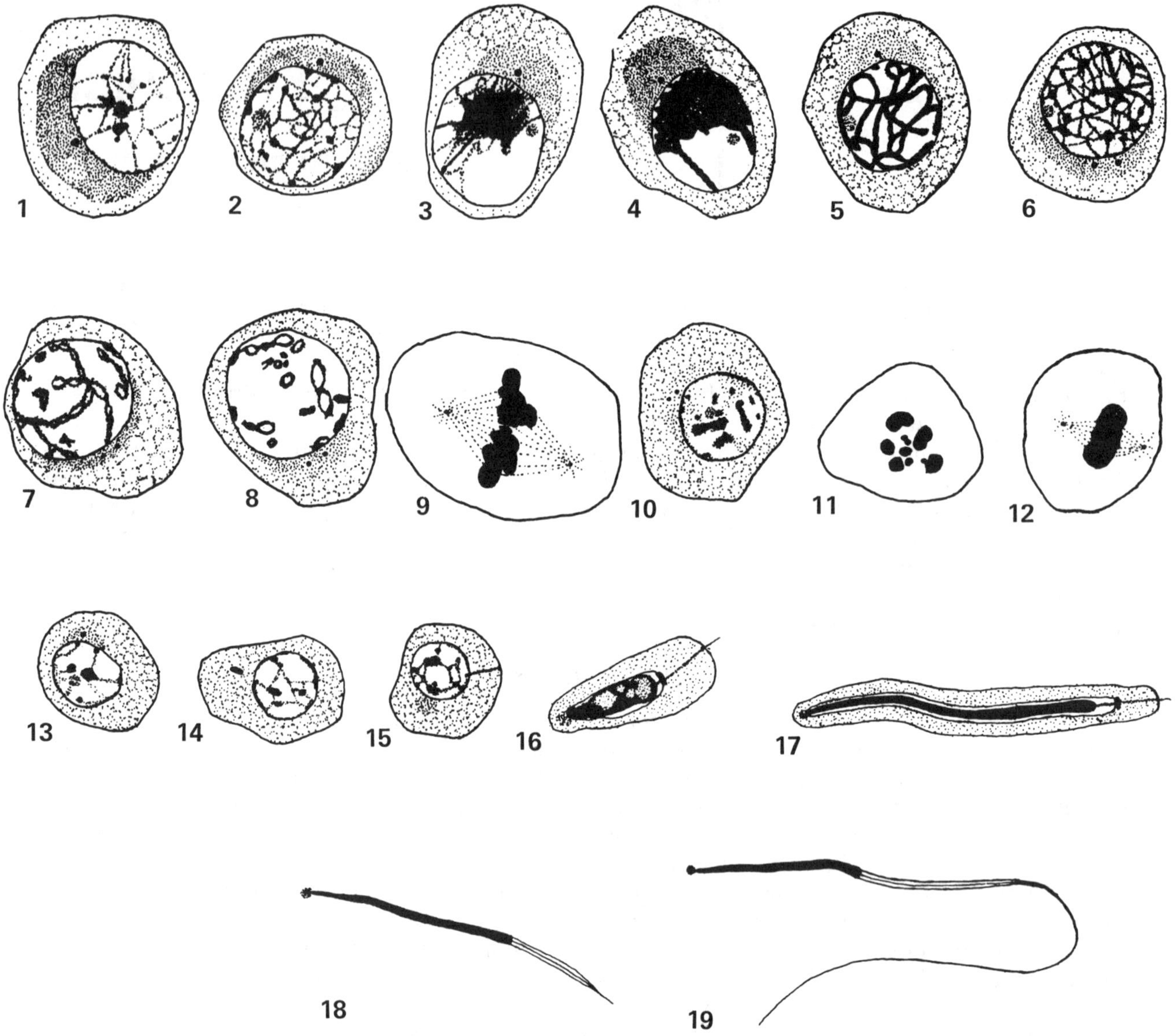

SPERMATOGENESIS IN THE ROOSTER

Fig. 1. Resting spermatocyte.
Fig. 2. Early spireme.
Fig. 3. Synaptene stage.
Fig. 4. Bouquet stage.
Fig. 5. Pachytene stage.
Fig. 6. Diffuse stage.
Fig. 7. Strepsinema breaking up into tetrads.

Fig. 8. Diakinesis.
Fig. 9. Primary spermatocyte division normal male, side view. Camera lucida drawings.
Fig. 10. Resting secondary spermatocyte.
Fig. 11. Secondary spermatocyte division polar view.
Fig. 12. Same, equatorial view.

Fig. 13. Resting spermatid.
Fig. 14-18. Several stages in the transformation of the spermatid into the spermatozoon. (In Figs. 13 and 14, only part of the tail filament is represented.)
Fig. 19. Spermatozoon free in lumen of a seminiferous tubule. Camera lucida drawings.

From Richard A. Miller. "Spermatogenesis in sex-reversed female and in normal males of the domestic fowl, *Gallus domesticus*." *Anat. Rec.* 70:155-189, 1938.

CHICKEN OVARY

PHOTOGRAPH OF LIVING OVARY OF HEN SHOWING, IN THE CENTER, SAC FROM WHICH AN EGG HAD RECENTLY BEEN OVULATED AND, SURROUNDING THIS, 13 OTHER EGGS IN VARIOUS STAGES OF GROWTH.

Courtesy Dr. Marlow Olsen, U.S. Agric. Res. Center, Beltsville, Md.

oöcytes. As these early oöcytes accumulate their *yolk* (deutoplasm), they come to lie closer to the surface of the ovary. The cell membranes of these early oöcytes are generally obscured by the very numerous cuboidal *follicle cells* which surround each oöcyte. No vitelline membrane is as yet developed. Closely wrapped around the follicle cells and early oöcyte is a layer of compact, fibrous connective tissue (possibly containing some muscle fibers) known as the *theca interna.* Outside of this layer is the more spongy tissue of the ovary, still encircling the egg cell, known as the *theca externa.* The rest of the ovary is made up of a loose ovarian stroma in which are embedded blood vessels and nerves. The main difference between the stroma and the theca externa is one of position.

Next locate the largest and, consequently, the most mature of the eggs in the sections and study the surrounding layers under highest magnification. The *vitelline membrane* should now be present between the *egg cell* membrane and the surrounding *follicle cells.* In the hen, however, the vitelline membrane is rather thick and is traversed by numerous minute radiating canals which probably convey nutrient materials from the surrounding areas to the egg itself. Due to these canals, the region is designated as the zona radiata, a term used in place of the vitelline membrane. Only the inner margin of the vitelline membrane can be made out. The numerous follicle cells have been stretched and flattened by the enlarging *oöcyte* and form a ring of cells outside of the zona radiata. Since the nuclei of these follicle cells are highly granular, the layer is known as the *zona granulosa.* Note the changes in both the *theca interna* and *theca externa,* due largely to the increase in egg size. The term *theca folliculi* refers to all of the connective tissue layers which enclose the developing egg.

In the mature hen's ovary, from which eggs are being liberated almost daily, one would find that the large, yolk-laden and mature egg has stretched the surrounding ovarian tissues (of the theca folliculi) so that there is a peripheral region of the ovary which appears to be devoid of blood vessels and even of the stroma. This is the region of the ultimate rupture of the follicle, with liberation of the egg into the body cavity. In the hen's ovary this *cicatrix* (stigma) appears at a region directly opposite to the position of the germinal vesicle, hence, at the vegetal pole of the egg. Obviously the cicatrix can be seen only in the whole ovary and in relation to a fully grown egg. It can be seen better in the kitchen than in the laboratory! Nuclear maturation is not completed until after insemination.

In preparing the ovarian tissue for sectioning and mounting on slides, it is difficult to avoid shrinkage of the yolk areas away from the adjacent regions so that artificially formed clear spaces will appear between the yolk and the egg cell membrane and between the yolk and the germinal vesicle. Occasionally this shrinkage so expands the vitelline membrane that its radiating lines seem to become a reticulum or network of fibers. (See Patten, *Chick,* page 19.)

The Hard-boiled Hen's Egg

If a fresh hen's egg is held in one position for 3 to 5 minutes, the blastodisc (embryonic mass) will float around to the uppermost position. If this position is marked on the shell with a pencil and the egg is then lowered into water and held in this position while bringing it to a boil (10 minutes), the coagulated embryonic blastodisc will be found directly under the pencil marking. When the egg is cool, break the shell to one side and remove the entire contents without breaking the yolk. Note the air chamber toward the blunt end of the shell and the whitish membrane within the shell. Gently try to peel the egg white off of the yolk and note that it has been deposited in spiral layers. Can you see any direction to the spiral? As you get closer to the yolk note the thin vitelline membrane which encloses it.

With a very sharp scalpel cut through the yolk by cutting first through the lighter colored disc on the dorsal side where the shell has been marked. If this is done without shearing, you will be able to identify the uppermost blastodisc region, beneath it the nucleus of Pander (lighter in color than most of the yolk), an enlongated connection of similar light-colored yolk, known as the neck of the latebra, and toward the center of the yolk the more liquid latebra. Under proper lighting, you may be able to see that the surrounding yolk had been put down in layers, alternately lighter and darker yellow. It is believed this may be correlated with the alternate day and night deposition of the two kinds of yolk.

The Raw Hen's Egg

The Mature Ovum at Metaphase of the Second Oöcyte Stage (Demonstration)

A fresh egg may be opened into a fingerbowl of appropriate (0.9% NaCl) salt solution. Without attempting to make any drawings, the student should note the following.

1. EGG SHELL—study fragments of the dried shell and note that it is made up of three layers: the inner mammillary layer, the intermediate spongy layer, and the thin outer cuticular membrane.
2. SHELL MEMBRANES—double membrane within the shell, the two separated from each other at the region of the air space at the blunt end of the egg. The outer shell membrane is the thicker and remains adherent to the shell and may be seen even at the time of hatching.

MATURATION, FERTILIZATION, AND EARLY CLEAVAGE OF THE CHICK EGG

1. Section of young ovarian follicle with large, centrally located germinal vesicle.
2. Blastodisc from an egg, taken from the isthmus, showing the first cleavage furrow.
3. Four-cell stage taken from the isthmus.
4. Sixteen-cell stage taken from the uterus.
5. Thirty-two cell stage taken from the uterus.
6. Approximately 100-cell stage, 74 central and 24 marginal.

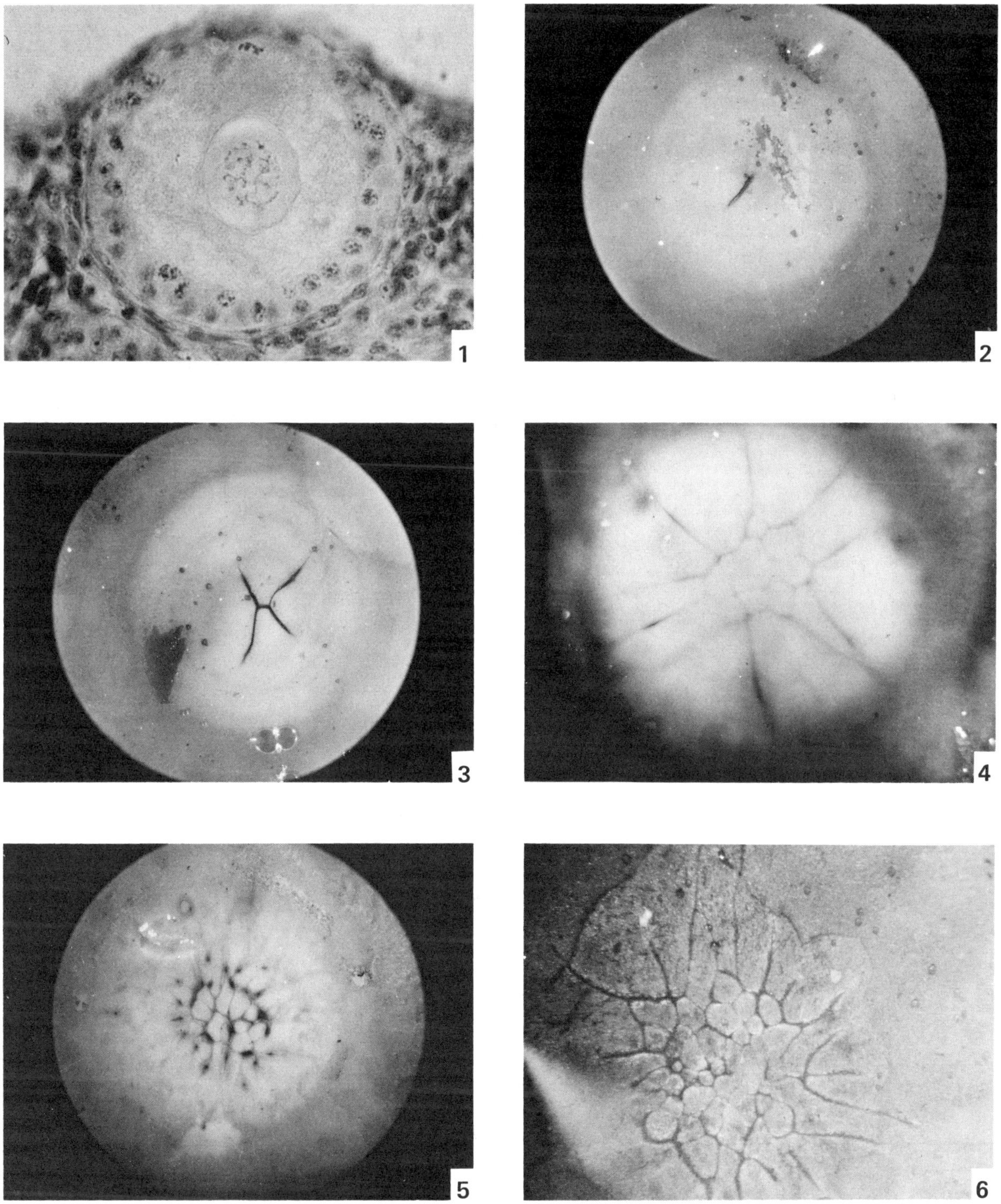

These remarkable photographs have been reproduced through the kind permission of Dr. M. W. Olsen of the Maryland Agricultural Experiment Station and appeared first in his paper of 1942: *Jour. Morphol.* 70:513-33.

3. THIN OUTER ALBUMEN—easily separated from the chalazae and yolk mass.
4. CHALAZAE—the inner and heavier layer of albumen around the yolk, twisted ropelike strands on either side of the spherical yolk mass.
5. DEUTOPLASM—the yolk mass generally floats around to the uppermost position. This *blastodisc* represents the region of earliest embryonic development and is relatively free from yolk, therefore, lighter in reflected light. It is instructive to examine a drop of fresh chick egg yolk under the microscope.

Frequently the blastodisc may contain a blood clot. This is proof either of the presence of an early embryo which developed for a while, then died (such an egg is still edible unless it is otherwise stale), or it could also be the incorporation of a blood clot from the oviduct prior to the addition of the layers of albumen. Such a clot, when found, does not indicate it is a "bad egg" or that it should be discarded, unless it is obviously a dead embryo.

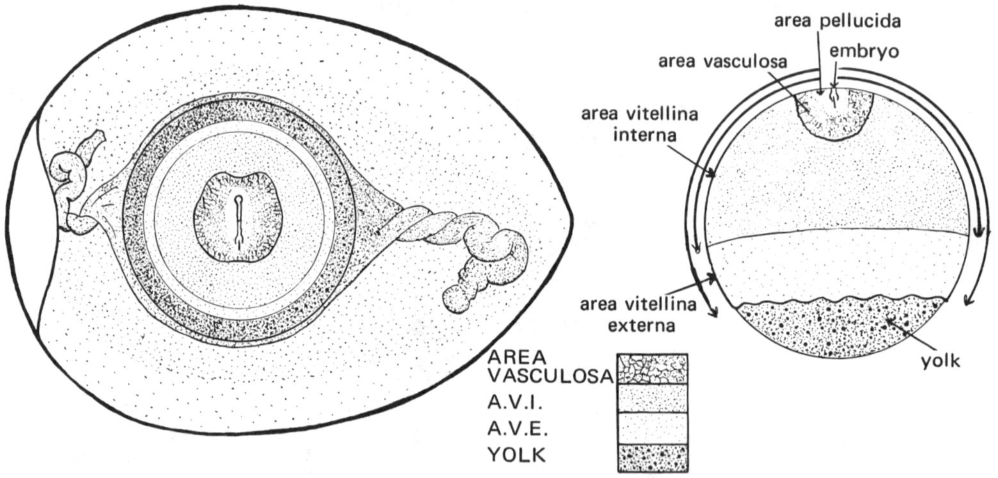

24-HOUR CHICK EMBRYO

THE 50-HOUR CHICK EMBRYO
ALBUMEN REMOVED

REFERENCES

Arey: 37, 38, 63-71
Balinsky: 4, 449
Hamilton-rev. of Lillie; Chapter I
Huettner: 192-201
Lillie: 1-68
McEwen: Chapter VIII
Nelsen: 3-278
Patten: *Chick*, 1-36
Patten (1964): Chapters 6, 7, and 8
Rugh: *Vert. Emb.* 105-17
Shumway: 15-18, 28, 33, 51

"As a physiologist I am especially impressed by the common ignorance of bodily organs and their functions. It seems to me now, as it seemed to Robert Boyle nearly three hundred years ago, that it is 'highly dishonorable for a Reasonable Soul to live in so Divinely built a Mansion as the Body she resides in, altogether unacquainted with the exquisite structure of it'."

W. B. Cannon, 1945, *The Way of An Investigator*, W. W. Norton Co., N.Y.

DRAWINGS: THE CHICK EGG

Exercise 17

THE CHICK EMBRYO:

Primitive Streak Stage

THE CHICK EMBRYO: PRIMITIVE STREAK STAGE

In the study of all chick whole mounts, it is important that the student use the dissection microscope or the lowest magnification of the compound microscope, because many of the embryos are mounted under elevated coverslips which will not allow the focal distance needed for the highest magnifications. All sectioned material is to be examined first under the low-power objectives and then, in a study of cellular detail, under high-power objectives.

Drawings are to be made so as to represent accurately the relative sizes of all organs. The student will, in all cases, first study the photographs of an entire embryo and then make specified drawings of sections. The exact location of each section drawn is to be indicated on the appropriate photograph or drawing. The sectioned material may be drawn with high-grade colored pencils, without any attempt at cellular detail, except when specifically required. The colors to be used are:

BLUE for ectoderm and derivatives.
YELLOW for endoderm and derivatives.
RED for mesoderm and derivatives.
GREEN for notochord.
ORANGE for yolk.

Ink may be used for blood vessels but only after pencil sketches have been approved by the instructor. Use of the above standard color designations will make it unnecessary to label the germ layers, but, as usual, the student is expected to label all organs and structures that can be identified.

WHOLE MOUNT

The entire blastoderm of the chick is mounted so that the student is able to locate the central clear area known as the *area pellucida.* It is within this area that the primary embryonic structures will appear. Peripheral to the area pellucida, locate the mottled and darkly stained region known as the *area opaca.* Since the peripheral blastoderm is in direct contact with the underlying yolk and is therefore dark, it is properly named the area *opaca.* By 20 hours of incubation this area may already be subdivided into an inner portion which takes the heavy mottled stain, the *area vasculosa* (in which blood vessels develop from blood islands) and a more peripheral area known as the area *vitellina,* where there are no blood islands.

Toward the center of the area pellucida locate a clear line flanked by parallel dark lines which terminate anteriorly in a slight bulge and a darkened margin. This is the *primitive streak,* and it extends almost to the posterior boundary of the area pellucida, where it flares out to merge with the area pellucida, this region being called the primitive plate. The clear line is actually a groove *(primitive groove)* between two ectodermal elevations *(primitive folds),* the groove terminating in a rather deep anterior pit, called *primitive pit,* surrounded by the dark *Hensen's node* (also called primitive knot). This entire primitive streak is considered by many to be the coalesced lips of the blastopore, Hensen's node representing the homolgue of the dorsal lip of the amphibian blastopore. The primitive pit may be a vestige of the nuerenteric canal. The embryo develops anterior to the primitive streak and is itself lengthened concurrently with the shortening of the primitive streak until this structure finally disappears entirely (about the third day). Study photographs of embryos from 18- to 96-hour stages.

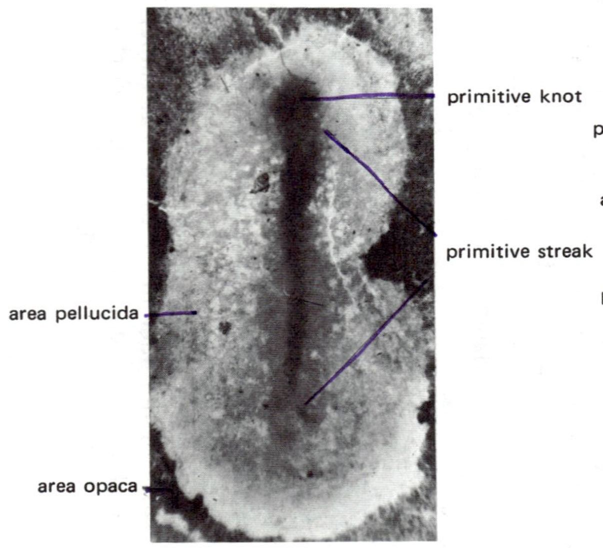

WHOLE CHICK EMBRYO AT 14 HOURS

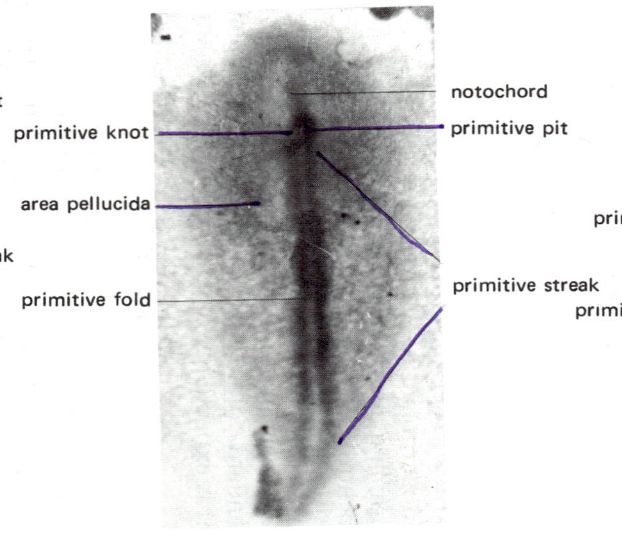

WHOLE CHICK EMBRYO AT 15 HOURS

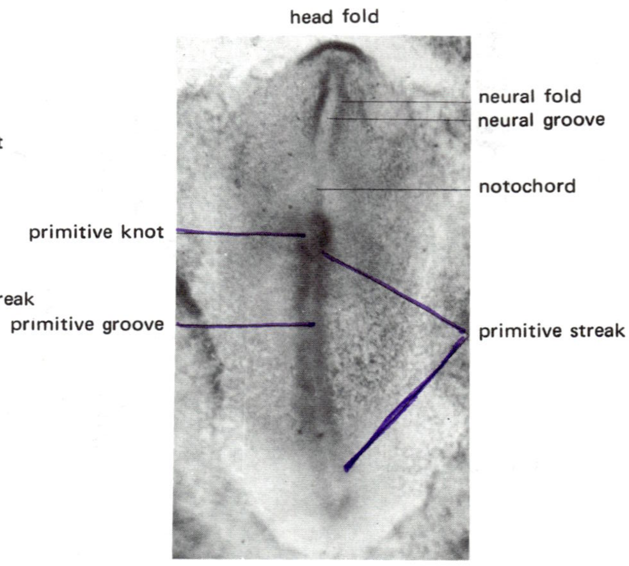

WHOLE CHICK EMBRYO AT 16 HOURS

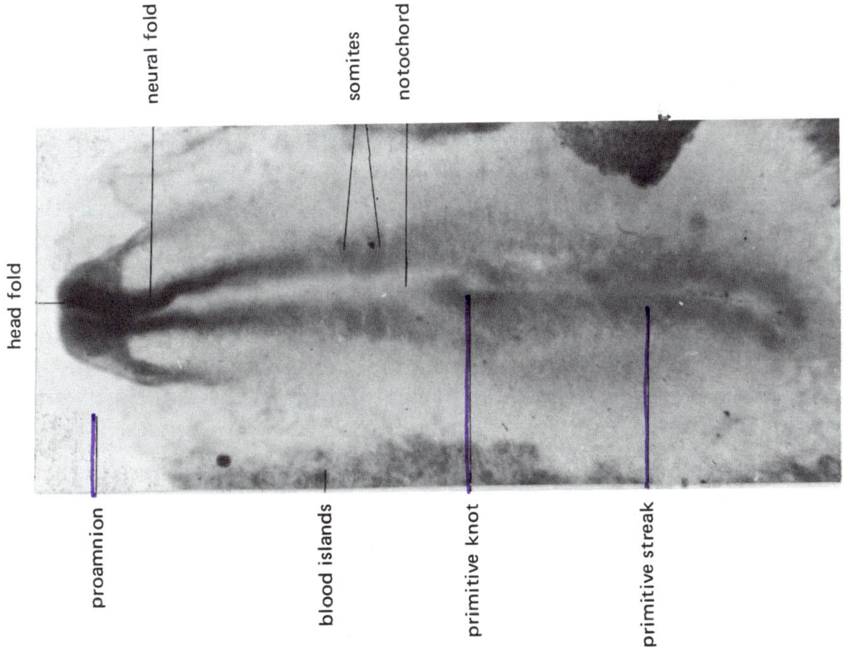

WHOLE CHICK EMBRYO AT 22 HOURS

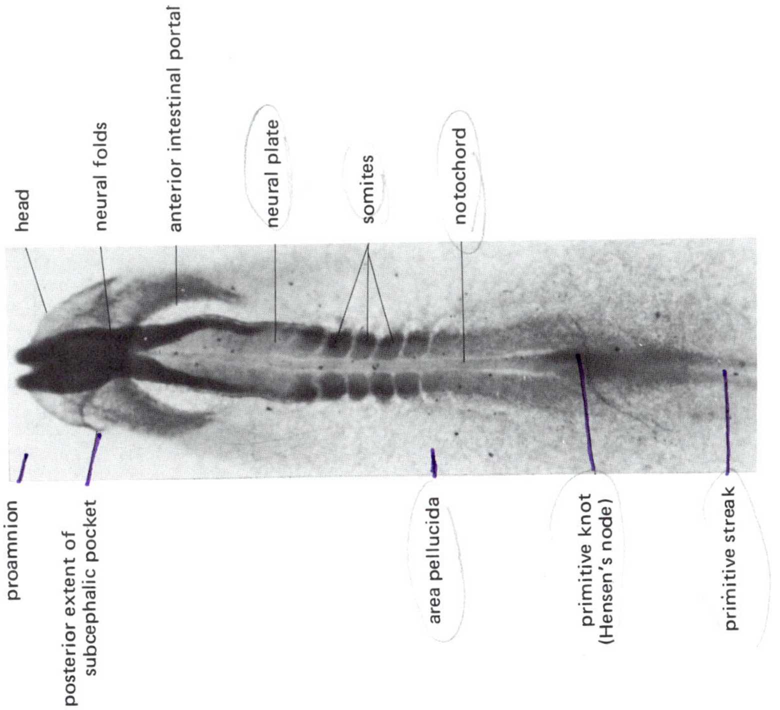

WHOLE CHICK EMBRYO AT 27 HOURS

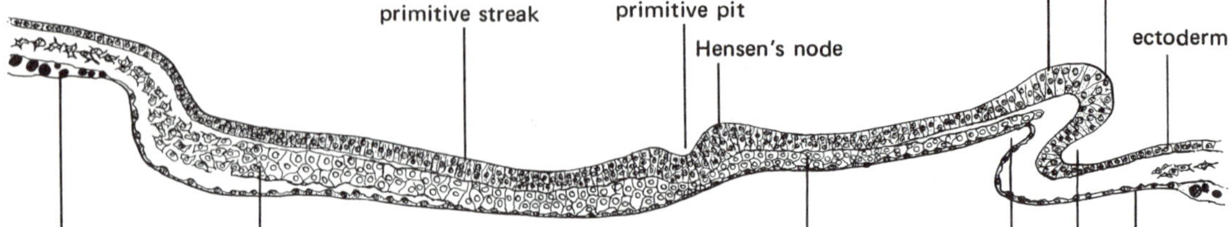

HEAD PROCESS AND PRIMITIVE STREAK OF THE CHICK EMBRYO
SAGITTAL SECTION

Depending upon the actual incubation age and the stage of development at the time of egg laying, slides labelled "primitive streak" may show various degrees of development, particularly anterior to the streak itself. The emphasis in this study is on the primitive streak, and this structure is essentially the same during the first 24 hours. Note that the axis of the embryo is in line with, but not directly anterior to, the line of the primitive streak. The slightly darkened triangular area just anterior to Hensen's node represents the first part of the embryo proper, specifically the notochord. This grows anteriorly from the primitive knot itself but is probably derived from or concurrently with the mesoderm. Flanking either side of this notochord, you may find indications of folds of ectodermal tissue, the beginnings of the *neural folds* with a *neural groove* between. It will appear as though the notochord were lying within the neural groove (although it is below) due to the transparency of the embryonic structures at this stage.

Even before the neural folds are well developed you will find an anterior transverse semicircular fold of tissue, called the *head fold*, representing an actual fold in the anterior ectoderm and endoderm which will give rise to the early head structures. As the head fold develops, there will appear, in consequence, a space beneath known as the *subcephalic pocket*. Anterior to this head fold locate a clear area (the *proamnion*), which is relatevely transparent because it is still free from mesoderm. The mesoderm is derived from the primitive streak and migrates anteriorly between the ectoderm and endoderm and out over the area pellucida, but for a time does not invade the region of the proamnion. (Actually this is a misnomer, for the true amnion is made up of ectoderm and mesoderm.) The posterolateral limit of the clear proamnion is the anterior limit of the *mesodermal wings,* which are growing anteriorly from the primitive streak.

The central portion of the area pellucida, that which contains the neural folds and other clearly defined embryonic structures, is known as the *embryonic area* in contradiction to the clearer more peripheral portion of the area pellucida, which is known as the *extraembryonic area.* Actually, all material, except the centrally located embryonic structures, represent extraembryonic areas. While these areas give rise to important structures such as blood vessels, act as channels of transport of nutrition from yolk to the embryo, and aid somewhat in the protection of the embryo, they are not incorporated into the body of the developing embryo. They are, therefore, permanently extraembryonic.

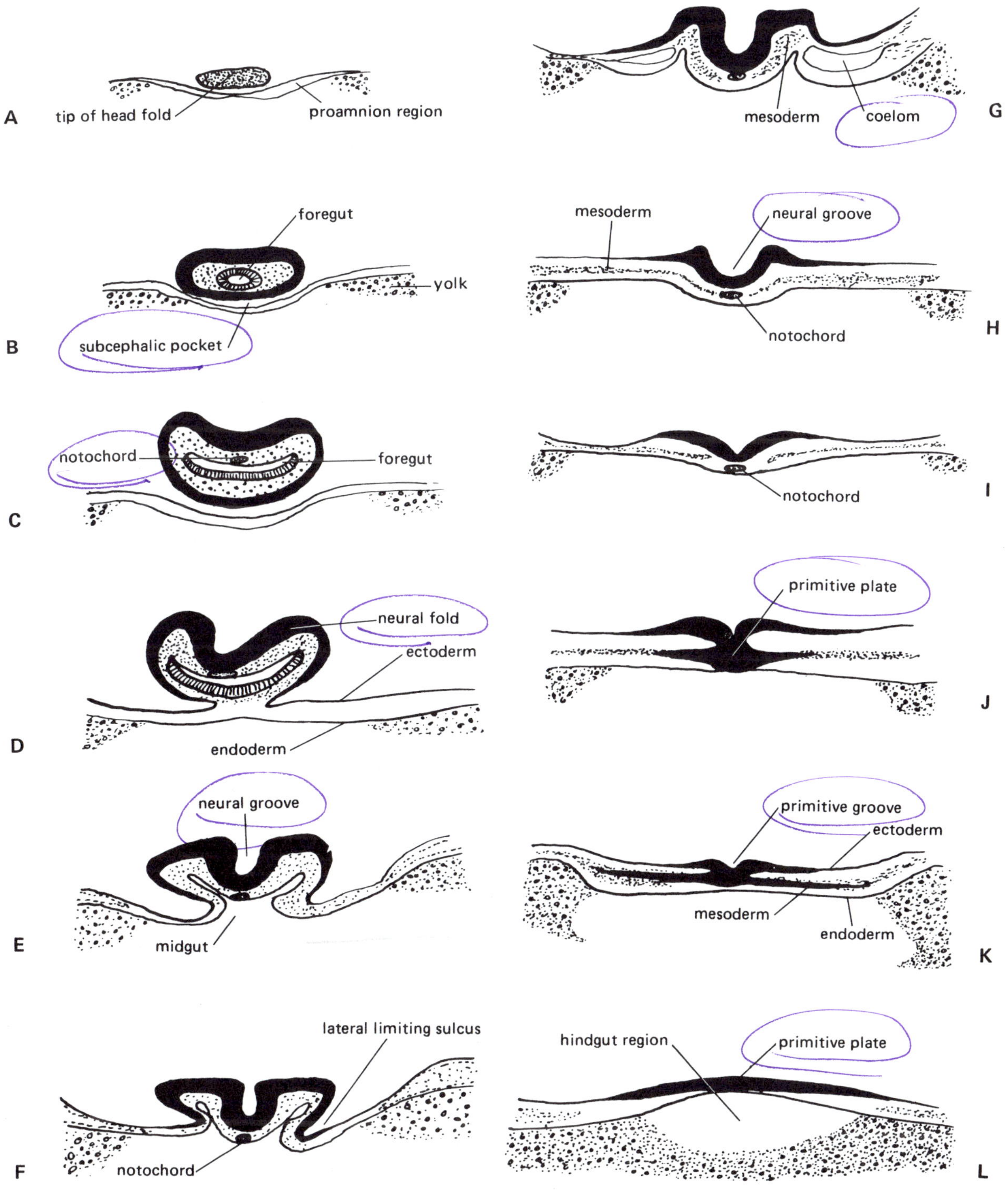

TRANSVERSE SECTIONS OF 18- TO 20- HOUR CHICK EMBRYO

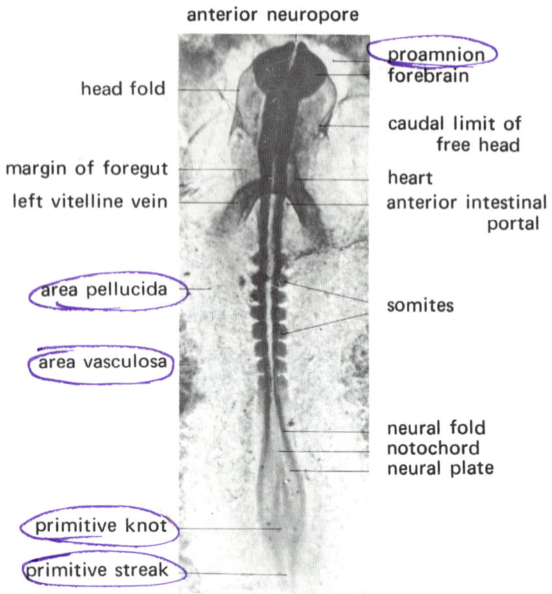

WHOLE CHICK EMBRYO AT 30 HOURS

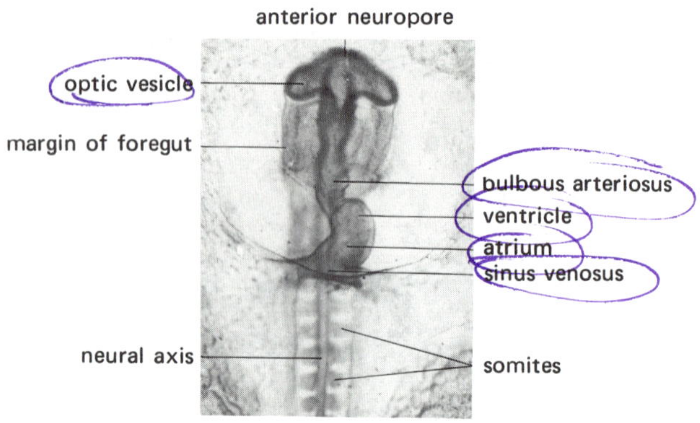

CHICK EMBRYO AT 33 HOURS (12 SOMITES) SHOWING MOST PRIMITIVE TUBULAR HEART

Transverse Sections

With the structures of the primitive streak whole mount clearly in mind study the serial transverse sections of an embryo of 18 to 20 hours of incubation, with the object of being able to distinguish at all times between sections taken through the streak and through the embryo proper. (See drawings on page 135 and Plate 1, page 361.)

First identify the marginal regions where the blastoderm is in direct contact with the large yolk granules beneath. This is the region of the *area opaca.* Midway between two such areas locate the *primitive folds* within the *area pellucida.* The sections should be studied with the yolk ventrally placed, i.e., toward the observer as seen through the microscope. The most dorsal germ layer is ectoderm, which shows considerable thickening toward the center of these transverse sections. The thickening may be solid *(Hensen's node)* or may be depressed into a groove *(primitive groove)* on either side of which are folds of ectoderm *(primitive folds).* It is of utmost importance that the student learn to distinguish between the primitive streak and the embryonic structures. There are anterior to the streak, *neural folds* and a *neural groove.* There are several ways to identify the primitive streak in transverse section and to avoid confusion with the neural structures. In the first place, there is no notochordal tissue beneath the primitive streak. In the second place the mesoderm of the primitive streak region is never organized into somites, is more concentrated at the streak, and the mesodermal cells seem to be coming directly from the streak (ectoderm) cells. In the third place, the ectoderm, mesoderm, and ventral endoderm all merge at the streak, while in sections through the embryo proper, the germ layers are separate.

Lateral to the primitive fold locate the rather loosely scattered cells which seem to be migrating outward between the upper layers of ectoderm and the lower layer of very thin endoderm. These cells represent the mesoderm, which originates from the primitive streak, and migrate anteriorly to take up positions lateral to the embryo itself, in the area opaca, where blood islands are to appear, and close to the neural folds where the somites will appear.

The space beneath the embryonic structures cleared of yolk and lying below the thin endodermal layer, is known as the *primitive gut.* This cavity is continuous beneath both embryonic and primitive streak regions. The missing floor of this primitive gut consists of yolk which was left behind when the blastoderm was removed for mounting.

DRAWINGS: THE CHICK EMBRYO

THE CHICK EMBRYO:

Twenty-Four-Hour Stage

Exercise 18

WHOLE MOUNT

Under the dissection microscope, or under the lowest magnification with the compound microscope, become thoroughly acquainted with the topography of the 24-hour-incubated chick embryo. A great deal can be made out with the naked eye, especially since the blastoderm has been dissected away from the underlying yolk. The identifying features of this stage consist of the presence of four pairs of distinct somites lateral to the open neural groove. The first of these somites will later be found just posterior to the otic (auditory) capsule and will be incorporated in the head of the adult bird. This indicates that by 24 hours of incubation only the most anterior structures of the chick embryo have been delineated.

The *head process*, consisting of ectoderm and endoderm, continues to grow forward as a distinct *head fold* lying above the clear *proamnion*. If one could manipulate this portion of the embryo it would be possible to grasp this head process with forceps or to pass a probe beneath it into the *subcephalic pocket*. Projecting into this head process from the posterior is a blind pocket of endoderm, the beginning of the *foregut*, which extends forward but not as far anterior as the extent of the neural folds. Locate first the lateral margins of this foregut and find where they pass out (posteriorly) over the yolk as undifferentiated archenteron. At this level the foregut lacks a floor except for the underlying yolk, and the margin of endoderm between the formed and unformed foregut marks the *anterior intestinal portal* (actually an opening into the foregut). As development proceeds, this anterior intestinal portal will come to lie more posteriorly simultaneously with the forward growth of the differentiated foregut. This portion of the gut thus lengthens in two directions. Locate this portal on the photographs and your slides.

The *neural folds* are very prominent at this stage, but they have come together only in the vicinity of the future midbrain. Toward the anterior end they flare apart at the level of the beginnings of the *optic vesicles*. At the most anterior tip of the head process, locate a small opening between the neural folds, known as the *anterior neuropore*. The pronounced thickening of the neural folds is reduced posteriorly until about the level of the most posterior somite, where they gradually begin to fade out until only faint incomplete folds are seen to diverge around what remains of the primitive streak. This latter region is known as the *sinus rhomboidalis*. The midline of the neural groove may be indistinct due to the presence beneath of the *notochord*. This structure extends cephalad as far as the anterior limit of the foregut (to the region of the future tuberculum posterius).

The mesoderm of the 24-hour chick embryo is abundant, all of it being derived from the region of the primitive streak and migrating forward to take up appropriate positions within the embryonic and extraembryonic areas. The mesoderm closest to the lateral neural folds is almost immediately organized into well-formed *somites* which appear in pairs at the rate of about one pair per hour after the twentieth hour of incubation, until about the fortieth hour, when the rate decreases. At first, they appear to lie ventral to the neural folds, but as the folds come together the correct lateral relation of the somites will be indicated. These are the first structures that show metamerism characteristic of vertebrates in general. In turn corresponding neural and skeletal metamerism will also appear.

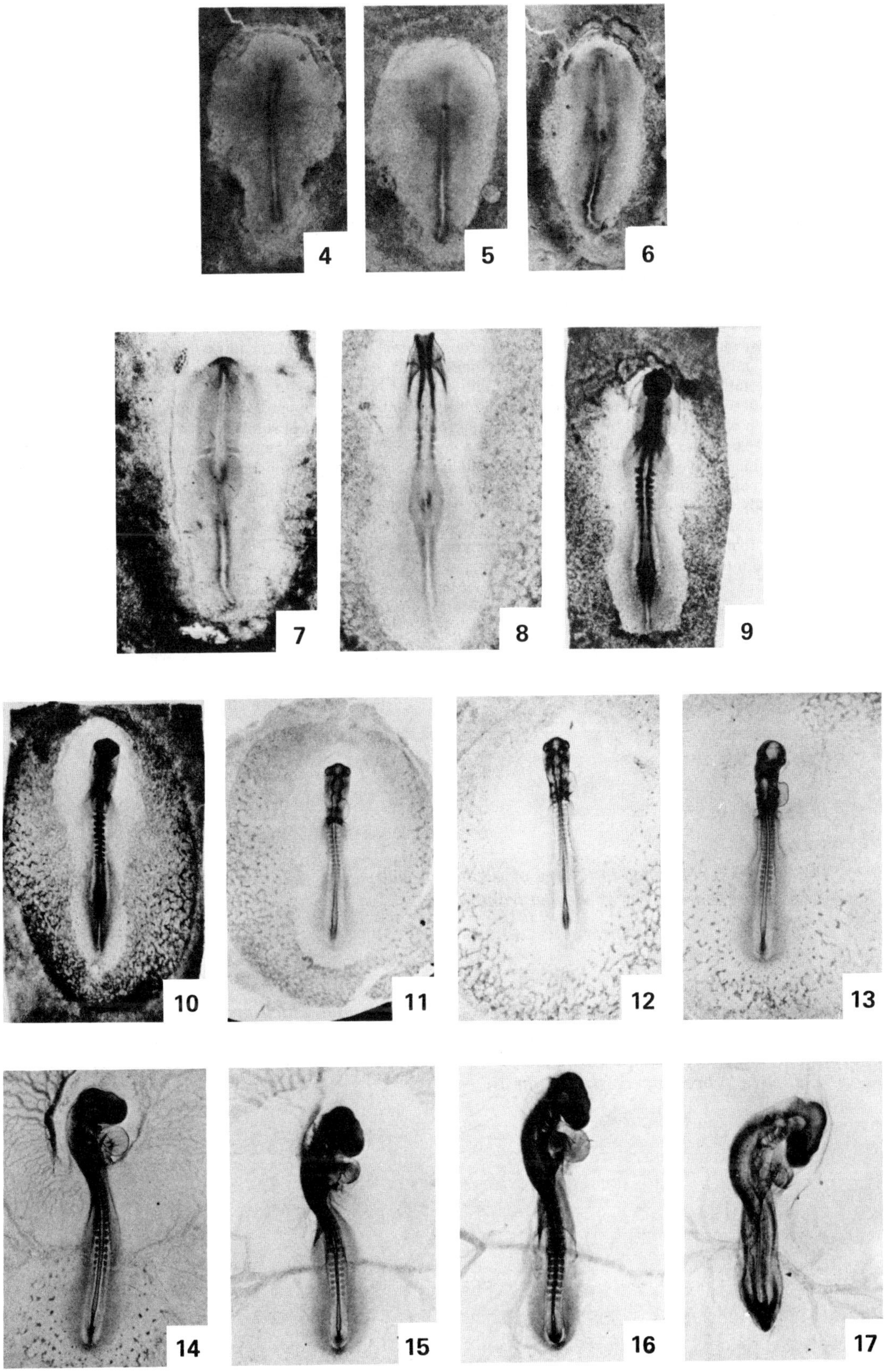
4
5
6
7
8
9
10
11
12
13
14
15
16
17

CHICK EMBRYOS

Stage 4: Definitive streak; 18-19 hours. Streak at maximal length, approx. 1.88 mm, showing primitive groove, primitive pit, Hensen's node; area pellucida pear shaped. Streak extends over about 75% total length.

Stage 5: Head process; 19-22 hours. Head process forming by notochord rod extending forward in dense mesenchyme anterior to Hensen's node. No head fold as yet. Notochord continues to lengthen.

Stage 6: Head fold; 23-25 hours. Definite fold in blastoderm anterior to tip of notochord marks anterior extremity of head; no lateral somites have formed.

Stage 7: One somite; 23-26 hours. Neural folds in head region.

Stage 8: Four somites; 26-29 hours. Neural folds touch at level of mid-brain; blood islands appear in posterior blastoderm.

Stage 9: Seven somites; 29-33 hours. Optic vesicles first delineated, paired heart primordia begin to fuse.

Stage 10: Ten somites; 33-38 hours. Most anterior or first pair of somites begins to disappear and is not included in further counts. Cranial flexure begins, three primary brain vesicles form; optic vesicles begin to constrict; heart slightly bent.

Stage 11: Thirteen somites; 40-45 hours. Cranial flexure definite, five pairs of neuromeres lateral to hindbrain; anterior neuropore closing; optic vesicles further constricted; heart bent to the right.

Stage 12: Sixteen somites; 45-49 hours. Head turning to left; anterior neuropore closed; telencephalon forming; optic stalk forward; auditory vesicle deep but open; heart S-shaped; head fold amnion covers forebrain level.

Stage 13: Nineteen somites; 48-52 hours. Head turned to left; cranial and cervical flexures distinct, telencephalon enlarged; atrio-ventricular canal indicated; headfold amnion covers to level of spinal cord; no hypophysis yet.

Stage 14: Twenty-two somites; 50-53 hours. Forebrain and hindbrain at right angles forming cranial flexure; cervical flexure a broad curve; body rotated back to somites 7-9; slight indication of lumbar flexure; visceral arches and clefts 1 and 2 formed; optic vesicles begin invagination with lens placode forming; auditory pit opening constricted; Rathke's pocket can be seen; ventricular loop of heart ventral to A-V canal; amnion folded back to level of somites 7-10.

Stage 15: Somite number becomes indistinct, hence variations must be expected for the following categories. Other features will be used to determine age; 50-55 hours. Lateral body folds extend to wing buds; mesoderm condensation in wing bud areas; somites 24-27; amnion folds to somites 7-14; forebrain and midbrain at acute angle forming cranial flexure; trunk distinct; rotation of body to somites 11-13; visceral arch and cleft # 3 distinct; optic cup formed and double at level of iris.

Stage 16: Somites 26-28; 51-56 hours. Lateral body folds between limb buds; wing bud as thickened ridge more prominent than leg bud; amnion fold to somites 10-18; all flexures accentuated; rotation of body to somites 14-15; tailbud; forebrain lengthened and interventricular constrictions forming in brain.

Stage 17: Somites 29-32; 52-64 hours. Lateral folds circumferential; wing and leg buds bulging from body; amnion enclosing embryo except through oval opening about somites 28-36; rotation to somites 17-18; epiphysis a knob; nasal pits.

Reprinted from H. L. Hamilton. *Lillie's Development of the Chick.* 1952. Holt, Rinehart & Winston, New York.

"The species is contained in the egg of the hen as completely as in the hen, and the hen's egg differs from the frog's egg as the hen from the frog."

C. O. Whitman

"Between vitalism and mechanism there is a middle ground which may be called 'Organizationism' or 'Emergence', which holds that life, differential sensitivity and reactivity, fitness and psychic phenomena, are results of increasing organization, these properties 'emerging' as it were, by a process of creative synthesis."

E. G. Conklin, 1944

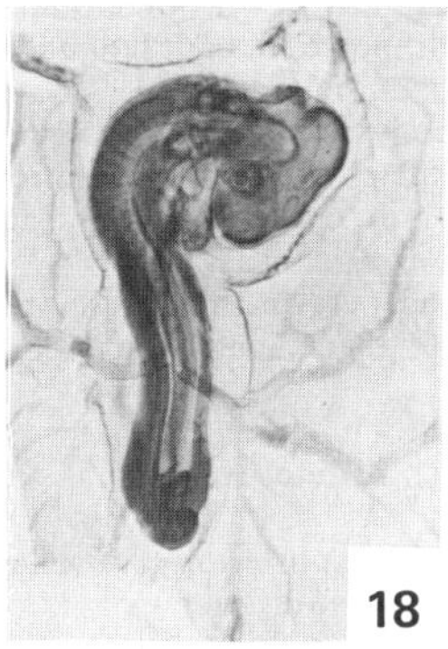
18

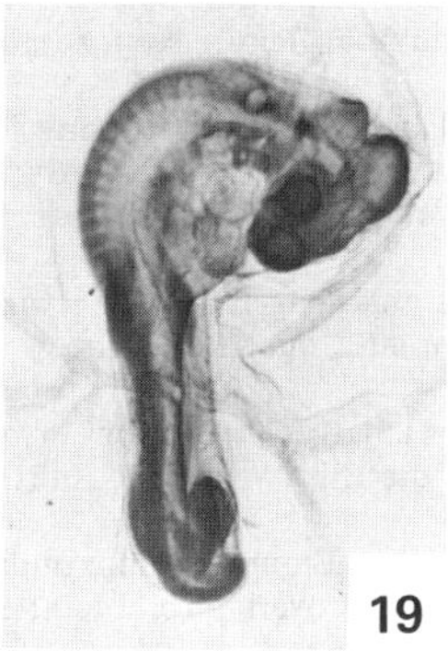
19

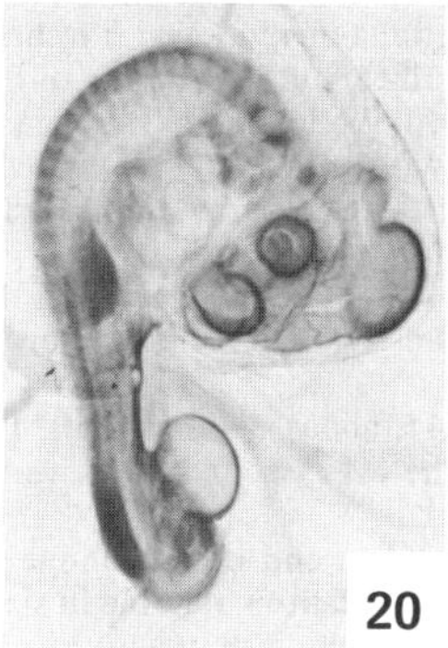
20

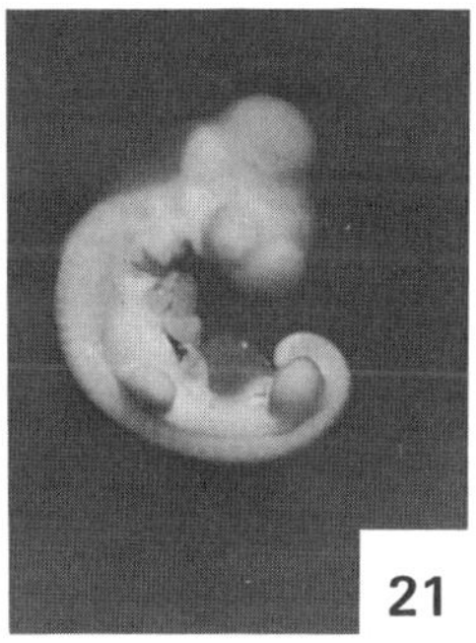
21

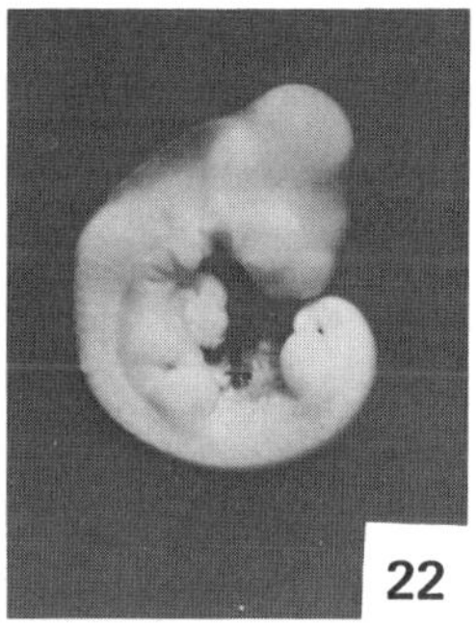
22

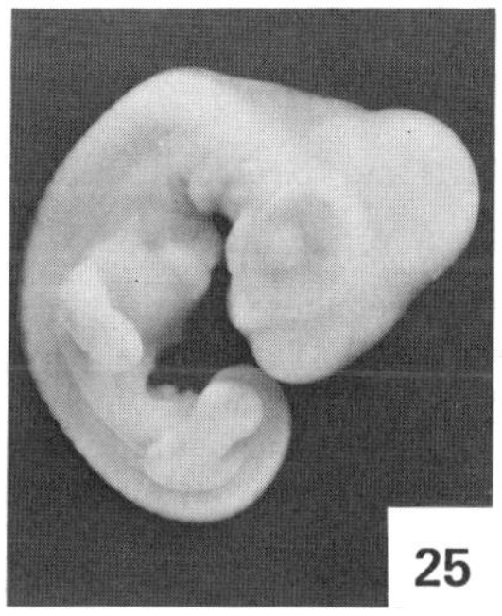
25

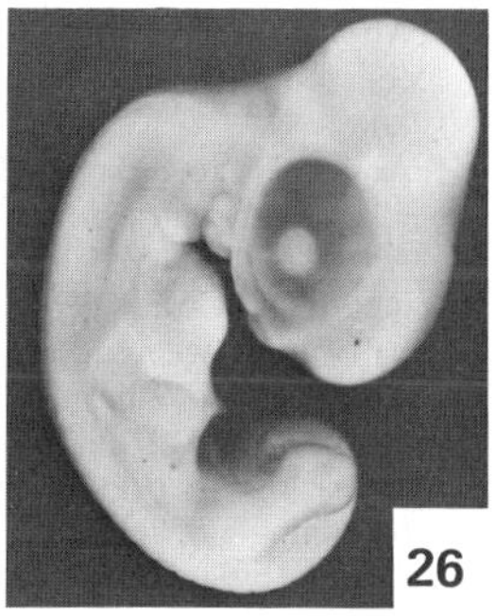
26

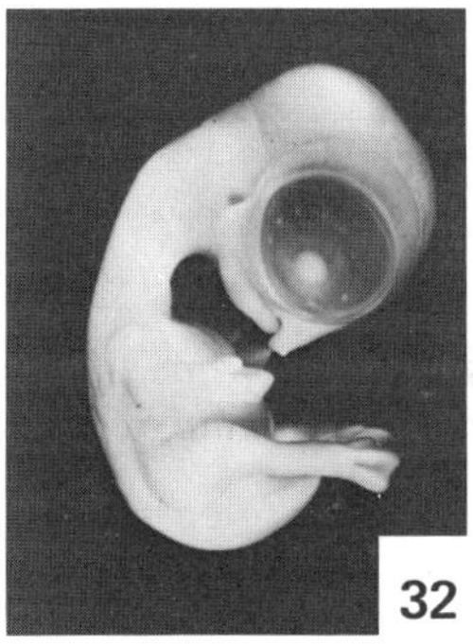
32

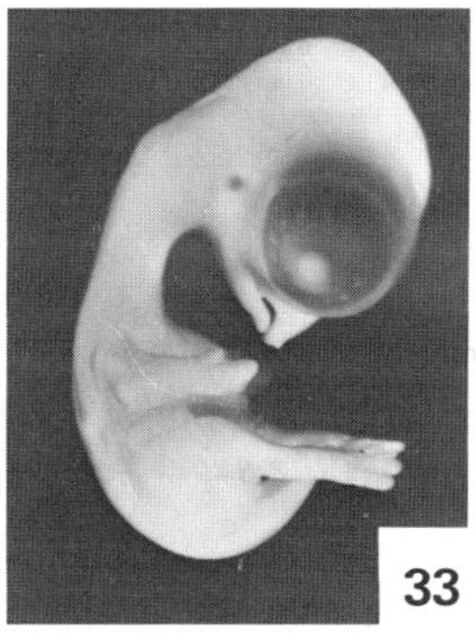
33

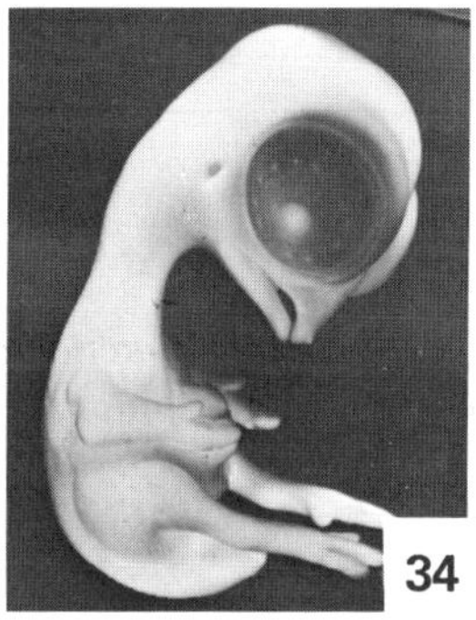
34

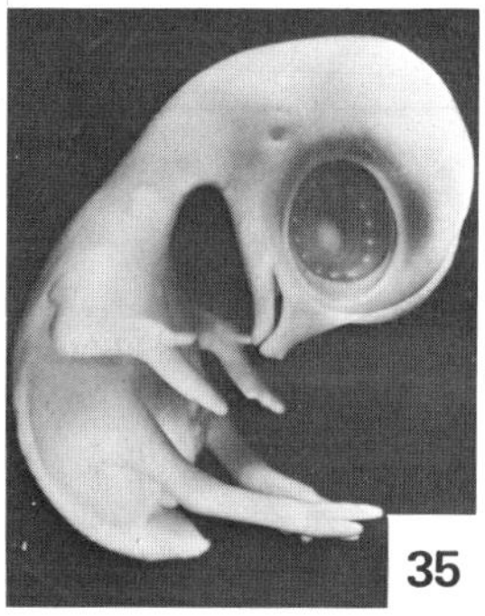
35

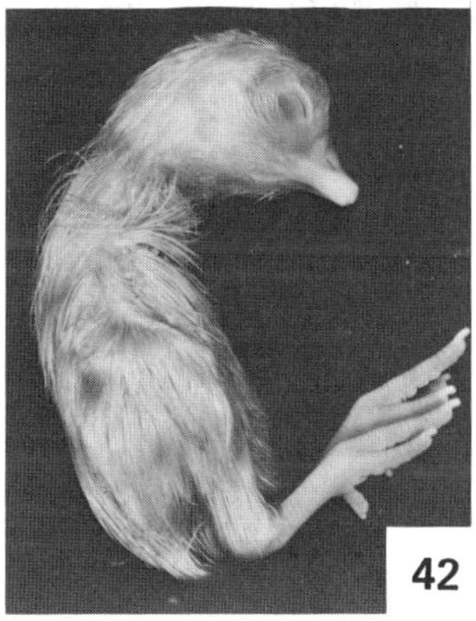
42

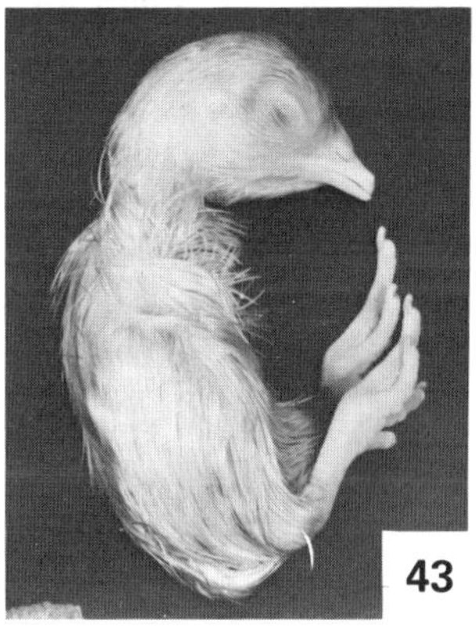
43

44

Chick Embryos (Continued)

Stage 18: Somites 30-36, 72 hours ±. Limb buds enlarged with leg buds slightly larger than wing buds; amniotic folds usually closed at raphe; medulla at right angles to body; lumbar flexure seen; rotation to posterior end; tailbud turned to right; maxillary process indistinct; allantois begins to form.

Stage 19: Somites 37-40, 72-84 hours. Leg buds enlarged and bulkier than wing buds; somites into unsegmented tail; medulla forms acute angle with trunk; tailbud curved with tip pointing anteriorly; maxillary and mandibular processes formed; vestige of first visceral cleft as narrow dorsal slit; fourth cleft formed but does not reach pharynx; allantois a pocket; eyes still without pigment.

Stage 20: Somites 40-43, about 3½ days. Leg buds enlarging faster than wing buds; cervical flexure more accentuated; tail bends toward lumbo-sacral region; rotation complete; maxillary process larger than mandibular; allantois vesicular; faint eye pigment.

Stage 21: Somites 43-44, about 3½ days plus. Limbs enlarged; posterior curvature to lumbo-sacral region; maxillary process extends to level of the eyes, second arch overlaps the third ventrally; fourth arch distinct and cleft as slit; allantois enlarging; eye pigment faint.

Stage 22: Somites extend to end of tail, about 96 hours. Elongated limb buds pointing posteriorly; maxillary process still enlarging; 4th cleft as slit; allantois variable in size but may overlap forebrain; eye pigment distinct.

Stage 25: About 4½ to 5 days. Elbow and knee joints formed, digital plate in wing; maxillar process lengthened; dorsal part of 3rd arch still visible; third and fourth clefts reducing.

Stage 26: About 5 days. Limbs lengthened; digital plates rounded; toe markings seen; contour of maxillary process broken; 3rd and 4th clefts gone.

Stage 32: About 7½ days. All digits and toes lengthened, rudiment of 5th toe gone; webs between digits thinning; sizes of digits and toes variable; anterior tip mandible reaches beak; eleven or more rows of feather germs at level of legs; one distinct row on tail.

Stage 33: About 7½ to 8 days. All digits and toes lengthened; mandible and neck conspicuously lengthened; tail has 3 distinct rows of feather germs.

Stage 34: About 8 days. Second and third digit and toe excessively grown; web contours concave and arched; mandible and neck lengthen; feather germs extensive; those on thigh protrude conspicuously; scleral papillae 13-14; nictitating membrane halfway between outer rim of eye (lid) and scleral papillae.

Stage 35: About 8½ to 9 days. Webs on appendages become inconspicuous; phalanges in toes distinct; beak lengthens further; eye more distant from beak; feather germs more conspicuous with mid-dorsal line standing out; ventral feather germs appearing; nictitating membrane approaches outer scleral papillae; eyelids begin to overgrow eyeball; eye circumference becoming ellipsoidal.

Stage 42: About 16 days. Beak pronounced, measuring 4.8 mm anterior angle of nostril to tip of upper bill; third toe 16.7 mm long.

Stage 43: About 17 days. Beak 5 mm long; labial grooves as white granular crust at edge of each jaw; third toe 18.6 mm long.

Stage 44: About 18 days. Beak length 5.7 mm, with transparent periderm covering but starting to peel off proximally; third toe 20.4 mm long.

"It is deserving of emphasis that the function of imagination is not merely the conception of mythical creations, but also, and quite particularly, the presentation, to the mind, of realities. Hence, imagination plays an important role in the exact Sciences."

A. J. Lotka, 1925

"I am unwilling to accept the defeatism of the vitalist, so long as means of investigation by experiment are available."

R. G. Harrison, 1945

"Theory without fact is fantasy, but fact without theory is chaos. Divorced, both are useless; united, they are equally essential and fruitful."

C. O. Whitman

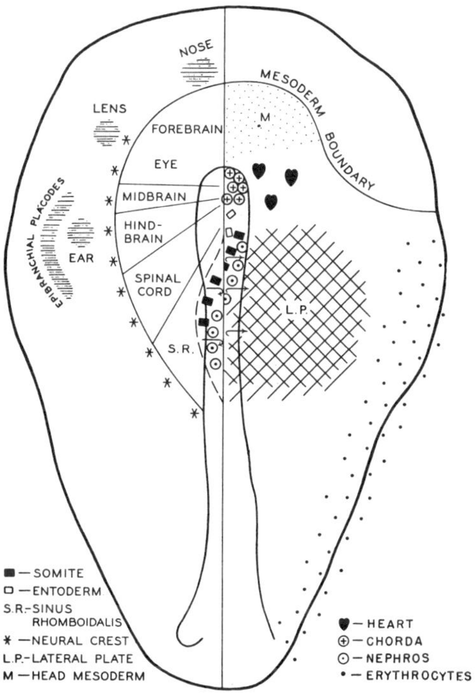

Prospective areas in the definitive primitive streak blastoderm of the chick. The superficial layer is shown at the left while the invaginated material is seen in the right half. The interrupted line on the left side of the anterior streak region marks the boundary between the ectoderm and the still uninvaginated mesoderm. This is on the assumption that invagination is as yet incomplete and that future invagination will be limited to material destined to form the embryo proper. All mesodermal boundaries need accurate experimental verification.

This composite map is drawn by Rudnick (1944) *Quart. Rev. Biol.* 19:187, and is based largely on work of Pasteels. In addition, there have been contributions from the work of Hunt (endoderm cells), Wolff (morphogenesis of trunk and tail), and Yntema (ectodermal placodes).

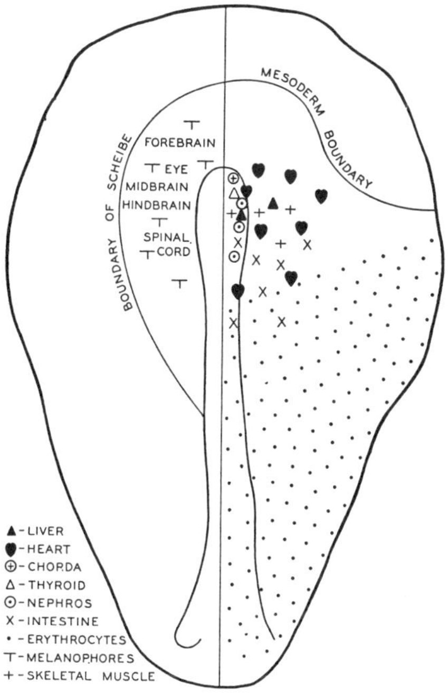

DISTRIBUTION OF POTENCIES IN THE DEFINITIVE PRIMITIVE STREAK BLASTODERM, TESTED UNDER VARIOUS EXPERIMENTAL CONDITIONS

Ectodermal potencies shown on left, mesodermal and entodermal on right; these have not been tested separately. Posterior and lateral extent of mesodermal potencies has not been specified.

From Rudnick 1944: *Quart. Rev. Biol.* 19:187

REFERENCES

Arey: 66, 67, 76-79, 505-10, 553-57
Balinsky: 205-40
Bodemer: 221-28, 256
Hamilton revision of Lillie: Chapter II
Huettner: 205, 211-12
Lillie: 69-90
McEwen: 300-31
Nelsen: 210-453
Patten: 37-74
Patten (1964): Chapters 9, 10, and 11
Rugh: *Vert. Emb.* 117-26
Shumway: 81-84, 93-98

"The hen does not produce the egg, but the egg produces the hen and also other hens. . . . We know that the child comes from the germ cells and not from the highly differentiated bodies of the parents and, furthermore, that these cells are not made by the parents' bodies but these cells have arisen by the division of antecedent cells . . . Parents do not transmit their characters to their offspring, but these germ cells, in the course of long development, give rise to adult characters similar to those of the parent."

E. G. Conklin, 1918.

The darkly staining triangular areas ventrolateral to the neural folds and located between the margin of the proamnion and the endodermal line of the anterior intestinal portal are known as the *amniocardiac vesicles.* These mesodermal cavities are continuous with the extraembryonic body cavities (coelom) and will subsequently grow in beneath the foregut to form the single pericardial cavity. At the posterior margins of the vesicles, it may be possible to discern the first indications of the large *vitelline* (yolk) *veins* which form between the cavities and the endoderm of the anterior intestinal portal. Later, these veins, running from the yolk sac into the embryo, will be known as the *omphalomesenteric veins.*

The area opaca at this stage is clearly divided into an inner, mottled *area vasculosa* and an outer, lighter *area vitellina. Blood islands* (dark aggregations of cells) will be abundant within the area vasculosa, and in slightly older embryos it may be possible to see the early formation of extraembryonic blood vessels. These blood vessels develop apart from the embryo but soon become continuous with the omphalomesenteric (embryonic) veins previously described. The area vitellina is made up of ectoderm and endoderm and the underlying adherent yolk. Between these two areas, encircling the area vasculosa, there will soon develop a large blood vessel, known as the *sinus terminalis.* The area vitellina will expand until it all but encompassed the yolk; the area vasculosa spreads more slowly. (See drawing on page 128.) There may be a slight difference in the area vitellina, designated as externa (ectoderm only) and interna (ectoderm and endoderm), but this will not be seen in excised blastoderms.

Transverse Sections

In studying the serial transverse sections of the 24-hour chick embryo the student should locate the most anterior sections first and then proceed to examine the series posteriorly, identifying each structure as it appears and disappears, making frequent reference to the whole-mount photograph and indicating in this picture exactly where each detailed transverse section was taken. The structure seen in a typical series of sections through a 24-hour chick embryo will be described briefly, in the order in which they should appear, from anterior to posterior. (Compare with drawings on page 128; also see illustrations on pages 146-47 and Plate 2, pages 362-63.)

The most anterior sections will show the anterior tips of the lateral *neural folds* which are not yet fused around the *anterior neuropore.* Directly below is the double-layered *proamnion,* made up of ectoderm and endoderm and limited laterally by the adherent yolk cells of the marginal *area opaca.* There are cellular differences in the ectoderm and endoderm which make them always distinguishable. Locate the *germ wall* (the expanding and proliferating margin of the blastoderm) and, more laterally, the earliest *blood islands.* These cell aggregates develop into blood vessels.

A few sections posteriorly, the *head process* will consist of the dorsal (head) ectoderm and the dorsally approximated *neural folds.* This thick-walled, ectodermal tube is now the *neural tube,* which will be seen to bulge laterally (evaginations) to form the *optic vesicles,* at the level of the *prosencephalon.* The central cavity is the *prosocoel.* Immediately ventral to the neural tube, there will shortly appear a group of cells which, more posteriorly, surround a cavity. This is the most anterior pocket of the *foregut*, later to be designated as the *preoral pit* (Seessel's pocket). Between the foregut and the neural tube locate the small circular aggregation of cells, the anterior tip of the *notochord.* Scattered cells may be found between the neural tube and the head ectoderm. They are embryonic mesoderm cells, and in this region they are designated as *head mesenchyme* cells. The space between the head process and the underlying double-layered proamnion is the space of the *subcephalic pocket.*

As the germ layers of the head process are reflected into the underlying proamnion, more posteriorly one will find that a pair of large spaces enclosed in a new double-layered mesoderm appear between the lateral ectoderm and endoderm. These spaces seem to come in under the foregut region as the primordia of the pericardial cavity, the embryonic part of the *amniocardiac vesicles.* The mesoderm has grown in toward the embryo, but before reaching the embryo it has formed this large cavity by splitting into an outer *(somatic)* and an inner *(splanchnic)* layer of mesoderm. These spaces are most obvious at about the level of the *anterior intestinal portal,* i.e., the region where the foregut is not yet a closed tube but opens out over the underlying yolk. The yolk is not present on these slides, having been removed when the blastoderm was excised. Note that the mesial portion of the splanchnic mesoderm (i.e., that portion closest to the gut region) is relatively very thick. This tissue will give rise to the muscle portion of the heart, known as *myocardium.* Note also that the *area opaca* at this level is closer to the embryo, that the neural tube is still open dorsally, and that there is abundant mesenchyme lateral to the neural folds where the somites are to appear. The notochord still marks the mid-dorsal line of the archenteric roof and the space below (between the layer of endoderm and the yolk) is later to be included in the foregut, with the posterior migration of the anterior intestinal portal.

The *amniocardiac vesicles* are large and will be seen through many posterior sections, but the *archenteric roof* opens out flat over the yolk so that there seems to be no indication of any gut structure except a roof. Other changes to note are the open neural tube, the early differentiation of *mesoderm* into the dorsal (*epimeric*) or segmental *mesoderm* (the somites), the *intermediate* (*mesomeric*) or nephric *mesoderm,* and the split lateral (*hypomeric*) *mesoderm.* This most lateral mesoderm is split into the dorsal *somatic* and ventral *splanchnic layers,* with the intervening *coelomic space.* The term coelom is properly reserved for the mesodermal cavity *within* the body of the embryo. Since most of the above-mentioned cavities will remain outside of the embryonic body, they are called the exocoels, or *extraembyonic body cavities* e. e. b. c.). The line of demarcation shifts with age, the coelom and exocoel remaining continuous for many hours. Locate *blood islands* in the splanchnic layer only, in the area opaca.

Examine sections further posteriorly until it becomes difficult to differentiate between the notochord and the neural tube, even under high magnification. As the notochord disappears one approaches the position of the receding *Hensen's node* and the *primitive streak.* The *neural folds* will flatten out and merge into the *primitive folds,* more posteriorly, in succeeding sections. Instead of formed somites there will appear large masses of mesoderm which seem to be merged with (growing out from) the ectoderm cells of the primitive streak. No notochord will be seen, but beneath will be the endodermal roof of the primitive gut. A section through Hensen's node will show it as an aggregation of cells, posterior to which is an angled depression, the *primitive pit.* At this level also identify the *area pellucida;* the *area opaca* where the yolk granules are abundant; and the *germ wall,* which consists of the inner margin of *zone of junction* about to be added to the area pellucida.

"The foundation of all embryonic development consists of the processes by which cells become differentiated from each other and from their common progenitors. These processes of cellular differentiation, though basic to an understanding of embryogeny, remain largely unknown, although two major factors are generally recognized as important in governing the course of cell development. First, the genetic makeup of cells defines the limits and the potentialities of their development, and, second, the diverse cellular environments of the embryo elicit the specific developmental responses in cells which lead to their maturation into the wide variety of cell types that characterize the adult. These local and highly specific cell environments are produced by cellular activity and are constantly changing as the responsible cells change. Thus, a continuously evolving, dynamic interchange between embryonic cells directs their differentiation into specialized adult cells and finally into senescent cells incapable of sustaining the life of the organism." **Markert and Silvers, 1956** ***Genetics 41: 429-50,*** **modified by Ebert, 1960, in "Aging," ed. N. W. Shock, A.A.A.S. Pub.**

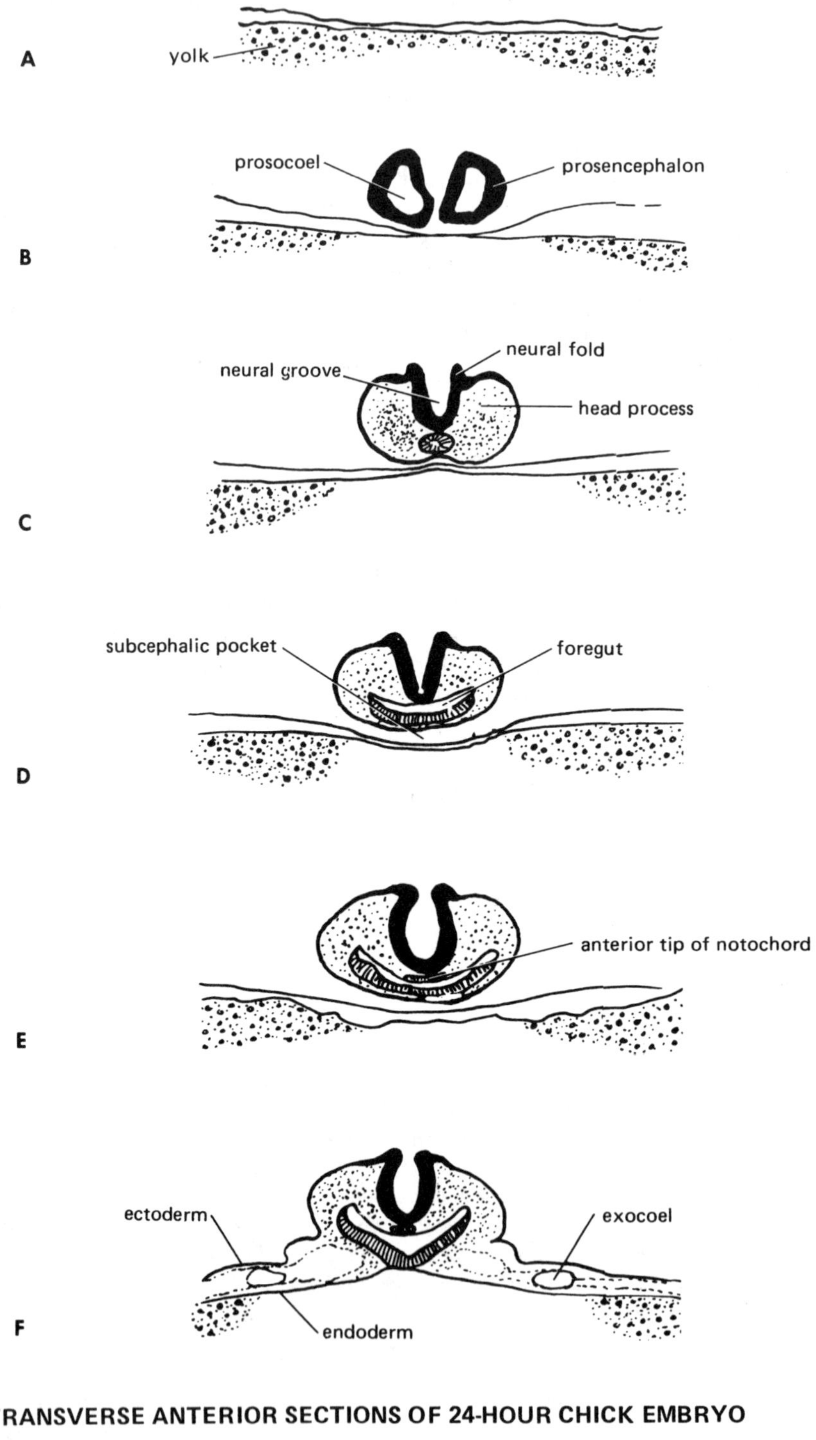

TRANSVERSE ANTERIOR SECTIONS OF 24-HOUR CHICK EMBRYO

THROUGH POSTERIOR PART OF NEURAL GROOVE

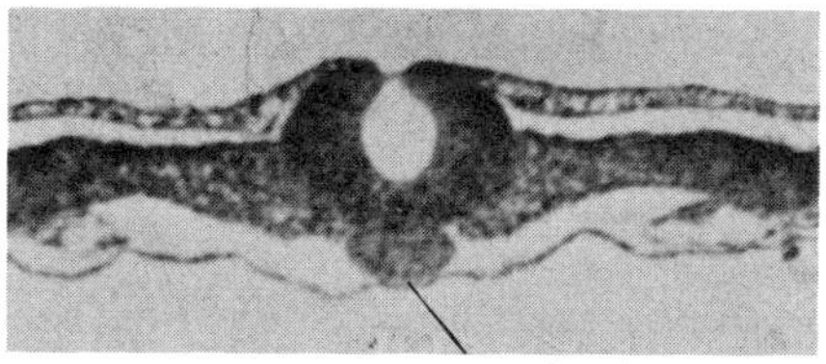

THROUGH ANTERIOR PART OF PRIMITIVE STREAK

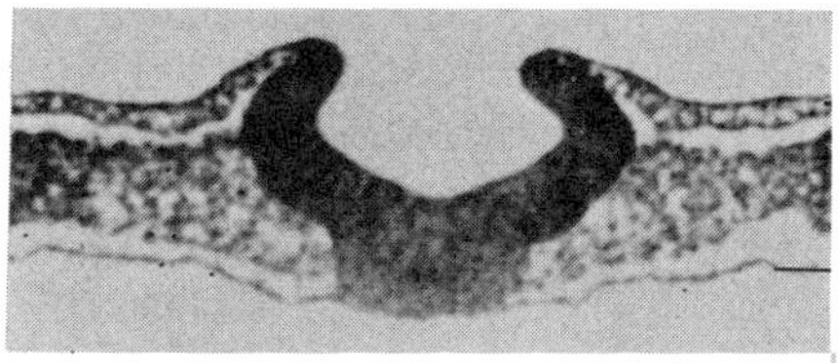

THROUGH ANTERIOR PART OF PRIMITIVE STREAK

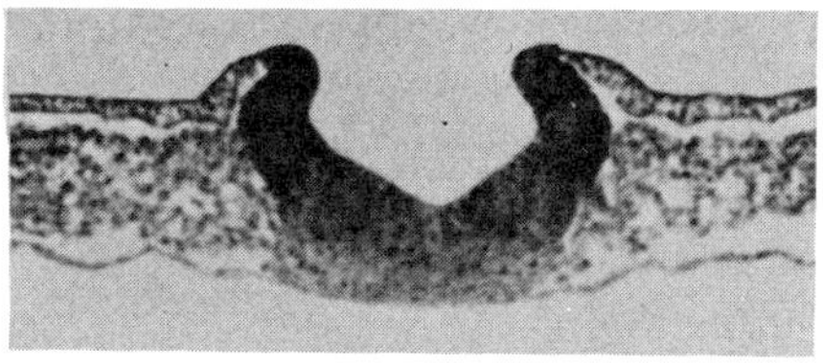

THROUGH PRIMITIVE STREAK

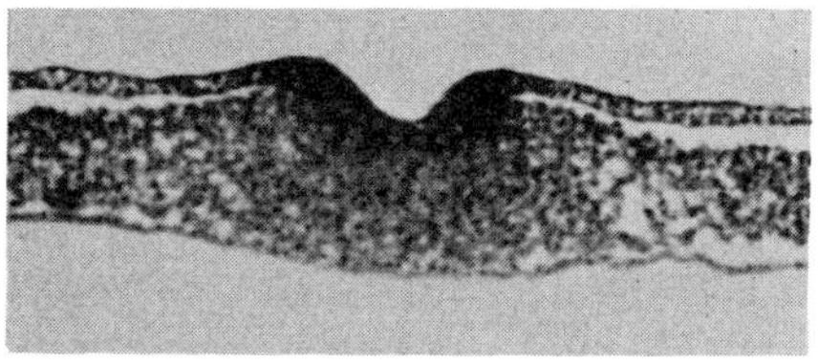

THROUGH PRIMITIVE PLATE

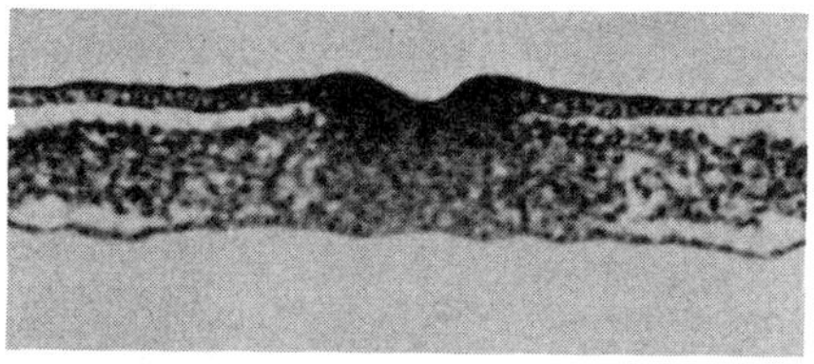

CHICK EMBRYO AT 24 HOURS
TRANSVERSE POSTERIOR SECTIONS

Notice that, in sections through the embryo proper, there is always some notochord below the neural groove and that this is absent beneath the primitive streak. Also, lateral to the neural groove, the mesoderm is being organized into somites, while it is diffuse lateral to the primitive streak. The primitive streak does not close over as a tube (neural tube). These are photographs of actual sections.

SAGITTAL SECTION

The main value of a sagittal section at this stage of development is to show the relation of the head fold to the other embryonic and extraembryonic areas, such as the heart, anterior intestinal portal, primitive streak, and germ wall. The head process or fold is represented by the anteriorly projecting ectoderm and associated endoderm which begin to grow forward and upward, away from the underlying yolk.

First, locate the most median of the sagittal sections, i.e., the one in which the maximum of *brain vesicles* and *neurocoel* show. Identify the three primary brain vesicles and note that the *prosencephalon* has already begun to bend ventrally over the tip end of the *notochord.* As in the frog, the fulcrum of the flexion is the region of the *tuberculum posterius.* The foregut may be seen as a definite tube extending anteriorly into the head about as far as the notochord. Posteriorly, this *archenteron* has only a roof, but this roof extends even beneath the primitive streak. The margin of the anterior gut opening, where the gut endoderm bends ventrally and anteriorly to become endoderm of the *proamnion,* is the *anterior intestinal portal,* or opening into the formed foregut.

Beneath the foregut and just anterior to the intestinal portal, locate small circular mesodermal cavities which represent parts of the *heart.* It should be possible to identify the *endocardium, myocardium,* and *pericardium* (the lining, muscular, and covering layers of the heart), the *posterior sinus venosus,* and the *anterior ventral aorta* (especially if the embryo happens to be about 30 hours of incubation age). The heart tissues are mesodermal and must not be confused with the *proamnion* which extends anteriorly beneath the *subcephalic pocket* and which is made up of ectoderm and endoderm. In some (older) embryos there may appear a fold of anterior ectoderm and somatic mesoderm folded back over the head process, called the *head fold* of the *proamnion.* More anteriorly locate the *area opaca* with its *area vasculosa* and *blood islands* and the double-layered mesoderm, which is growing in toward the embryo between the layers of the proamnion. With the invasion by this mesoderm the true amnion (and chorion) will shortly develop.

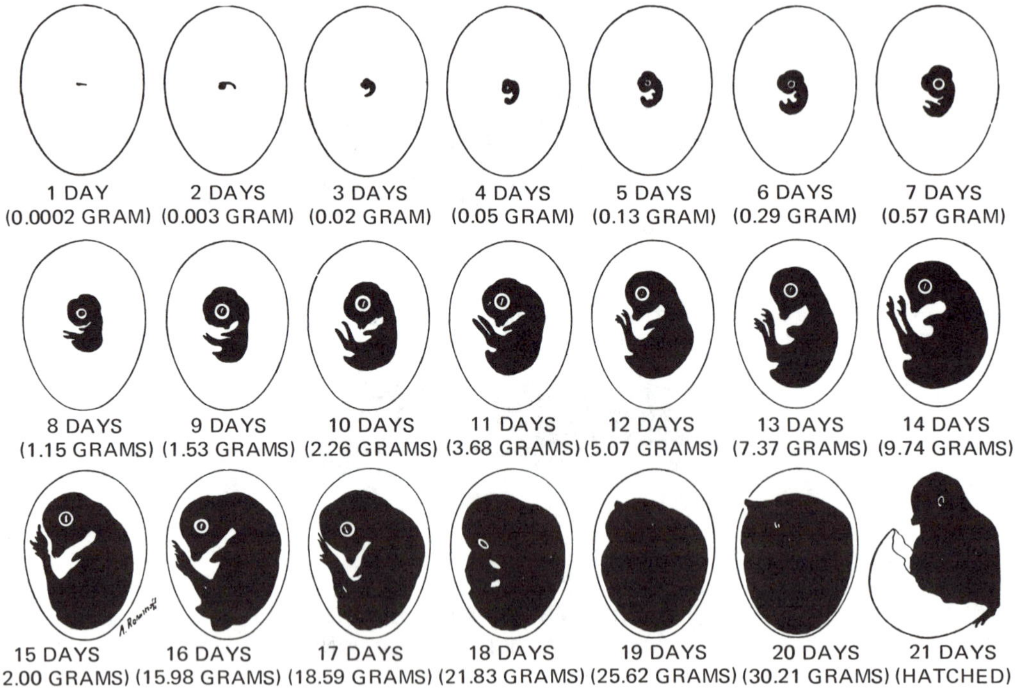

DAILY CHANGES IN WEIGHT AND THE FORM OF THE DEVELOPING CHICK EMBRYO (WHITE LEGHORN)

Drawing from the New York State College of Agriculture.

In sections slightly lateral to these median sections, locate the well-formed somites, and posteriorly identify parts of the *primitive streak*. It is particularly important that the student distinguish between the various germ layers in these sections. (See Patten, *Chick,* p. 99.)

REFERENCES

Arey: 558-66
Balinsky: 49-280
Bodemer: 193-213
Hamilton revision of Lillie: Chapters III and IV
Huettner: 218-24
Lillie: 91-122
McEwen: 300-31
Nelsen: 516-54
Patten (1957): 130-45
Patten (1964): Chapters 6-11
Rugh: *Vert. Emb.* 117-26
Shumway: 107-10, 119, 285-87

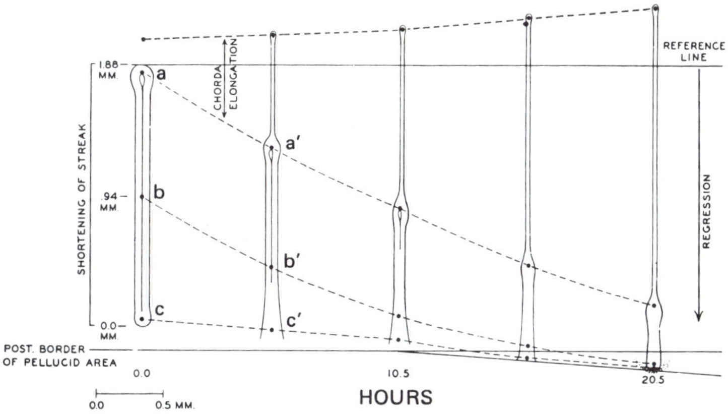

REGRESSION OF THE CHICK PRIMITIVE STREAK

Scale-drawn summary diagram illustrating regression and shortening of the primitive streak and elongation of the notochord during about 20-hours explanation *in vitro.* The slope of the broken lines in relation to the fixed reference line is an approximate measure of the rate of movement of the marked cell groups (the black dots). In later stages of regression the streak frequently extends behind the posterior border of the pellucid area. The diagram is based on the analysis of 195 marked, living blastoderms. None of these shows a significant variation from the course of events depicted here.

From Spratt 1947: *Jour. Exp. Zool.* 104:69.

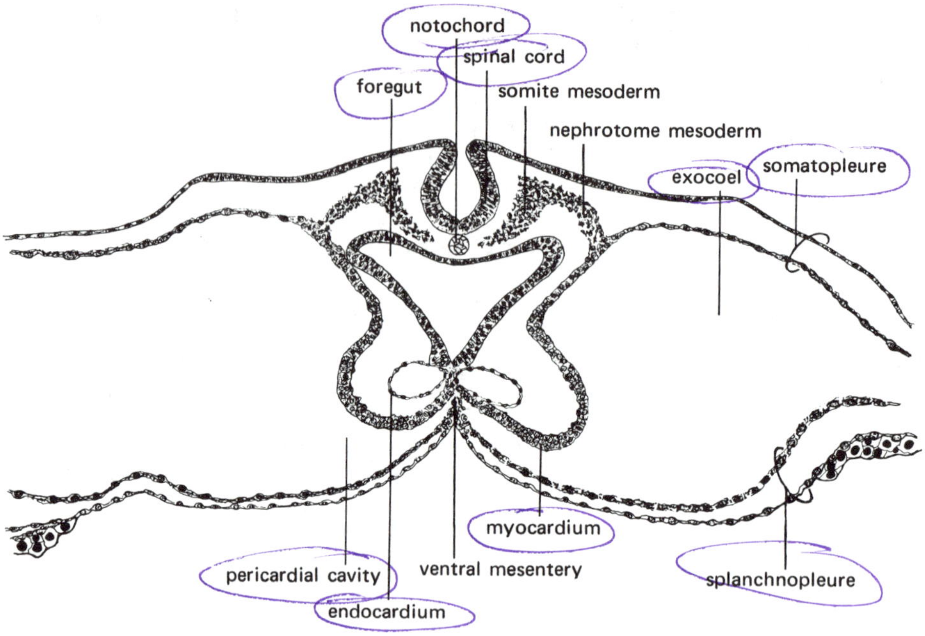

CHICK EMBRYO OF SEVEN SOMITES (29-33 HOURS)

TRANSVERSE SECTION JUST ANTERIOR TO ANTERIOR INTESTINAL PORTAL

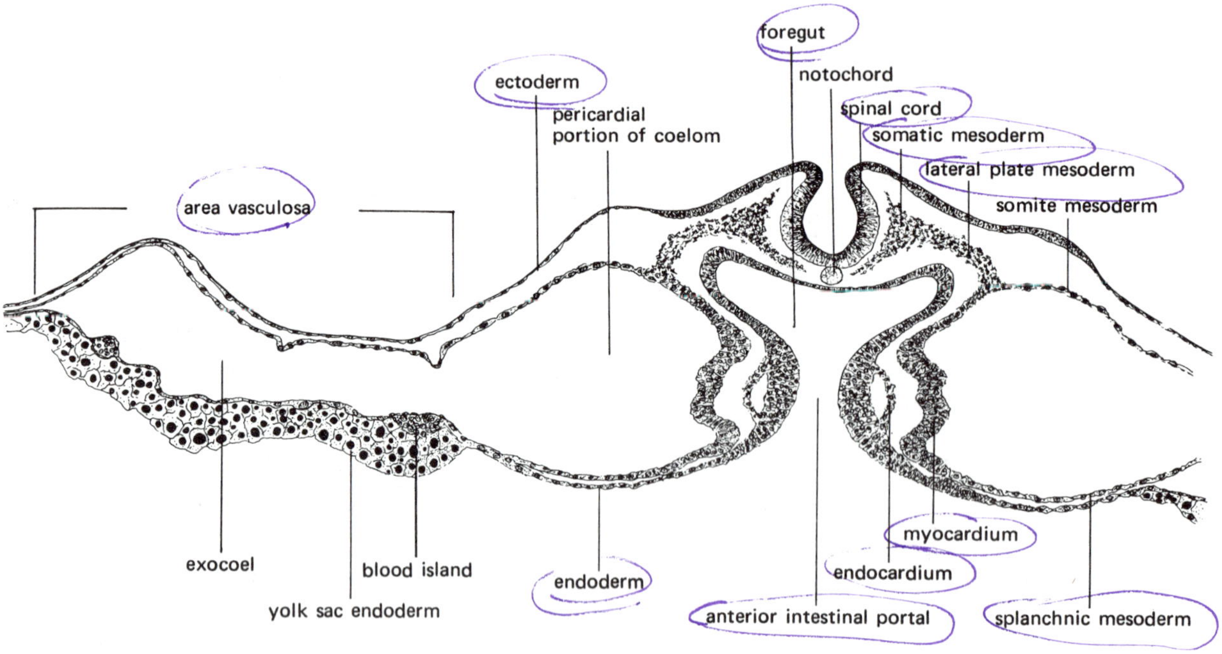

CHICK EMBRYO OF SEVEN SOMITES (29-33 HOURS)

TRANSVERSE SECTION AT LEVEL OF ANTERIOR INTESTINAL PORTAL

DRAWINGS: CHICK EMBRYO: 24-HOUR STAGE

THE CHICK EMBRYO:

Thirty-Three-Hour Stage

Exercise 19

WHOLE MOUNT

By 33 hours of incubation the embryo will measure 4 mm in total length, although a portion of this is represented in a *ventral flexion* of the head process, that part of the body around the anterior end of the notochord. Obviously this flexion cannot progress very far because of the underlying inert yolk so that in later stages there will develop a *dextral torsion* (head turns with its left side toward the yolk) also beginning at this anterior end. Note the large *optic vesicles;* the constriction between the two portions of the forebrain (prosencephalon); the single midbrain vesicle (mesocoel); and several vesicles (metacoel and myelocoel) arranged linearly which comprise the hindbrain (rhombencephalon). This cephalic portion of the central nervous system is known as the *encephalon* to distinguish it from the more posterior spinal cord (spinal cord level) known as the *myelocoel* (cavity of the myelencephalon). The brain consists of eleven segments or *neuromeres:* three in the forebrain, two in the midbrain, and six in the hindbrain. Is there any evidence of such metamerism of the encephalon? Due to the presence of 12 to 13 pairs of somites at this stage, the neural tube may also appear to have metameric bulges. Posteriorly there remains but a slight vestige of the primitive streak and the *sinus rhomboidalis.*

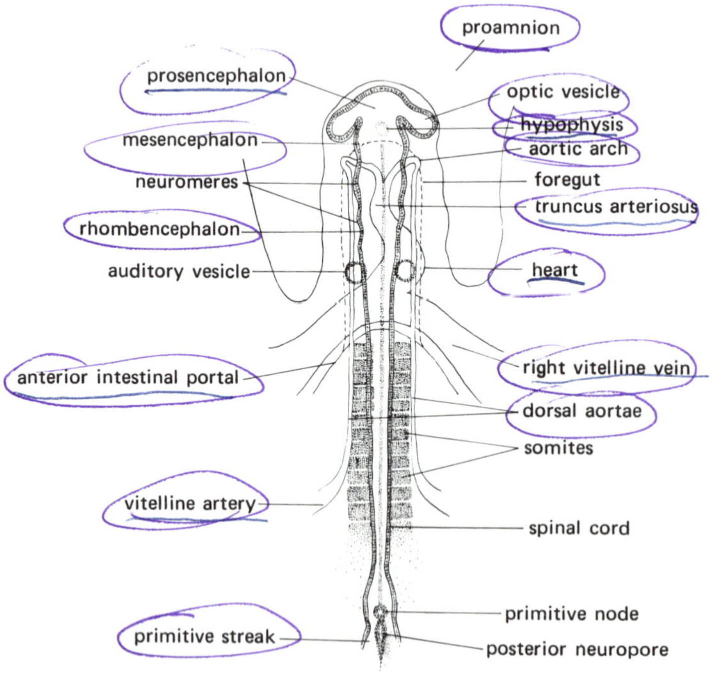

THE 31-HOUR CHICK EMBRYO

SCHEMATIC DRAWING

At the posterior margin of the median vesicle of the forebrain, locate a semicircular depression, which is the *infundibulum.* This ventral evagination of the brain ectoderm will later join the ectoderm of *Rathke's pocket* to form the *pituitary body.* Just posterior to the infundibulum locate the anterior tip of the notochord. It may be possible to trace this notochord posteriorly to the shortening *primitive streak.* The head ectoderm now clearly marks off the head structures from the underlying *proamnion,* and the *subcephalic pocket* is almost immediately ventral to the level of the heart.

The *heart* at this stage is a thin-walled, S-shaped, single, tubular sack which bugles to the right, apparently projecting outside of the body. Locate the two large omphalomesenteric veins which join (*sinus venosus*) and empty into the posterior part of this bulbous heart (*atrium*). Anteriorly the heart wall may appear somewhat thicker, indicating the material of the future *ventricles.* These cavities lead forward to the ventral *aortic roots* and thence to the *aortic arches* which pass dorsally around the foregut. The first evidences of the paired *dorsal aortae* may be seen running posteriorly beneath the *somites* and growing out onto the yolk about the level of the most posterior somites. In the *area vasculosa* note the further development of the *blood islands* and *blood vessels* and the enlarging *sinus terminalis.* The areas *pellucida, opaca, vasculosa,* and *vitellina* are all easily distinguishable at this stage of development.

TRANSVERSE SECTIONS

(Examine with the naked eye to gain perspective and understand how the sections are mounted. See drawings on page 156 and Plate 3, pages 364-65.)

Again begin the study at the most anterior of the serial transverse sections of the 33-hour chick embryo and locate the *opticoel* with the connecting *optic stalks,* all associated with the *diocoel* (cavity of the diencephalon). Dorsal to the opticoel, you will find abundant *head mesenchyme.* The *subcephalic pocket* and the *proamnion* are much the same as they were in the 24-hour stage, but laterally the *exocoels* have grown in toward the head *process* so that there is a clearer demarcation between the *somatopleure* and the *splanchnopleure.* The former is made up of ectoderm and somatic mesoderm, and the latter is composed of splanchnic mesoderm and endoderm. Laterally the *area opaca* is identified by the adherent underlying yolk, at the region of the *germ wall.* The splanchnic mesoderm of the area vasculosa is now supplied with abundant blood islands, some with contained corpuscles.

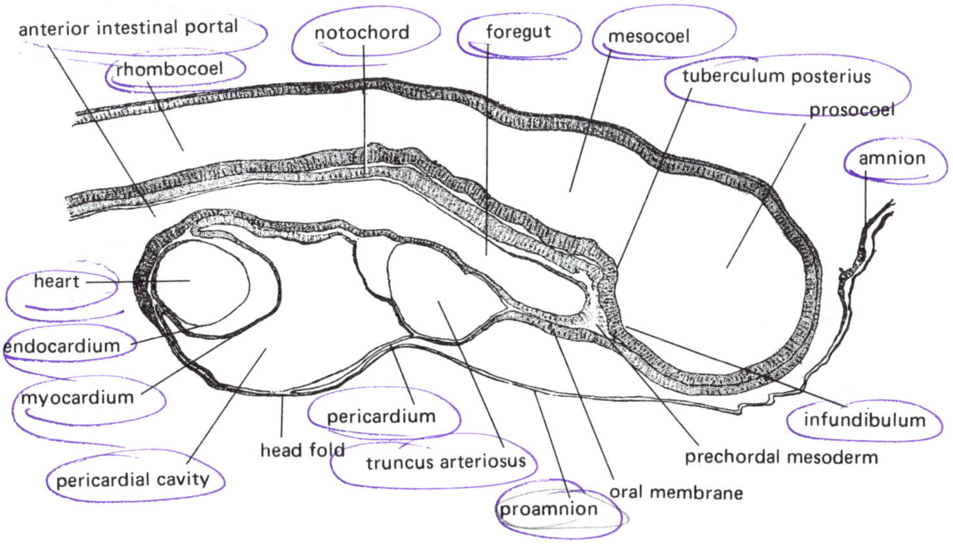

THE 33-HOUR CHICK EMBRYO—SAGITTAL SECTION

From H. L. Hamilton. *Lillie's Development of the Chick,* 1952. Holt, Rinehart & Winston, New York.

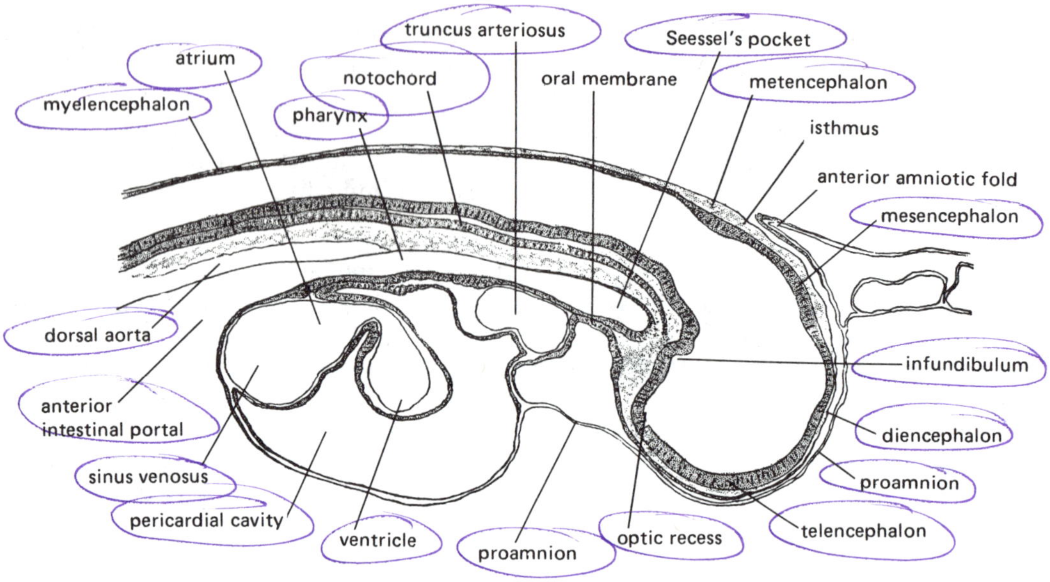

SAGITTAL SECTIONS SHOWING BRAIN AND HEART DEVELOPMENT IN CHICK EMBRYO

From H. L. Hamilton. *Lillie's Development of the Chick.* 1952. Holt, Rinehart & Winston, New York.

Several sections posteriorly the *opticoels* will appear to be separated from the *diocoel,* indicating a slightly posterior direction to the optic evaginations from the forebrain. Just as the opticoels disappear, you will find a pair of small blood vessels, *the paired anterior cardinal veins,* close to the *mesocoel.* Directly ventral to these veins and much larger are the *dorsal aortic roots,* the anterior projections of which become the carotids. At about this level a median ventral invagination appears as the foregut comes into view. The fused ectoderm and endoderm of this region comprise the *oral plate,* the invagination being the ectodermal *stomodeum.* As the *pharynx* widens, you will find either a pair of *ventral aortic roots* or some connections between the dorsal and the ventral aortae, which represent the rudiments of the *aortic arches.* Attempt to identify these aortic arches by number, counting from the most anterior. On some slides the *amniotic folds* will have closed over the embryo dorsal to the head process, indicating the beginning of the true double-layered *amnion* and *chorion.* It must be emphasized that, while the layer next to the embryo is always ectoderm, the second layer of the amnion (the outer one) is mesoderm. The chorionic layers are oppositely arranged. Posterior to the first appearance of the *notochord,* which is between the *mesocoel* and the *pharynx,* locate the *ventral aorta* as a single, large, thick-walled vessel ventral to the pharynx. (This will become the *bulbus arteriosus.)*

Following the bulbus posteriorly, locate the thin inner lining of the heart (*endocardium*) and the thicker muscular part *(myocardium).* The thick, muscular portion immediately associated with the bulbus is the embryonic *ventricle.* The heart mass is suspended to the floor of the gut by a transient double-layered membrane known as the *dorsal mesocardium.* The coelom in this vicinity will become the *pericardial cavity,* derived from the original amniocardiac vesicles. The enclosing mesodermal membrane is the *pericardium.* Within the *splanchnopleure* it will be possible to identify the vitelline *blood vessels,* but it may not be possible to distinguish between veins or arteries. A few sections posteriorly, however, the *omphalomesenteric veins* will be seen to converge from their more lateral positions to enter the *atrium* by way of the *sinus venosus.* Some of the epimeric mesoderm at these levels may be organized into *somites.*

Posterior to the heart, the gut is open over the yolk, the margin (*anterior intestinal portal*) having moved somewhat posteriorly since the preceding stage. The floorless gut at this level will, in all probability, later become the midgut. The *omphalomesenteric veins* now appear to be far out in the *splanchnopleure,* lateral to the *limiting sulcus,* which is the line of (future) demarcation between the embryonic and extraembryonic structures. In some sections it will be possible to discern this partial constriction which separates the *coelom* from the *exocoel* or extraembryonic body cavity (e.e.b.c.). The *dorsal aortae* are very large. The small, paired *posterior cardinal veins* may be seen just dorsal to the *mesomeric* (nephric) *mesoderm.* Well-formed *somites* should be apparent, and, above these, in the angle between the body ectoderm and the neural tube identify the *neural crests* (in the same position as in the frog).

More posteriorly, as the neural groove opens and the somites disappear in favor of solid mesodermal areas, note that the notochordal tissue is displaced by a larger mass of compactly arranged cells representing *Hensen's node.* The lateral mesoderm here seems to be continuous with the Hensen's node material. Dorsally the groove has flattened and disappeared entirely. This is about all that remains of the original primitive streak.

REFERENCES

Arey: 567-77
Balinsky: 346-66
Bodemer: 214-17
Hamilton revision of Lillie: Chapters III and IV
Lillie: 91-127
Mc Ewen: Chapter X
Nelsen: 516-54
Patten (1957): 84-108, 146-68
Rugh: *Vert. Emb.* 105-54
Shumway: 121-25: 288-91

"A foolish consistency is the hobgoblin of little minds, adored by little statesmen and philosophers and divines. To be great is to be misunderstood."

Emerson in *Self Reliance*

"It would appear, when the marine ancestors of terrestrial vertebrates emerged from the sea and adventured upon dry land, they packed, as it were, a portion of their saline environment in their baggage, and took along with them on their excursion as an essential part of their internal milieu . . . we ourselves carry about with us in our arteries and veins, if not a portion of the actual ocean, at least a roughly approximate replica of its brine . . . There are strong indications that the fluids of the highest animals are really descended from sea water."

L. J. Henderson

"Since the Stone Age the average size of man has increased by five centimeters (2 inches). When members of the Scottish nobility wanted to present a pageant for Queen Victoria during her wedding tour, they found that the historic suits of armour were too small. The descendants had outgrown the armour of their ancestors."

F. Kahn, 1944

"Countless human beings are being starved to death for the need of nitrogen (as protein) even though they are living in an atmosphere of 79% nitrogen, which they cannot use. This is probably the greatest paradox of terrestrial life, and at the same time the greatest tragedy of mankind."

F. Kahn, 1944

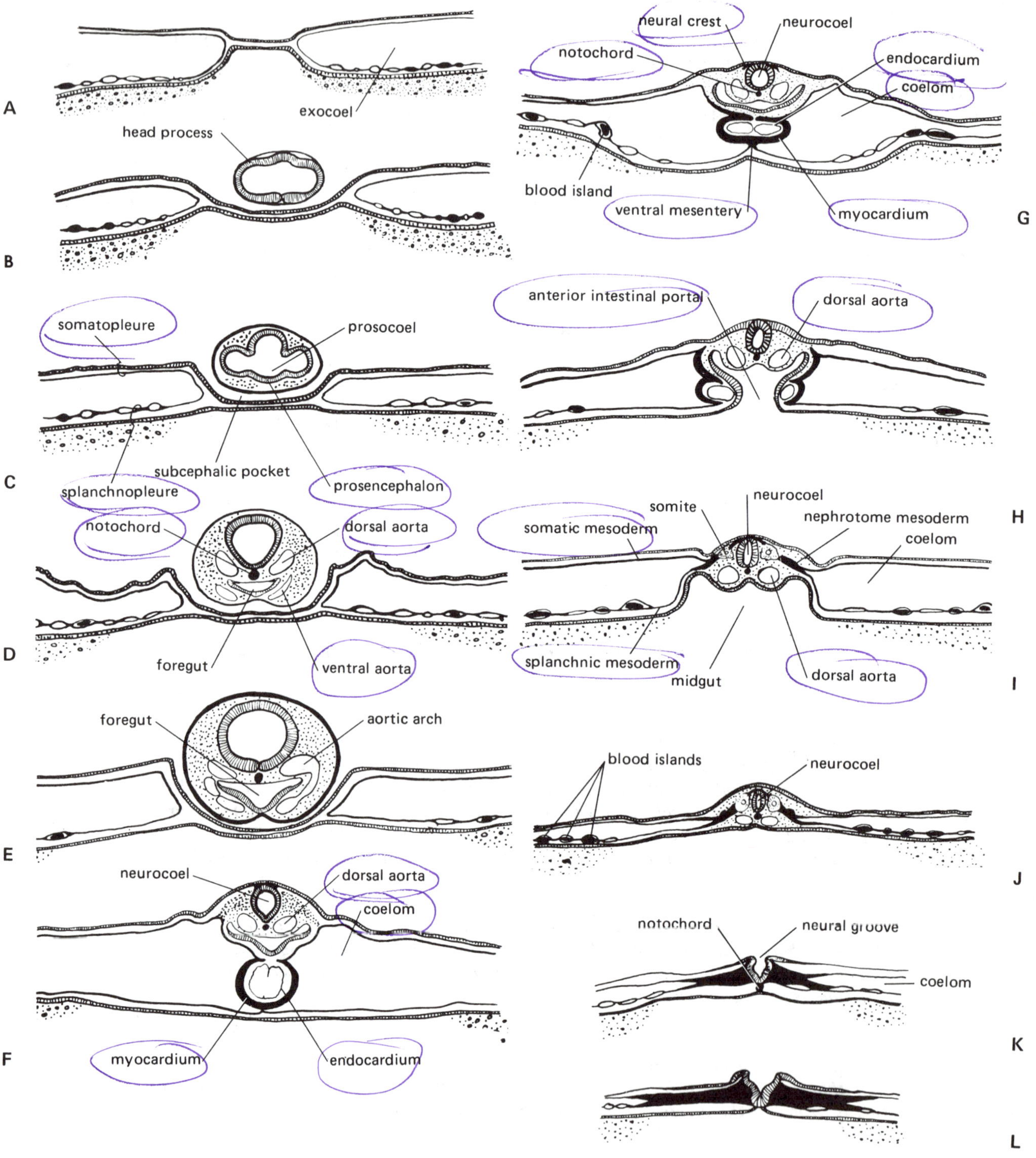

33-HOUR CHICK EMBRYO

Courtesy Dr. A. C. Scott, Colby College.

THE GERM LAYER CONCEPT

It has long been known that vertebrate embryos begin at an early stage to develop distinctly separate cell layers: one outer, one inner, and one between, known, respectively, as ectoderm, endoderm, and mesoderm. No animal or embryo is aware of these distinctions, but, through experimental embryology, we have learned that with the separation of these cell layers under normal conditions each will be destined to produce certain precise organs and organ systems. However, such destinations are not absolute, because we have learned that, even though the layers become distinct and even separable, they can be interchanged; parts can be exchanged, and yet the organism, as a whole, will develop normally with all organs in their proper places. This means that, under normal conditions of development, we can map out various regions on the three cell layers which will give rise to the eyes, the brain, the gut, the kidneys, etc., but, if we shift the cell layers at an early stage, this does not produce an abnormal embryo and organism. Thus, there must be an integrating force which protects the organism, as a whole, through these early stages, insuring normal development no matter what shifts occur in the primary cell layers. But, if the cell layers have been established for some time, then they acquire imprints that cannot be nullified, and transplantation of parts of one cell layer to another will produce an abnormal embryo. It could have extra appendages, eyes, or eyes in the wrong place, or inversion of organs, etc. Thus, the cell layers are the germinal orgins for various organ systems, under normal conditions, and are, therefore, called "germ layers." Before you accept this interpretation, let us suggest that each original cell is "totipotent," meaning it is capable of becoming any kind of cell in the body of the organism in which it is found. This complete potential is gradually lost in favor of a very specific potential, appropriate in time and/place for the formation of a normal organism, as a whole. Thus, a pigmented cell of the eye could come from cells of any of the three germ layers, ectoderm, endoderm, or mesoderm, but it normally comes from ectoderm, which is found to be in the proper place in the eye at the proper time for normal development. From totipotency to complete differentiation is a process in development to exercise one's cerebral cortex in contemplation.

It is a great convenience for us students of embryology to derive the germ layer concept, because, no matter what embryo we study, the organs and organ systems whixh derive from each of these three germ layers are exactly the same for all vertebrate embryos.

GERM LAYER DERIVATIVES OF THE 33-HOUR CHICK

Derivatives of the Ectoderm

NEURAL TUBE—elevation and fusion of paired lateral neural folds.

OPTIC VESICLES—paired evaginations ventrolateral from diencephalon.

INFUNDIBULUM—single evaginations from floor of diencephalon just anterior to tip of notochord.

AUDITORY PLACODE—thickened evaginations dorsolateral to myelenecephalon.

Derivatives of the Endoderm

FOREGUT—head and lateral folds approach each other beneath the head process to form forward pocket of endoderm.

PHARYNGEAL MEMBRANE—fusion of foregut endoderm with ventral head ectoderm to form double membrane later to break through as opening to the pharynx, the mouth.

Derivatives of the Mesoderm

NOTOCHORD—origin uncertain but probably from early migrating cells of anterior end of primitive streak, regresses posteriorly.

SOMITES—paired condensation of cells anterior to segmental plates.

PRONEPHRIC CORDS—paired solid cords of cells arising from nephrotome.

PRONEPHRIC DUCT–fusion of tips of pronephric cords that bend posteriorly, later to become tubular.

HEART–from splanchnic mesoderm, paired primordia composed of both endo- and epimyocardium.

BLOOD VESSELS–defferentiate in situ from hemangioblasts, and include aortae, first aortic arches, vitelline veins and arteries.

BLOOD VESSELS, EXTRAEMBRYONIC–from blood islands of endothelium outside of embryonic body, which fuse together to form network.

BLOOD CELLS–from blood islands in entraembryonic splanchnic mesoderm in area vasculosa (erythroblasts).

"Scientists are the most important occupational group in the world today. At this moment, what they do is of passionate concern to the whole of human society."

Sir Chas. Snow, *Science* 133:256, 1961

"Research is to see what everybody else has seen and think what nobody has thought."

Dr. Albert Szent-Györgyi

Exercise 20

THE CHICK EMBRYO:

Forty-Eight-Hour Stage

WHOLE MOUNT

Along with the rapid growth of the anterior portion (cephalization) of the chick embryo, the 48-hour stage shows both *ventral flexion* (bending) and the *dextral torsion* (twisting) of the anterior end so that about half of the total 7-mm length of the embryo is lying on its left side while the posterior half remains in its original position. The *cranial flexure* (at midbrain level) is quite pronounced, while the *cervical flexure* (near the junction of the hindbrain and spinal cord) is just indicated. The torsion involves all of the head, and this torsion accommodates the further extension of both the cranial and the cervical flexures. At the posterior end a tail fold begins to develop in exactly the same manner as did the anterior head fold.

Since the anterior portion of the embryo is lying on its left side, the bilateral nature of the organ systems is less evident, but many of the hitherto hidden structures become more distinct. The *amniotic fold* has grown posteriorly until it covers about half of the embryo, and this thin membrane, consisting of both *amnion* and *chorion,* gives a rather hazy appearance to the covered anterior structures. Locate the posterior margin of the amniotic fold.

The cranial (cephalic) flexure makes its bend around the anterior tip of the *notochord* so that the *tuberculum posterius* becomes the most anterior projection of the floor of the brain. This is used as a "landmark," and a line drawn from the tuberculum to the *dorsal isthmus* separates the *mesencephalon* from the *metencephalon.* Another imaginary line from the tuberculum to the anterior limit of the dorsal thickening separates the *prosencephalon* from the *mesencephalon,* so that the dorsal thickening represents the extent, both anteriorly and posteriorly, of the midbrain (*mesencephalon*). The *optic vesicles* have now become two-layered cups, each open ventrally by the *choriod fissure.* The inner layer of the cup is the thicker nervous *retinal layer,* and the outer layer (i.e., the more meisal) is the thinner, *pigmented* layer of the retina. The *lens vesicle* should be visible because from this lateral view one looks directly into the lens invagination of the right eye.

Between the position of the eye and the tuberculum posterius, locate the *infundibulum* as a depression in the floor of the forebrain. Close to the infundibulum, and consisting of integumentary ectoderm, anterior to the oral plate, locate the invagination, known as *Rathke's pocket*, which projects toward the infundibulum. This is the homologue of the frog's stomodeal hypophysis. The ear *vesicles* (otocysts) mark the position of the hindbrain, just posterior to which you will find the most anterior of the *somites* in lateral view. The *neural tube* still appears to be open posteriorly although the dorsal neural folds have come together.

Next locate the three *visceral* (branchial or gill) *clefts* which open from the outside into the *pharynx.* The most anterior is the *hyomandibular cleft* (or pouch), followed by the *second, third,* and finally the *fourth visceral pouches* (later to be opened as clefts). The hyomandibular (endodermal) pouch evaginations fuse with the invaginating ectoderm only at the dorsal ends, to become the spiracular clefts found prominently in the elasmobranchii. The second and third clefts form completely, but the fourth is somewhat reduced or does not open at all. Both anterior and posterior to each cleft, the mesenchyme is concentrated to become the *visceral arch,* containing important blood vessels, the aortic arches. The first arch is the *mandibular* (anterior to the hyomandibular cleft); the second is the *hyoid;* followed by the *third, fourth,* and *fifth visceral* arches, also known as the first, second, and third branchial arches, respectively. The distinction relates to those arches which will be related to gill rudiments. In some embryos it may be possible to make out a much reduced sixth visceral (fourth branchial) arch.

The *heart* by this time has elongated so much that it has become twisted upon itself. The *ventricle* later comes to lie posterior to the position of the *atrium* (auricle), both outside of the body of the embryo. The most posterior and bulbous portion of the heart, found within the body, is the *sinus venosus,* into which empties the *ductus Cuvieri* (common cardinal on each side) and the *omphalomesenteric veins* (from the vitelline circulation). Leading anteriorly from the ventricle is the *bulbus arteriosus* (really the ventral aorta), which almost immediately breaks up into paired *aortic arches.* These aortic arches pass through the substance of the (visceral or) *branchial arches* to connect with the dorsal aortae, which carry blood both to the head and to the body. Between the ventral margins of the second pair of visceral arches, you may find the median ventral evagination of endoderm, which is the single *thyroid primordium rudiment.*

Most posteriorly the *somites* may be clearly seen, along with the *intermediate mesodermal mass* (mesomere), the *neurocoel,* which terminates in an open *posterior neuropore,* and finally the dense mesodermal mass comprising the *tailbud.*

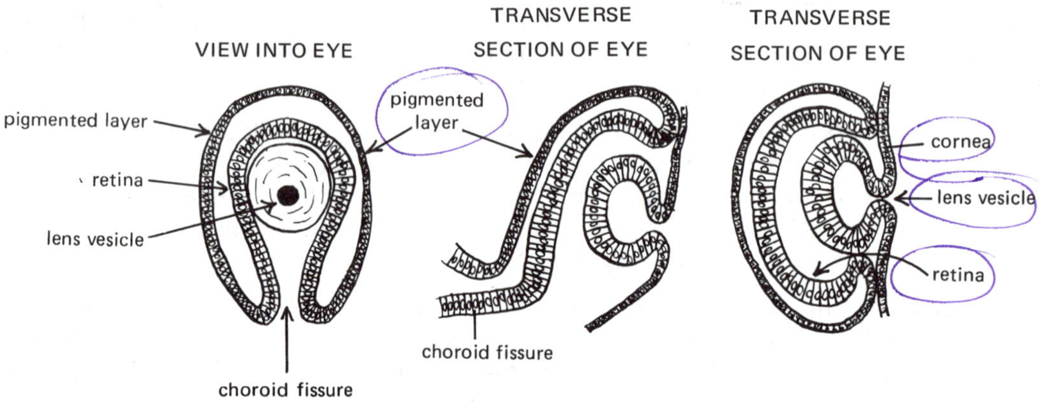

THE 48-HOUR CHICK EMBRYO: SCHEMATIC SECTIONS OF EYE

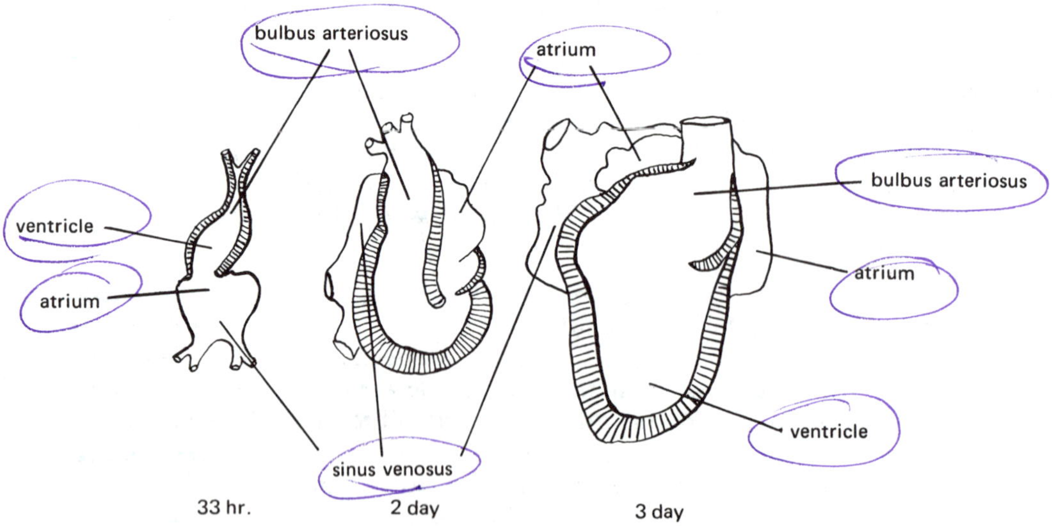

CHICK HEART DEVELOPMENT
VENTRAL VIEWS

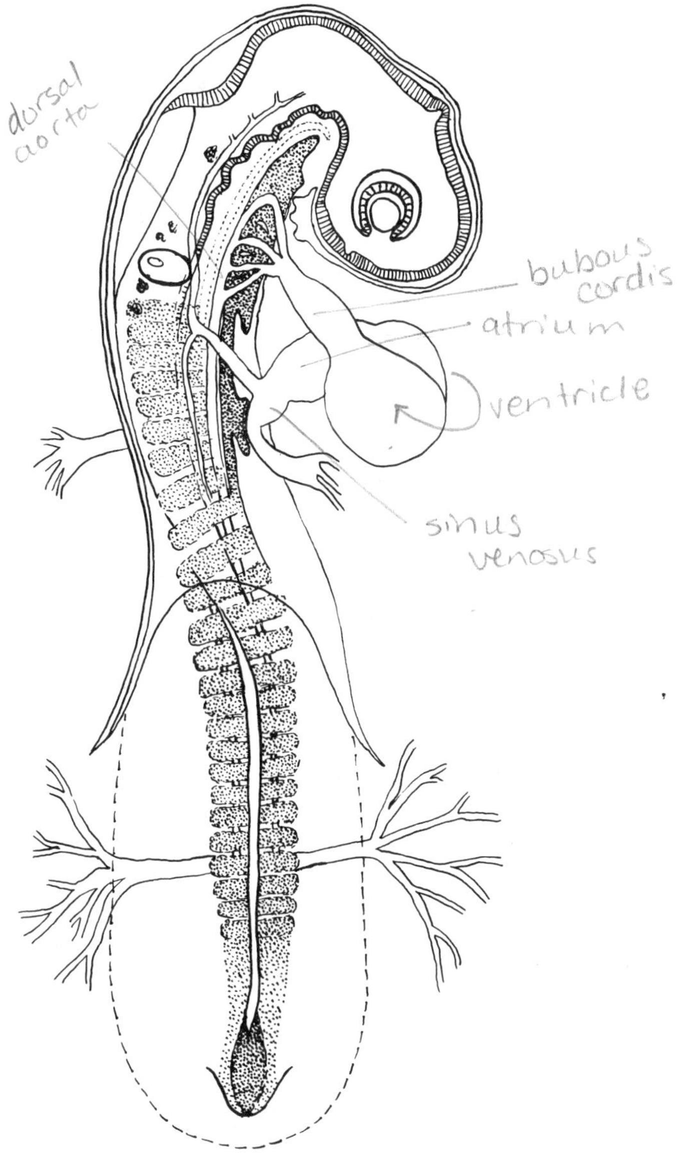

THE 43-HOUR CHICK EMBRYO

(Add all labels)

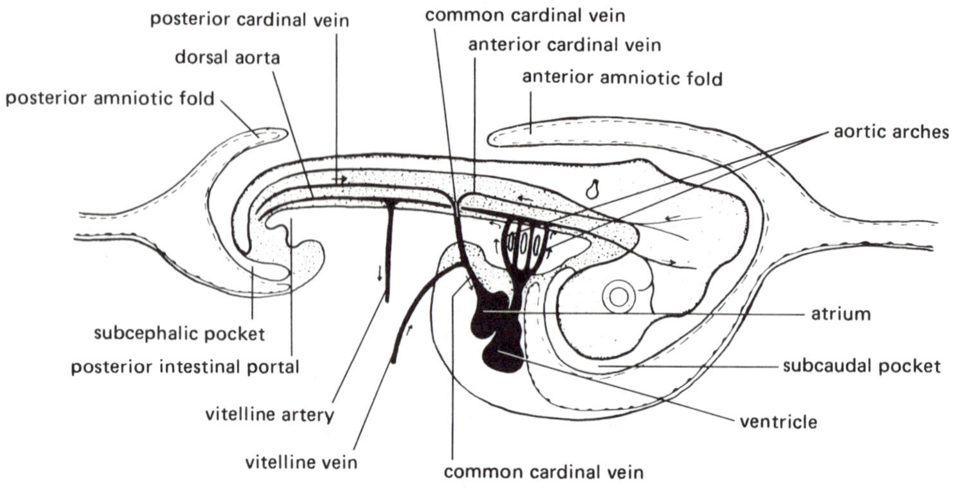

LATERAL VIEW OF VASCULAR SYSTEM OF 48-HOUR CHICK EMBRYO

TRANSVERSE SECTIONS

It is important to remember, when studying embryos beyond the 33-hour stage, that the cranial and cervical flexures plus dextral torsion of the anterior end of the embryo make it impossible to secure isolated transverse sections of the head region comparable to those of earlier studies. A transverse section at the anterior end may include, for instance, both the telencephalon and the myelencephalon. In sections just anterior to the tip of the notochord the structures are arranged exactly as one would expect in frontal sections. Many of the anterior sections will, therefore, include both the head and the body as apparently separate items, and the student must learn to distinguish between them. The second point of major importance is to learn to distinguish between the *serosa* (chorion) and the *splanchnopleure*. By this stage a good portion of the embryo is contained within the embryonic membranes, the inner double membrane being the *amnion* and the outer double one being the corollary, the *chorion* (or serosa). The third membrane, intimately associated with the yolk and full of blood vessels, is known as the *yolk sac splanchnopleure*. All of these membranes are double, one layer always being mesoderm. In making drawings, the student should always place the yolk sac splanchnopleure toward the observer. In the anterior transverse sections it will be possible to identify the *lateral folds* of the amnion and the space between these folds and the embryo, known as the *lateral limiting sulcus*. (See Plate 4, pages 366-71.)

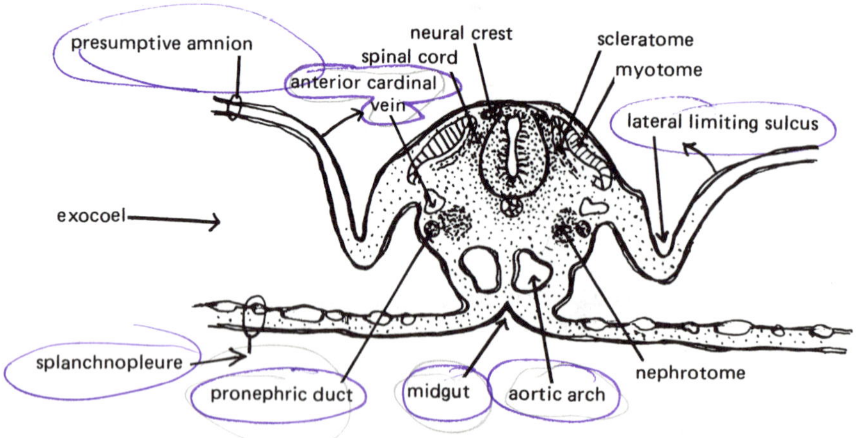

THE 50-HOUR CHICK EMBRYO
TRANSVERSE SECTION

GERM LAYER DERIVATIVES OF THE 48-HOUR CHICK

Derivatives of the Ectoderm

OPTIC CUPS—invaginations of paired optic vesicles as they touch outer head ectoderm, forming outer pigmented and inner sensory layers of the eyes.

OPTIC (CHOROID) FISSURES—invagination of the floor of optic stalks.

NASAL PLACODES—paired thickenings of head ectoderm adjacent to telencephalon.

CEREBRAL HEMISPHERES—paired lateral evaginations of telencephalon.

ACOUSTIC GANGLIA—paired cell aggregates from otic vesicles to brain.

SEMILUNAR AND FACIAL GANGLIA—paired cell aggregates from neural crests and epibranchial placodes.

GLOSSOPHARYNGEAL GANGLIA—paired aggregates of neural crest cells (9th cranial).

RATHKE'S POCKET—single evagination from roof of stomodeum in association with infundibulum.

Derivatives of Endoderm

AMNION (ECTO- AND SOMATIC MESODERM)—inner double membrane formed from anterior and later folds which under cut the embryo.

CHORION (ECTO- AND SOMATIC MESODERM—outer part of folds which formed the amnion.

FIRST THREE (PAIRED) PHARYNGEAL POUCHES—paired lateral evaginations from foregut (3).

THYROID GLAND—single ventral evagination from foregut anterior to ant. int. portal.

LIVER PRIMORDIA—ventral evaginations from foregut posterior to thyroid extension of gut primordia into tail.

HINDGUT—entention of gut primordia into tail.

Derivatives of the Mesoderm

MESONEPHRIS TUBULES—hollowed out chords of mesoderm (paired) which join the mesonephric (pronephric) ducts.

MESONEPHRIC DUCTS (WOLFFIAN)—remnants of old paired pronephric ducts.

HEART—parts of developing heart, single (sinus venosus, atrium, ventricle, and bulbus cordis).

AORTIC SAC—fused portion of ventral aortae.

AORTIC ARCHES—differentiation in situ within branchial arches (II and III).

DORSAL AORTA—fusion of originally paired aortae (descending aorta).

INTERNAL CAROTID ARTERIES—anterior extensions of paired dorsal aortae into head.

VEINS—differentiation in situ within embryonic body (common cardinals, precardinals, postcardinals).

SOMITES—differentiating into outer dermatome, median myotome, and inner sclerotome).

The most anterior sections which contain an enlarged brain cavity are generally cut through the *diencephalon* and the *mesencephalon,* showing the spacial as well as tissue connection between the two. The diencephalon may be identified by the evaginating *optic vesicles.* The mesencephalon, which has a thick roof (*dorsal thickening*), may be distinguished from the *rhombencephalon,* which has a thin roof, later to be designated as the posterior choroid plexus when invested with numerous blood capillaries. Several sections posteriorly these two brain cavities will be separated by the intervening *tuberculum posterius* and still more posteriorly by the anterior tip of the *notochord.* In the vicinity of the diencephalon locate the anterior margins of the *optic vesicles.* The head mesoderm is still in the form of loose *mesenchyme.* These anterior sections illustrate well the fact that the ectodermal wall of the brain shows extreme variations in thickness and that there are already indications of convolutions, constrictions, and bulges.

At the level of the eye sections, locate the paired blood vessels close to and on either side of the brain vesicle. These are the *anterior cardinal veins* which, if followed posteriorly., will be seen to empty into the *sinus venosus* by way of the *ductus Cuvieri.* Small branches of these veins may be seen scattered about the brain.

As the *preoral gut* or *pit* (i.e., Seessel's pocket, the anterior extension of the *foregut*) enlarges, you will find *visceral clefts* cutting in lateral to the pharynx and separating the *visceral* and (contained) aortic arches. The large paired blood vessels which now appear between the pharynx and the brain are the *dorsal aortae.* The aortic arches, within the (visceral or) branchial arches, connect the *ventral aortic roots* with the dorsal aortae. Between the preoral pit and the *forebrain* find the enlarged and ectoderm-lined *Rathke's pocket,* generally triangular in shape and representing the homologue

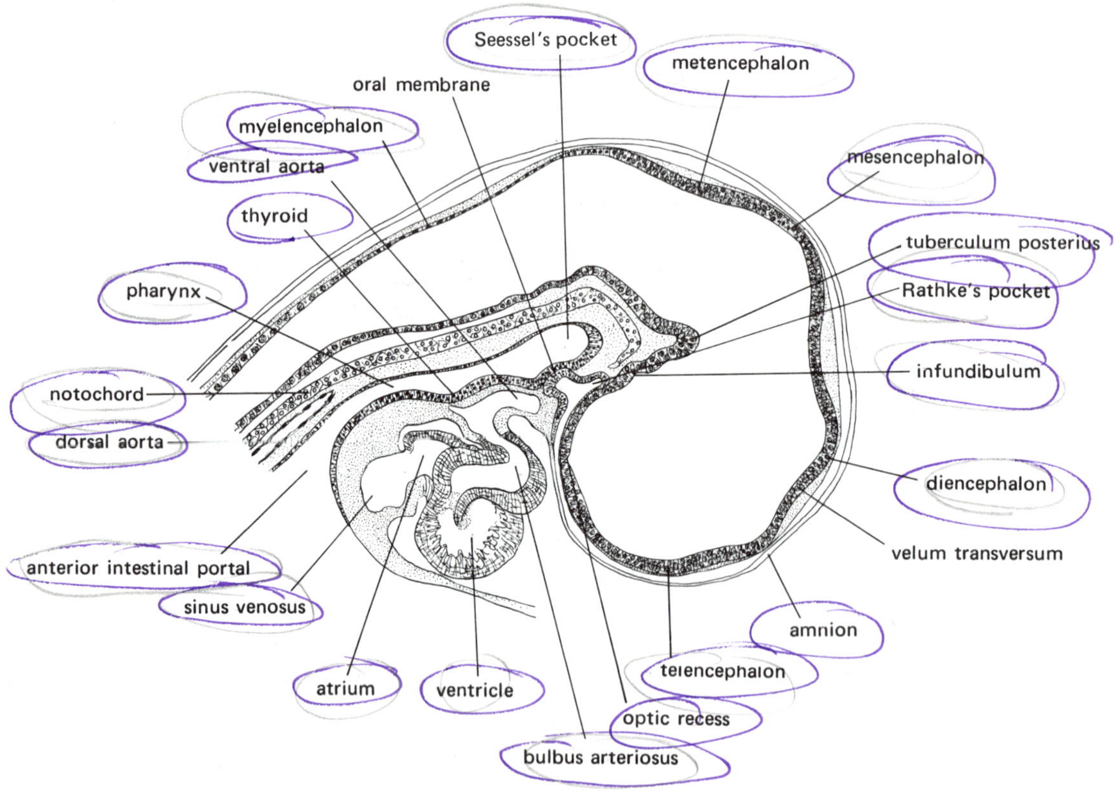

CHICK EMBRYO OF 22 SOMITES OR 50-53 HOURS SAGITTAL SECTION

From R. Rugh. *Vertebrate Embryology.* 1964. Harcourt, Brace and World.

of the frog's stomodeal hypophysis. Rathke's pocket is external to the embryo, and its cavity is continuous with the *amniotic cavity.* Note the close association of Rathke's pocket with the *infundibulum,* this latter cavity already partially constricted off from the rest of the *diocoel.* Locate the diocoel, *mesocoel, auditory vesicles* (possibly with opening from the outside), and the *optic cups* with the *lens vesicles.* The head *mesenchyme* (at these levels) is not organized into somites. The position of the *notochord* may be used to distinguish between *prosencephalon* and *metencephalon.* Locate also the *stomodeal invagination* and the *oral plate.*

Additional visceral arches and clefts are found posteriorly until the level where the head is separated entirely from the body. The student should have before him the whole mount drawing of this stage to understand completely the topography of the sections at this level, as well as the relationship of the various membranes to the embryo.

Aortic Arch	Visceral Arch	Time of Appearance	Disposition
I	I (mandibular)	10 somites (30 hours)	disappears at 3 days
II	II (hyoid)	19 somites (40 hours)	disappears at 4 days
III	III	26 somites	common carotid arteries
IV	IV	36 somites	systemic arch disappears
V	V	96 hours	disappears
VI	VI	120 hours	pulmonary arteries

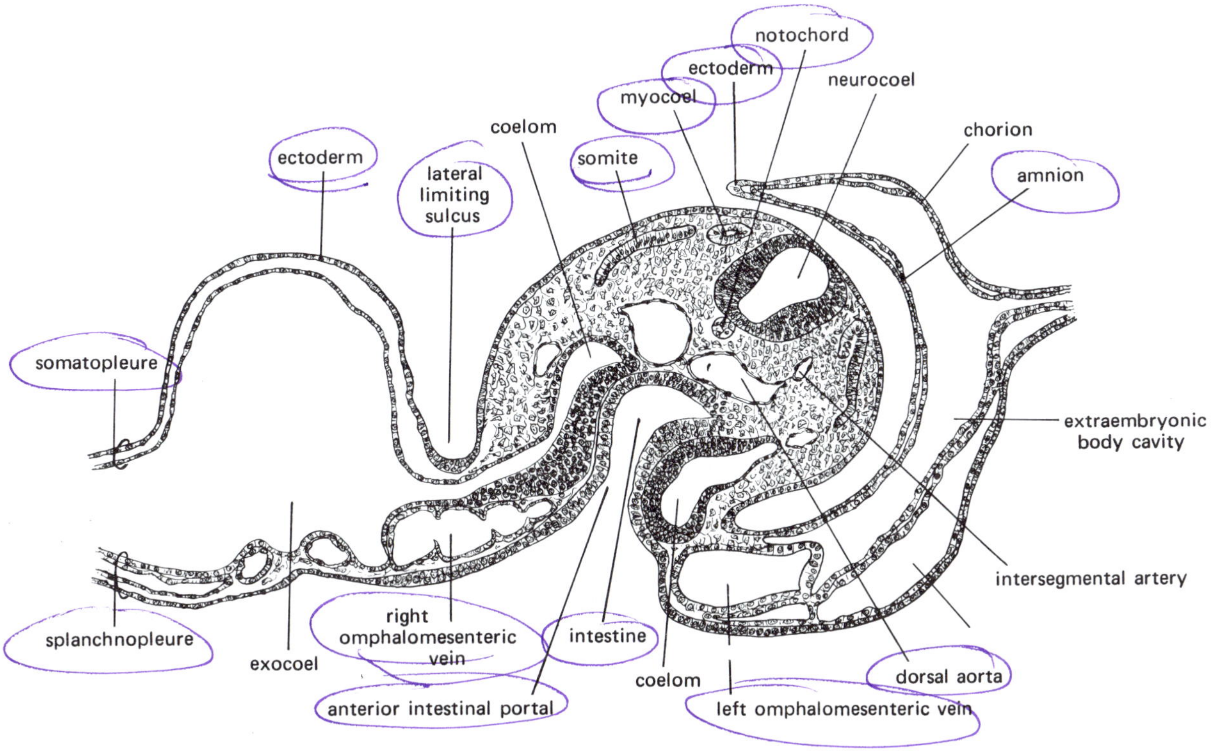

CHICK EMBRYO IN 23-SOMITE STAGE

TRANSVERSE SECTION THROUGH LEVEL OF 5TH SOMITE AND ANTERIOR INTESTINAL PORTAL

From Rugh. *Vertebrate Embryology.* 1964. Harcourt, Brace & World, New York.

Still further posteriorly locate the *ventral bulbus arteriosus,* which is connected with a pair of *aortic arches.* The floor of the pharynx in this vicinity will show the thick, single evagination of the *thyroid primordium,* the tubular connection of this future endocrine gland with the foregut being called the thyroglossal duct. The portion of the heart first seen is the thick-walled ventricle with its *endothelial lining,* continuous with the *bulbus arteriosus.* Sections at certain angles will show the tubular connection of the ventricle, through the bulbus, into the aortic arches, and possibly even to the dorsal aortae. Locate the rather large *anterior cardinals* (future jugulars) dorsolateral to the dorsal aortae at these levels.

As the visceral clefts disappear and the thyroid primordium is no longer visible, the *foregut* is seen to have a thin roof and a thick floor. A median ventral groove appears in this floor, extending for quite a few sections, and leading eventually to the origin of the lung buds. This is known as the *laryngotracheal groove.* At these levels the heart appears to have no connection with the body. It will be possible to identify *endocardium, myocardium,* and even blood corpuscles. The more posterior sections of the heart, where the walls become thin, represent portions of the *atrium.* Locate the groove between the body and the amnion known as the *lateral limiting sulcus;* the *somites,* which will later be incorporated into the occipital region of the head; and the single, large *dorsal aorta.*

Now examine sections posteriorly until the apparently isolated heart again establishes connections with a region of the embryo ventral to the gut. While viewing these sections, observe also the

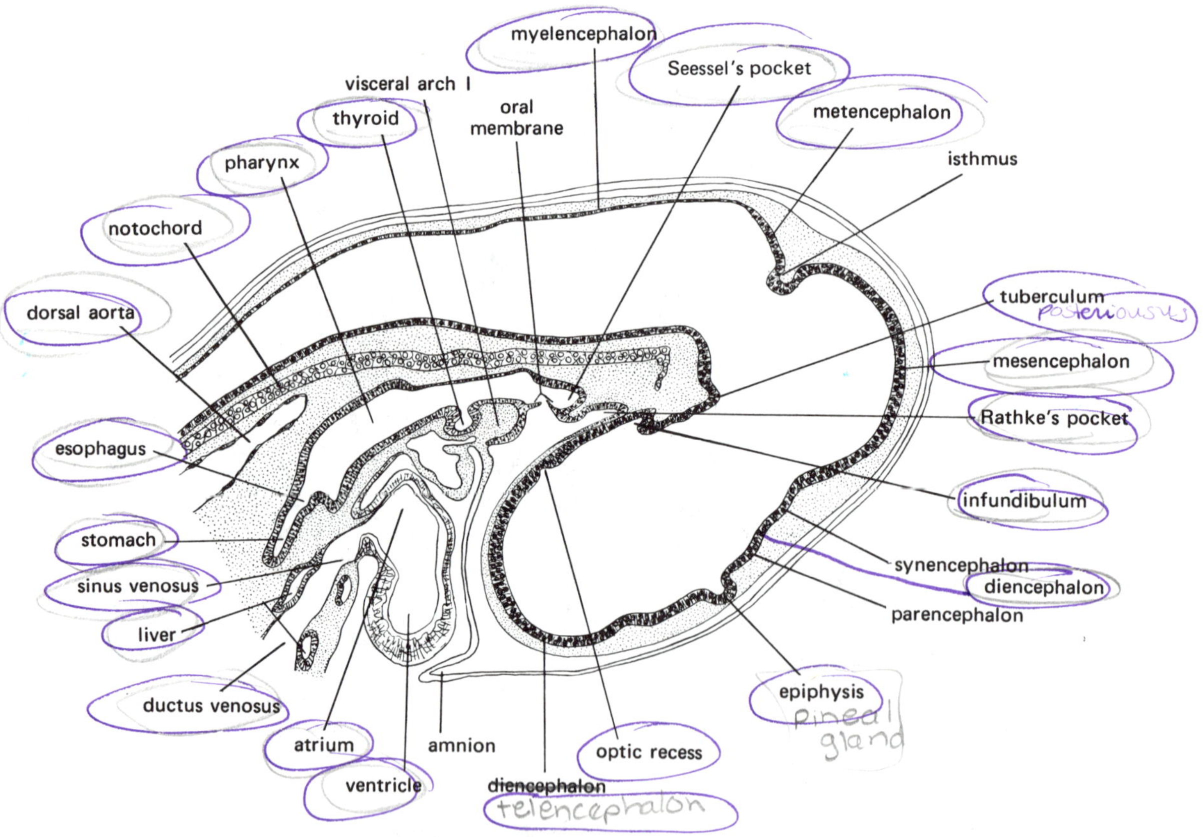

CHICK EMBRYO OF 39 SOMITES

MEDIAN SAGITTAL SECTION

From Rugh. *Vertebrate Embryology.* 1964. Harcourt, Brace & World, New York *anterior cardinal veins,* which before were lying lateral to the dorsal aortae, and note that they gradually approach and finally enter the heart, ventral to the gut. The vertically directed veins are really the *common cardinals,* which empty into the *sinus venosus* by way of the *ductus Cuvieri.** In order to reach the heart these vessels must cut off portions of the *coelom* from the *exocoel,* establishing for the first time a derivative of the coelom, the *pleural cavities* into which the lung buds will subsequently grow. The posterior limit of these pleural cavities, formed by the junction of the tissue around the heart and the ductus, is known as the *septum transversum,* a transverse partition. The *sinus venosus* itself is best identified as that portion of the heart into which the ductus empties, from either side of the embryo. Posteriorly it can be readily established that the two large *omphalomesenteric veins* bring nutrient (vitelline) blood directly to the heart by way of the sinus venosus.

Just anterior to the fused pair of omphalomesenteric veins locate a median ventral diverticulum of the foregut known as the *liver primordium.* If this is followed posteriorly, in its place will appear the *anterior intestinal portal,* or the region of the floorless midgut. Lateral to the anterior intestinal portal find the very large blood vessels in the *splanchnopleure,* here called the *vitelline veins,* continuous with the omphalomesenteric veins within the body. It is well for the student to reidentify the various cavities that appear, such as the extraembryonic body cavity, or *exocoel;* the intraembryonic body cavity, or *coelom;* the pleural cavities derived from the *coelom;* the *amniotic cavity;* and the *archenteron,* or gut.

In the future midgut region the somite material has differentiated into *dermatome, myotome, myocoel,* and *sclerotome.* The *posterior cardinals,* near the nephrogenic or *mesomeric material,* the *Wolffian ducts,* and the *neural crests* are also distinct. Note that the embryo at these levels is no longer lying on its side and that the bilaterality of structures is again quite obvious. The line of dorsal fusion of the amniotic folds, called the *seroamniotic junction* (or amniotic raphe), is found at this level. It will be pissible to follow the *dorsal aorta* posteriorly until it bicomes divided, and finally the main branches of the paired aortae leave the body to pass out along the splanchnic mesoderm to the yolk as the *vitelline arteries.* The embryo is separated from the underlying yolk only by the rather thin roof in the midgut, and yet none of the yolk material passes as food into the formed gut. Observe changes in size and cellular structure of the developing central nervous system, of the somites, and of the general body form in progressively posterior sections.

OH REALLY NOW

More posteriorly, at the levels where the notochordal cells become merged with the floor of the neural tube, identify the remaining portions of the originally prominent *primitive streak.* This will be the level of the *sinus rhomboidalis* as seen on the whole mount. Locate *Hensen's node, primitive folds,* and ventral *primitive gut.* Note the mesodermal derivatives and the fact that the splanchnic mesoderm is very vascular, particularly at the inner margin of the *area vasculosa.*

REFERENCES

Arey: 577-91
Balinsky: 321-66
Bodemer: 217-24, 311-29
Hamilton revision of Lillie: Chapters V, VI
Huettner: 239-71
Lillie: Chapter VI
McEwen: 332-69
Nelsen: 516-54
Patten: 108-46
Patten (1957): 169-89
Rugh: *Vert. Emb.* 126-53
Shumway: Chapters VII-IX, pp. 291-94.

"Science does not explain anything. Science is less pretentious. All that falls within its mission is to observe phenomena and to describe them and the relations between them."

A. J. Lotka, 1925

"The elements of our own behavior are found in all organisms."

E. G. Conklin, 1944

*The junction of the anterior and posterior cardinals is considered to be the common cardinal and the passage of this into the sinus venosus as the ductus Cuvieri, although many authors consider these as synonyms.

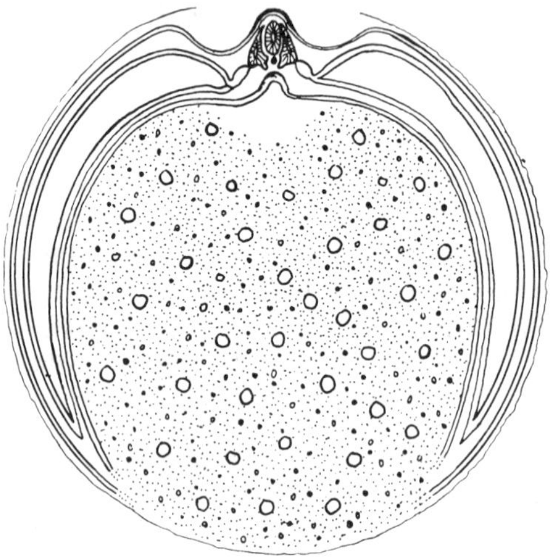

THE 3-DAY CHICK EMBRYO

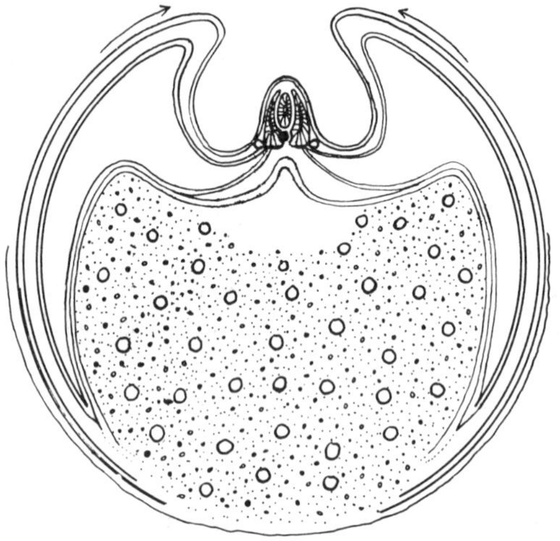

THE 4-DAY CHICK EMBRYO

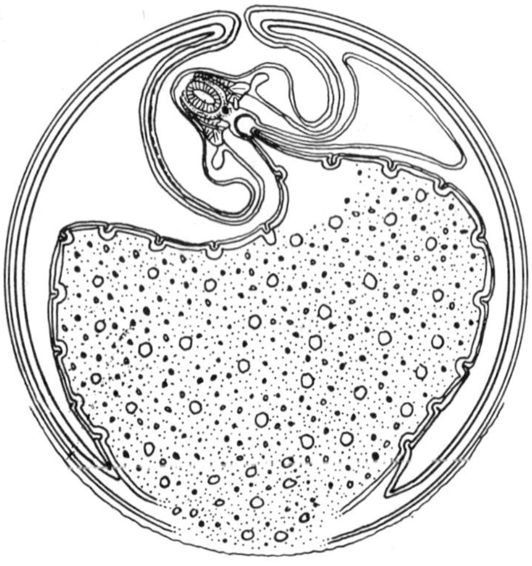

THE 5-DAY CHICK EMBRYO

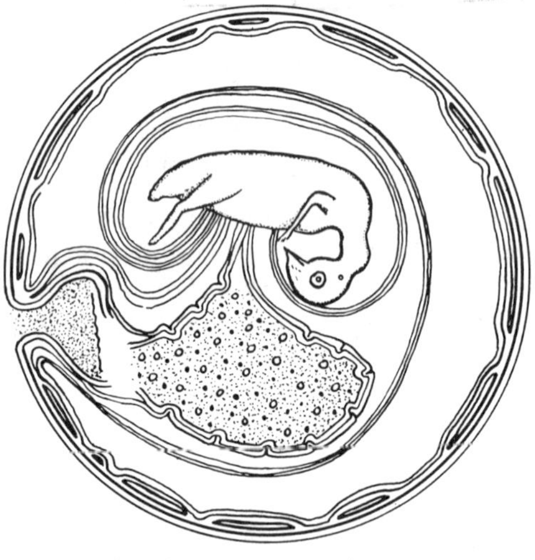

THE 9-DAY CHICK EMBRYO

THE DEVELOPMENT OF CHICK EMBRYONIC MEMBRANES

(IDENTIFY AND LABEL ALL MEMBRANES)

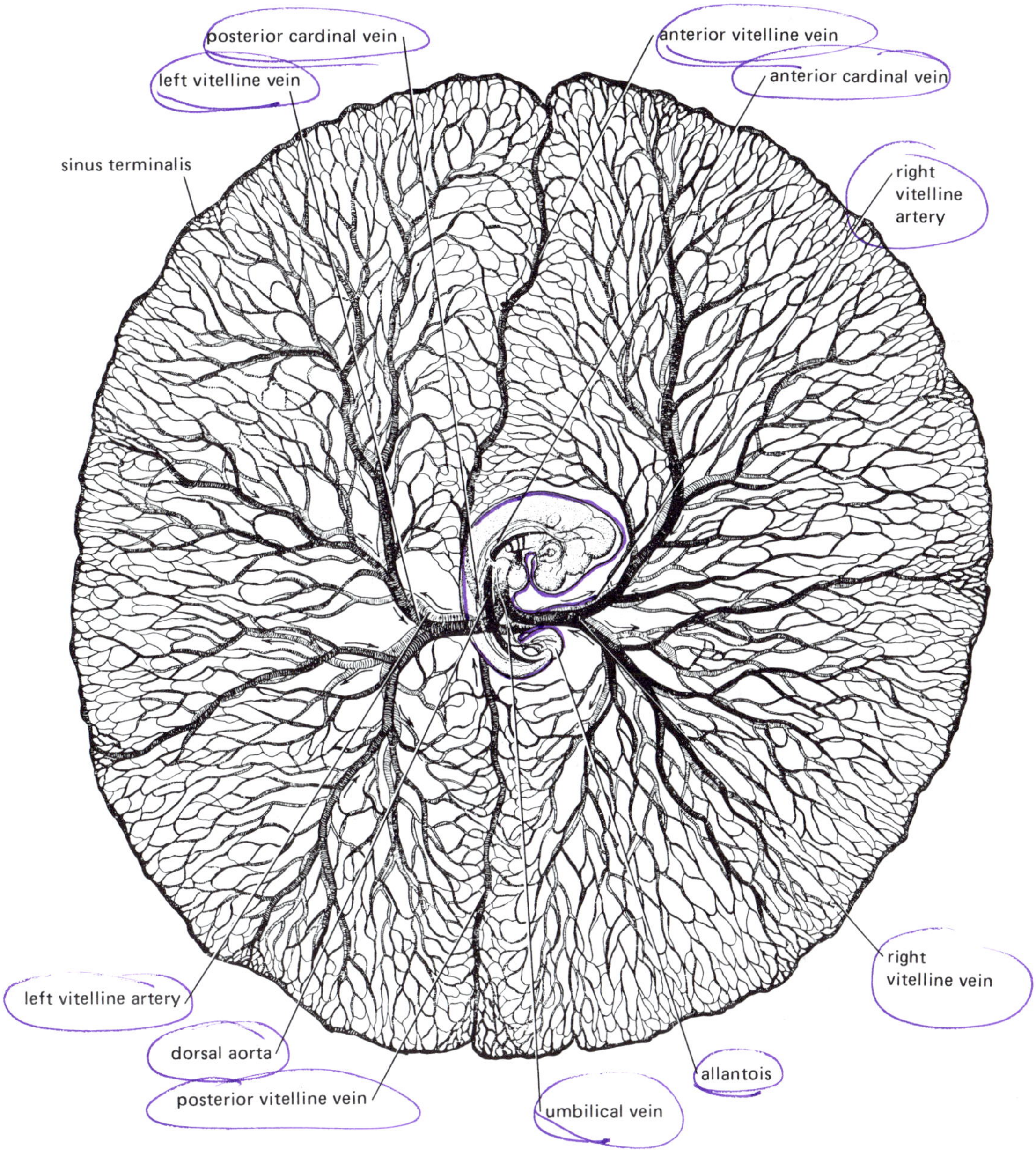

DORSAL VIEW OF VASCULAR SYSTEM OF 96-HOUR CHICK EMBRYO

From A. F. Huettner. *Fundamentals of Comparative Embryology of the Vertebrates.* Rev. ed. 1949. Macmillan, New York.

SUMMARY OF DEVELOPMENT OF CHICK AT 48 HOURS

Membranes

1. AMNION – ectoderm and somatic mesoderm separated from body by amniotic cavity.
2. CHORION (serosa) – ectoderm and somatic mesoderm enclosing all embryonic parts. The space between this membrane and the amnion is known as the seroamniotic cavity or exocoel. Very little blood evident in the chorion.
3. YOLK-SAC SPLANCHNOPLEURE – endoderm and splanchnic mesoderm, closely associated with the yolk and abundantly supplied with vitelline blood vessels.

Ectodermal Derivatives

1. EPIDERMIS – the entire covering of the embryo, single layer thick.
2. BRAIN AND SPINAL CORD –
   - TELENCEPHALON – most anterior brain with lateral vesicles; later associated with nasal placodes.
   - DIENCEPHALON – portion of the brain and vesicles associated with the optic stalks and epiphysis.
   - MESENCEPHALON – brain posterior to diencephalon and anterior to isthmus; thick walls, and thick roof (i.e., dorsal thickening). Cavity is the mesocoel.
   - METENCEPHALON – posterior to the isthmus, roof becomes the cerebellum.
   - MYELENCEPHALON – brain posterior to metencephalon, very thin roof; level of the paired auditory vesicles.
   - SPINAL CORD – thick lateral walls and narrow tubular neurocoel, extending full length of the embryo to region of disappearing Hensen's node.
3. SPECIAL SENSE ORGANS –
   - OLFACTORY (nasal) PLACODE – paired and ventrolateral to the telencephalon.
   - OPTIC CUP and LENS – lateral to diencephalon, associated with it by optic stalk.
     - Optic cup – double-walled cup formed by invagination of the optic vesicle of the lateral diencephalic wall previously evaginated; joined to brain by optic stalk.
     - Lens – thick-walled invagination of outer ectoderm following contact with wall of optic vesicle; comes to lie within the optic cup, paired.
   - OTIC (auditory) VESICLE – lateral to myelencephalon as thick-walled invagination.
4. CRANIAL NERVES –
   - SEMILUNAR GANGLION of trigeminal nerve (V) – most anterior of large cranial ganglia.
   - ACOUSTIC GANGLION (VIII) of auditory nerves, ventroanterior to otic vesicles.
   - SUPERIOR GANGLION OF GLOSSOPHARYNGEAL (IX) – small mass of nerve cells posterior to each otic vesicle.
5. STOMODEUM – ectodermal invagination which meets the endodermal evagination from the foregut, together to form the oral plate.
6. RATHKE'S POCKET – ectoderm, anterior to the stomodeum, which invaginates toward the infundibulum. This is the stomodeal hypophysis of the future pituitary gland.

Endodermal Derivatives
(linings of the various organs described)

1. PREORAL GUT or Seessel's pocket – forward projection of foregut, anterior to oral plate.
2. VISCERAL CLEFTS – inner portion lined with endoderm; three pairs open by 48 hours.

3. THYROID GLAND – median ventral evagination between the second pair of visceral arches, communicating with the pharynx by thyroglossal duct. Well-formed columnar cells.

4. PHARYNX – chamber into which visceral clefts and thyroid open.

5. LIVER – two narrow median ventral diverticula just anterior to the anterior intestinal portal, flanked on either side by the large omphalomesenteric veins. Liver cells will invade the walls of these veins to form a frilled organ, the liver primordium.

6. ANTERIOR INTESTINAL PORTAL – rather sudden loss of the floor of the foregut just posterior to the liver diverticulum. Gut opened out over the yolk.

7. MIDGUT – roofed archenteron, no floor; from anterior to posterior intestinal portals.

8. HINDGUT – closed portion of the gut, posterior to the posterior intestinal portal, which extends into the tail process.

9. PRIMITIVE GUT – ventral to tail process, remnant of primitive streak area.

Mesodermal Derivatives

1. MESENCHYME – loose stellate cells of embryonic tissue which will give rise to the bulk of the connective tissue of the adult, most abundant as mesenchyme in the head.

2. NOTOCHORD – this may be derived from the mesoderm. At least it is formed at the same time and place as the early mesoderm. It is the embryonic precursor of the animal axis, found ventral to the neural tube. There is no subnotochordal rod as is found in the amphibian embryo.

3. EPIMERE – the most dorsal condensed mesoderm, lying lateral to both sides of the notochord and the neural tube.
   a. DERMATOME – most superficial cellular mass of epimere, giving rise to the integumentary mesoderm.
   b. MYOTOME – from inner cell mass of somite, giving rise to striated muscle.
   c. SCLERETOME – loose mesoderm from epimere which migrates to a position around both nerve cord and notochord, later to give rise to the axial skeleton. (The epimeric mesoderm comes from the primitive streak, but as it reaches the embryo proper, it is segmentally divided into blocks of mesoderm known as somites, each somite becoming differentiated into the three parts enumerated above.)

4. MESOMERE – paired nephrogenic mesoderm, continuous from anterior to posterior and found ventrolateral to the somites. The cavity within is the nephrocoel which may become the tubules connecting with the pronephric or Wolffian duct, depending upon the level of the section and the stage of development.

   The pronephric tubules are best seen at about the level of the fifteenth somite, but they extend from the fifth to the sixteenth somite. Some of these tubules unite to form the pronephric duct which grows posteriorly. The tubules appear first at 36 hours and are thick-walled contorted vesicles which open into the nearby coelom by ciliated nephrostomes. The duct is made up of similar cells throughout and is circular in cross section. There is neither glomus nor glomerulus in association with the pronephros. The mesonephric tubules are distinct, extending from the twentieth to the thirtieth somite, and are more numerous than the pronephric tubules (three or more at the level of each somite). In the 48-hour chick these mesonephric tubules are just beginning to appear, emptying into the posterior prolongation of the pronephric or Wolffian duct. The mesonephros functions from 5 to 11 days of incubation.

5. HYPOMERE – paired lateral plate mesoderm, intraembryonic, but continuous with the lining of the exocoel. This mesoderm is never segmented, as are the somites, but remains as sheets of

mesoderm with an outer (somatic) and an inner (splanchnic) layer and a cavity (coelom) between.

6. COELOM – cavity between layers of lateral plate mesoderm, not yet separated into intra-embryonic and extraembryonic body cavities (exocoels) by the undercutting of the lateral body wall. Continuous, at this stage, with the developing pericardial cavity and the presumptive pleural cavity.

7. CIRCULATORY SYSTEM – the heart begins to beat and circulate fluid at 42 hours.

THE HEART – a bent tubular organ which is lined with endothelium (endocardium) and consists principally of the muscular myocardium derived from splanchnic mesoderm.

MEATUS VENOSUS – fused vitelline (omphalomesenteric) veins entering the sinus.

SINUS VENOSUS – thin-walled chamber receiving meatus venosus and ductus Cuvieri (common cardinals) and joined to the atrium.

ATRIUM – thin-walled, free, contorted portion of heart (posterior at this stage).

VENTRICLE – thick-walled, most lateral portion of the heart. Actually, this portion of the heart is outside of the body at this stage.

BULBUS ARTERIOSUS – circular in cross section, posterior part just outside of the body. Anterior part (ventral aorta) bifurcates to become the several aortic arches (paired).

THE ARTERIES (All paired):

INTERNAL CAROTID – dorsal portion of third aortic arch and anterior remnant of the dorsal aortae growing into the head. External carotids are the anterior ventral aortae.

FIRST AORTIC ARCH – joins dorsal and ventral aortae through the first visceral or mandibular arch, later to disappear.

SECOND AORTIC ARCH – blood vessels which run through the second visceral arch in close proximity to the thyroid promordium.

THIRD AORTIC ARCH – blood vessels in the third pair of visceral arches.

DORSAL AORTAE – double at the level of the aortic arches; single, posterior to the level of the heart. Situated dorsal to the gut and beneath the notochord. Receives all aortic arches and splits again in the posterior part of the body to give off the large vitelline arteries to the yolk and to form the caudal arteries.

VITELLINE (OMPHALOMESENTERIC) ARTERIES – paired arteries arising after the bifurcation of the dorsal aorta, and passing out onto the yolk by way of the splanchnopleure. Leave body at about level of seventeenth somite.

SEGMENTAL ARTERIES – lateral branches from the dorsal aortae which pass between the somites.

THE VEINS:

CARDINAL VEINS – the cardinal system forms the letter "H," the cross bar being the common cardinals which empty into the ductus Cuvieri, while the upper pair of prolongations are the anterior and the lower pair the posterior cardinals. The anterior cardinals drain from the head, lying ventrolateral to the brain vesicles, and carry the venous blood to the heart along a route lateral to the dorsal aortae. The posterior cardinals drain blood from the epimeric and the nephrogenic tissue along the full length of the embryo, while the common cardinals receive blood from both the anterior and the posterior branches and empty it into the sinus venosus. The common cardinals (ductus Cuvieri) join the sinus venosus.

VITELLINE (OMPHALOMESENTERIC) VEINS – these large veins bring nutrient blood from the yolk to the sinus venosus. Out on the yolk (splanchnopleure) the contributing vessels are known as the anterior, posterior, and lateral vitelline veins, all associated with the circumferential sinus terminalis.

LABELS FOR CHICK SECTIONS

AA – aortic arch
ACV – anterior cardinal vein
AIP – anterior intestinal portal
ALV – allantoic vein
AMN – amniocardiac vesicle
ATR – atrium

BA – bulbus arteriosus
BLV – blood vessel

CC – common cardinal vein
CH – chorion
CL – cloaca

DA – dorsal aorta
DC – ductus Cuvieri
DERM – dermatome
DIEN – diencephalon
DM – dorsal mesentery

EEBC – extraembryonic body cavity (also exocoel)
END – endocardium

HYAR – hyoid arch
HYCL – hyomandibular cleft

INF – infundibulum

JUG – jugular vein (ant. card.)

LE – lens
LGB – lung bud
LLS – lateral limiting sulcus
LTG – laryngotracheal groove
LV – lens vesicle

MAND – mandibular arch
MES – mesencephalon
MET – metencephalon
MYEL – myelencephalon
MYOC – myocardium
MYOT – myotome

NC – neurocoel (neural canal)
NEM – neuromere
NOT – notochord

OLF – olfactory placode
OPS – optic stalk
OT – otocyst (ear vesicle)
OV – omphalomesenteric vein

PCH – pharyngeal pouch
PCV – posterior cardinal vein
PH – pharynx
PLC – pleural cavity

RAP – Rathke's pocket

S1 – first somite
S2 – second somite
S3 – third somite
SCLER – sclerotome
SEPT – septum transversum
SOMA – somatopleure
SPLAN – splanchnopleure
SUBC – subcardinal vein

TA – truncus arteriosus
TEL – telencephalon
TH – thyroid primordium
UR – ureter

VENT – ventricle
VIA – vitelline artery
VIC – visceral (or branchial) cleft
VIP – visceral (or branchial) pouch
VITA – vitelline artery
VM – ventral mesentery
VV – vitelline vein

WD – Wolffian duct (mesonephric duct)

III – oculomotor nerve
V – trigeminal nerve or ganglion
VII-VIII – acoustico-facialis ganglion
IX – glossopharyngeal nerve

7 to 10 – neuromeres

"We have seen that all organisms are composed of essentially like parts, namely cells: that these cells are formed and grow in accordance with essentially the same laws; hence that these processes must everywhere result from the operation of the same laws."

Schwann

ANTERIOR

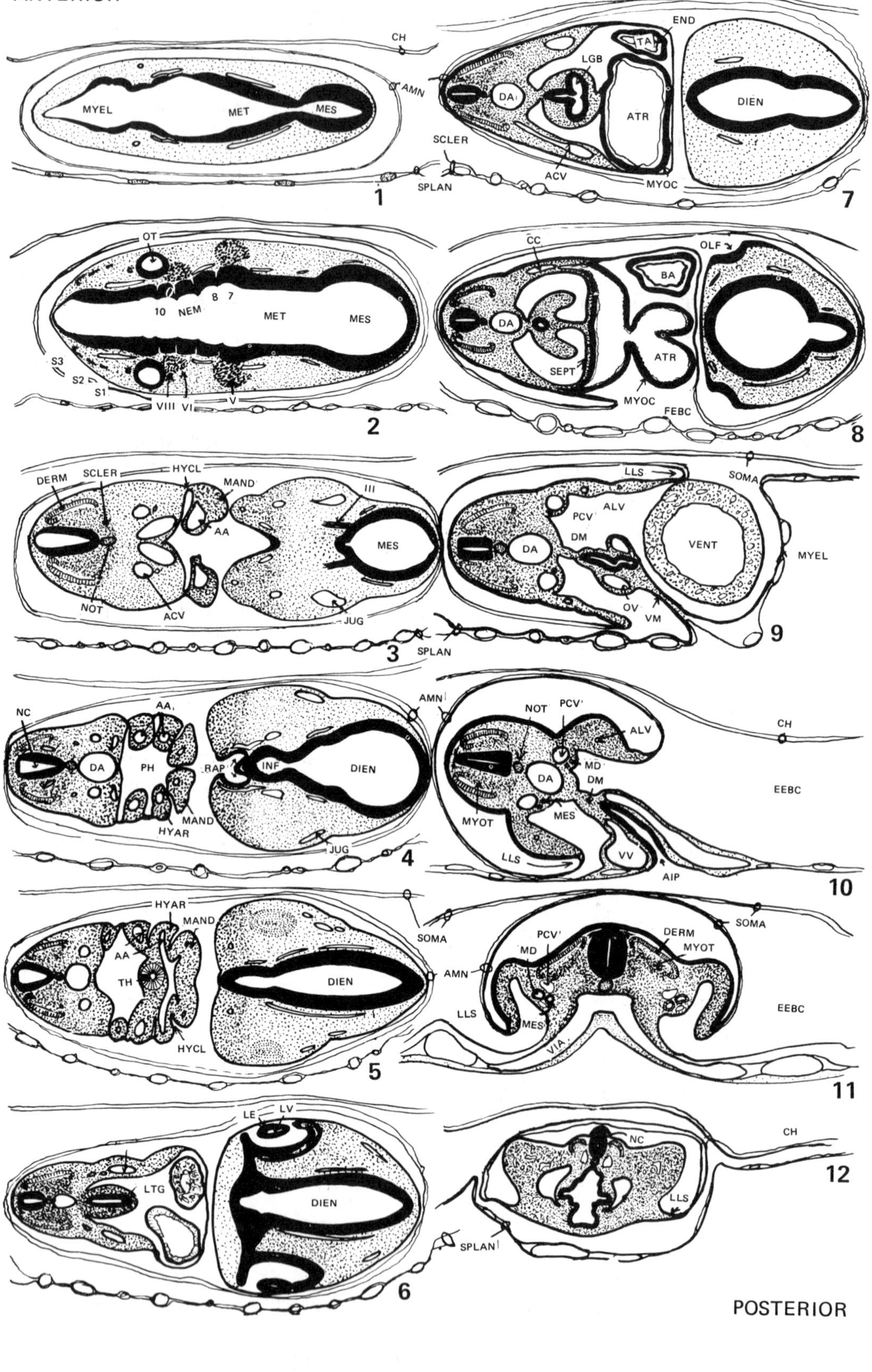

POSTERIOR

60-HOUR STAGE

DRAWINGS: CHICK EMBRYO: 48 HOURS

Exercise 21

THE CHICK EMBRYO:

Seventy-Two-Hour Stage

WHOLE MOUNT

By this stage the head fold of the amnion has been carried posteriorly to a point about the level of the omphalomesenteric (vitelline) arteries, while the tail fold, consisting of the same layers, has progressed anteriorly so that only an oval opening in the *amniotic membrane* can be discerned. This gradually closing oval opening will be found above the level of the hindgut, and the final closure occurs above the thirty-first somite. The process has the appearance of a drawstring closing the membranes. There are *lateral folds* continuous with the *head* and *tail folds*, so that the margin of the amniotic membrane is an unbroken line.

The *cranial* flexure has bent the head so that the eyes are pointing posteriorly, and the accentuated *cervical flexure* carries the *myelencephalon* and the *spinal cord* out to the left from the original midline of the embryo. The general shape of the entire embryo at this stage may be compared to a reversed question mark: It lies on its left side to a point posterior to the heart and the anterior intestinal portal. The portion of the embryo posterior to the amniotic fusion is still in the dorsoventral position.

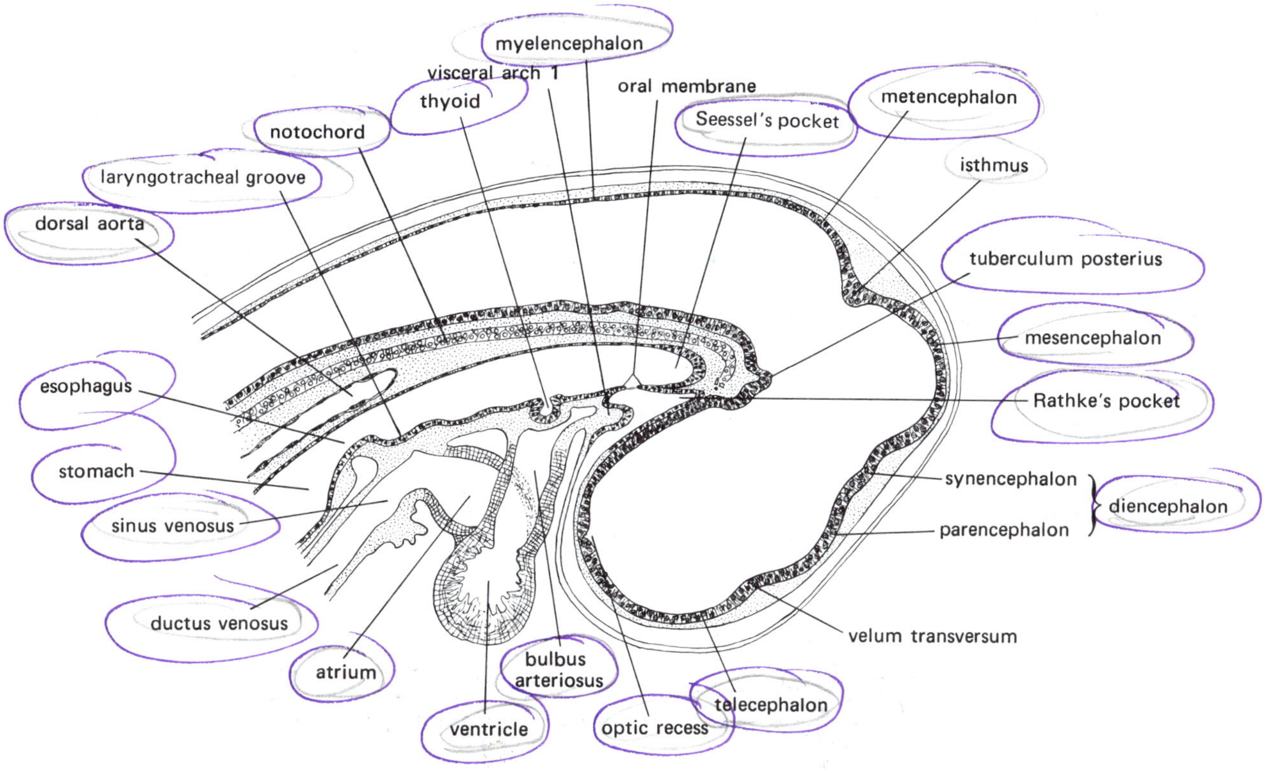

CHICK EMBRYO OF 30 SOMITES OR ABOUT 72 HOURS

MEDIAN SAGITTAL SECTION

From R. Rugh. *Vertebrate Embryology*. 1964. Harcourt, Brace & World, New York.

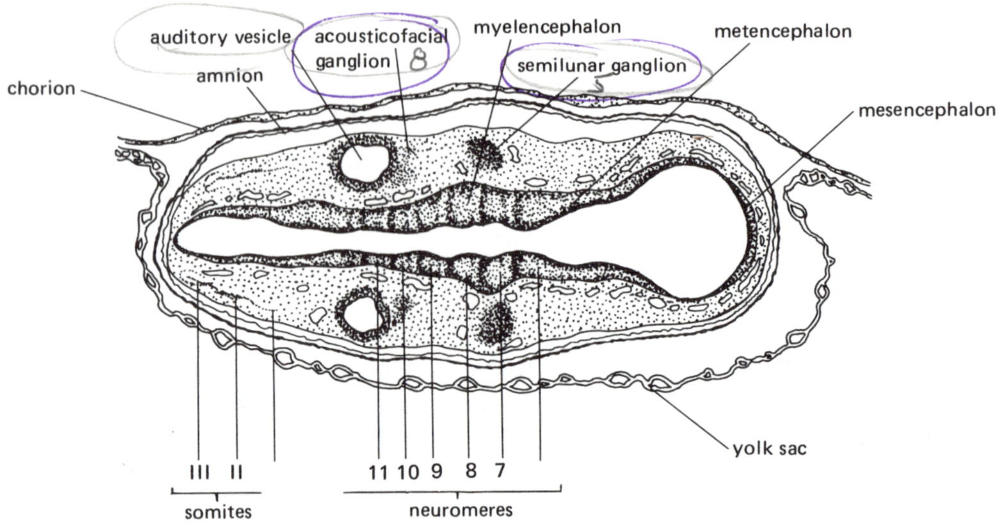

FRONTAL SECTION THROUGH HINDBRAIN REGION OF 36-SOMITE CHICK EMBRYO

The brain at this stage shows considerable enlargement, with thickenings and thinnings of the walls at various levels. The *epiphysis* can be made out distinctly for the first time in a whole mount. Locate the previously described "landmarks" of the major brain vesicles, the five parts of the brain now being clearly defined. Note particularly the further development of the eye and ear vesicles.

Anterior and dorsal to the first visceral cleft locate the darkened mass of the fifth (V) *cranial ganglion*. More dorsally and just anterior to the *auditory* (otic) *vesicle* find the *seventh* (VII) cranial ganglion and possibly the *eighth* (VIII), close to the auditory vesicle itself. Posterior to this ear vesicle find the smaller *ninth* (IX) and *tenth* (X) *cranial ganglia*. These will be particularly clear in the more transparent specimens and consist of only placodes and crest material, which will again be located in the transverse sections. All ganglia are paired.

Other changes to note in the whole mount are:

a. Lengthening of the foregut toward thc infundibular region.
b. Further development of the visceral clefts.
c. Further twisting of the heart and enlargement of the aortic arches.
d. First undercutting of the posterior end of the embryo to form the subcaudal pocket, the hindgut, and the posterior intestinal portal.

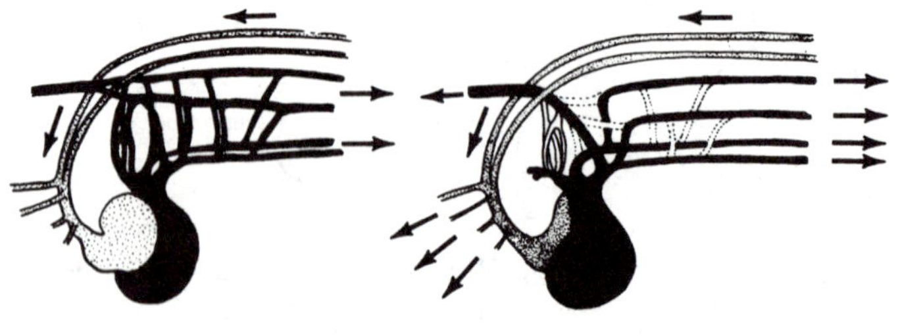

FATE OF CHICK AORTIC ARCHES

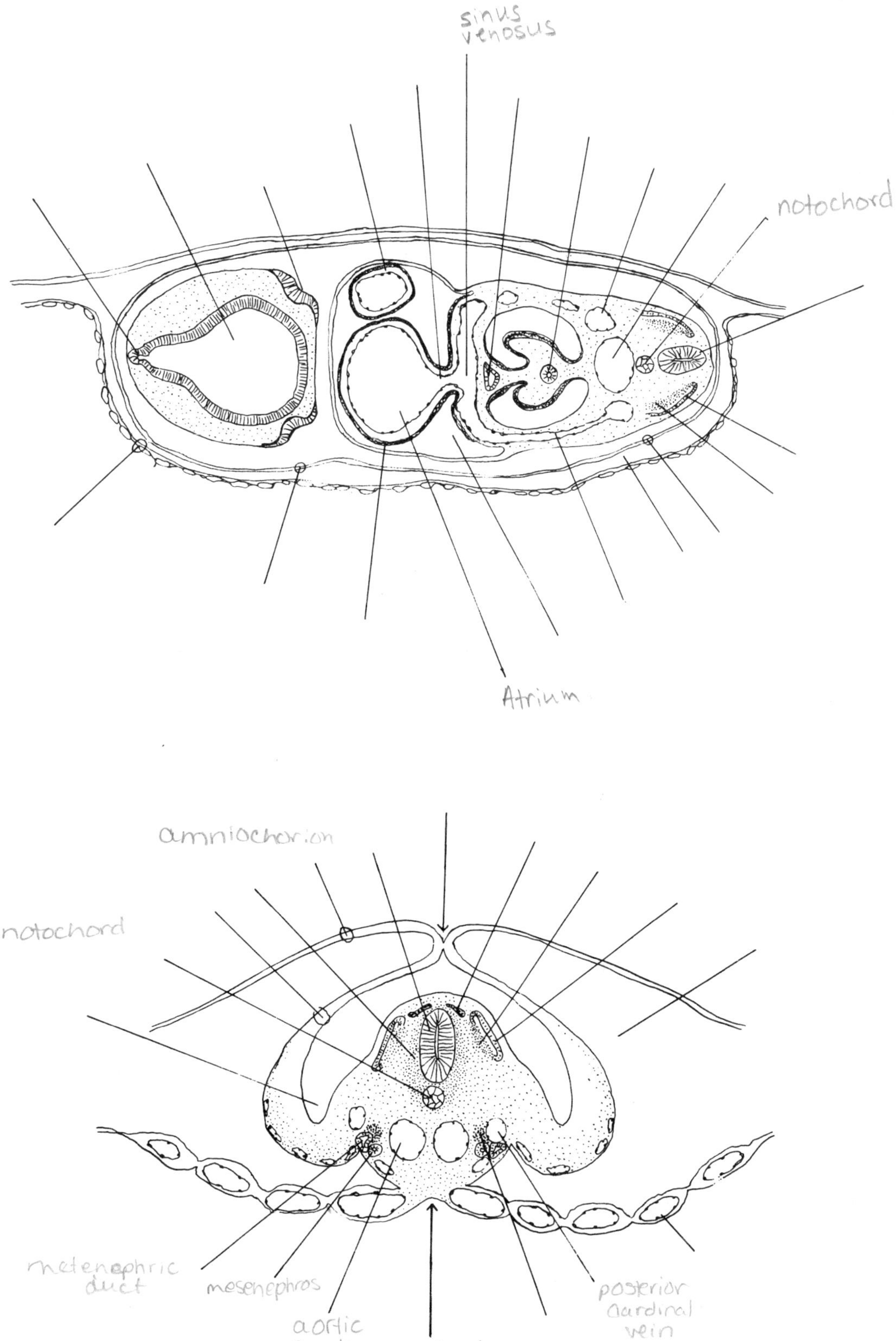

CHICK EMBRYO AT 72 HOURS

TRANSVERSE SECTIONS

(Add all labels)

TRANSVERSE SECTIONS

In the most anterior of the serial sections you will utilize the varying thicknesses of the brain wall and the proximity of the sense organs (eye and ear vesicles) and the cranial nerves to identify the exact location of any particular section. Aside from the *choroid fissure* of the optic stalk, within which will develop the optic nerve, and the undeveloped olfactory (I) nerves, the most anterior cranial nerves to be observed are the very small pair of *oculomotor* (III) *nerves.* Embryonic ganglia are nerve condensations out of which nerves and nerve trunks (extensions) develop. This pair of cranial nerves is derived from the floor of the mesencephalon and will be found in only a few sections anterior to both the eye and the ear vesicles, just where the notochord makes its first appearance (see page 174, Fig. 3). The next to be seen is the very large *Gasserian ganglia* (V) of the trigeminal nerves. Closely associated with the ear vesicles but just anterior to them, find the facial (VII) and the auditory (VIII) *ganglia.* These originate together and may appear as single large masses at this stage, but by 96 hours they are completely separate. Locate the cranial *ganglia* (IX and X) posterior to the *ear vesicles.* Note that the walls of the ear vesicles are of uneven thickness, and locate in the same or nearby sections some of the *aortic arches, viserceral arches,* paired and fused *aortae,* the *carotids,* and the *anterior cardinals.* (See, for comparison, Plate 5, 86-hour chick embryo, pages 372-77.)

The most anterior tip of the foregut is the endodermal Seessel's pocket *(preoral gut),* which must be distinguished from the ectodermal *Rathke's pocket,* close to the *infundibulum,* and also from the *exocoel* between the head and the body of the embryo. At the level of the visceral clefts the pharynx is now very spacious. The *thyroid* will have grown almost apart from the foregut, the groove joining the two being the *thyroglossal duct.* The position of the thyroid identifies the related second pair of visceral clefts. Just ventral to the thyroid find the *truncus arteriosus* (ventral aorta) as a single, large blood vessel leading from the heart around the pharynx by way of the *aortic arches* to the *dorsal aorta.* Posterior to the thyroid the gut cavity changes in shape so that it appears as a vertical slit which shortly widens (ventrally) to form the two *lung bud* evaginations. This foregut groove, which leads to the lung primordium, is known as the *laryngotracheal groove.* Since these lung buds appear and become pinched off in progressively posterior sections, it is proper to assume that they are growing posteriorly. They will ultimately give rise to the lungs and air sacs of the bird. The buds are also growing laterally into the preformed *pleural cavities.* The *ductus Cuvieri* flanks these pleural cavities to bring blood from the *common cardinals* to the large, single *sinus venosus.* At about this level, you will also find the anterior tip of the brain *(telencephalon),* along with the *nasal placodes* and the *nasal vesicles.* Posterior to the level of the lung buds there appears a much more extended ventral evagination of the foregut, namely the *liver diverticulum.* This is closely associated with the *lateral omphalomesenteric veins,* because these veins are being encroached upon by the growth of the liver and become broken up in the rapidly growing liver tissue, as blood sinuses. That portion of the omphalomesenteric veins which previously led into the sinus venosus will become the hepatic vein carrying blood from the liver to the heart. Just posterior to this level the gut will open out over the yolk at the level of the *anterior intestinal portal,* a position in the embryo somewhat posterior to that in the 48-hour stage.

At the midgut level find the lateral *amniotic folds* which have not yet fused dorsally, forming an umbilicallike oval ring and the rather thick lateral wings of the embryo proper. These are the potential *wing buds,* but at this stage they are mere concentrations of mesoderm containing some distinct blood vessels, possibly the beginnings of the *allantoic veins.* Within the body find the *Wolffian duct,* the *nephrogenic tubules* and *nephrostomes,* the *neural crests,* and the *dorsal aortae.* Find the section in which these dorsal aortae send the *vitelline* (omphalomesenteric) *arteries* out onto the yolk by way of the *splanchnopleure.* Note changes in the *dermatome, myotome,* and *sclerotome,* and posteriorly find the paired lateral thickenings which are the beginnings of the *hindlimb buds.*

Locate the level of the *posterior intestinal portal* where the hindgut is being separated from the floorless midgut. The roof of this hindgut is very thin, but the floor is very thick and heavily stained. Shortly the floor will be seen to give off a large vesicular evagination with prominent blood vessels on either side. The vessels are the *allantoic arteries*, derived from the dorsal aortae, and the vesicle is the *allantois.* The large space on either side of the allantois is the posterior *coelom,* which has been incorporated within the tail level of the body along with the development of the posterior intestinal portal and the tail fold. The coelom is still continuous with the exocoel. Posterior to the allantoic vesicle is the cloacal portion of the hindgut. Finally locate the last vestiges of the hindgut, the extension into the tail fold known as the *postanal gut.* Identify the covering membranes at all levels.

REFERENCES

Arey: 591-605
Balinsky: 321-536
Bodemer: 225-331
Hamilton revision of Lillie: Chapter VI
Huettner: Chapter XV
Lillie: Chapters VI and VII
McEwen: Chapter XI
Nelsen: 524-44
Patten(1957): 190-206
Rugh: *Vert. Emb.* 154-80
Shumway: Chapters X, XV, XVI

"The ground motive of science is a high order of curiosity, led on by ambition to overcome obstacles."

H. F. Osborne

"In science you must not talk before you know."

Ruskin

"It is deserving of emphasis that the function of imagination is not merely the conception of mythical creations, but also, and quite particularly, the presentation, to the mind, of realities. Hence, imagination plays an important role in the exact Sciences."

A. J. Lotka, 1925

"The world is moving so fast these days that the man who says it can't be done is generally interrupted by someone doing it."

Elbert Hubbard

"The common facts of today are the products of yesterday's research."

Dean Duncan MacDonald

"The value of a scientific hypothesis depends, it seems to me, first on the possibility of testing it by direct observation, or by experience; second, on whether it leads to advance; and lastly, on its elimination of certain possibilities."

T. H. Morgan

GERM LAYER DERIVATIVES OF 72-HOUR CHICK

Ectodermal Derivatives

NERVES:

I OLFACTORY–axons grow from neuroblasts of olfactory epithelium toward telencephalon on both sides.

III OCULOMOTOR–axons grow from neuroblasts in floor of mesencephalon to four pairs of ultimate eye muscles, of each eye.

V TRIGEMINAL–dendrites grow from neuroblasts of semilunar ganglia to first branchial arches and the eyes; maxillary, mandibular, and ophthalmic branches.

SPINAL NERVES–growth from dendrites of sensory neuroblasts in spinal ganglia (dorsal root), but motor neurons arise from neuroblasts in ventral root of spinal cord.

SPINAL GANGLIA–paired aggregations of neural crest cells at level of each somite.

LENS EPITHELIUM–arise from outer wall of lens vesicle.

LENS FIBERS–differentiate in inner wall of lens vesicle.

OLFACTORY PITS–invaginations of paired olfactory placodes.

EPIPHYSIS (PINEAL GL.)–single evagination of roof of diencephalon.

BRANCHIAL GROOVE IV–invagination of surface ectoderm toward fourth paired pouches (evaginating from endoderm).

Endodermal Derivatives

SEESSEL'S POCKET–foregut anterior to pharyngeal membrane.

MOUTH–inner part is endodermal part of foregut, while outer part is ectoderm of stomodeum.

PHARYNX–foregut associated with pharyngeal pouches.

PHARYNGEAL POUCH IV–dorsolateral evaginations from endoderm of pharynx between branchial arches IV and V.

LARYNX–ventral evagination from foregut just behind pouches.

GLOTTIS–slitlike opening from larynx into foregut.

TRACHEA–ventral evagination from foregut just behind pouches, distal portion only.

LUNG BUDS–primary branchi paired distal evaginations from trachea.

ESOPHAGUS–posterior to foregut, dorsal half.

STOMACH–gross enlargement of foregut posterior to esophagus.

DUODENUM–level of foregut to which liver and pancreas attach.

PANCREATIC RUDIMENT–broad evagination from dorsal wall of duodenum.

LIVER RUDIMENTS–both anterior and posterior evaginations from the duodenum.

INTESTINE, SMALL–foregut between duodenum and anterior intestinal portal.

INTESTINE, LARGE (COLON)–hindgut anterior to the cloaca.

CLOACA–hindgut associated with colon, mesonephric ducts, allantois, and tail gut.

*Most organs are made up of some mesoderm, so that, when we state here that the stomach is derived from endoderm, we mean the lining only, which determines the shape of the organ.

CLOACAL MEMBRANE—fusion of floor of cloaca and ventral ectoderm, therefore, both ecto- and endodermal.

ALLANTOIS—single, ventral evagination of hindgut endoderm and associated with it the splanchnic mesoderm.

TAIL GUT—posterior end of hindgut, posterior to cloaca.

Derivatives of the Mesoderm

DORSAL MESENTERY—fusion of splanchnic mesoderm above the various organs (esophagus, stomach, duodenum, intestine) suspending all to the dorsal wall.

MESOESOPHAGUS
MESOGASTER
MESODUODENUM
MESENTERY

AORTIC ARCH I—lose their connections to doral aorta, degenerates.

AORTIC ARCH IV—differentiate in situ in IV branchial arches.

ARTERIES:

INTERSEGMENTAL—paired outgrowths from dorsal aorta toward sides of neural tube.

MESENTERIC—unpaired branches of dorsal aorta, ventral.

VEINS:

INTERSEGMENTAL—paired, dorsal outgrowths from both anterior and posterior cardinal veins, which unite with the capillary plexi of the intersegmental arteries.

UMBILICAL IALLANTOIC—arise in somatic mesoderm in situ but also from splanchnic mesoderm of allantois.

DUCTUS VENOSUS—fusion of proximal ends of vitelline veins as they approach the heart.

SUBCARDINAL—arise in situ beneath mesonephric ducts.

KIDNEYS—pronephric (mesonephric) ducts hollow out and join hindgut and blocks of nephrotome influenced by these ducts transform into the tubules which are connected with the ducts—embryonic kidneys.

MESONEPHC RIDGE—bulge of mesonephric kidneys into body cavity.

GERMINAL EPITHELIUM—thickened coelomic epithelium on mesonephric ridges.

GENITAL RIDGES—thickened germinal epithelium bulging into body cavity.

UROGENITAL RIDGES—mesonephric plus genital ridges.

WING AND LEG BUDS—condensations of somatic mesoderm covered with skin ectoderm, include both mesoderm and ectoderm, two pairs, lateral.

PLEURAL CAVITY—body cavity surrounding lung buds and lined with mesoderm.

PERICARDIAL CAVITY—body coelom which surrounds the heart, lining only.

DRAWINGS: CHICK EMBRYO: 72 HOURS

DRAWINGS: CHICK EMBRYO: 72 HOURS

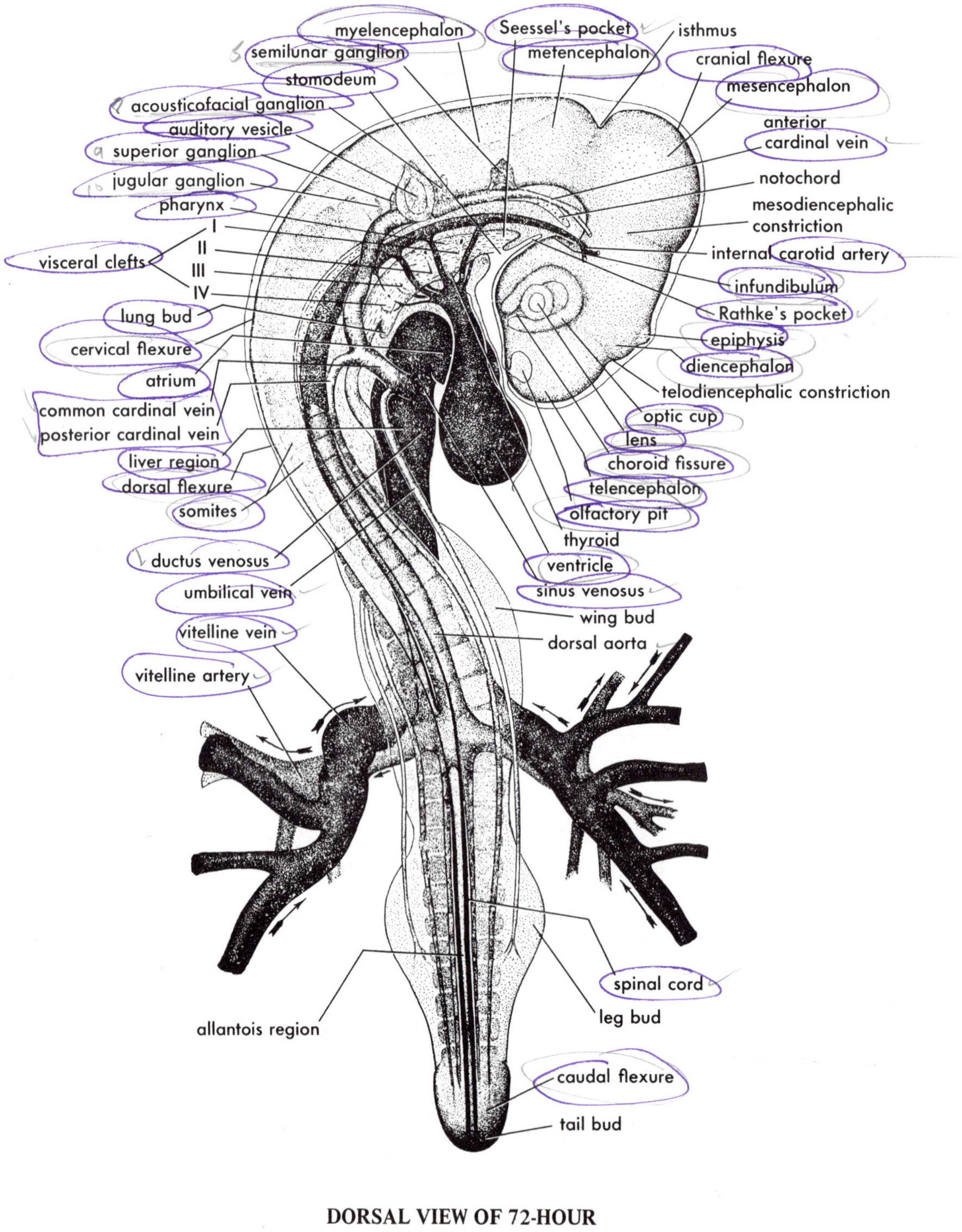

DORSAL VIEW OF 72-HOUR CHICK EMBRYO

From A. F. Huettner. *Fundamentals of Comparative Embryology of the Vertebrates.* Rev. ed. 1949. Macmillan, New York.

Exercise 22

THE CHICK EMBRYO:

Ninety-Six-Hour Stage

WHOLE MOUNT

The "apparent" length of the 96-hour chick embryo is not appreciably greater than that of the 72-hour stage, because the *cranial, cervical,* and *caudal flexures* have compensated for most of the actual cephalocaudal growth. The main body divisions of *head, trunk,* and *tail* are better defined and the *appendages,* the *allantois,* and the very large *vitelline veins* and *arteries* can easily be located. The *wing* (forelimb) *buds* appear at the level of the seventeenth to nineteenth somites, while the *leg* (hindlimb) *buds* appear at the level of the twenty-sixth to thirty-second somites, near the *tailbud.*

The head is so bent that the *telencephalic vesicles* now lie behind the pharynx in the section sequence with the *olfactory* (nasal) *pits* practically against the *heart.* The *otic* (ear) *vesicles* lie dorsal to the pharynx. The telocoels will give rise to the lateral ventricles; the roof of the telencephalon will become the cerebrum; the floor will give rise to the preoptic part of the hypothalamus; and the most anterior part becomes the olfactory lobes, associated with the first pair of cranial nerves and the olfactory organs. The *diocoel* is the next large vesicle, identified by the forward evagination of the *epiphysis,* later to become the pineal. From this portion of the forebrain are derived the third ventricle, the infundibulum, the mammillary portion of the hypothalamus, the thalamus, and, dorsally, the thin anterior choroid plexus. The earliest and most prominent derivatives of the diencephalon are the optic vesicles. The *mesocoel,* derived beneath the dorsal thickening, has a relatively thicker roof than the rest of the brain and is separated from the *metacoel* by the very deep isthmus. The cavity becomes the aqueduct of Sylvius; the roof becomes the corpora bigemina (centers of vision and hearing), or optic lobes, while the floor becomes the brain stem (crura cerebri), through which neurons connect the cerebral hemispheres with the spinal cord. The third and fourth cranial nerves arise from this portion of the brain. The small *metencephalon* is thick roofed and gives rise to the cerebellum, the metacoel, and the pons Varolii (a bundle of axons between the two sides of the cerebellum and the cerebellar peduncles). The myelencephalon is thin roofed and narrows to the spinal cord. Its roof becomes the posterior choroid plexus; the floor and sides become the posterior part of the brain stem; and the cavity becomes the fourth ventricle. The fifth to twelfth cranial nerves inclusive emerge from the hindbrain. The *ear* (otic) *vesicles* are found at the level of the myelencephalon. The spinal cord is closed at 55 hours, except for the posterior *sinus rhomboidalis.* The primordium of the upper jaw, known as the *maxillary process,* is now identified as the thick, mesodermal mass growing forward from the most dorsal portion of the first visceral arch. The bulk of this arch will, however, give rise to the lower jaw, or the *mandibular process.* At least three *visceral clefts* will be seen, the fourth beginning to form at 60 hours but never really opening. The first two clefts close during the fourth day, the upper part of the first (hyomandibular) giving rise to the Eustachian tube. Note how the dorsal portion of this cleft is bent forward. The second cleft is associated with the *thyroid primordium,* while the third and the fourth clefts are associated with the primordia of the *thymus gland.* Adjust the lighting and it may be possible to see the thyroid even in the whole mount. Also locate Seessel's pocket and the large *dorsal aorta.*

While part of the *heart* is still outside of the embryonic body, this fact is less apparent since it now lies between the telencephalon and the liver. The intervening space is obliterated. The heart is still a looped tube. The auricles will be differentiated from ventricles. These chambers are not yet divided into right and left. The *vitelline* (omphalomesenteric) *veins* and *arteries* are close together now, entering and leaving the body at the level of the midgut. Distinguish between the two by structure, if possible.

The *somites* now extend well into the tail, numbering 42 pairs. Since the embryo has turned completely on its left side, the bilateral arrangement of these somites will not be so apparent. Further, there are so many somites that they appear to be fused together with only slight indications of metamerism.

There also appears at this stage a *tail fold* (comparable to the head fold), resulting from the lifting of the tail process off of the underlying yolk. Subsequently there is a forward growth of the amniotic tail fold. In the curve between the body and the tail, locate the evaginating *allantoic vesicle* and the nearby posterior *limb buds*. In the umbilical stalk you will also find the large *vitelline* (and allantoic) *veins*.

TRANSVERSE SECTIONS

It is important that the student have before him an accurate whole-mount drawing or a mounted specimen while studying the serial transverse sections. The flexions and torsions are so great that practically none of the sections are truly transverse in the sense that very few of them cut across the embryonic axis at right angles. The most anterior sections will appear to be frontals. The sections at the level of the oral plate are apt to be the most confusing, and the sections through the level of the tail fold will be reminiscent of those through the earlier head fold. (See, for comparison, Plate 5, 86-hour chick embryo, pages 372-77.)

The majority of the structures seen in the 96-hour stage have already been identified in earlier stages, so that a mere listing of the structures of this stage will be given under the headings of their germ layer derivatives. Some of these structures may be difficult to locate in any particular series of sections, but each student should become familiar with the list of derivatives presented below.

Membranes

AMNION – inner layer ectodermal, outer layer mesodermal; together forming the innermost sac which encloses the embryo. Later acquires smooth muscle fibers which help to keep the embryo in more or less constant motion, preventing adhesions.

CHORION – inner layer somatic mesoderm, outer layer ectoderm; formed simultaneously with, and reciprocally to, the amnion. This membrane will appear as a thin, nonvascular layer dorsal to the embryo and superficial to the amnion. Syn., serosa.

YOLK SAC SPLANCHNOPLEURE – splanchnic mesoderm and ventral endodermal layer comprise this membrane which, peripheral to the yolk sac umbilicus, separates the embryo from the underlying yolk. There are numerous blood vessels in the splanchnic mesoderm of this layer. In the periphery lateral to the vascular area, locate the junction of this splanchnopleure and the chorion, the space between being the exocoel or extraembryonic body cavity. This cavity is surrounded by somatic and splanchnic mesoderm.

ALLANTOIS this membrane is made up of a bladderlike median ventral diverticulum of the hindgut endoderm, covered with splanchnic mesoderm. It connects with the hindgut, which will be found only in sections posterior to the posterior intestinal portal at this stage of development. Ultimately, this double membrane will fill the exocoel, and its outer layer of mesoderm will fuse with mesoderm of the chorion and the amnion and finally with the splanchnic mesoderm of the yolk sac splanchnopleure. Its function in the chick is related to respiration and excretion.

Ectodermal Derivatives*

EPIDERMIS – the entire outer covering of the chick embryo is of ectodermal origin and is made up largely of squamous epithelium but will later also include horny scales, feather germs, quills and barbs, claws, beak coverings, and a temporary eggtooth.

By invaginations from the surface, the linings of the following structures are also derived

*See Balinsky, 367-423.

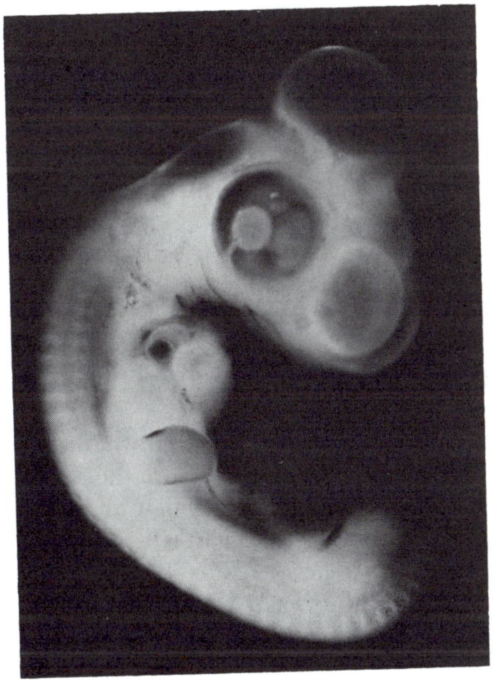

CHICK EMBRYO–STAGE 21 (3.5 DAYS)

Courtesy J. W. Saunders Jr., State Univ., N.Y.

Ectodermal Derivatives (Continued)

from ectoderm: the mouth (stomodeal portion and stomodeal hypophysis); cloaca (proctodeal portion); visceral clefts (peripheral halves); nostrils; eye chamber and lens; otic vesicles; and external auditory meatus.

BRAIN VESICLES AND RELATED STRUCTURES:

PROSENCEPHALON (Forebrain)

TELENCEPHALON – cerebral hemispheres with thick lateral walls, called the corpora striata; torus transversus, which becomes the anterior commissure; paraphysis, which comes from a dorsal evagination of the median telencephalon, the anterior choroid plexus anterior to the velum; olfactory lobes from the ventral evaginations associated with the first cranial nerves; telocoels, which become the lateral ventricles and join the third ventricle by way of the foramina of Monro. First neuromere.

DIENCEPHALON – epiphysis becomes the frilled pineal; posterior commissure is the thickened roof just posterior to the epiphysis; optic thalami from thickened lateral walls; optic chiasma from optic nerve fibers; infundibulum plus Rathke's pocket forms the pituitary gland; and the diocoel becomes the posterior part of the third ventricle. (There are two telencephalic and one diencephalic choroid plexus.) Second and third neuromeres.

MESENCEPHALON (Midbrain)

The dorsolateral thickenings of this portion of the brain become the optic lobes; the ventral and lateral thickenings become the crura cerebri associated with the third and fourth cranial nerves; the roof becomes the corpora bigemina; the cavity (mesocoel) will be the narrow aqueduct of Sylvius which joins the third and the fourth ventricles. Fourth and fifth neuromeres.

Ectodermal Derivatives (Continued)

RHOMBENCEPHALON (Hindbrain)

METENCEPHALON – the cerebellum comes from the roof, the pons Varolii from the ventrolateral walls, and the cavity is the metacoel. This portion of the hindbrain is separated from the mesencephalon (anteriorly) by the isthmus, a dorsal depression. Sixth neuromere.

MYELENCEPHALON – the roof becomes the thin and vascular posterior choroid plexus; the floor and sides thicken to become the brain stem and medulla associated with the fifth to twelfth cranial nerves inclusive; the cavity is the myelocoel or part of the fourth ventricle. Seventh to eleventh neuromeres.

SPINAL CORD –

This mid-dorsal tubular extension of the brain ectoderm and cavity (neurocoel), with 38 pairs of spinal ganglia and related sympathetic ganglia, extends into the tail fold. The spinal ganglia are derived from the neural crests. The sympathetic ganglia come from the same source but migrate toward the dorsal aorta.

Endodermal Derivatives

FOREGUT:

PREORAL PIT – Seessel's pocket, present in the embryo only.

MOUTH CAVITY – contiguous with the stomodeum.

PHARYNX – the dorsoventrally flattened cavity just posterior to the mouth cavity; related to the visceral pouches and clefts.

VISCERAL CLEFTS – the paired lateral evaginations of the pharyngeal cavity which meet and fuse with the invaginations of head ectoderm to form the visceral clefts. Three and sometimes four pairs present in the chick embryo.

THYROID GLAND – median ventral evagination from the pharyngeal floor near the base of the second pair of visceral arches.

LARYNGOTRACHEAL GROOVE – median ventral groove in the floor of the foregut posterior to the thyroid primordium and leading to the origin of the trachea. The connection with the gut is the glottis.

ESOPHAGUS – thick-walled oval tube anterior to origin of lung buds.

PRIMARY BRONCHI – the tracheal evagination grows caudad and bifurcates to form the lung buds, also associated with the air sacs later.

STOMACH and DUODENUM – section between the origin of the lung buds and a point just anterior to the liver and pancreatic rudiments.

LIVER – median ventral diverticulum of foregut which appears early in development and invades the large omphalomesenteric veins, causing them to become broken up into sinuses.

PANCREAS – three rudiments, one dorsal and the others ventral to the duodenum.

ANTERIOR INTESTINAL PORTAL – immediately posterior to the liver diverticulum the foregut opens out over the yolk, determining the anterior margin of the unformed midgut.

MIDGUT:

Portion of the undifferentiated gut bounded by the anterior and the posterior intestinal portals,

Endodermal Derivatives (Continued)

therefore, merely a roof over the yolk. Has thick lateral walls but no floor. Gives rise to intestines.

HINDGUT:

ALLANTOIS – bladderlike median ventral evagination from the floor of the hindgut; endodermal lining. This sac eventually fills the exocoel, its outer layer of mesoderm fusing with the inner lining of the exocoel.

CLOACA – portion of hindgut posterior to the origin of the allantois and that region which receives the mesonephric ducts at the dorsolateral margins; later receives the reproductive ducts and functions also as a rectum. The common outlet for the three systems.

POSTANAL GUT – extension of hindgut into tail fold, comparable to preoral gut.

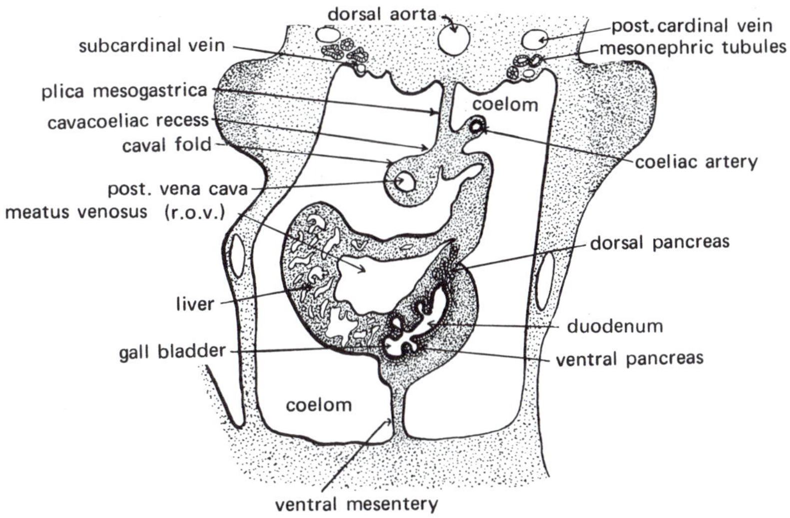

DUODENAL REGION AT 96 HOURS

Mesodermal Derivatives

NOTOCHORD – embryonic precursor of centrum of axial skeleton, composed of vacuolated cells. Precise origin is doubtful.

MESENCHYME – loose embryonic mesoderm, found most concentrated in the head.

EPIMERIC MESODERM – (dorsal derivatives)

DERMATOME – dorsal mesodermal mass, continuous with dermatome; precursor of striated musculature. This includes cranial, axial, and appendicular muscles.

SCLEROTOME – loose mesenchymal cells from the somites which migrate around the nerve cord and notochord to give rise to the axial skeleton.

MESOMERIC MESODERM – (intermediate portion)

NEPHROTOME – primordium of excretory system, both tubules and ducts. The mesonephros may be seen at 72 hours and the metanephros at 96 hours. Portions give rise to gonoducts.

Mesodermal Derivatives (Continued)

HYPOMERIC MESODERM – (lateral plate mesoderm)

SOMATOPLEURE – mesodermal sheet combined with embryonic ectoderm.

SPLANCHNOPLEURE – mesodermal sheet combined with embryonic endoderm, very vascular. Mesodermal portion gives rise to visceral musculature.

COELOM – space within the body between somatic and splanchnic mesoderm, from the walls of which develop the gonads. Subdivided into pericardial, pleural, and peritoneal cavities.

GONADS – primordia of sex glands from dorsal germinal epithelium (mesoderm) appearing about the eighth day, near dorsal mesentery.

VISCERAL ARCHES – (all paired)

MAXILLARY PROCESS – mesoderm in head which will form upper jaw.

MANDIBULAR PROCESS – first visceral arch, anterior to hyomandibular pouch which will form parts of lower jaw.

HYOID PROCESS – second visceral arch posterior to hyomandibular pouch.

THIRD VISCERAL ARCH – (also called first branchial) posterior to second visceral cleft.

FOURTH VISCERAL ARCH – (also called second branchial) posterior to third visceral cleft.

CIRCULATORY SYSTEM:

ARTERIES

DORSAL AORTAE – paired vessels dorsal to the pharynx which fuse to form a single large vessel posteriorly. Give off the following branches:

INTERSEGMENTAL ARTERIES – paired dorsolateral branches between somites.

BRACHIAL ARTERIES – branches of dorsal aorta to forelimb rudiments.

MESENTERIC ARTERY – many branches to the developing gut.

VITELLINE ARTERIES – paired vessels which leave the body at about the level of twenty-second somite to course out over the yolk.

ALLANTOIC ARTERIES – numerous small vessels from the iliac artery to the developing allantois.

RENAL ARTERIES – branches of dorsal aorta to the nephrogenic tissue.

ILIAC ARTERIES – lateral extensions of the dorsal aortae at the level of the posterior limb buds.

CAUDAL ARTERIES – extensions into the tail of the fused dorsal aortae.

PULMONARY ARTERIES – extensions of the original sixth pair of aortic arches, growing with the development of the lungs and air sacs.

EXTERNAL CAROTIDS – small paired vessels extending from the base of the third pair of aortic arches toward the mandibular process.

INTERNAL CAROTIDS – anterior extensions of the paired dorsal aortae into the head.

VEINS

CARDINAL VEINS – paired vessels forming letter "H" from dorsal view.

ANTERIOR CARDINALS – carry venous blood from the head to common cardinals.

POSTERIOR CARDINALS – carry venous blood from dorsal nephrogenic tissue toward the common cardinals. Posterior portion later degenerates while the anterior portion becomes the subclavians.

SUBCARDINALS – smaller veins located ventral to the nephrogenic tissue, later to fuse to form part of the posterior vena cava.

COMMON CARDINALS – fused junction of anterior and posterior cardinals which empty into the sinus venosus through the ductus Cuvieri.

Mesodermal Derivatives (Continued)

PULMONARY VEINS – ventral to the primary bronchi, carry blood to the left auricle.

VITELLINE VEINS – very large paired vessels from the yolk sac, called the omphalomesenterics as they enter the embryonic body. These veins converge as the meatus venosus before entering the sinus venosus. The meatus venosus later becomes the hepatic vein, and those portions of the omphalomesenterics posterior to the liver and the newly formed dorsal mesenteric vein become the hepatic portals.

SUBINTESTINAL VEIN – small vessel appearing in the ventral mesentery.

DORSAL MESENTERIC VEINS – develop late and assume functions of vitelline veins about the time of hatching.

INTERSEGMENTAL VEINS – small metameric veins which lead from the intersomitic regions to either the anterior or the posterior cardinals.

ALLANTOIC VEINS – formed along with the development of the allantois, but reach the ductus Cuvieri by way of the lateral body walls of the embryo rather than within the body proper. Junction later changes to the hepatic vein near the heart, connections with ductus degenerating.

HEART

This is a tubular organ attached to the body at the level of the bulbus arteriosus anteriorly and the sinus venosus posteriorly, the intermediate portion remaining somewhat outside of the body.

SINUS VENOSUS – receives common cardinals through ductus Cuvieri; also receive the meatus venosus which is the fused omphalomesenteric veins.

ATRIUM – (primordium of the) auricles, ventral to the sinus venosus; thin walled.

VENTRICLE – projected farthest from the embryonic body; ventral to atrium; very thick-walled.

BULBUS ARTERIOSUS – (also conus) junction between ventricle and the aortic arches; circular in transverse section and identical with the ventral aorta.

AORTIC ARCHES – these vessels pass through the visceral arches from the bulbus arteriosus (ventral aorta) to the dorsal aortae; located directly dorsal to the pharynx. Since the aortic arches run through correspondingly designated visceral arches (previously identified) a total of five (from II to VI) may be located. The last two may be fused.

REFERENCES

Arey: 605-08
Balinsky: 324-41
Bodemer: 225-31
Hamilton revision of Lillie: Chapter VII
Huettner: Chapter XV
Lillie: Chapters, VI, VII
McEwen: Chapters, XI, XII
Patten (1957): 207-46, Chapter XIII
Rugh: *Vert. Emb.* 180-220
Shumway: Chapters XV, XVI

"It is the thought transmitting propotency of the human species, more than any other, that gives it a superlative lead over all the creatures of the globe" and "Man is the only animal who, in any considerable measure, bequeathed to his descendants the accumulated wisdom of past generations."

Lotka, 1952

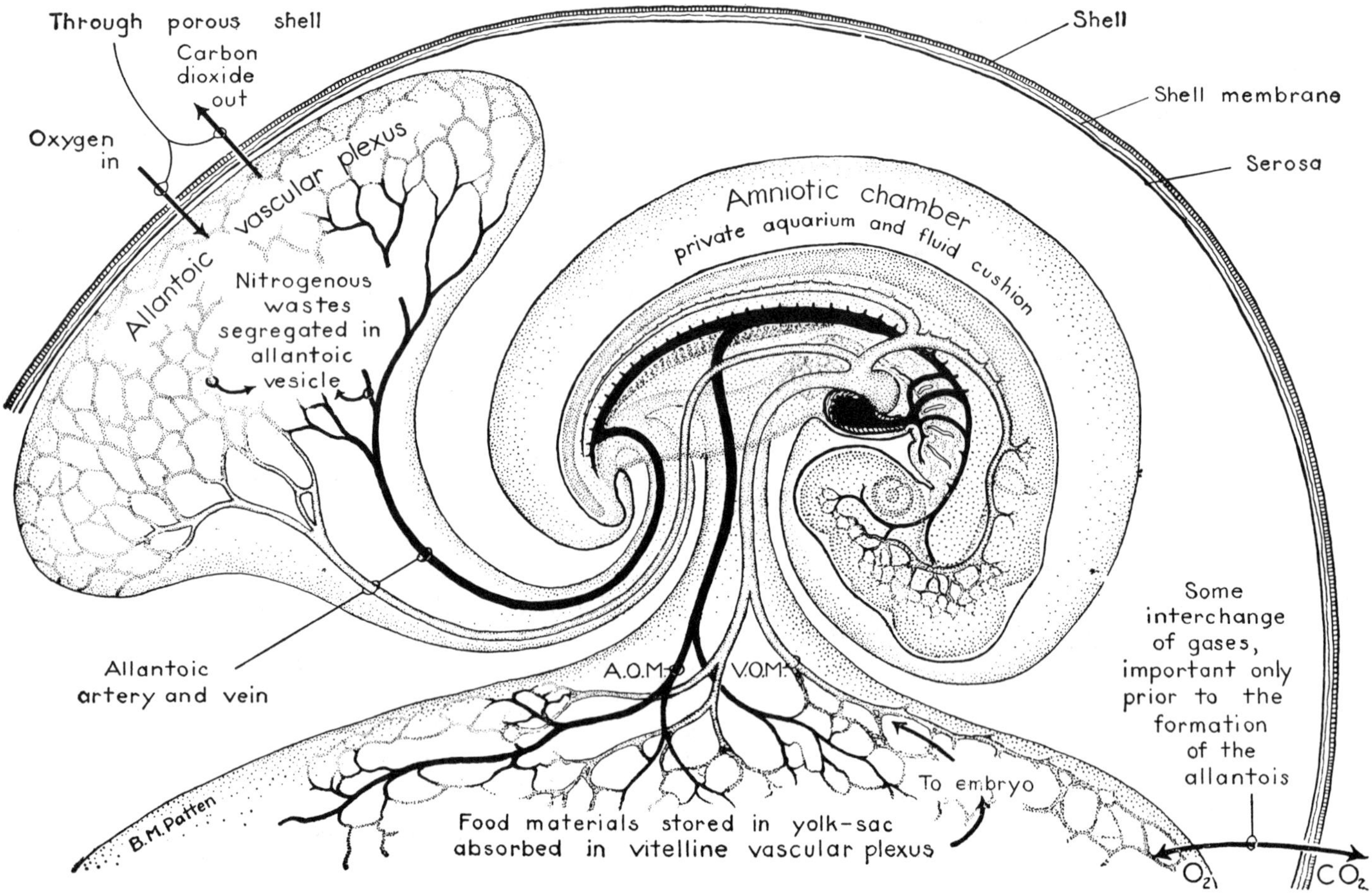

Schematic diagram showing the arrangement of main circulatory channels in a young chick embryo. The sites of some of the extraembryonic interchanges important in its bioeconomics are indicated by the labeling. The vessels within the embryo carry food and oxygen to all its growing tissues and relieve them of the waste products incident to their metabolism. Abbreviations: A.O.M., omphalomesenteric artery; V.O.M., omphalomesenteric vein.

From Patten. 1951. Am. Nat. 39:225 (with permission).

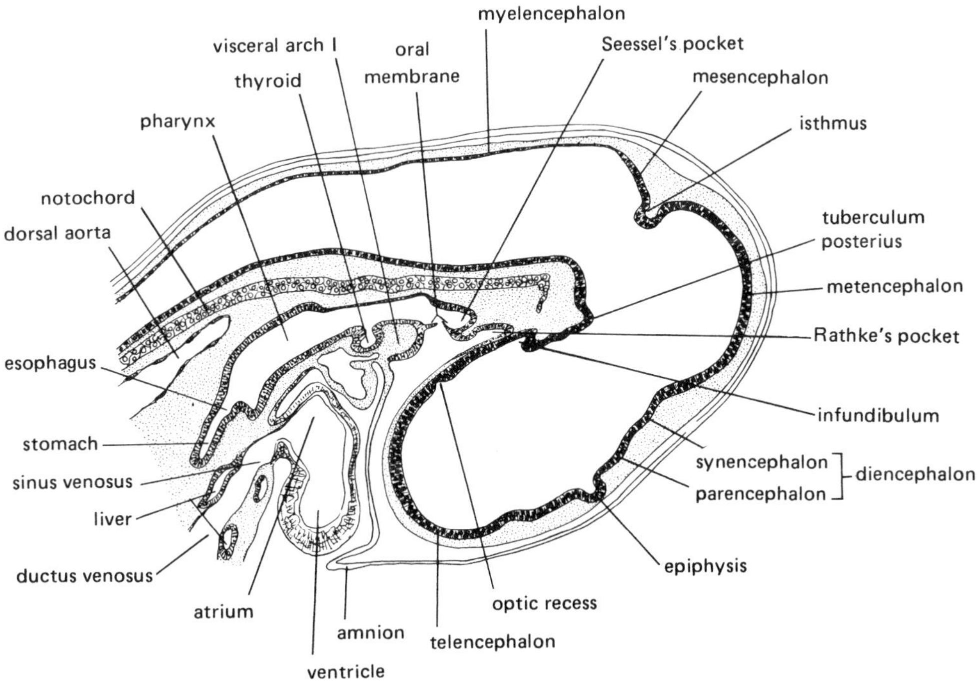

SAGITTAL SECTION OF BRAIN AND HEART REGION OF 39-SOMITE CHICK EMBRYO

From R. Rugh. *Vertebrate Embryology*. 1964. Harcourt, Brace & World, New York.

DRAWINGS: CHICK EMBRYO: 96 HOURS

THE CHICK EMBRYO:

Four to Nine Days

Exercise 23

4-DAY (96-HOUR) CHICK EMBRYO: SAGITTAL SECTION

Each student is provided with a single median sagittal section of a 4-day chick embryo in which he or she should attempt to locate each of the following:

CENTRAL NERVOUS SYSTEM:

- Telocoel and telencephalon
- Diocoel and diencephalon
- Mesocoel and mesencephalon
- Metecoel and metencephalon
- Myelocoel and myelencephalon
- Neurocoel and spinal cord
- Tuberculum posterius
- Infundibulum
- Recessus opticus
- Epiphysis
- Optic chiasma
- Isthmus
- Dorsal thickening

DIGESTIVE TRACT AND RELATED PARTS:

- Rathke's pocket
- Seessel's pocket
- Stomodeum and oral plate
- Mandibular arch
- Pharynx
- Laryngotracheal groove
- Thyroid
- Thyroglossal duct
- Esophagus
- Stomach
- Liver
- Pancreas (3)
- Midgut
- Allantoic stalk
- Allantois
- Proctodeum
- Postanal gut

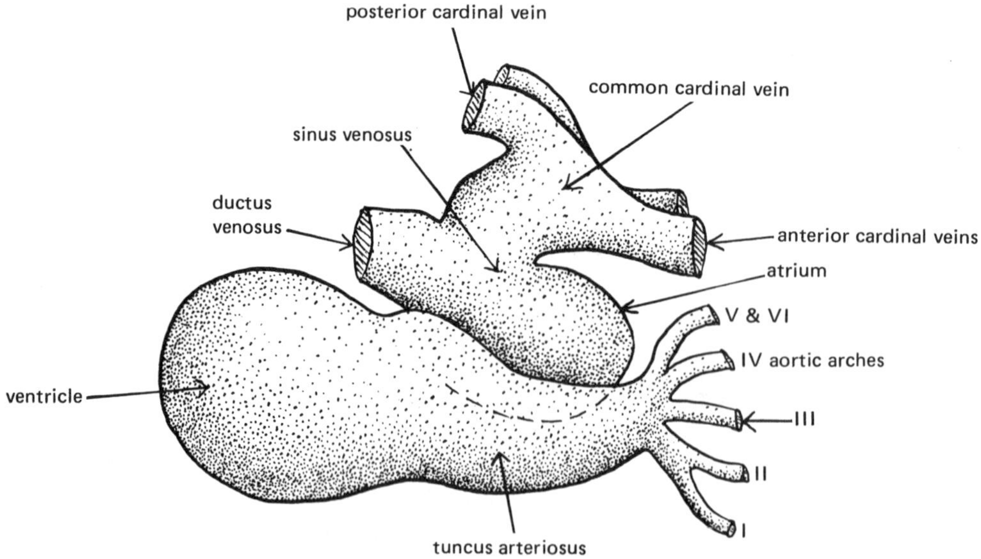

THE 96-HOUR CHICK EMBRYO HEART

BLOOD VASCULAR SYSTEM AND RELATED PARTS:

- Dorsal aorta
- Ventral aorta
- Descending aorta
- (Anterior double dorsal aorta)
- Aortic arches
- Atrium
- Ventricle
- Sinus venosus
- Allantoic vein
- Omphalomesenteric vein and artery

MISCELLANEOUS STRUCTURES:

- Subcephalic pocket
- Subcaudal pocket
- Amnion
- Chorion
- Yolk-sac splanchnopleure
- Exocoel
- Notochord
- Tailbud
- Mesonephros
- Mesodermal somite
- Mesenchyme
- Coelom

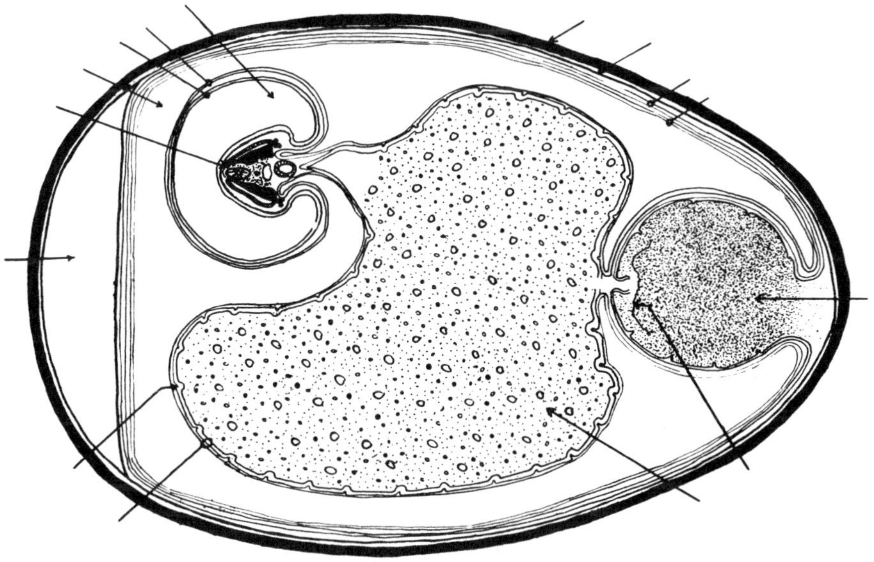

THE 14-DAY CHICK EMBRYO – SCHEMATIZED
(Add all labels)

"Of course, I am not forgetting that development and evolution are in the main epigenetic processes by which the more complicated end stages are built upon the less complicated earlier ones, but I also refuse to forget that these earlier stages are also complex, that the egg or Paramecium are complex organisms and that development is endogenetic as well as epigenetic. Both epigenesis and endogenesis are involved in all development and evolution."

E. G. Conklin, 1944

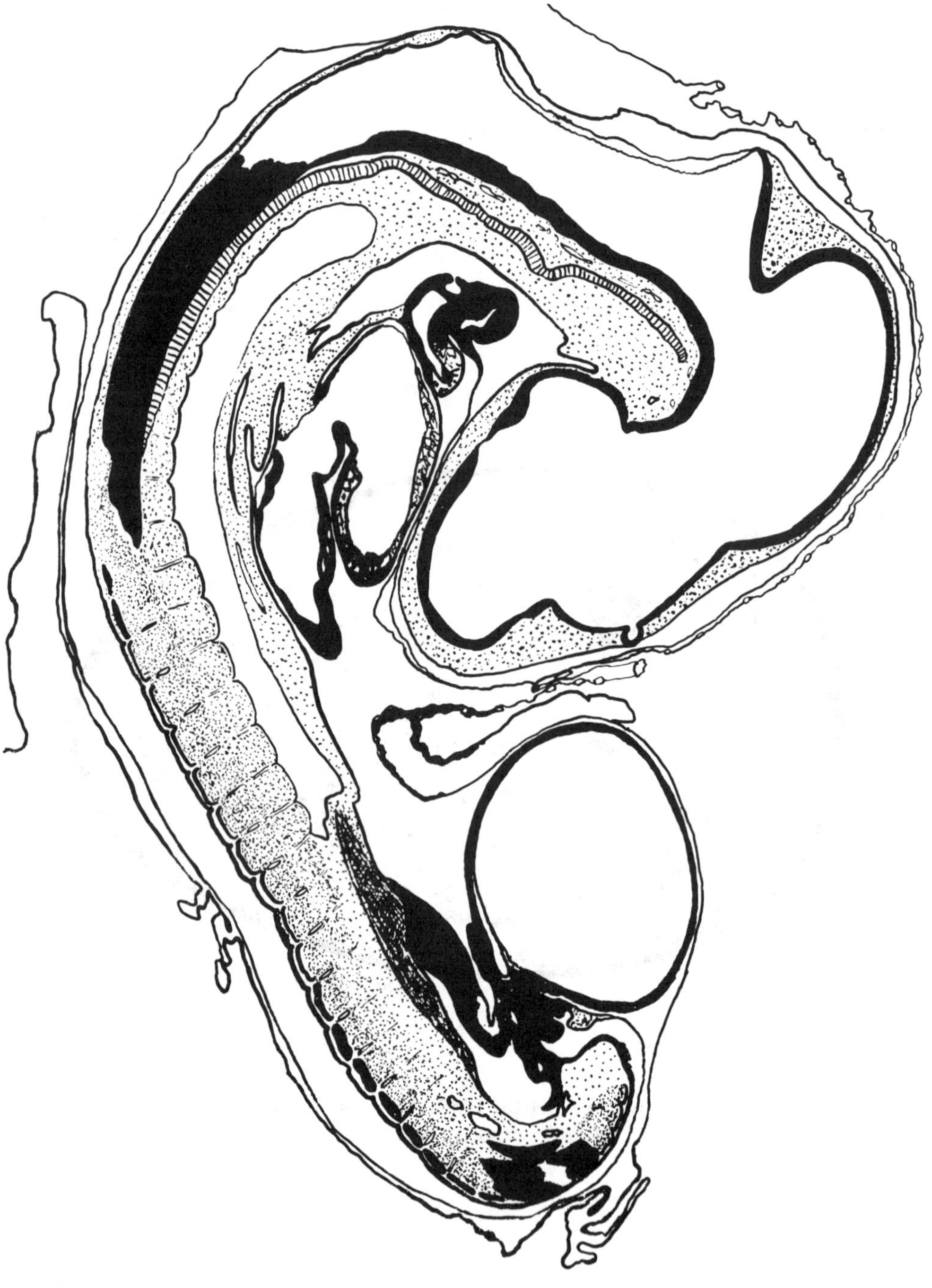

CHICK EMBRYO 96-HOUR STAGE
SAGITTAL SECTION

Drawn from actual specimen.

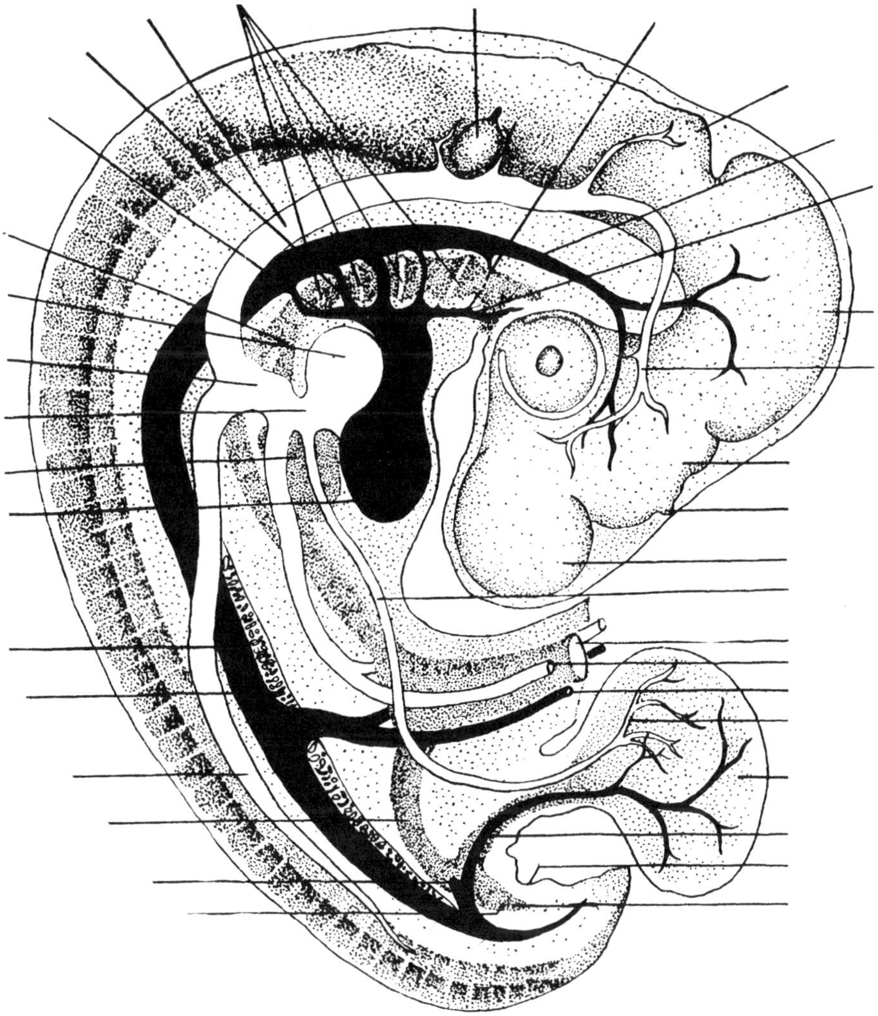

THE 4-DAY CHICK EMBRYO

From B. M. Patten. *Embryology of the Chick 1957.* Copyright The P. Blakiston Company, Philadelphia.

(After the student labels this diagram in pencil, he or she should check it against the diagram on page 208 for corrections.)

THE CENTRAL NERVOUS SYSTEM OF THE 5-DAY CHICK EMBRYO

A single transverse section cut through the midbody level of a 5- to 6-day chick embryo is provided each student. Under high magnification study the central nervous system.

The epithelial cells of the neurocoel are stretched to the outer margin of the spinal cord. The cells in the roof and the floor are rather low. As the wall of the neural tube grows in thickness, these epithelial cells are stretched continuously, but their nuclei remain near the central canal. These cells soon become ciliated and form the ependymal layer. The more distal portions of these cells then form the framework of the spinal cord, later to support the nervous and other elements. During the third day, neuroblasts and glia cells begin to form the gray matter, while peripherally there appear areas of white matter consisting of myelinated axons. The original epithelial cells give rise to both neuroblasts and glia cells. Locate the following:

WHITE MATTER – dorsal, lateral, and ventral; above, between, and below the spinal roots.
NEUROBLASTS – ventrolateral to ependyma; will give rise to ventral roots.
ANTERIOR COMMISSURE – white fibrous tract ventral to the neurocoel connecting two masses of neuroblasts.

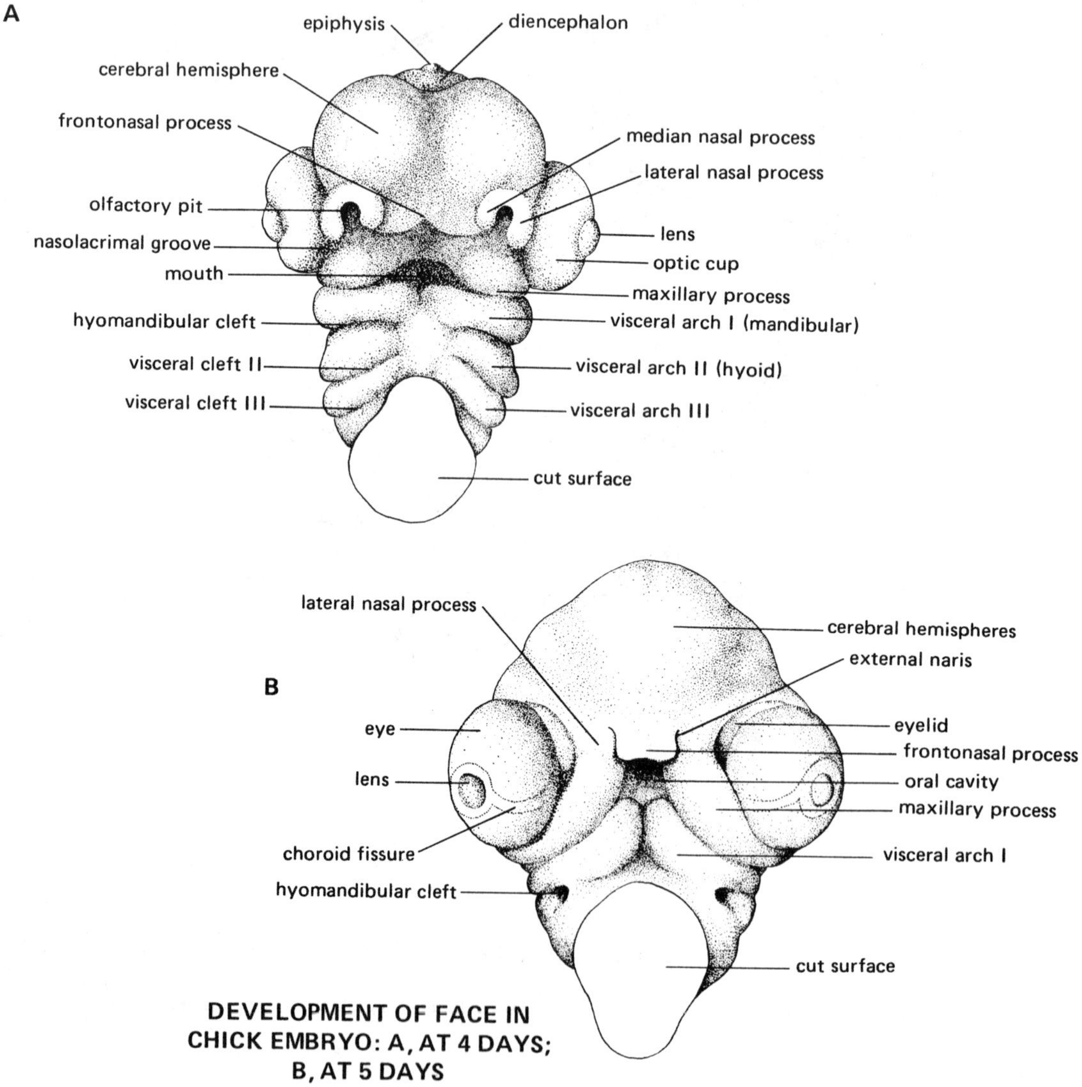

DEVELOPMENT OF FACE IN CHICK EMBRYO: A, AT 4 DAYS; B, AT 5 DAYS

From R. Rugh. *Vertebrate Embryology.* 1964. Harcourt, Brace & World, New York.

EPENDYMA – the bulk of the spinal cord at this time; cells have distinct nuclei and are generally crowded around the central canal. Cilia may project from ependyma into the canal.
BLOOD VESSELS – these may be seen between the neuroblasts and ependymal cells.

THE EYE OF THE 8-DAY CHICK EMBRYO

A single section through the center of the chick eye is provided. With the naked eye and also under low magnification, locate the following parts:

LENS – with cavity (remnant of lens vesicle) and fibrous structure.
CHAMBERS – anterior chamber is between the cornea and the lens, and posterior chamber is between the iris and the suspensory ligament of the lens. The large cavity of the eye, posterior to the lens, is filled with vitreous humor.
RETINA – the thick cellular layer surrounding the vitreous humor and terminating in front of the lens at the iris. Lying thinly over the face of the retina are the neuron extensions of the optic nerve. The section may show the region of entrance of the optic nerve.
PIGMENTED LAYER – this is the thin, black outer layer of the retina which is closely surrounded by the choroid coat.
CHOROID COAT – a mesodermal layer of rather loose connective tissue, between the pigmented layer and the sclerotic cartilage.
SCLEROTIC CARTILAGE – the thin, cartilaginous layer next to the choroid coat which completely surrounds the eye cup.
SCLEROTIC COAT – this layer is continuous over the front of the eye where, together with the ectodermal covering, it is known as the cornea.
CORNEA – this portion of the eye consists of the layers covering the anterior chamber, ectoderm, and mesoderm.
CILIARY PROCESS – muscle fibers which control the iris (diaphragm).
EYELIDS – both upper and lower eyelids may have developed by this stage.
NICTITATING MEMBRANE – fold of ectoderm between the lower eyelid and eyeball.

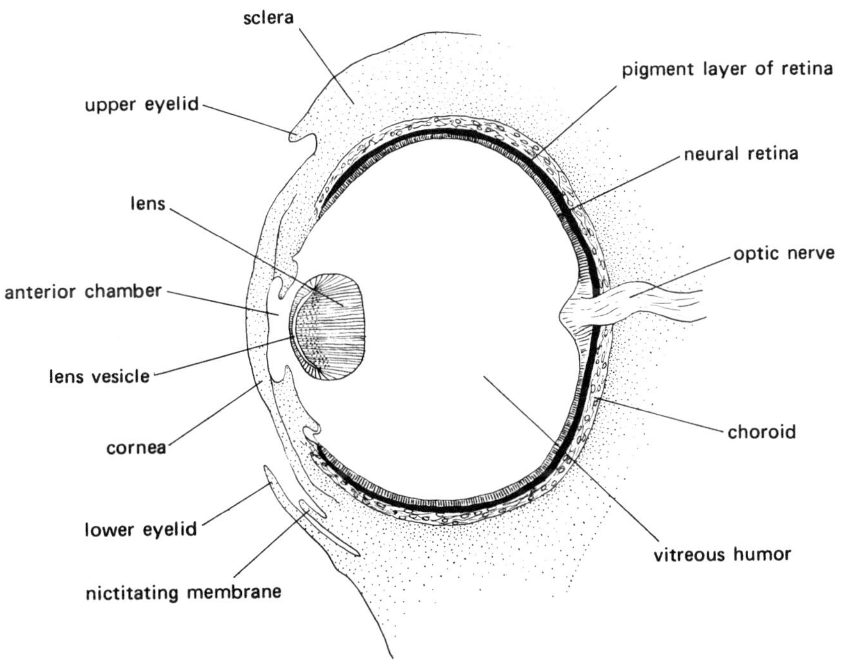

SCHEMATIC SECTION OF 9-DAY CHICK EMBRYO

THE GONAD PRIMORDIA OF THE 8- TO 9-DAY CHICK EMBRYO

The purpose of this study is to acquaint the student with the origin and the location of the gonad primordia. Also, the sections are very satisfactory for the study of an advanced stage in embryonic development. Serial transverse sections of such an embryo would occupy some 100 slides; hence, a single selected section is offered for study. Identify the following:

NERVOUS SYSTEM:

- Ependymal cells
- Neuroglia
- Neuroblasts
- White matter
- Connective tissue around cord
- Dorsal root and ganglion
- Ventral root and spinal nerve
- Ramus communicans
- Sympathetic ganglion
- Sympathetic nerve to dorsal aorta

SKELETAL SYSTEM:

- Notochord (to become centrum)
- Perichordal sheath around notochord
- Loose sclerotome
- Sclerotome-forming cartilage
  - Spinous process
  - Transverse process of neural arch
  - Region of future centrum
- Cartilage in limb bud
- Ilium

DIGESTIVE SYSTEM:

- Liver – a frilled organ
- Stomach (proventriculus)
- Gizzard – large, thick-walled cavity
- Pancreas – three rudiments
- Intestine – sections of coiled tube

BLOOD SYSTEM AND RELATED PARTS:

- Dorsal aorta
- Coeliac artery
- Inferior vena cava (directly beneath dorsal aorta)
- Umbilical (allantoic) vein
- Ventricle and other parts of the heart
- Posterior cardinal vein
- Subcardinal vein (between mesonephros and dorsal aorta)

EXCRETORY SYSTEM AND RELATED PARTS:

- Mesonephros–with tubules and Malphigian corpuscles; largest, nephrogenous mass
- Metanephros–dorsolateral to mesonephros, and made up of cortical tubules from intermediate nephrogenous mesoderm
- Mullerian duct–thick-walled tube dorsolateral to the mesonephros
- Wolffian duct–thin walled tube, ventral to the Müllerian duct; appears like large uriniferous tubule

GONADS:

- Gonad rudiments – mesial to mesonephros, bilateral in both males and females at this stage of development; appear as solid ovoid masses; examine for cellular detail under high magnification

MISCELLANEOUS:

- Coelom and derivatives – pleural, pericardial, and peritoneal cavities, each with mesodermal epithelial mesothelial lining
- Lungs and abdominal air sacs–the latter being large, thin-walled cavities
- Dorsal mesentery
- Embryonic membranes

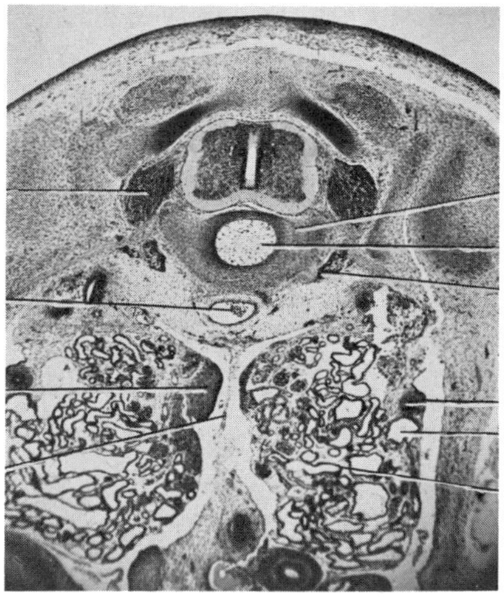

CHICK EMBRYO AT 8 DAYS
(X-SECTION)

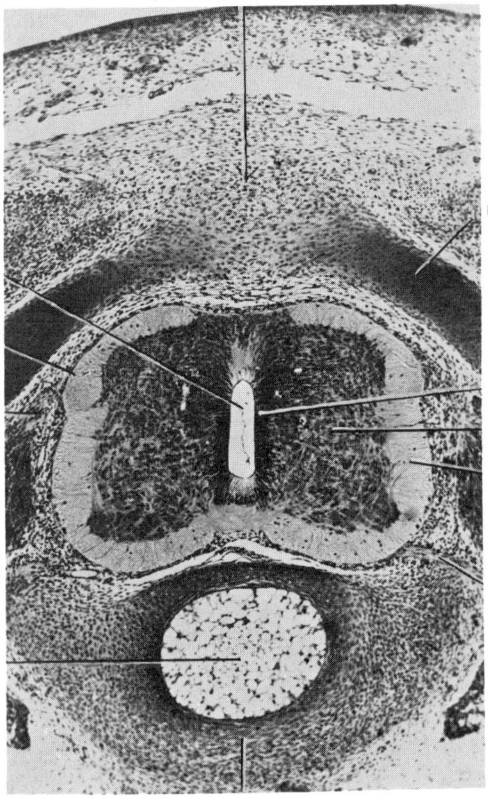

CHICK EMBRYO AT 8 DAYS
(AXIAL SKELETON)

TRANSVERSE SECTION THROUGH THE LUNGS OF A CHICK EMBRYO OF 11 DAYS

a. th. A. S., Anterior thoracic air sac. Ao., Aorta. Aur. d., s., Right and left auricles. B. d., s., Right and left ducts of Botallus. F., Feather germs. Li., Liver. P. C., Pericardial cavity. p. p. M., Pleuroperitoneal membrane. P. V., Pulmonary vein. Par'b., Parabronchi. Pl. C., Pleural cavity. Pt. C., Peritoneal cavity. R., Rib. Sc., Scapula. V. d., s., Right and left ventricles.

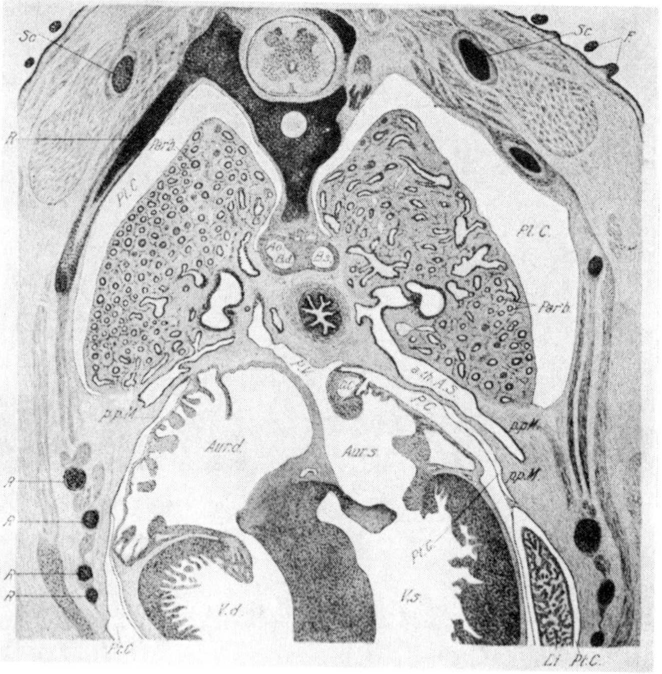

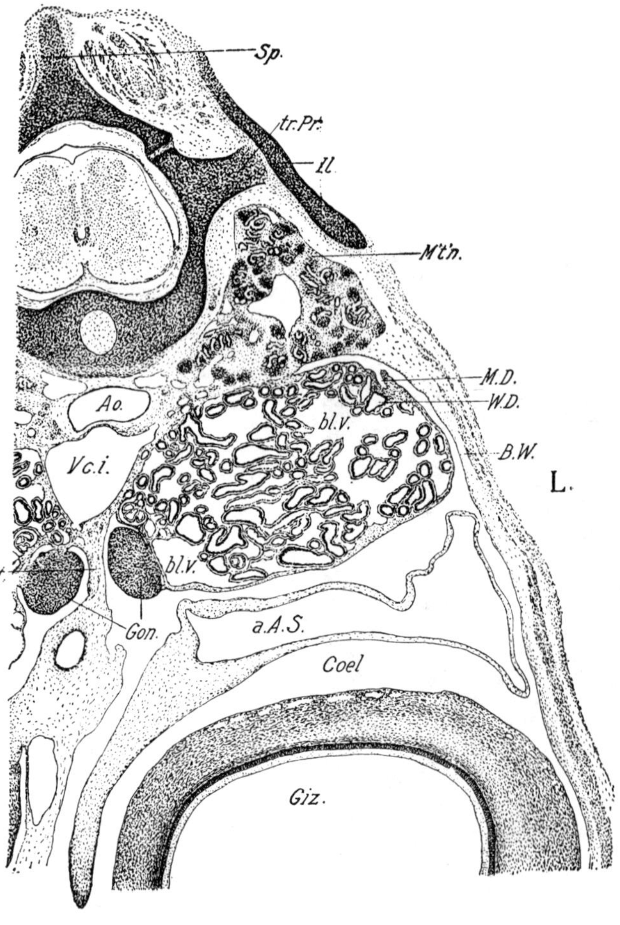

L.

TRANSVERSE SECTION THROUGH THE METANEPHROS, MESONEPHROS, GONADS, AND NEIGHBORING STRUCTURES OF AN 11-DAY MALE CHICK

a. A. S., Abdominal air sac. Ao., Aorta. B. W., Body-wall. Coel., Coelome. Giz., Gizzard. Il., Ilium. M. D., Remains of degenerating Mullerian duct. M's't., Mesentery. M't'n., Metanephros. Sp., Spine of neural arch. tr. Pr., Transverse process of the neural arch. V.c.i., Vena cava inferior. W. D., Wolffian duct.

Above pictures reproduced from *Development of the Chick* by Frank R. Lillie; Copyright, 1908, 1919 by Henry Holt and Company.

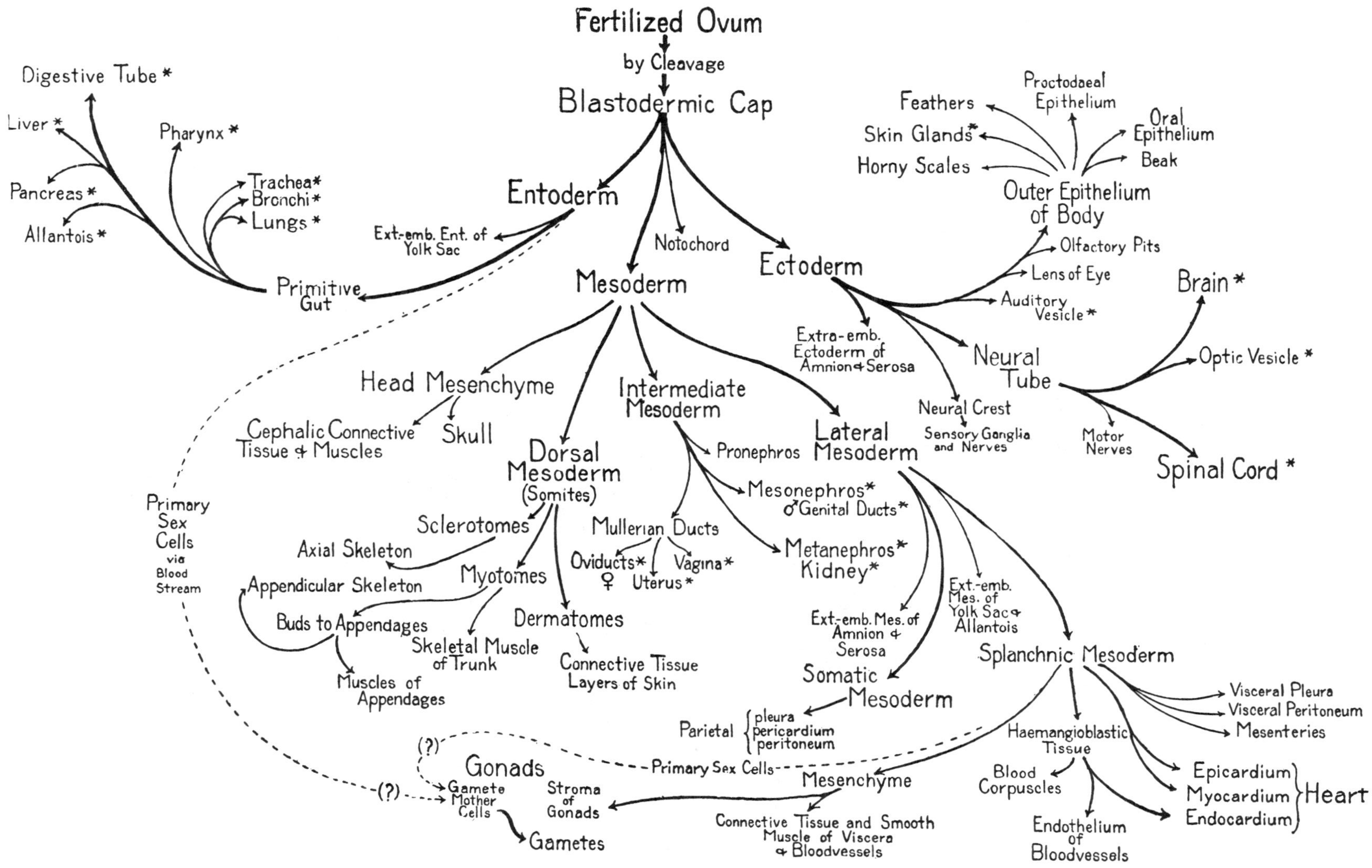

*Chart indicates origin of epithelial part of organ. All starred organs have secondary supporting investment of mesodermal origin.

Chart showing the derivation of various parts of the body by progressive differentiation and divergent specialization. Note especially how the origin of all the organs can be traced back to the three primary germ layers.

From B. M. Patten: *Embryology of the Chick*. 1957. Copyright The P. Blackiston Company, Philadelphia.

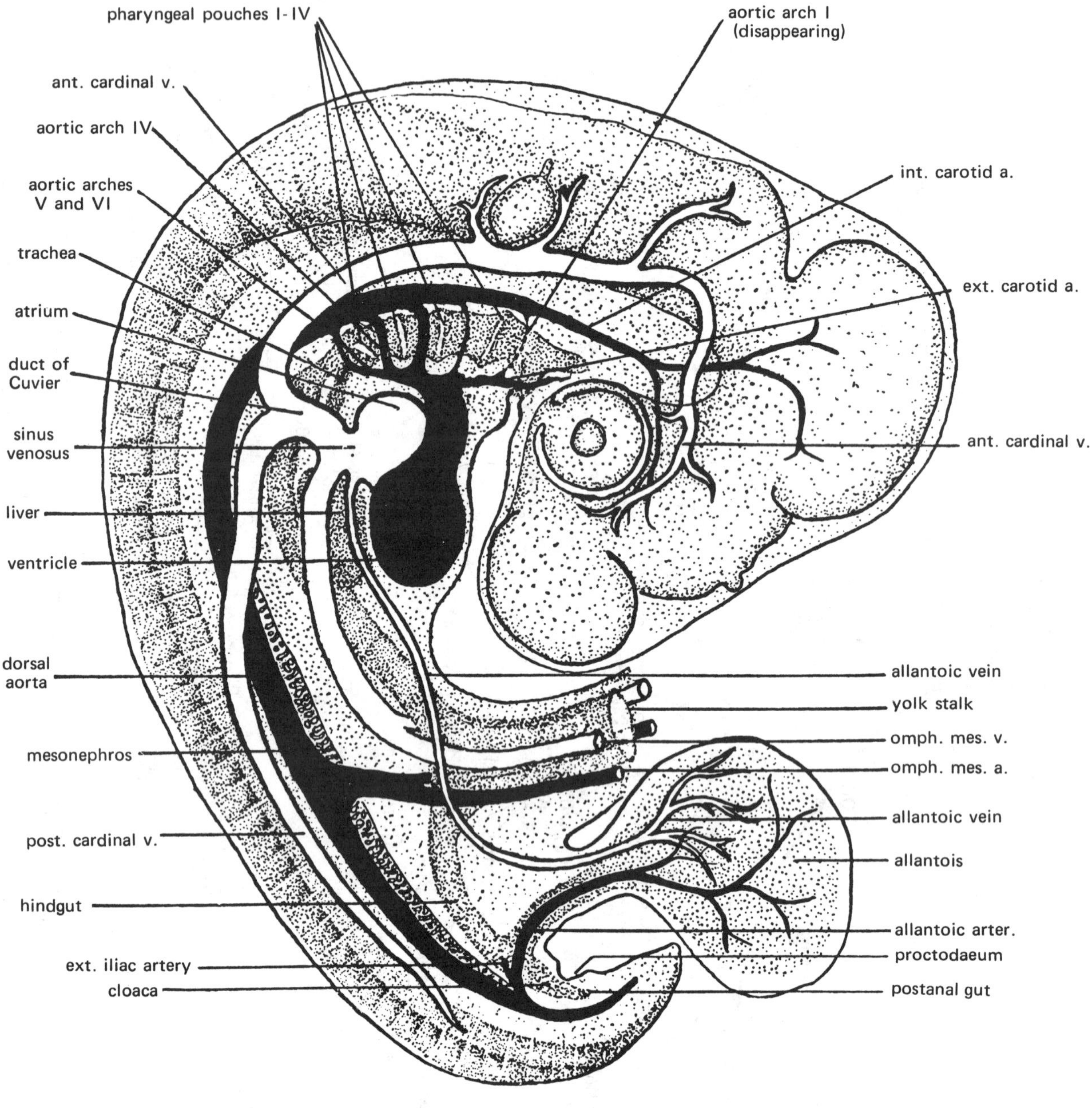

Simplified plan showing the location and relations of the main vascular channels of the 4-day chick. Except for the omphalomesenteric arteries and veins paired structures are represented only on the side toward the observer.

From B. M. Patten. *Embryology of the Chick.* 1957. Copyright The P. Blackston Company, Philadelphia.

See: Patten (1957): 207-46
Nelsen (1953): 541
Rugh: *Vert. Emb.* 180-231
Rugh: *Exp. Emb.* 405-52

METHOD FOR STUDYING LIVING CHICK EMBRYOS

Fresh fertile chick eggs may be kept at 20°C for as long as a week before the beginning of incubation at 103°F. There are various types of incubators, but the two important variables that must be controlled are the temperature and the humidity. Rotation of eggs is not important during the first 5 days of incubation. In calculating incubation time for eggs temporarily stored at a lower temperature, an interval of about 6 hours must be allowed before the egg temperature reaches that of the incubator.

It is suggested that the student begin with the easier 72- or 96-hour chick embryo and work back to the 24-hour stage, which is considerably more delicate, hence fragile.

THE 24- AND 48-HOUR STAGES (1 and 2 days)

The earlier stages are a bit more difficult to dissect out cleanly and transfer to a microscope for observation. It may prove of more value to transfer these stages to Syracuse dishes; by gradually reducing the amount of salt solution in the dish and pipetting fluid onto the embryo, a very nice flat demonstration may be secured. The heartbeat begins at about 38-40 hours and the entire circulation functions at about 44 hours. Note particularly the general body shape and the degree of flexure and the beginning of torsion at the anterior end.

The 24-hour embryo will appear as a white circular area on the dorsal side of the ovum and cannot be positively identified until after dissection and removal to a microscope for examination. The earlier primitive streak stage is the most difficult to identify in the living. Some technicians drop fixative directly onto the egg, whereupon the blastoderm changes color and may then be seen and dissected out. However, it is very instructive to study the living embryo at these stages, when the head process, proamnion, somites, and area pellucida are so distinct. For fixation, the method used for the 72-hour stage is recommended.

THE 72-HOUR STAGE (3 days)

Make up a normal salt solution consisting of 0.75% NaCl in distilled water and place in incubator for several hours or heat gently to about 37.5°C. Hold the egg in your hand for a few moments (to allow the blastoderm to float to a dorsal position) and then gently crack the underside of the egg against the edge of the fingerbowl containing the warm salt solution, separating the shell into two halves to allow the contents to flow out *into the salt solution.*

Examine the egg shell and locate the air chamber at the broader end. This air space is found between the two shell membranes, but elsewhere these two membranes are so close together that they appear as one. The shell itself is made up of three layers, consisting largely of calcareous spicules (see page 213).

Examine the egg and locate the twisted cords of heavy albumen known as the chalazae. With a pair of forceps, one may manipulate the egg by grasping the chalaza. The yolk mass (vitellus) is spherical and represents the true ovum. It is surrounded by the vitelline membrane, which may be seen against a dark background.

The blastodisc (living portion of the early egg) floats around to a dorsal position and in the 72-hour stage one can easily distinguish the embryo, the area pellucida, area vasculosa, and the sinus terminalis. Grasp the chalaza on one side with forceps and swing the egg around until it is convenient to insert very sharp pointed scissors near the position of the forceps but just outside of the area vasculosa. By cutting (not too deeply) through the blastodisc and at the same time moving the ovum by the chalaza (by means of the forceps), it is possible to quickly cut around the entire living blastodisc and remove it from the underlying yolk without injury to the embryo. Float it freely in the salt solution, away from the yolk mass. Shake it free from the vitelline membrane. Now, lower a clean microscopic slide beneath the salt solution and gently slide the blastodisc up onto the slide, making sure that the yolk side of the blastodisc is uppermost. By lowering and raising the slide and maneuvering the blastodisc, it will be possible to get it into position, flattened and spread out on the slide. Now, to hold it in position; place a small square of filter paper on the edge of the blastodisc and against the slide as both are raised out of the fingerbowl. This embryo should be studied immediately under a binocular microscope, or it may be preserved by pipetting onto it some Bouin's fluid or Klinenberg's picro-sulfuric acid solution.

In this 72-hour stage, the entire circulatory system can be made out. Also note the relatively tremendous brain vesicles; the various flexures and torsions; the heart, which is almost outside of the body; and the point of emergence and exit of the vitelline veins and arteries. Particularly instructive is the study of the beating of the various parts of the heart and variations in circulatory rate in different regions of the embryo and the area vasculosa.

THE 5- OR 6-DAY EMBRYO

By this time the embryonic tissues have almost completely enveloped the vitellus (yolk) so that care must be exercised to avoid tearing the embryo, which tends to stick to the surrounding shell membranes. These embryos should be studied while they are submerged in the warm salt solution, and, even with the naked eye, the rocking movements due to contractions of the smooth muscle fibers of the amnion can be noted. The allantois is now a large fluid-filled sac which almost envelops the embryo but is rather vascular as compared with the amnion. Look for indications of the yolk-sac septa or areas where the yolk is being invaded and digested.

With the naked eye identify various parts of the embryo proper. Then find the anterior and posterior cardinal veins, the duct of Cuvier, the dorsal aorta, and the various parts of the extraembryonic circulation.

Now, carefully rupture the chorion with fine scissors and expose the embryo lying within the amnion. By means of a blunt instrument, rotate the embryo and locate the yolk-sac umbilicus which attaches the embryo to the underlying vitellus, or yolk. Since all of the extraembryonic blood vessels enter the embryonic body through the umbilicus, it should not be necessary to cause any hemorrhage up to this point.

After identifying all possible structures and making appropriate sketches, grasp the amnion with forceps, cut through the umbilicus, and remove the embryo to a Syracuse dish containing warm salt solution. The embryo will remain active for some time even though it has been separated from the yolk and its circulatory system has been cut off from the extraembryonic areas. Dissect off the amnion to get a better view of the embryo itself.

PERMANENT MOUNTS OF CHICK EMBRYOS

For fixation either Bouin's or Klinenberg's picro-sulfuric has been used successfully. The important step is to have the entire blastoderm fixed in a completely flat position and free from yolk. This may be accomplished by cutting an opening out of a piece of filter paper just a bit larger than the blastoderm to be fixed and laying this over the wet blastoderm on the slide. Wash off adhering yolk by gently pipetting clean tap water or distilled water over the embryo and letting the excess water drain off. The fixative may then be gently pipetted onto the embryo and the whole gently lowered into a larger container of the same fixative. The filter paper may later be removed very easily, but such embryos, if left for 24 hours or more in the fixative, will generally remain flat.

After proper fixation the fluid on the embryo is gradually changed to 70% alcohol to which a small amount (2%) of NH_4OH (ammonia) has been added. This will remove all of the yellow stain, though the process may take several days or several changes of the alkaline alcohol. Just before staining, the fluid must be changed gradually back to distilled water and thereby neutralized.

For either whole mount or sectioned material the best stain is Conklin's modification of the haematoxylin stain. This consists of:

Harris's haematoxylin	1 part
Water	4 parts

Klinenberg's picro-sulfuric
acid, 1 drop per cc of above mixture

Time for staining:

1. 24 hours — 1½ minutes
2. 48 hours — 3 minutes
3. 72 hours — 5 minutes
4. Older stages may be overstained for several hours and then later destained to the proper degree in acidified 70% alcohol. However, whenever destaining by acidification, there must follow a neutralizing by thorough washing in pure 70% or better, the addition of a slight amount of ammonia in the 70% alcohol.

GENERAL REVIEW QUESTIONS ON CHICK EMBRYOLOGY

1. Describe the processes of maturation and fertilization in the chick.
2. Draw and label a schematic vertical and sagittal section through the early segmenting blastoderm as the segmentation cavity is being formed.
3. With a series of sketches illustrate gastrulation in the chick. How does this compare with gastrulation in Amphioxus, the fish, and the frog?
4. What are the reasons, pro and con, for homologizing the frog blastopore and the chick primitive streak?
5. How does the original mesoblast form? The first somite? The head mesenchyme?
6. Describe the origin of the coelom and its derivatives.
7. What are the major changes noted in the whole mount from 20 to 48 hours in incubation?
8. What is the relationship of Hensen's node to the future embryo? Is there a neurenteric canal in chick development? What is the positional relation of the future anus to the primitive streak?
9. What is the relationship of the neural tube to the notochord at the anterior and the posterior extremities?
10. Exactly what is meant by germ wall, zone of overgrowth, marginal periblast, zone of junction, area vitellina, and area vasculosa?
11. What and where are the yolk nuclei?
12. What is the origin of the blood vascular system, both in the embryonic body and in the extraembryonic areas?
13. Where and how do the heart rudiments develop? Describe through the fourth day of incubation.
14. Sketch a lateral view of a 72-hour chick embryo in outline only, and then draw in the entire circulatory system of that age, both embryonic and extraembryonic.
15. What are the changes in the blood vascular system at the time of hatching?
16. Be prepared to sketch either a transverse or a sagittal section through the level of the yolk stalk of a 3-day chick embryo.
17. Be prepared to sketch the development of the embryonic membranes, illustrating particularly the relationship of the allantois to the other membranes.
18. Has the frog any structures which might be considered as homologous to the chick extraembryonic membranes? How are the functions of these membranes achieved in the frog embryo?
19. What is the basis for the fundamental differences in development of the frog and the chick?
20. What evidences are there in chick development in support of the biogenetic law?
21. What are the developmental relations of the pro-, meso-, and the metanephric units in the chick?
22. What is the significance of the neuromere? Evidence in the chick? What are crests, placodes, ganglia, ependyma, and neuroglia?
23. Be prepared to enumerate all of the embryonic derivatives of each of the germ layers.
24. What are the major similarities and differences in the origin and development of the three major sense organ systems: olfactory, optic, and auditory?
25. What is the developmental relation of the sclerotome to the skeleton, to the skull, to the branchial skeleton, and to the jaw parts?
26. What is the fate of the branchial clefts and the visceral arches?

27. Describe the origin and early development of the lungs and air sacs.
28. Explain the interrelation of the liver, blood system, and the pleura.
29. What is the origin of the reproductive system (primary and secondary) in the chick?

"Continuity of life is the great end of living; survival its greatest good; extinction its greatest evil."

Conklin, 1944.

DRAWINGS: CHICK EMBRYO
LATE STAGES

TECHNICAL PROCEDURES FOR CHICK EMBRYO EXPERIMENTATION*

For some time it has been possible to extirpate, transplant, explant, and vital stain parts of chick embryos; this enables us to learn a great deal more about the developmental processes of this warm-blooded form than was previously possible. The choric-allantoic membrane of the 8½-10-day chick embryo has proven to be an excellent substratum for the development of structures isolated from other and younger embryos and examined after 9 to 10 days. The inherent developmental potencies of such grafts can be determined without interference from the normal inductive influences of the host, since the host's nutritious vascular bed alone is used. We will briefly outline here some of the possibilities for the chick embryo so that the qualified and interested student may be initiated into the rewards of such experimentation.

The source of fertile eggs should be reliable, and every egg should be examined by candling before use. Such eggs should not be stored for more than 6 days, kept at 55°F or 12-13°C. The incubator should preferably have forced air ventilation and be kept at about 103°F (or 37.5°C). Humidity control is even more important than is exact temperature control, with an optimum at 60%. The developing eggs should be turned twice daily, and the incubation period is normally 21 days.

Eggs of almost any breed, which are obtained directly from a farm where there are roosters, are best if used between February and June. Good fertile eggs are available at other seasons, the lowest production periods being during the heat of the summer or the extreme cold of the winter. During incubation there are two periods of relatively high mortality, the first at about days 3-4 when the membranes are beginning to function actively and the second between days 18-20 when the membranes begin to dry up and the chick must begin to rely more upon its own innate physiology.

THE METHOD OF CHORIO-ALLANTOIC GRAFTING

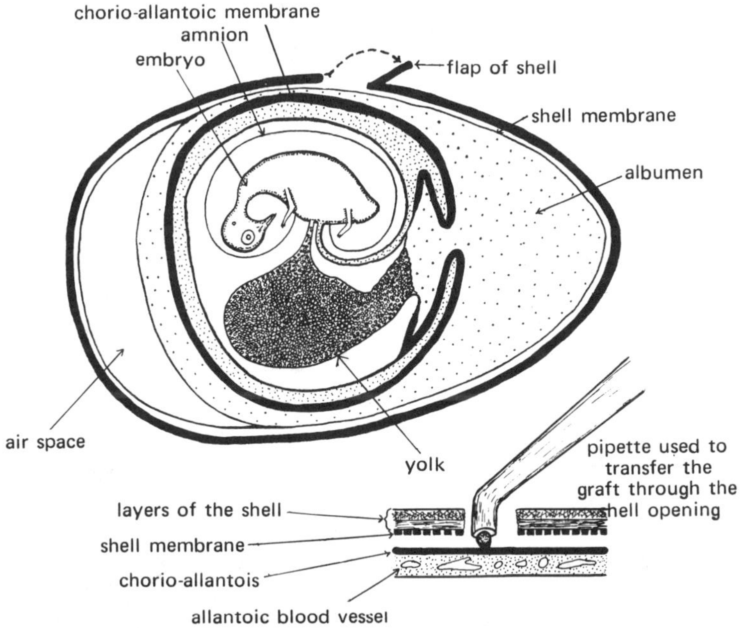

Diagram illustrating the method of chorio-allantoic grafting. Host embryo is generally incubated for 8 to 9 days. The graft may be any primordium (e.g., the limb bud of 72-hour stage). The graft is generally recovered at about the nineteenth day, just before the time of hatching of the host when the membranes begin to dry up.

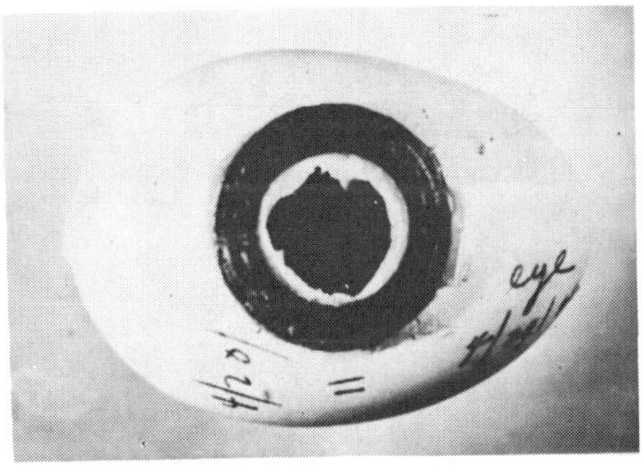

Chick egg with window above the developing embryo. Black washer is sealed to the egg shell and to the covering coverglass with paraffin. Irregular margin of torn shell membrane visible through the window. Embryo can develop and hatch.

*See final exercise in this Guide, where techniques and procedures are described which will aid in planning and carrying out experimental procedures with chick embryos.

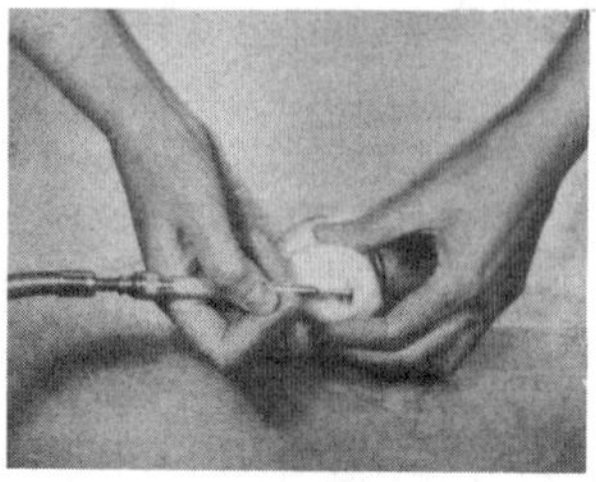

Sawing shell open
(Donor)

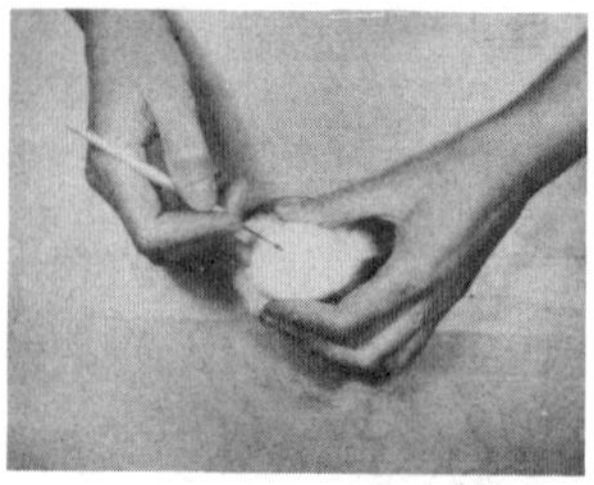

Picking away shell

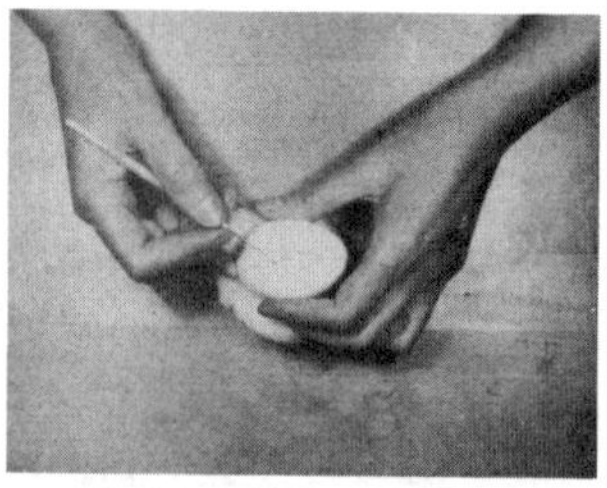

Shell partially removed

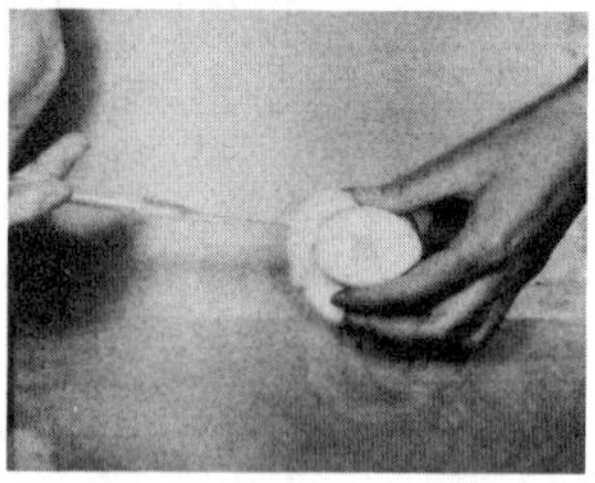

Moistening shell membrane

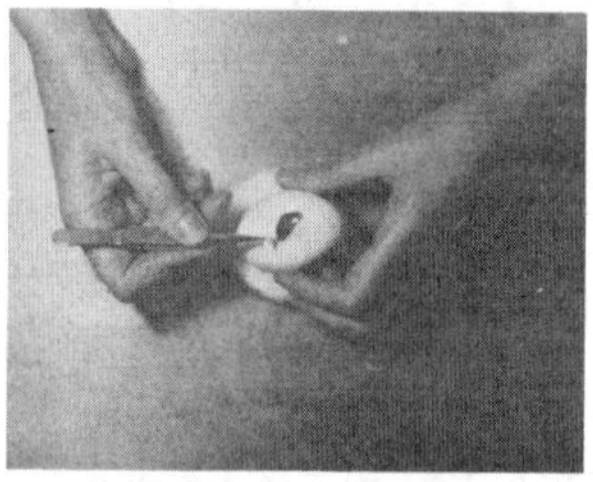

Stripping shell membrane

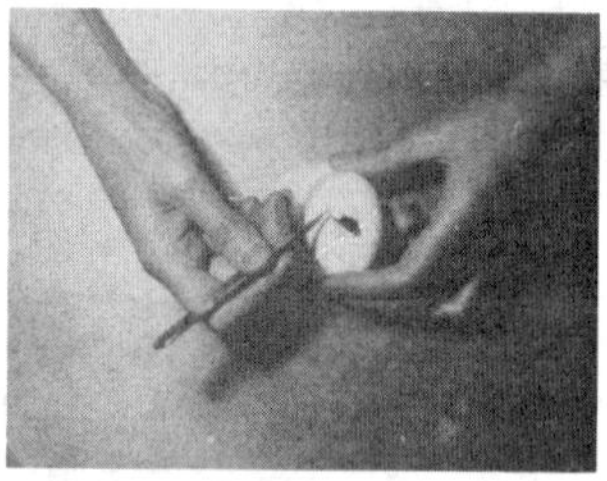

Stripping shell membrane

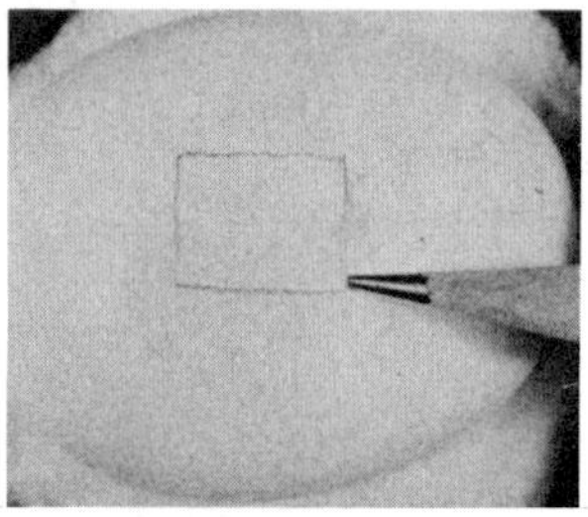

Marking window (host)

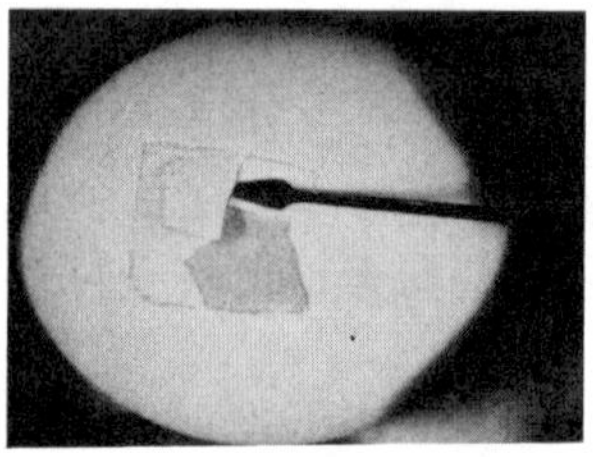

Picking off shell after window
is outlined with dental saw

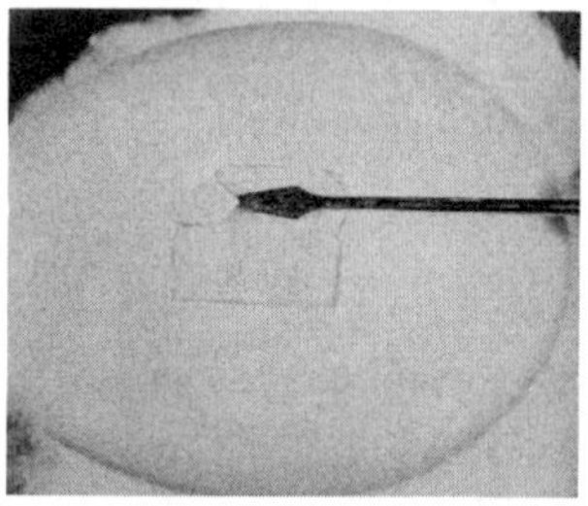

Shell membrane exposed

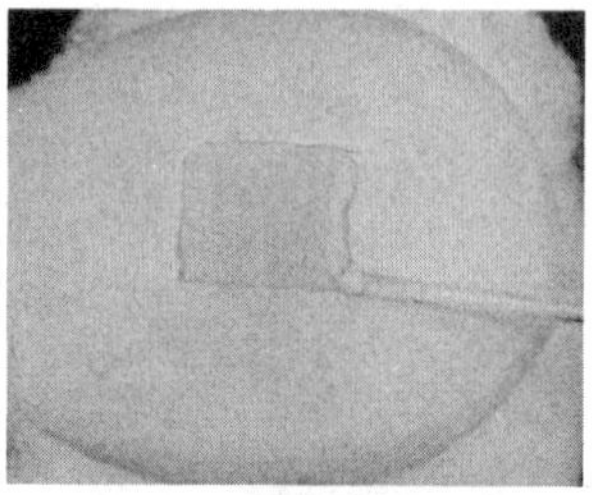

Moistening shell membrane

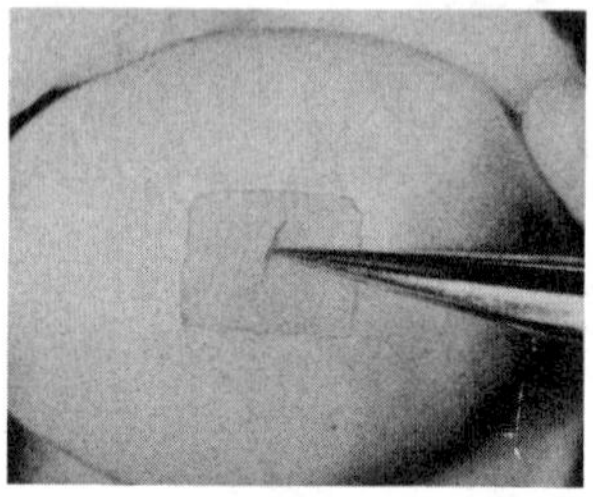

Making initial slit
in shell membrane

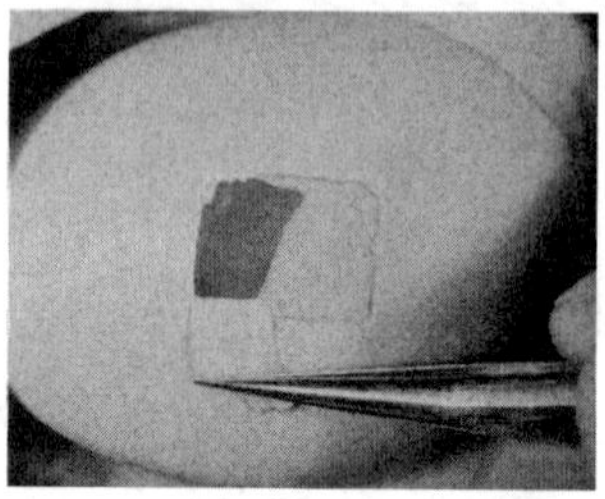

Stripping off shell membranes

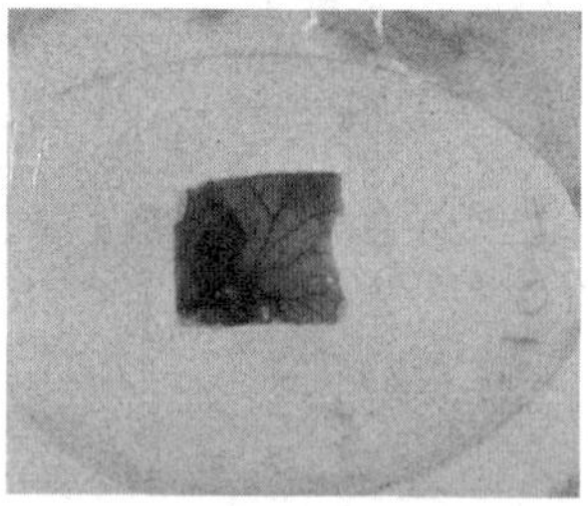

Donor embryo exposed for
operation

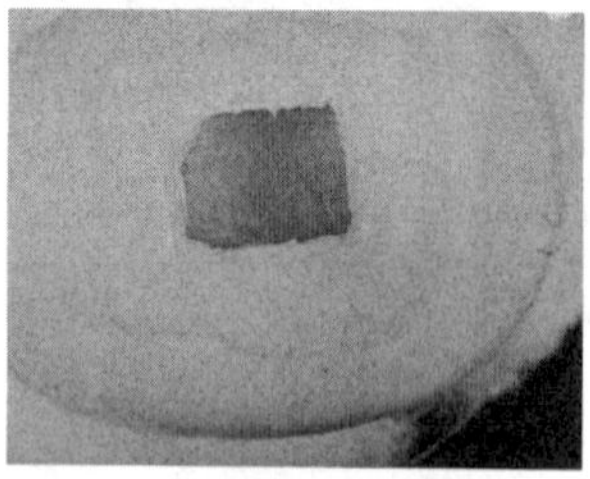

Paraffin sealed window
after operation

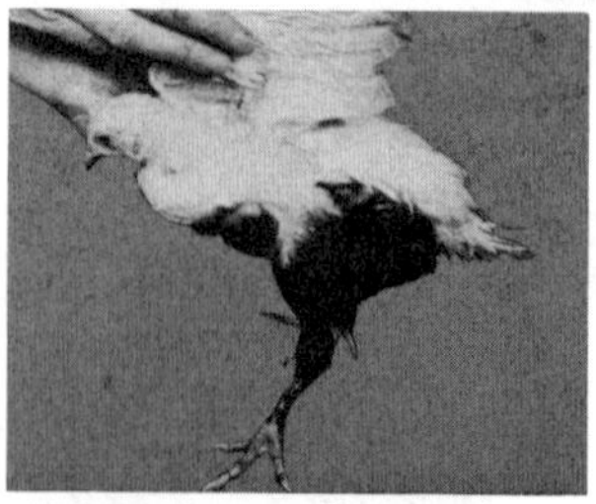

White leghorn with neural crest
graft from black donor

CHICK TRANSPLANTATION TECHNIQUES

Courtesy Dr. Frances Dorris Humm.

Candling equipment can be made easily by inverting a metal wastebasket over a 100-watt light held in a firm socket and cutting an oval hole in the bottom of the wastebasket not quite large enough to allow an egg to fall through. If the basket is then painted black and a black cloth attached to its circular base, a light-proof viewing area is provided through which one can quickly determine whether a viable embryo exists within any particular egg. In a matter of seconds the embryo will float around to the top. The yolk sac circulation can be seen at 2½-3 days; rocking movements of the embryo within its amniotic sac are visible during the next few days, and the expanding vitelline circulation can be followed. Should the early embryo die, a blood coagulum around the periphery of the area vasculosa becomes visible some time between days 3 and 7. The chorio-allantoic circulation becomes evident by day 4, and after day 13 the living embryo is so dense that its structures are difficult to discern even by candling.

The student who has become acquainted with the normal development of the chick embryo will enhance his knowledge by daily viewing the changes in the living egg via the candling method. In addition, it is a rather simple matter (under aseptic conditions) to open the shell and cut through its membranes on the upper side, as the egg is held in the horizontal position, and then substitute for the shell membrane and shell a thin circular coverglass held in place by a ring of melted paraffin. In this way and as long as the egg is held in the same position, the student will have a clear view through a window of the embryo developing within.

From days 8½ to 10 the chorio-allantois is so well developed and vascular that if it is exposed one can simply place upon it an eye primordium from a 33-hour chick donor or a limb primordium from a 2-3 day chick, cover with coverglass which is then sealed into place with paraffin, and then observe as long as the host remains alive, which may be until almost the normal time for hatching, 21 days.

With training and unlimited patience, it is possible even to transplant organs from a (black) chick embryo to another (white) chick embryo or vice versa and to produce a chicken with feathers, for example, of two colors. (Detailed descriptions are given in the author's *Experimental Embryology* or in Hamburger's *A Manual of Experimental Embryology.*)

The fact that the chick egg maintains a balanced and high temperature with great humidity and requires turning to avoid adhesions presents situations which lend themselves to experimentation. Alterations in temperature at critical times and to considerable degrees to determine tolerance of such variations; introduction of various drugs (e.g. thyroxine) into the air chamber for direct absorption into and effects on the embryo; placing minute particles upon the blastoderm to observe movements; and testing the tolerance of extirpations are all possibilities for the beginner in avian research.

> ***"I finally saw the blood, forced by the action of the left ventricle into the arteries, was distributed to the body at large, and its several parts, in the same manner as it sent through the lungs, impelled by the right ventricle into the pulmonary artery, and that it then passed through the veins and along the vena cava, and so around to the left ventricle ---- which motion we may be allowed to call circular."***
>
> **The origin of the term "circulatory system" by Harvey, 1628.**

EMBRYOLOGY OF THE PIG

Exercise 24

THE PIG

The student who has just completed a study of chick embryology will have little difficulty in understanding the early embryology of the pig. Depending upon the amount of available time, it is recommended that he examine the 6-mm and, if possible, the 10-mm pig embryo in serial sections and compare with the later stages in chick development. Following is a brief description which may be used in such a study, with a preliminary statement about the adult gonads and the maturation process. Fertilization, cleavage, gastrulation, and implantation are not available for study in the pig as they are in the mouse, for the following exercise.

THE REPRODUCTIVE ORGANS OF THE ADULT

The Female

The paired ovaries lie in the pelvic portion of the abdominal cavity, each gonad practically surrounded by the flared ostial opening of the oviduct. It is not known whether they are cilia in the vicinity of the infundibulum to aid in egg transport, but the positional relation of the oviduct and the ovary practically ensures proper transport of the eggs. If the eggs are fertilized within the bicronate uterus, they become attached and develop. The two branches of the uterus fuse and form the corpus which leads through the cervix into the vagina. The vagina serves both as an organ for copulation and a birth canal. Sows which fail to become pregnant after copulation come into heat again after an interval of about 21 days and maintain the cycle until it is interrupted by pregnancy.

The Male

The paired testes lie outside of the body in the scrotal sac; the body temperature is too high for the survival of the spermatozoa. Mature spermatozoa pass from their seminiferous tubules into the tubuli recti and then into the anastomosing rete testis. The efferent ducts collect these spermatozoa and convey them by way of the coiled epididymis into the large, heavy-walled vas deferens or ductus deferens. This duct leads back into the body of the boar and loops around the ureter to join the urethra at about the level of the prostate and Cowper's glands. Just before entering the urethra, the ductus deferens enlarges as the ampulla. This is the seminal vessel, and its walls are highly glandular. During coitus the sperm are carried into the vaginal canal along with secretions from the seminal vesicle and the prostate and Cowper's (bulbourethral) glands.

THE MATURATION PROCESS

OÖGENESIS: Relatively few mature ova are produced by the mammal, the bulk of the ovary consisting of stroma. Scattered egg nests are formed by the breaking up of ovigerous (sex) cords, with a few cells enlarging at the expense of all others, any of which might have become ova. The cells surrounding the one selected by chance for development will constitute the Graafian follicle, and, as the presumptive oocyte grows, the follicle expands and acquires a large cavity (antrum) filled with liquor folliculi. The maturing egg cell comes to lie within the follicular fluid and is attached to the wall of the follicle by a neck of cells known as the discus proligerus. The mature egg and follicle are always found peripherally in the ovary and in the more mature stages may be seen projecting from the surface ready to rupture into the limited body cavity, thence to be picked up by the infundibulum, the fimbriated funnel of the oviduct.

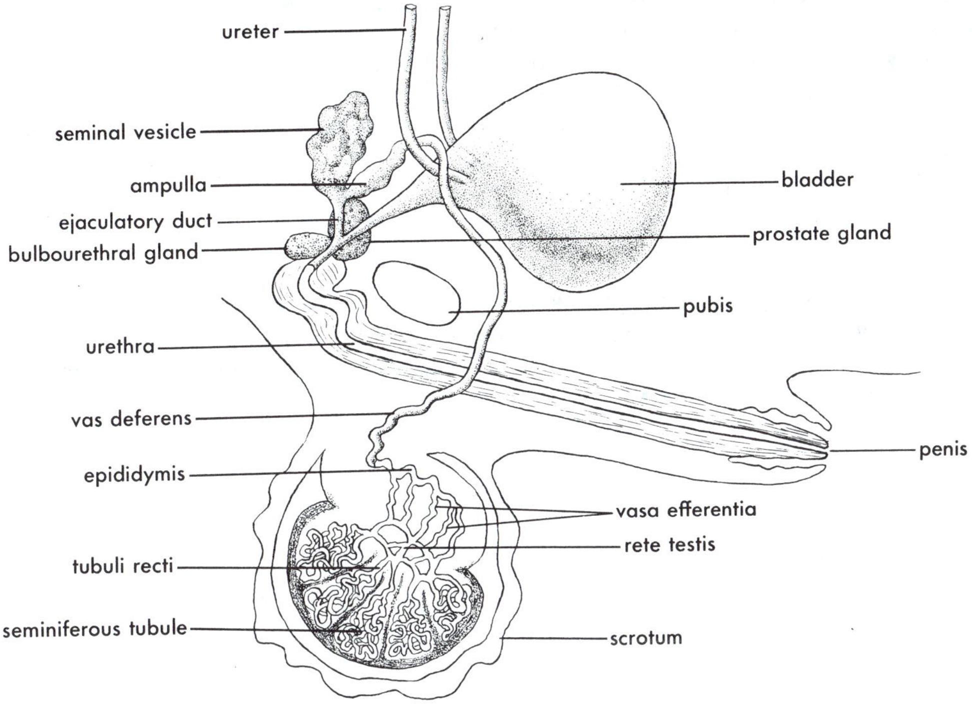

MALE (BOAR)

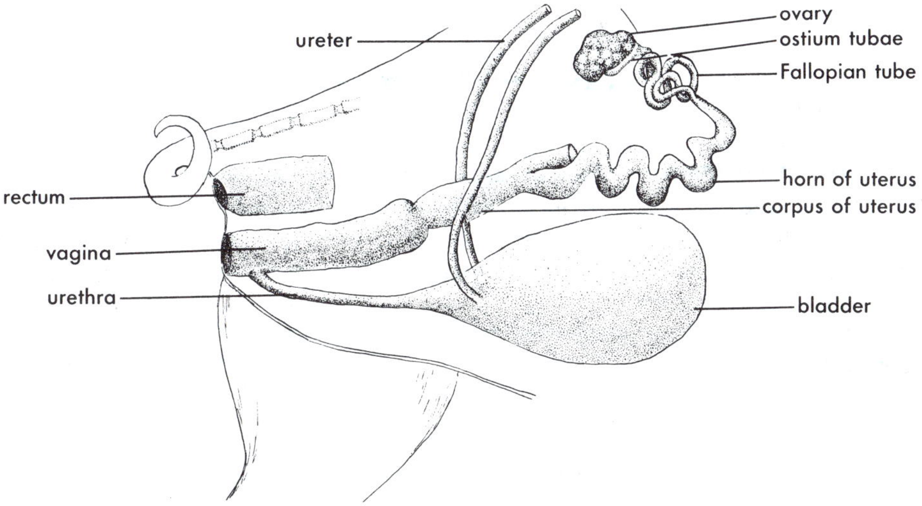

FEMALE (SOW)

UROGENITAL SYSTEMS OF THE MALE AND FEMALE PIG, BOTH LATERAL VIEWS

From B. M. Patten. *Embryology of the Pig.* 3rd ed. 1948 McGraw-Hill, New York.

The mature ovum shows a well-formed nucleus, cell and vitelline membrane, and a clear zona pellucida or zona radiata. The corona radiata, outside of the zona radiata, is continuous through the discus proligerus (or cumulus oöphorus) with the stratum granulosum (follicle cells) which make up the wall of the entire follicle surrounding the liquor folliculi in the antrum. Surrounding this (Graafian follicle) are the theca interna and the theca externa of the ovarian stroma, which are penetrated by blood vessels and nerves. These two connective tissue layers are collectively known as the theca folliculi. The causes of follicular rupture are not known but may be related to circulatory or muscular pressure or to a cytolytic enzyme in the follicular fluid, or all these. The first maturation division probably occurs at the time of ovulation.

Corpora lutea generally form from all ruptured follicles but remain only if the ova concerned are fertilized and implanted within the uterine mucosa. Corpora lutea of ovulation and of pregnancy are histologically similar but are not of equal survival. Eventually the corpus luteum becomes scar tissue of the ovary known as corpus albicans.

SPERMATOGENESIS: See Patten (1948), Chapter II.

THE SEX CYCLE

In mammals, there is a cyclic production of ova with concurrent changes in the physical, physiological, and psychic activity of the female. The changes are collectively known as the estrous cycle (21 days in the pig), made up of four stages:

a. *Estrus* – complete readiness for reproduction, high sexual desire.
b. *Post estrum* – regressive changes in reproductive tract due to failure of the ovum to achieve fertilization.
c. *Diestrum* – rest period, generally rather longer than the other periods.
d. *Proestrum* – changes in reproductive tract leading to estrus. In most mammals (except primates) copulation is allowed by the female only at times when fertilization can be accomplished. Changes in the vaginal and uterine tract are now recognized as accurate criteria for changes within the maturing Graafian follicle.

CLEAVAGE AND EARLY DEVELOPMENT

Contrary to expectations, the cleavage rates of eggs of warm-blooded mammals are very slow. The pig ovum reaches the 2-cell stage about 50 hours after fertilization, the 4-cell stage in about 60 hours, and the earliest blastula stage in about 4½ days. The noncellular zona pellucida becomes displaced by the cellular trophoblast, clinging to the wall of which is an inner cell mass suspended into the blastocoel. The cell mass becomes the blastocyst, the inner cell mass elongates and acquires a primitive streak (diploblastic) which shortly gives rise to mesoderm (triploblastic). Hensen's node, the notochord, and coelomic cavities are developed much in the same manner as in the chick, with the exception that yolk is lacking. The pig's primitive streak of 12 days is comparable to that of the chick at 19 hours. The neural groove appears on the thirteenth day and several somites on the following day. (Reference: Patten, 1948, Chapters IV and V.)

THE 6-MM PIG EMBRYO OF 18 DAYS

This stage of pig development corresponds well with that of the 72-hour chick embryo. The embryonic membranes, amnion, chorion, and allantois develop much in the same manner as they do in the chick, but their junctions are modified. There is no permanent yolk sac in the pig. The allantois fuses with the chorion to constitute the placenta. This organ aids in respiration, excretion, and nutrition of the embryo. (Besides the illustrations on the pages that follow, the 6-mm pig embryo is photographed in serial sagittal sections in Plate 6, pages 378-79, and in serial x-sections in Plate 7, pages 380-83.)

EXTERNAL FORM: The body of the 6-mm pig embryo is shaped much like an almost closed letter "C," with the telencephalic vesicle all but touching the flexed tail. The anteriorly protruding cephalic flexure (at level of mesencephalon) plus the cervical flexure (at the myelencephalon) tend

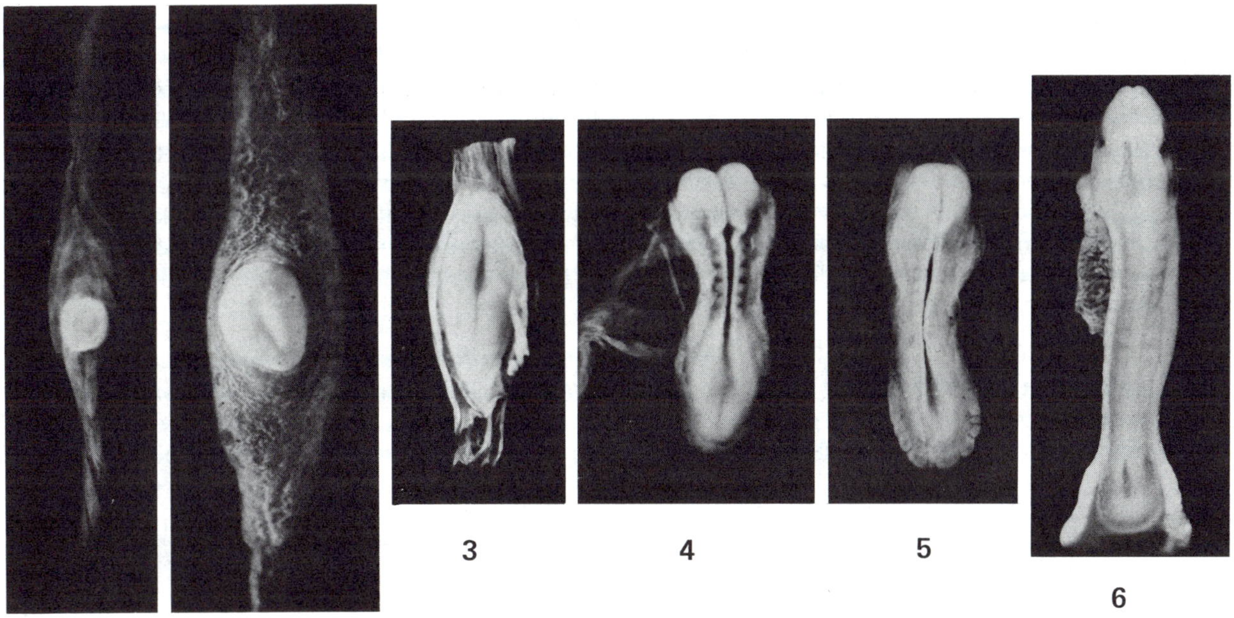

1 2 3 4 5 6

EMBRYO FROM PRIMITIVE GROOVE TO 3-mm STAGE

(DORSAL VIEW)

Carnegie Institute, Washington, D.C. Publ. #394; Contrib. Embr. #109.

Dorsal views of earliest available pig embryos. The first three show stages in primitive groove formation. The last three show neural groove, neural folds, a remnant of the primitive streak (posteriorly), and both anterior and posterior neuropores (Fig. 4), in an embryo measuring 3 mm and with seven somite pairs. Unfortunately, such early stages are not available for laboratory study.

to separate the head from the body. Serially arranged somites are seen from the cervical flexure extending posteriorly into the tail. The telencephalic olfactory bulbs lie midway between the eyes and the heart, dorsal to which are three branchial (visceral) arches with intervening pseudo-clefts. Between the arches and the body is a cervical sinus, a vertical cleft. The anterior atrium and the posterior ventricle remind one very much of the chick heart. Posteriorly the large liver bulge may be seen, and at about the midbody level is the anterior limb bud. Posteriorly to the limb bud is the huge and precocious mesonephros typical of the pig embryo. The hindlimb bud may be seen near the tail.

NERVOUS SYSTEM: The five regions of the brain are differentiated at this stage: the telencephalon, diencephalon, mesencephalon, metencephalon, and myelencephalon. The spinal cord extends into the tail, giving off numerous paired spinal nerves with ganglia. The third, seventh, and eighth to twelfth cranial nerves are all indicated. The sense organs consist of the thickened ectodermal olfactory pits, optic cups with related choroid fissures, lens with vesicles, and the otic vesicles with elongating endolymphatic ducts. CN 5, 7, 8, 9, 10

DIGESTIVE AND RESPIRATORY SYSTEMS: Rathke's pocket is the epithelial (ectodermal) hypophysis which has formed anterior to the oral membrane. The pharynx is a broad cavity with an anteriorly projecting diverticulum or blind pocket known as Seessel's pocket. Posterior to the large mandibular arch is the tongue, formed from two lateral components. Then comes the triangular elevation, the tuberculum impar about the level of the hyoid arch, posterior to which is the previously developed median thyroid diverticulum. The fourth and fifth visceral arches plus the tuberculum impar form the epiglottis. Parts of these same arches form the roots of the tongue. The thyroid gland lies between the ventral diverticula of the third pair of visceral (or first branchial) pouches.

The glottis groove, posterior to the epiglottis, leads into the laryngotracheal groove. At this level of the heart, a digestive tract derivative has separated as the trachea, which bifurcates into the primary bronchi and lung buds. These buds project into the pleural (coelomic) cavities. The short esophagus passes into the enlarged straight stomach, which later will twist to the left. The large liver diverticulum is next seen with the developing cystic duct, gall bladder, and common bile duct. The pancreas has both dorsal and ventral rudiments at this stage. The yolk sac is that portion of the midgut which extends almost out of the body, while the hindgut gives off the allantois from a posterior dilatation. Mesonephric ducts open into this cloaca. The cloacal endoderm fuses with the nearby ectoderm to form the cloacal membrane, which later breaks through to form the anus and urogenital apertures (See postanal gut).

THE URINOGENITAL SYSTEM: The mesonephros consists of many vascular glomeruli and tubules, and the mesonephric (Wolffian) duct runs along the ventral surface of the mesonephros, eventually entering the expanded cloaca. Near the cloaca the mesonephric duct gives off a dorsal diverticulum which is to become the permanent kidney or metanephros. The allantois is joined to

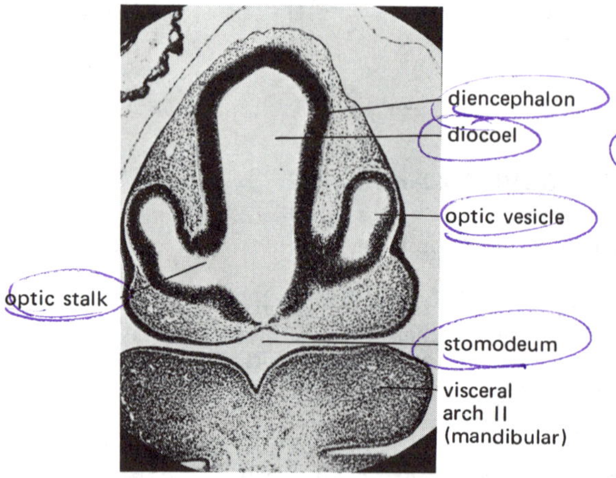

PIG EMBRYO AT 6-mm

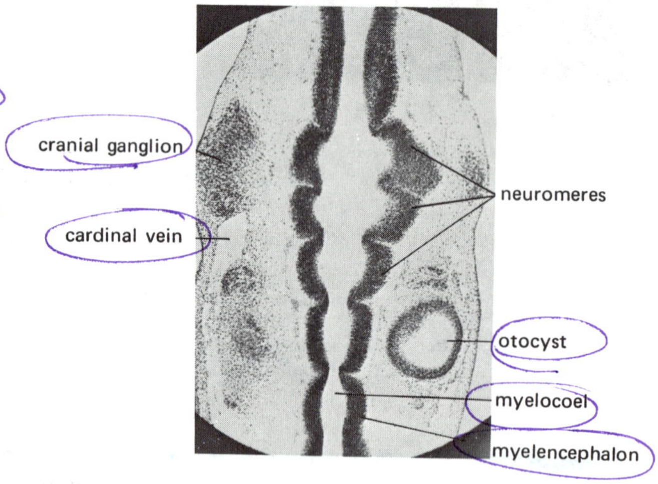

PIG EMBRYO AT 6-mm

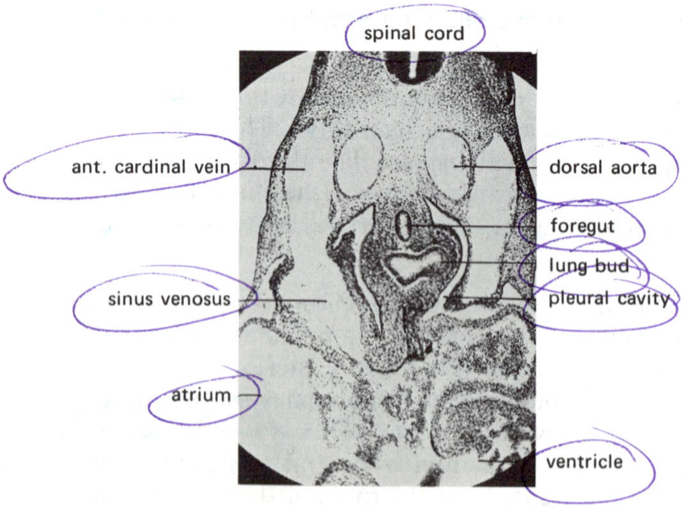

PIG EMBRYO AT 6-mm

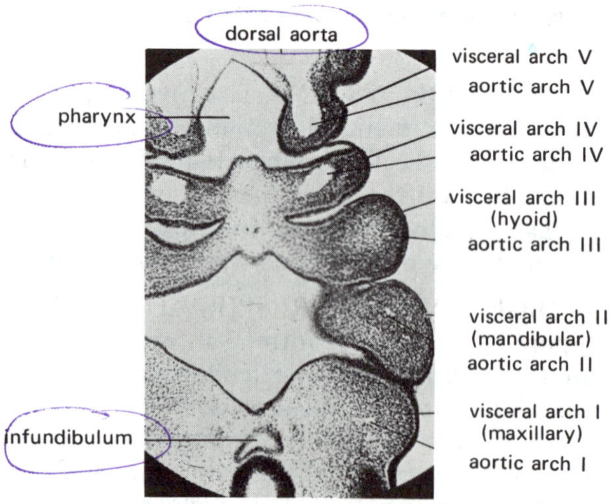

PIG EMBRYO AT 6-mm

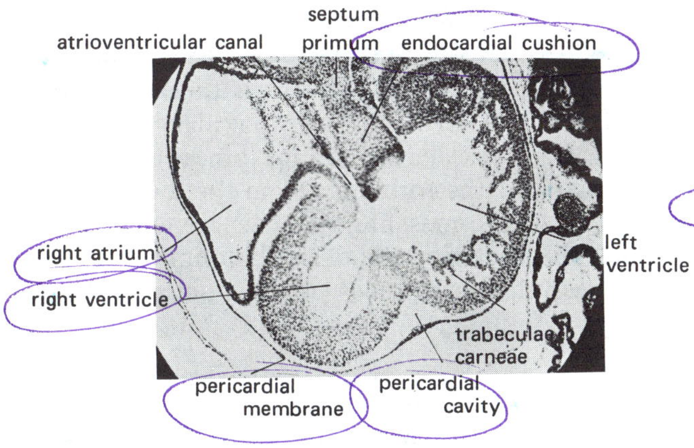

PIG HEART AT 6-mm

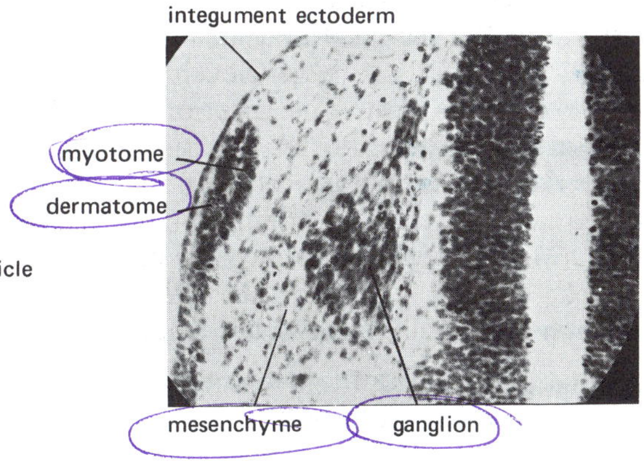

PIG EMBRYO AT 6-mm

the ventral wall of the cloaca. The genital folds are found along the median and ventral face of each mesonephros at this stage of development.

THE COELOM AND MESENTERIES: The originally continuous coelomic cavity has, by this stage, become subdivided into a single pericardial cavity, paired pleural cavities, and the large common peritoneal chamber. The pericardial cavity is constricted off from the coelom in part by the septum transversum, which will give rise to some of the diaphragm. The double splanchnic mesoderm which supports the gut may now be recognized as the greater omentum (from the stomach), mesoduodenum, mesentery, and mesocolon. Other mesenteries are the lesser omentum between the stomach and duodenum and the falciform ligament between the liver and the body wall.

THE VASCULAR SYSTEM: The thin-walled and paired atrial cavities may be easily distinguished from the thick-walled and paired ventricles. The smaller right ventricle is continuous with the ductus arteriosus and the dorsal aorta. The paired common cardinals and vitelline veins all empty into the sinus venosus, which in turn feeds the right atrium or auricle. The right vitelline vein later becomes the common hepatic, bringing blood from the umbilical veins, the yolk sac, gut and liver sinusoids to the heart. Auricular-ventricular canals and valves within the heart may be seen at this stage.

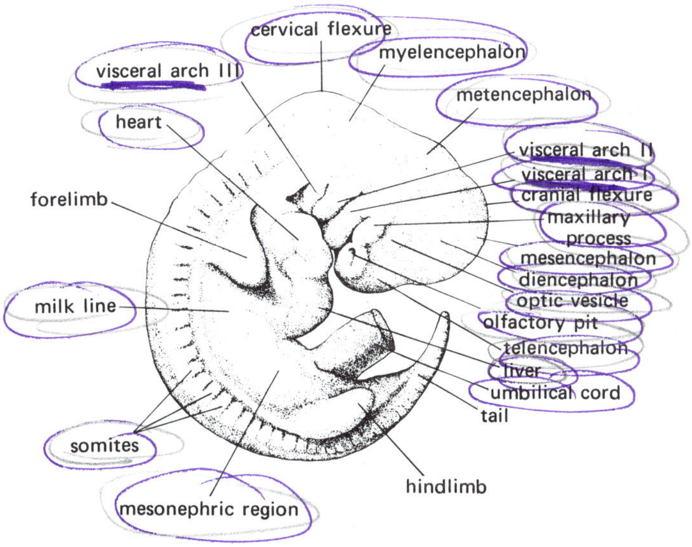

6-mm PIG EMBRYO 18 DAYS

From the ventral aorta five pairs of aortic arches run axially through the corresponding visceral arches and join the paired dorsal aortae. The apparent fifth is really the sixth, later to become the pulmonary, with the fifth never developing properly. Of these arches the fourth is the largest and the first two the smallest. The anterior extensions of the dorsal aortae into the maxillary processes are the internal carotids. Posteriorly the paired dorsal aortae (descending aortae) join, then give off intersegmental arteries between the somites; the subclavians to the forelimbs from the seventh pair of intersegmentals; renals to the mesonephros; the ventral to the mesenteries, which extend to the liver and stomach region, collectively known as the coeliac arteries. Near the hindlimb level the dorsal aorta divides, giving off paired umbilical arteries, then continues as the common iliacs, hypogastrics, and finally the caudals to the tail.

The originally paired vitelline veins are now fused, and, as they enter the body, they are joined by the superior mesenteric vein to form the hepatic portal vein. Actually, this consists largely of the right vitelline vein with contributions from the superior mesenteric. Within the liver, these vitelline veins break up into sinusoids. The original portion of the right vitelline vein between the liver and the sinus venosus joins the inferior vena cava. The umbilical veins originate in the chorion and allantois, the left one being the larger. These veins pass by way of the somatopleure until they enter the liver where the left one widens as the ductus venosus, which joins the right vitelline vein. The remnant of the right umbilical vein joins the portal vein in the liver. The anterior cardinals are ventrolateral to the hindbrain, draining blood from the head. These are joined by the external jugulars (also from the branchial arches) and together empty into the common cardinals and sinus venosus. The posterior cardinals are dorsolateral to the mesonephros and carry blood to the common cardinals. By ventromesial growth these posterior cardinals give rise to subcardinals located ventrally to the mesonephros. The right subcardinal, right posterior cardinal, and the right vitelline trunk (common hepatic vein) are all connected with the liver sinuses by a small vein which develops in the mesenchyme to the right of the mesogastrium, later to become the inferior vena cava.

REFERENCES

Arey: 609-47 Patten (1948): 1-91 Rugh: *Vert. Emb.* 305-28

THE 10-mm PIG EMBRYO

This is the most instructive stage in pig embryonic development and compares well with the mouse embryo at 14 days or the 12-mm human embryo of about 7 weeks. Its morphological details will be readily identified by those who have studied the 72- or the 96-hour chick embryos. (Besides the illustrations on the pages that follow, the 10-mm pig embryo is photographed in serial transverse sections in Plate 8, pages 384-87.)

THE EXTERNAL FORM

Compare the specimen to be studied with the accompanying drawing. Note that the head is bent sharply at right angles to the longitudinal axis of the body as a result of the cephalic flexure at the mesencephalon and the cervical flexure at the level of the myelencephalon. Distinguish the head, trunk, tail, and paired anterior and posterior limb buds.

Note that the head shows the typical brain vesicle bulges. On the ventral surface find the olfactory pits, bounded by lateral and median nasal processes. The optic cup and lens are well developed. The maxillary and mandibular portions of the first branchial or pharyngeal arch are very large, the former fusing with the nasal processes to form the upper jaw, while the mandibular arches fuse ventrally to form the lower jaw. The nasolachrymal groove marks the anterior boundary of the maxillary process. The hyomandibular groove, posterior to the mandibular arch, becomes the external acoustic meatus (ear). The hyoid (second visceral) arch and the third are largely hidden in the cervical sinus, which is formed by the cervical flexure.

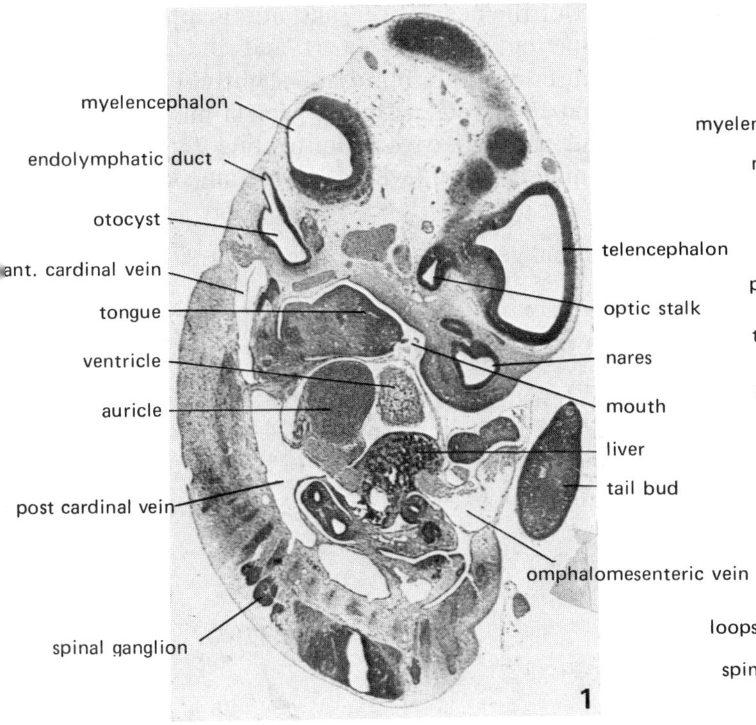

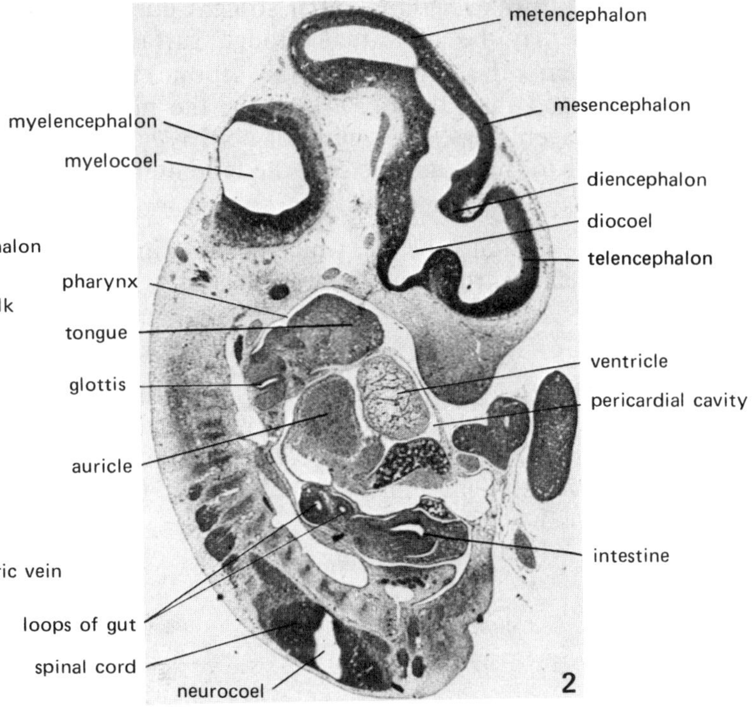

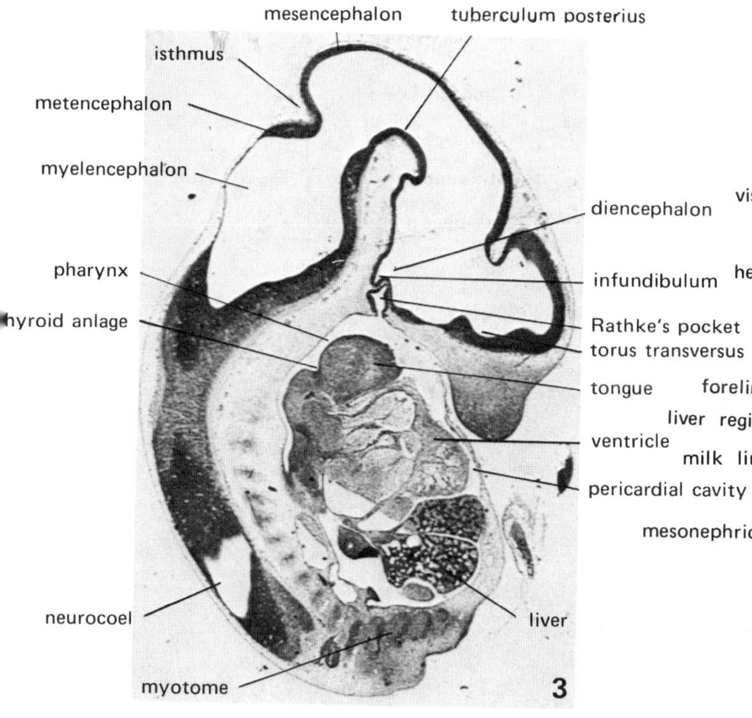

PIG EMBRYO AT 6-mm
SAGITTAL SECTIONS

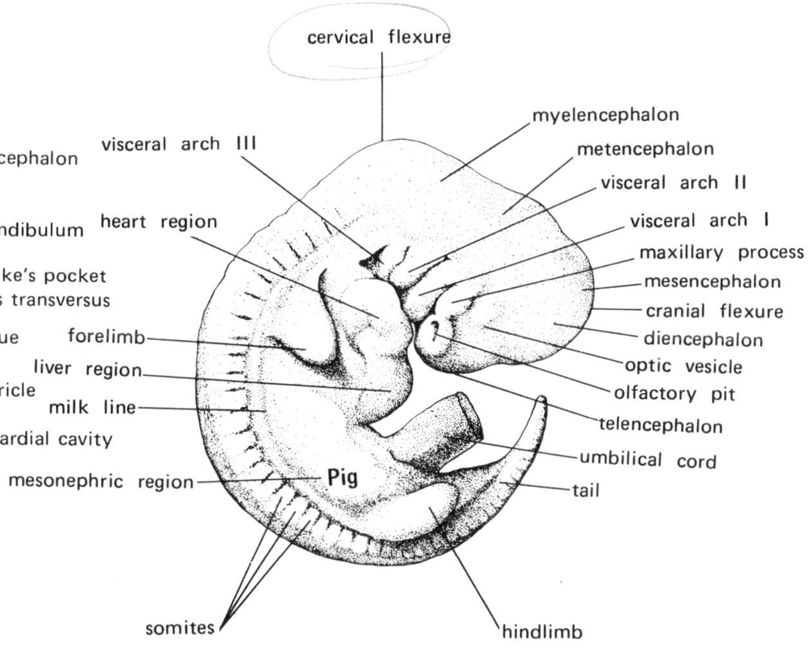

LATERAL VIEW OF 6-mm (18-day) PIG EMBRYO
From R. Rugh. *Vertebrate Embryology*. 1964. Harcourt, Brace and World, New York.

The trunk region is rather straight due to the development of the heart, liver, and mesonephros, which may be indicated through surface bulgings. The line between the heart and the liver represents the position of the septum transversum. The anterior limb buds are slightly more developed than the posterior pair; the metameric somites and their related spinal ganglia may be seen as bulgings; the umbilical cord is large and ventral; and parallel to the somites but slightly ventral to them, may be seen the faint milk line, where the mammary glands will later (15-mm stage) appear.

The genital tubercle, which is the primordium of the external genitalia, may be seen in the fold between the trunk and the tapering tail.

INTERNAL ANATOMY OF 10-mm PIG EMBRYO

(The following description is given in terms of organ systems rather than in terms of section levels. The student is advised to refer to the accompanying labelled drawings [pages 224-46] and to Patten's *Embryology of the Pig* for details relating to any specific section studied. It is important that the student understand the stage or degree of development of each of the various organ systems.)

REFERENCES

Balinsky: 367-423 Bodemer: 252-57 Rugh: *Vert. Emb.* 304-77

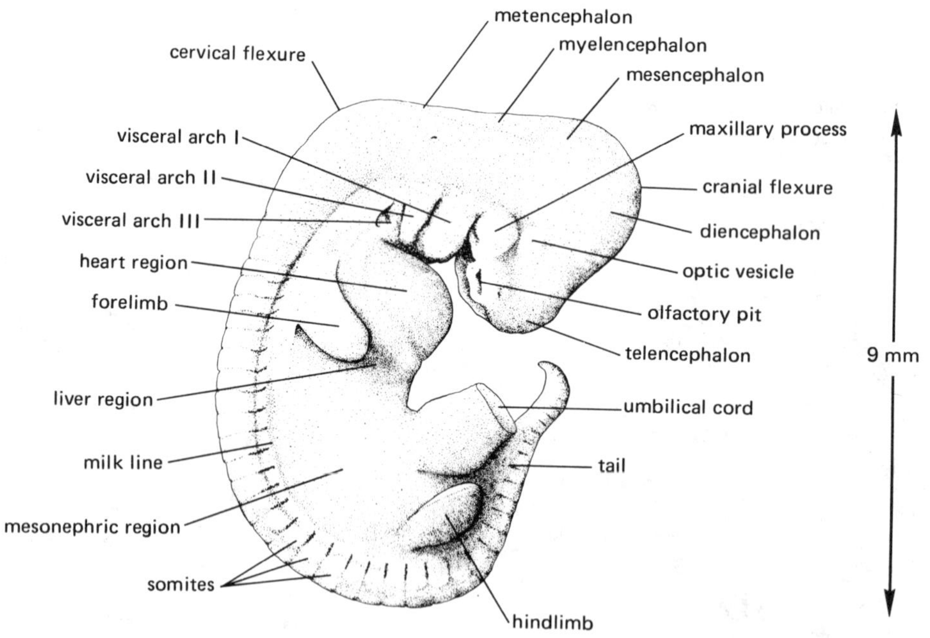

LATERAL VIEWS OF 9-mm PIG EMBRYO

From R. Rugh. *Vertebrate Embryology*. 1964
Harcourt, Brace & World, New York.

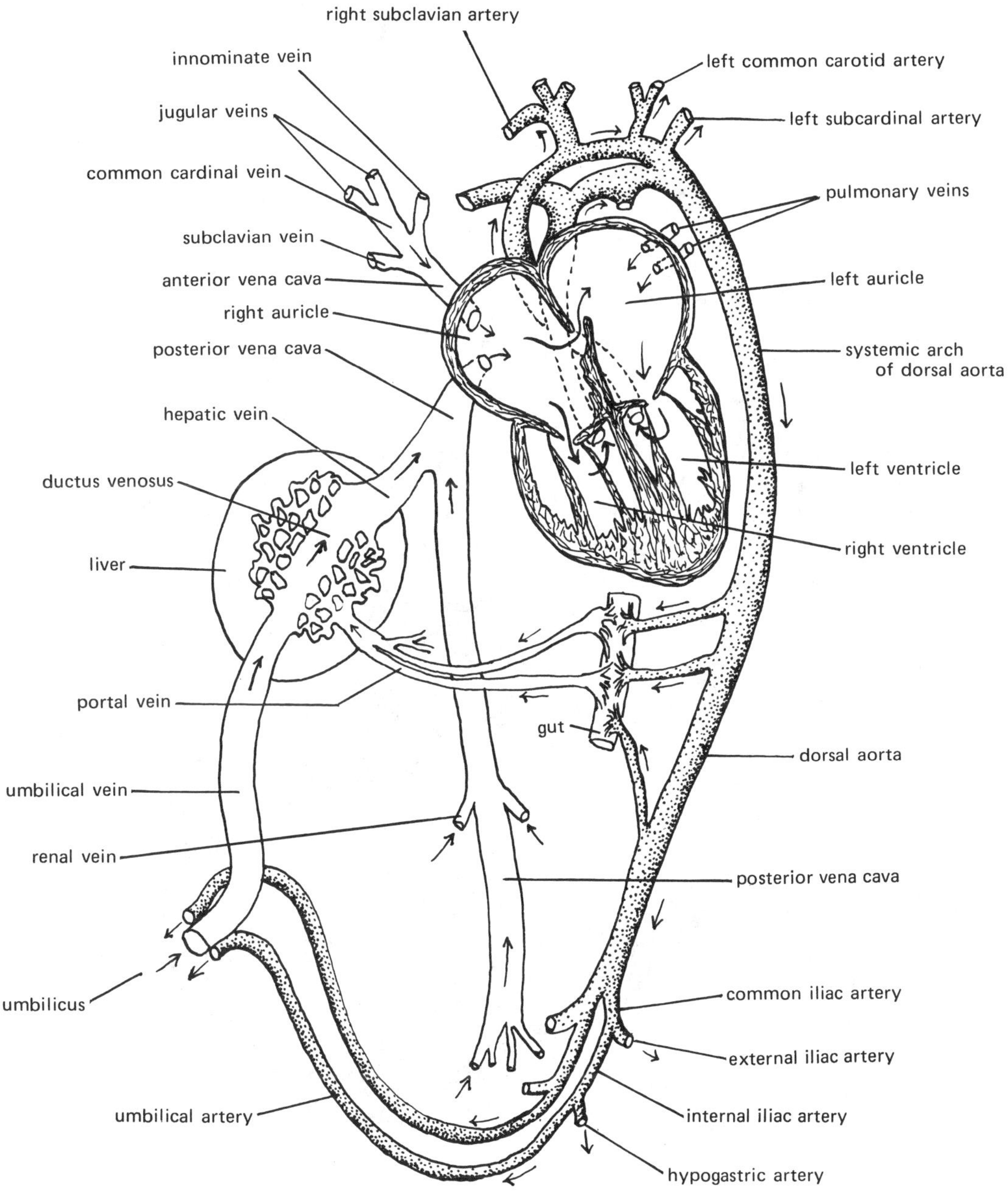

CIRCULATION LATE MAMMALIAN FETUS

Redrawn from Patten, 1968.

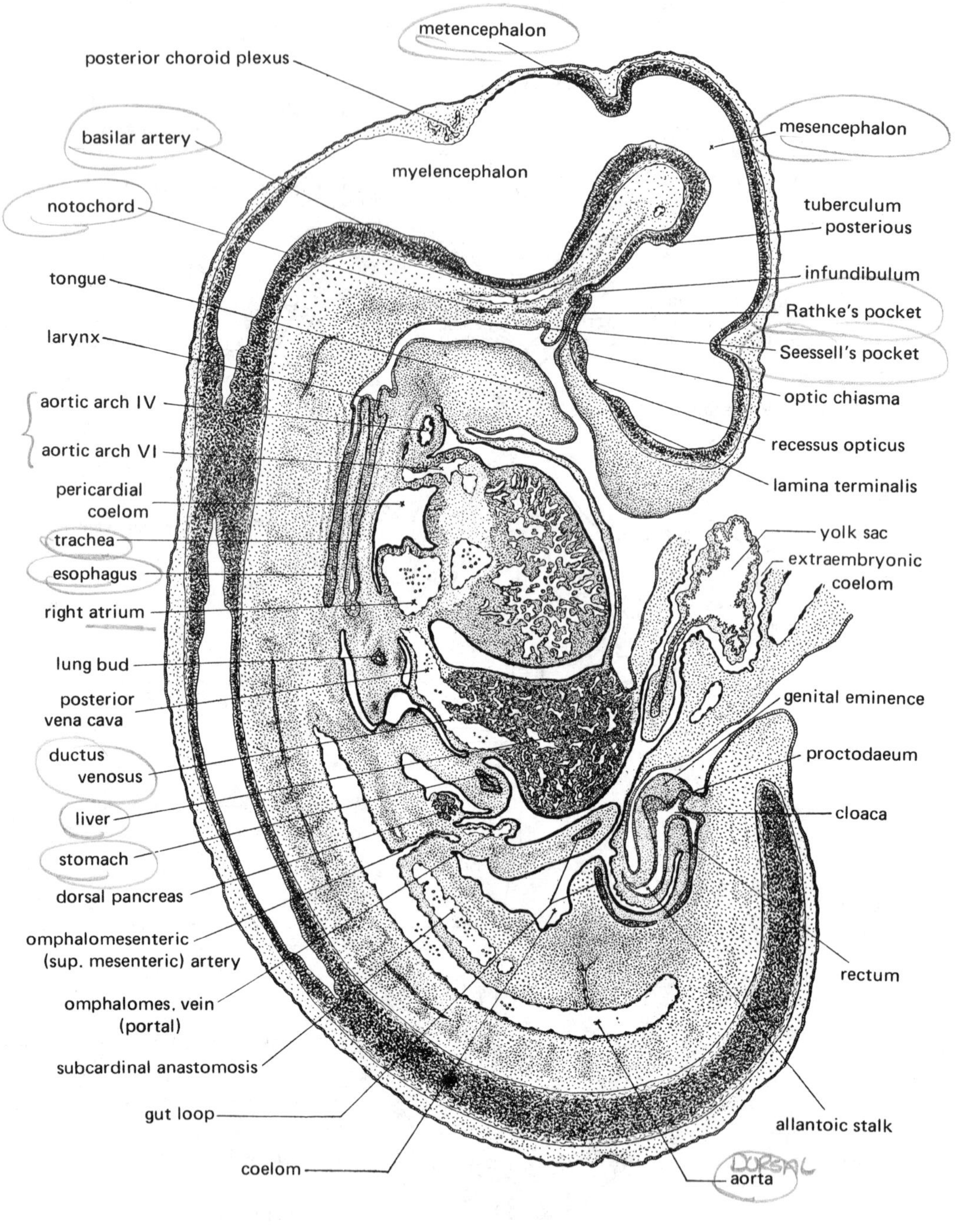

SAGITTAL SECTION OF 10-mm PIG EMBRYO

Compare with chick, page 200.

From B. M. Patten. *Embryology of the Pig.* 3rd ed. 1948. Copyright The P. Blakiston Company, Philadelphia.

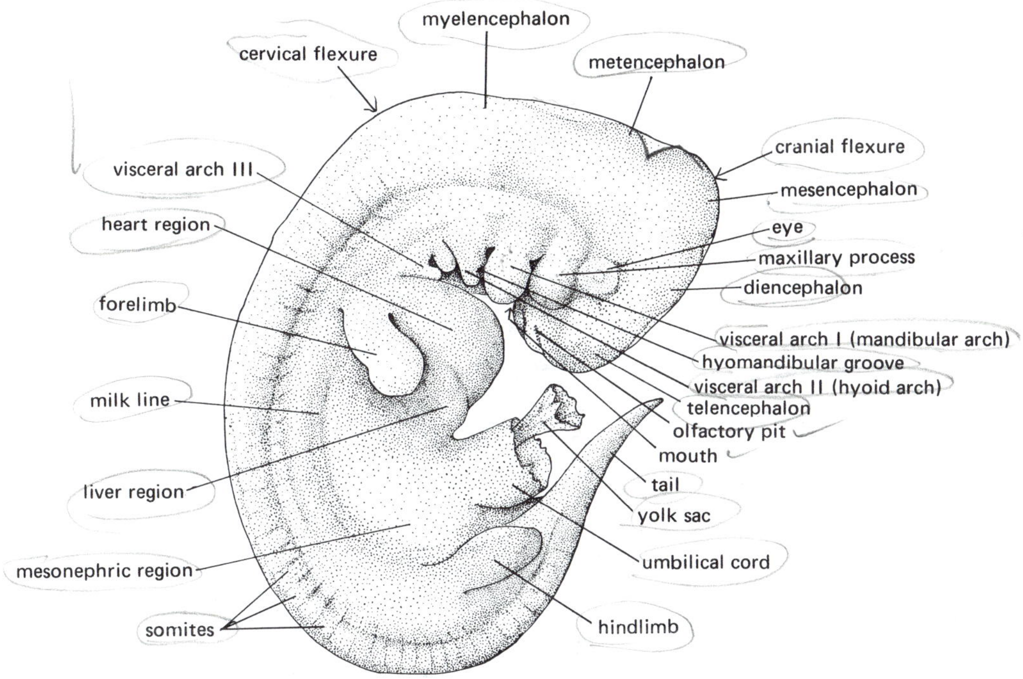

THE 10-mm PIG EMBRYO

THE ECTODERMAL DERIVATIVES IN THE PIG

EPIDERMIS: A single layer of ectoderm covers the entire body.

NERVOUS SYSTEM:

THE BRAIN – divided into five distinct regions –

#1. TELENCEPHALON with thick-walled, rounded, lateral outgrowths which become the cerebral hemispheres. It is separated from the diencephalon by two constrictions, the velum transversum in its roof and the optic recess in its floor. The telocoels (lateral ventricles) are known as the first and second ventricles and communicate with each other by way of the interventricular foramina with the third ventricle (foramina of Monro). First neuromere.

#2. DIENCEPHALON, associated with the ventrolateral optic stalks and the median ventral evagination (infundibulum) which becomes the posterior or neural lobe of the hypophysis (pituitary gland). The cavity is the diocoel or the third ventricle. Extends to dorsal thickening. Second and third neuromeres.

#3. MESENCEPHALON is at the point of the cranial flexure and is the original midbrain. It is thick-walled, rounded in cross section, and contains the mesocoel or aqueduct of Sylvius (cerebral aqueduct), which leads into the fourth ventricle. It is bounded posteriorly by the tuberculum posterius and dorsally by the isthmus. Fourth and fifth neuromeres.

#4. METENCEPHALON (anterior hindbrain or rhombencephalon) gives rise to the cerebellum dorsally and the pons Varolii ventrally and is separated by the isthmus from the mesencephalon. Its roof is thicker than that of the myelencephalon. Sixth neuromere.

#5. MYELENCEPHALON which is elongated and tapering toward the spinal cord, having a roof of thin, nonnervous ependyma. Neuromere markings may still be seen ventrolaterally. The cavity of the entire hindbrain is the fourth ventricle, or myelocoel. Seventh to eleventh neuromeres.

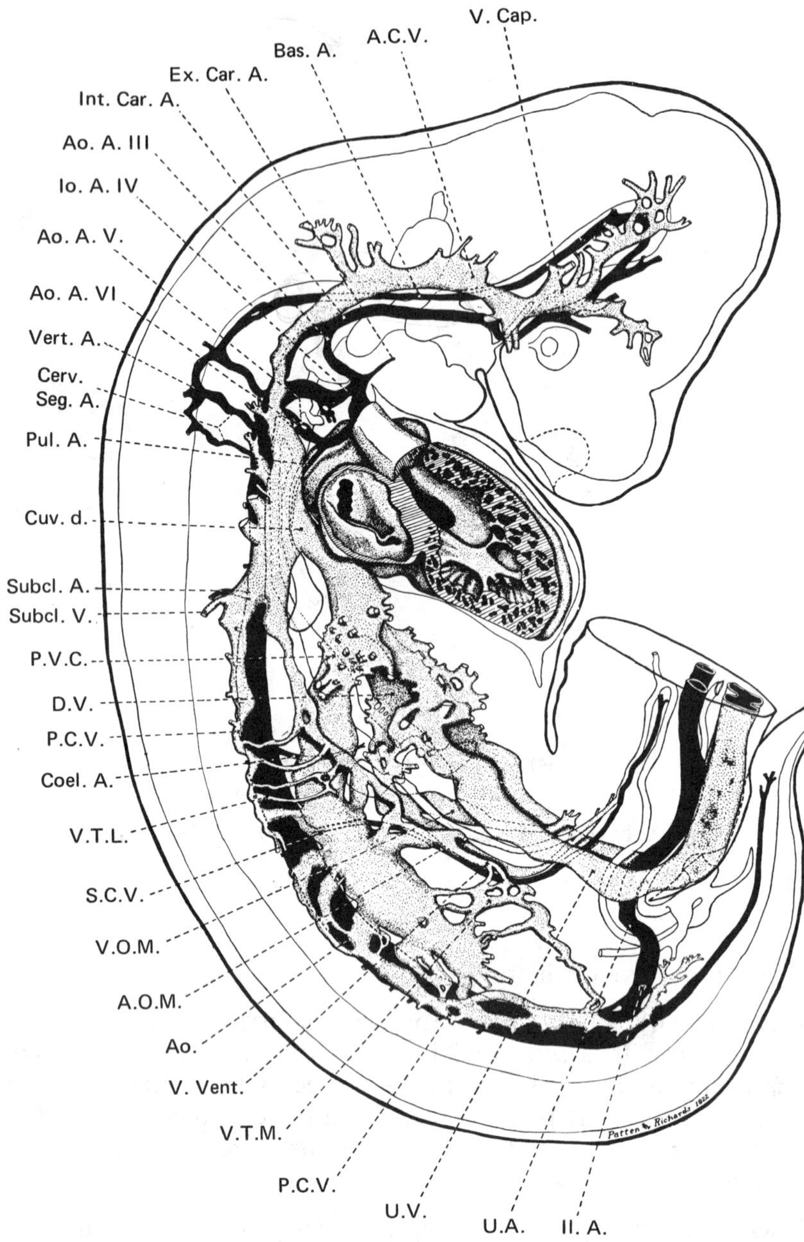

RECONSTRUCTION (X 14) OF THE CIRCULATORY SYSTEM OF A 9.4-mm PIG EMBRYO

ABBREVIATIONS

A. C. V. – anterior cardinal vein.
Ao. – aorta.
Ao. A. – aortic arch.
A. O.M. – omphalomesenteric artery.
Bas. A. – basilar artery.
Cerv. Seg. A. – intersegmental branches of aorta in cervical region.
Coel. A. – coeliac artery.
Cuv. d. – common cardinal vein (Duct of Cuvier).
D. V. – ductus venosus.
Ex. Car. A. – external carotid artery.
Il. A. – iliac artery.
Int. Car. A. – internal carotid artery.
P. C. V. – posterior cardinal vein.
Pul. A. – pulmonary artery.
P. V. C. – posterior vena cava.
S. C. V. – subcardinal vein.
Subcl. A. – subclavian artery.
Subcl. V. – subclavian vein.
U. A. – umbilical (allantoic) artery.
U. V. – umbilical (allantoic) vein.
V. Cap. – vena capitis (continuation of anterior cardinal vein).
V. O. M. – omphalomesenteric (portal) vein.
V. T. L. – lateral transverse veins of mesonephros.
V. T. M. – medial transverse veins of mesonephros.
V. Vent. – ventral vein of mesonephros.
Vert. A. – vertebral artery.

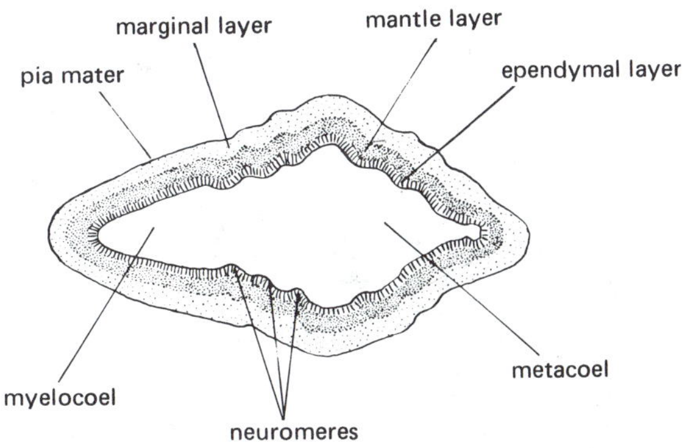

CELL LAYERS OF THE BRAIN OF PIG

THE CRANIAL NERVES – (all paired) –

I. OLFACTORY: grows from olfactory pit to telencephalon. Generally not developed at this stage.

II. OPTIC: grows from retina within the choroid fissure to the diencephalon. Generally not seen at this stage.

III. OCULOMOTOR: arises from the ventrolateral wall of the mesencephalon and supplies four eye muscles: inferior oblique, superior rectus, inferior rectus, and internal rectus.

IV. TROCHLEAR: arises from the ventrolateral wall of the mesencephalon but emerges dorsally at the isthmus. Difficult to trace. This will innervate the superior oblique muscle of the eye.

V. TRIGEMINAL (semilunar ganglion): arises from the broad anterior extremity of the myelencephalon and has three branches: the ophthalmic, maxillary, and mandibular. Syn. Gasserian ganglion.

VI. ABDUCENS: later arises from the ventrolateral wall of the myelencephalon to innervate the external rectus muscle of the eye.

VII. FACIAL (geniculate ganglion): arises along with the eighth cranial nerve but later separates to act as a sensory and motor nerve to the face (second visceral arch) and tongue.

VIII. AUDITORY (acoustic ganglion): arises just anterior to the otocyst, along with the seventh. Innervates the otic (auditory) vesicle and is purely sensory.

IX. GLOSSOPHARYNGEAL (dorsal superior ganglion and ventral petrosal ganglion): just posterior to the otocyst. Sensory and motor to the tongue and pharynx.

X. VAGUS (jugular and nodosal ganglia): sensory and motor, largely related to the viscera and parasympathetic system.

XI. ACCESSORY (froriep's ganglion): multiple origin, from lateral wall of posterior myelencephalon and associated with the first six cervical ganglia. It is motor in function and follows the vagus to sternomastoid and trapezius muscles.

XII. HYPOGLOSSAL: six points of origin from the ventral wall of the myelencephalon. Purely motor to the tongue muscles.

The cranial nerves which carry sensory fibers generally show conspicuous ganglia (V, VII, VIII, IX, X). All of these except VIII also carry motor fibers and are, therefore, mixed. There are no external ganglia for the purely motor nerves (III, IV, VI, XI, XII).

I olfactory
II optic
III occulomotor
IV trochlear
V semilunar
VI abducens
VII geniculate
VIII acoustic
IX superior
X jugular
XI Accessory
XII hypoglossal

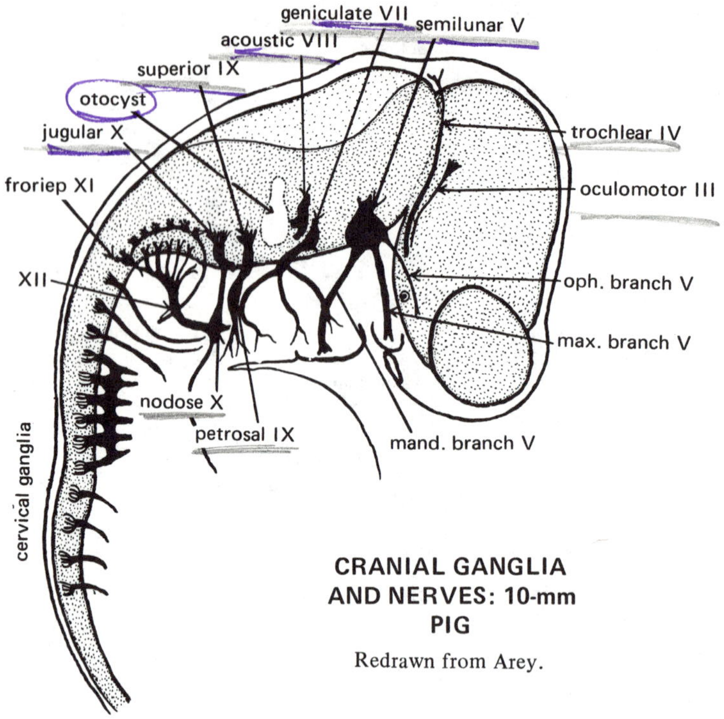

CRANIAL GANGLIA AND NERVES: 10-mm PIG

Redrawn from Arey.

THE SPECIAL SENSE ORGANS –

OPTIC VESICLE:

Lateral evagination of the diencephalon, drawn out into optic stalks and peripheral invaginated optic cups with:

- Inner layer of sensory nerve cells – the nervous layer of the retina.
- Outer layer of darkened cells – pigmented layer.
- Lens – almost solid, found within rim of optic cup.

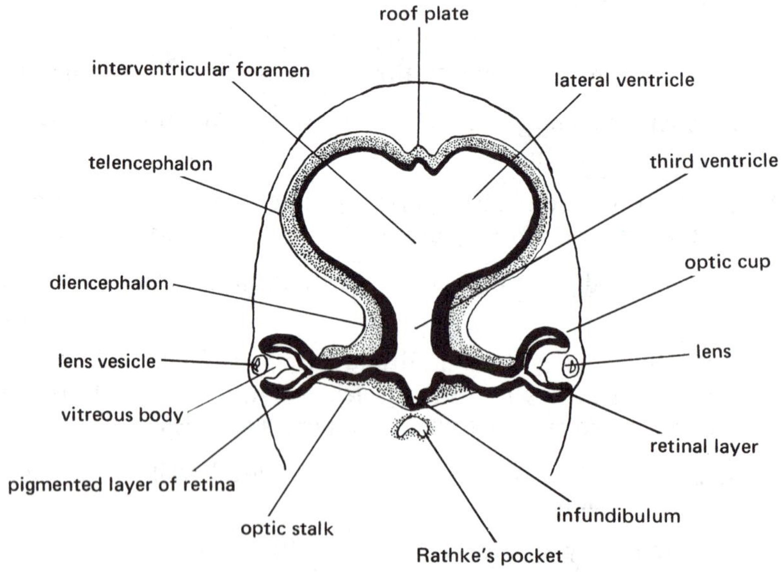

SECTION THROUGH EMBRYONIC FOREBRAIN

OLFACTORY VESICLE:

Ventrolateral surface of the head, close to the telocoels. Thick-walled invagination.

OTIC VESICLE:

Lateral to the myelencephalon. Walls unevenly thick, with slight constriction into dorsal utricle and ventral saccule. Between the vesicle and the myelencephalon is the endolymphatic duct which is circular in cross section and joins the saccule.

THE SPINAL CORD AND NERVES –

This part of the central nervous system begins without clear demarcation from the myelencephalon and extends into the tail, just dorsal to the notochord.

THE SPINAL CORD:

Marginal layer – outer white matter composed of axons and neuroglia.
Mantle layer – intermediate layer of neuroblasts and spongioblasts.
Ependymal layer – ciliated epithelium lining of the neurocoel.
Neurocoel – central canal or cavity of spinal cord.

THE SPINAL NERVES:

Dorsal root – bundle of afferent axons associated with dorsolateral wall of spinal cord.

Dorsal root ganglion – large mass of neurones on the dorsal root lateral to the spinal cord.

Ventral root – bundle of efferent axons emerging from the ventrolateral wall of the spinal cord.

Spinal nerve – formed by junction of dorsal and ventral roots, with three branches (rami).

- Dorsal ramus – grows dorsolaterally to the body wall to supply the skin and trunk muscles.
- Lateral ramus – grows laterally to the ventral body wall.
- Communicating ramus – grows ventrally and mesially to the sympathetic ganglia, or parasympathetic in the sacrol region.

THE BRACHIAL PLEXUS: The lateral rami of the spinal nerves at the level of the anterior limb buds are interconnected.

THE LUMBOSACRAL PLEXUS: The lateral rami of the spinal nerves at the level of the hind-limb bud are interconnected.

THE SYMPATHETIC GANGLIA: There are paired neuron masses lying dorsal to the aorta, seen best at the level of the anterior limb buds but also along the auterocoeliac and mesenteric ganglia.

RATHKE'S POCKET –

This is the ectodermal pocket just anterior to the stomodeum which gives rise to the epithelial hypophysis (pituitary).

THE ENDODERMAL DERIVATIVES IN THE PIG

By this stage, the embryonic gut or endodermal tube has become differentiated into the major parts of the digestive and respiratory systems. (See Balinsky: 505-36, Bodemer: 258-72, and Rugh: *Vertebrate Embryology*. 304-77.)

STOMODEUM – oral aperture, bounded by the maxillary and mandibular processes of the first visceral arch and lined with ectoderm.

ORAL CAVITY – area of junction of stomodeal ectoderm and the anterior end of the foregut (endoderm).

PHARYNX – this is the portion posterior to the oral cavity which is compressed dorsoventrally. It is bounded laterally by four pairs of visceral pouches and five pairs of visceral arches. The latter are aggregations of mesenchyme containing blood vessels (aortic arches) and nerves. Other structures relating to the pharynx are:

TONGUE – floor of the pharynx is a thickening, the tuberculum impar which forms the root of the tongue. On either side are elongate thickenings which rise from the mandibular arches, the three parts forming the pig's tongue.

THYROID – consists of branching epithelial cords located in the median ventral line midway between the third and fourth visceral arches and connected by the thyroglossal duct to the pharynx just posterior to the tuberculum impar. Duct may be closed. The thyroid gland has thus moved posteriorly from its point of origin near the second pair of visceral pouches.

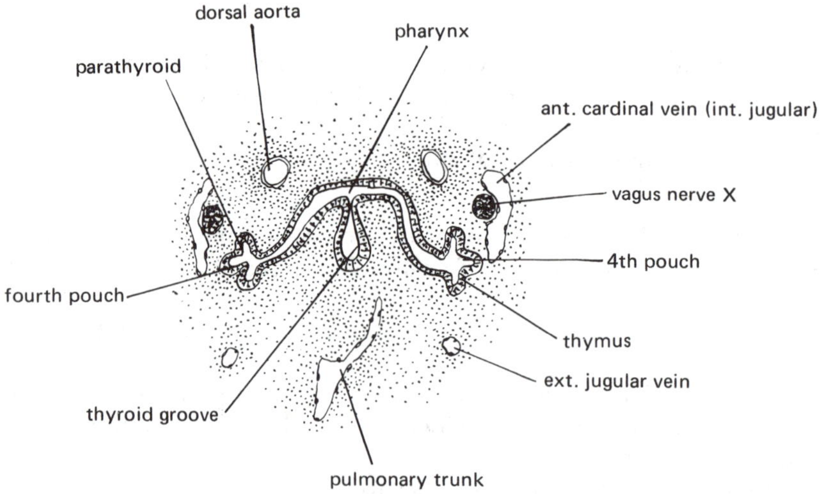

SECTION THROUGH LEVEL OF FOURTH VISCERAL POUCH 10-mm PIG

EPIGLOTTIS – ring around the glottis which develops from the bases of the fourth and fifth visceral arches, on each side of which is the arytenoid fold of the larynx.

PHARYNGEAL POUCHES: Each pouch now bears a dorsal and a ventral wing.

Auditory tube and tympanic cavity from first pouch.
Tympanic membrane includes an outer layer from ectoderm of the first visceral groove.
External auditory meatus, remnant of groove itself.
Palatine tonsils – from dorsal wings of third and fourth visceral pouches.
Parathyroid glands – from dorsal wings of third and fourth visceral pouches.
Thymus bodies – from ventral wing of third and fourth visceral pouches.
Ultimobranchial bodies – tubular outgrowth caudal to the fourth pouch.

LARYNX, TRACHEA, AND LUNGS –

LARYNGOTRACHEAL GROOVE – ventral median longitudinal groove in the floor of the foregut beginning just posterior to the fourth visceral pouch. Leads to origin of the trachea.

TRACHEA – ventral to the esophagus and circular in cross section.

PRIMARY BRONCHI – bifurcated ends of the trachea or the lung buds. There may be further subdivisions into two lobes of the left lung and a median and ventral lobe of the right lung.

ESOPHAGUS – section between pharynx and stomach, thick walled and circular in transverse section.

STOMACH – somewhat more expanded than esophagus, thick-walled and circular in section. Suspended by dorsal mesentery. Diverted posteriorly toward the right.

INTESTINE –

DUODENUM – from pyloric end of stomach. Extends from constriction of stomach to level of the bile duct. Circular and thick walled in section.

LIVER – four lobed, derived from the original hepatic diverticulum and fills the space between the heart, stomach, and duodenum. Hepatic duct connects liver tissue with original diverticulum.

COMMON BILE DUCT (ductus choledochus) – main tubular connection with the original liver diverticulum, leading to the distal end of the duodenum.

GALL BLADDER AND CYSTIC DUCT – lateral sacculation of the common bile duct.

PANCREAS –

Ventral pancreas arises from common bile duct near its origin.
Dorsal pancreas arises caudally and dorsally from the level of the duodenum.
The two pancreatic rudiments interlock, with the duct of the dorsal pancreas persisting as the functional pancreatic duct.

INTESTINE PROPER – posteriorly directed gut of small bore and thick wall, accompanied by vitelline veins and arteries.

INTESTINAL LOOP – major loop of small intestine extending out into the umbilical cord to connect with the yolk stalk.

JEJUNUM AND ILEUM – section of small intestine from duodenum to the caecum.

CAECUM – diverticulum on the caudal limb of the intestinal loop, the beginning of the large intestine (colon and rectum).

CLOACA – enlarged cavity at caudal end of colon, which divides into the rectum and the urogenital sinus.

ALLANTOIC STALK – duct opening ventrally into the cloaca. Passes through the body stalk to the allantois.

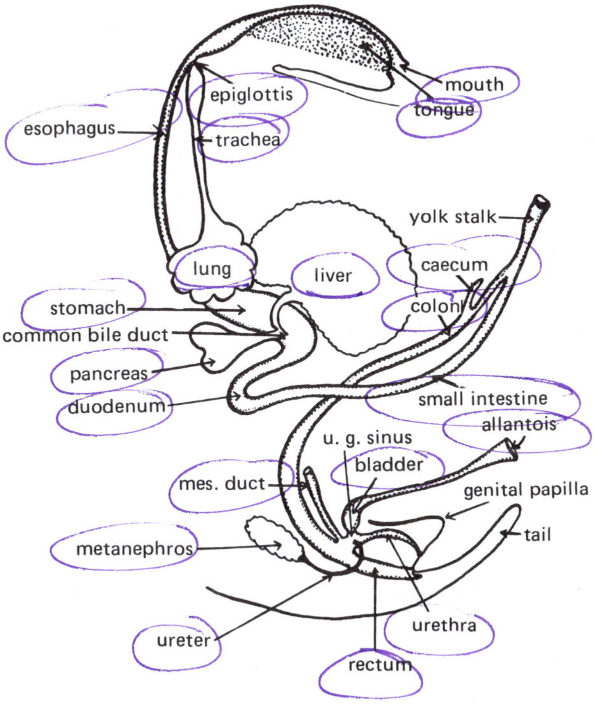

ALIMENTARY TRACT AND ITS DERIVATIVES

THE MESODERMAL DERIVATIVES IN THE PIG

The loose embryonic tissue which fills the spaces between developing organs is known as mesenchyme. This is the forerunner of mesoderm and all of its derivatives. Its cells are star-shaped with protoplasmic processes which extend from cell to cell to form a reticulum or network of cells. (See Balinsky: 424-504, Bodemer: 273-331, and Rugh: *Vertebrate Embryology* 304-77).

NOTOCHORD – in the mammal, this cylindrical rodlike structure, located ventral to the spinal cord from tuberculum posterius into the tail, is considered of mesodermal origin. It constitutes the primary skeletal axis of the embryo.

SOMITES – the 10-mm pig embryo has approximately 44 pairs of somites, each of which has undergone extensive differentiation into:

DERMATOME – solid cord of cells just beneath the epidermis lateral to the spinal cord.

MYOTOME – the more dense, dorsally placed aggregation of cells adjoining the dermatome but between it and the spinal cord. This gives rise to the skeletal muscles of the back.

SCLEROTOME – loose mesenchymal cells of the somite which migrate from it to encircle the spinal cord and notochord to form the skeletal axis of the adult.

MYOCOEL – generally obliterated cavity of the somite between the dermatome and the myotome.

NEPHROTOME – intermediate mesoderm or mesomere which gives rise to the excretory system.

PRONEPHROS – already degenerated except for a few anterior nephrotomes.

MESONEPHROS – very large functional kidneys of the 10-mm pig embryo. Located posterior to the liver; retroperitoneal in position.

MESONEPHRIC DUCTS – along ventral margins of mesonephroi leading posteriorly to the cloaca. Flattened oval in section.

MESONEPHRIC TUBULES – hollow coiled tubes laterally situated, which connect the Malpighian corpuscles with the mesonephric duct.

GLOMERULUS – knot of blood capillaries surrounded by Bowman's capsule.

BOWMAN'S CAPSULE – which is the double-walled cup at the end of the onephric tubule, surrounding the glomerulus.

METANEPHROS – begins to develop at this stage and may be found caudally between the roots of the umbilical arteries.

COLLECTING PORTION – from epithelium of the mesonephric duct, the proximal slender duct becoming the ureter and the distal dilatation the pelvis of the adult kidney.

SECRETORY PORTION – condensed mesenchyme of nonsegmented nephrogenic tissue posterior to the mesonephros which will differentiate into secretory tubules.

CLOACA* – common chamber for embryonic excretory, reproductive, and gut channels.

UROGENITAL SINUS – portion of the cloaca which receives the allantois, the mesonephric, and metanephric ducts.

RECTUM – portion of cloaca receiving the large intestine and postanal gut.

CLOACAL PLATE – the membrane formed by the fusion of the lateral walls of the terminal portion of the cloaca.

LATERAL PLATE MESODERM – hypomere. These lateral sheets of mesoderm extensions of the original nephrotomes, are differentiated into:

SPLANCHNIC MESODERM – associated with the viscera and also giving rise to the visceral linings of the various body cavities.

*Parts of these systems are also endodermal.

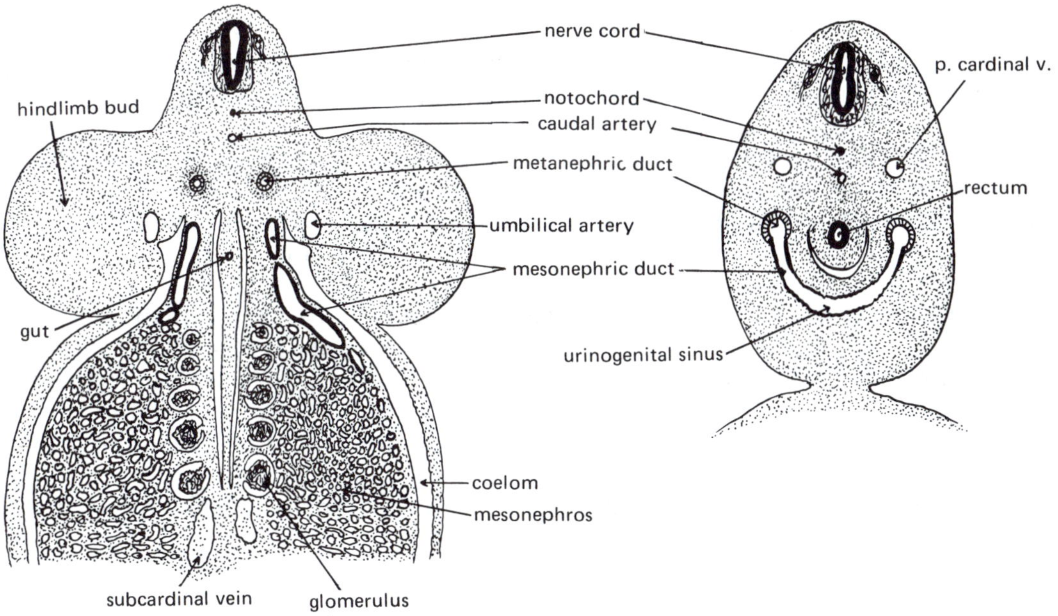

**URINOGENITAL DUCTS:
10-mm PIG**

COELOM – a continuous, communicating cavity which includes the pericardial and peritoneal cavities connected by pleural canals. The separation of these cavities begins at the 10-mm stage by the development of the septum transversum, between the heart and the liver.

MESENTERIES – the ventral mesentery breaks through* with the formation of the peritoneal cavity, but the double sheet of splanchnic mesoderm forming the dorsal mesentery persists and is named according to what it supports.

MESOGASTRIUM (or greater omentum) – suspends the stomach.**
MESODUODENUM – suspends the duodenum.
MESENTERY PROPER – suspends small intestines.
MESOCOLON – suspends the colon.

REPRODUCTIVE SYSTEM – thickened primordia (genital ridges) on the ventro-mesial side of the mesonephros.

GERMINAL EPITHELIUM – thickened splanchnic mesoderm which covers the genital ridges.

PRIMORDIAL GERM CELLS – with large vesicular nuclei directly beneath the germinal epithelium.

GONODUCTS – undifferentiated. The mesonephric duct in the male becomes the ductus deferens.

EXTERNAL GENITALIA – genital tubercle constitutes the only primordium of the external genitalia at this stage.

VASCULAR SYSTEM – by the 10-mm stage this system is already very complex.

*Note: The ventral mesentery persists only between the gut and the liver as the lesser omentum and between the liver and the body walls as the falciform ligament.
**Note: A saccular recess develops between the liver and the stomach due to a shifting in the position of the stomach, resulting in the blind pouch known as the omental bursa. This communicates with the coelom by way of the epiploic foramen, otherwise known as the foramen of Winslow.

THE HEART –

SINUS VENOSUS – receives blood from the right and left common cardinals and the common hepatic vein. Entrance into the right atrium guarded by valves temporarily joined as the septum spurium.

LEFT ATRIUM – receives single pulmonary vein and is separated from the right atrium incompletely by the septum primum, with an opening, the interatrial foramen.

RIGHT ATRIUM – dorsal wall develops septum secundum with opening known as the foramen ovale. Atrial septum finally formed by the fusion of these two septa and the sinus valve.

ATRIO-VENTRICULAR CANALS – valvular connections between atrium and ventricle on each side. Tricuspid valve on the right, bicuspid valve on the left.

INTERVENTRICULAR SEPTUM – partition separating ventricles, with temporary interventricular foramen.

BULBUS CORDIS – large vessel separating into ascending aorta and pulmonary artery, undivided at the heart.

TRABECULAE – thick and spongy walls of the ventricles, highly muscular, forming sinusoids.

THE ARTERIES –

AORTIC ARCHES – the pig embryo has the typical vertebrate plan of six pairs of aortic arches, but only the third, fourth, and sixth persist to the 10-mm stage.

III – with extensions of dorsal aortae into the head constitute the *internal carotids.* Near the brain vesicles. From the ventral aorta grow the paired *external carotids* into the mandibular arch.

IV – the largest of the arches and persist on the left side only as the arch of the systemic aorta.

VI – seen best at junction with dorsal aortae, extending as small *pulmonary arteries* to the lungs. The remnant of the arch which shrivels up at birth is the ductus arteriosus.

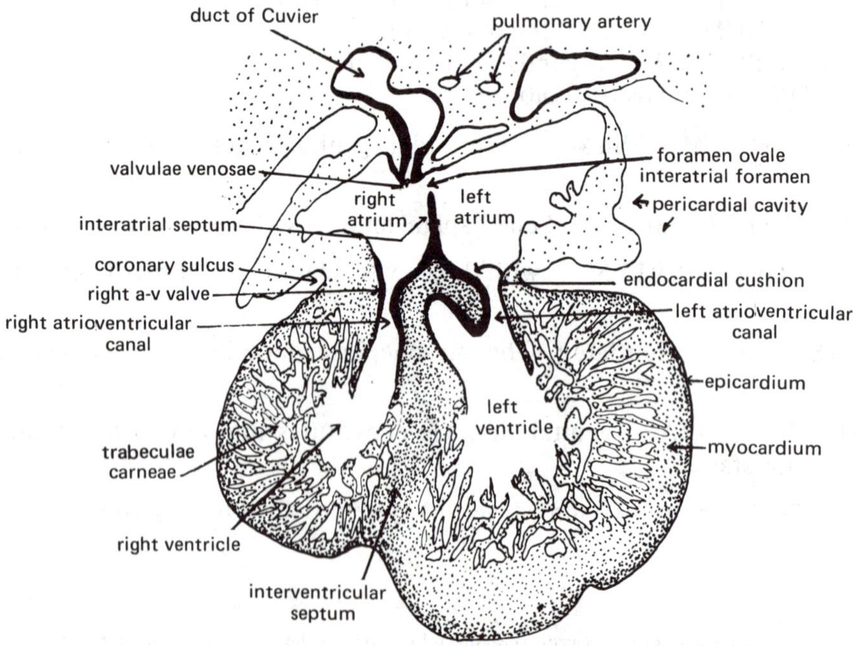

SECTION THROUGH 10-mm PIG HEART

BULBUS ARTERIOSUS – the thick-walled, single, large vessel emerging from the heart, which divides into the systemic trunk and pulmonary trunk almost immediately. The systemic trunk comes from the left ventricle, while the pulmonary trunk emerges from the right ventricle.

DORSAL AORTAE – united opposite the eighth segment and continuous posteriorly as the *descending aorta.* There are dorsal, lateral, and ventral branches.

- INTERSEGMENTAL ARTERIES – small paired vessels which arise from the dorsolateral margins of the aortae and pass upward between the somites.
- SUBCLAVIAN ARTERIES – arise from point of fusion of the dorsal aortae and pass into the forelimb buds.
- VERTEBRAL ARTERIES – arise along with subclavians and pass cephalad into the head. These are continuous with the unpaired basilar artery lying beneath the brain. The basilar artery is an extension of the internal carotids by way of the circulus arteriosus near the hypophysis.
- COELIAC ARTERY – from ventral side of aorta to stomach.
- SUPERIOR MESENTERIC ARTERY – (primitive vitelline) from aorta to the small intestine and out along the yolk stalk.
- INFERIOR MESENTERIC ARTERY – from aorta to large intestine.
- UMBILICAL ARTERIES – lateral vessels from aorta at level of origin of common iliacs, which run along the allantoic stalk into the umbilical cord.
- ILIAC ARTERIES – extensions of the aorta into the hindlimb buds.
- CAUDAL ARTERIES – posterior extension of the aorta into the tail.

THE VEINS – the venous system of the 10-mm pig embryo is quite complicated but is so similar to that of the advanced chick stages that it should be readily understood. The cardinal system drains the more solid body tissues; the vitelline and allantoic veins bring nutriment from the yolk sac and the placenta; and the posterior vena cava brings blood from the viscera.

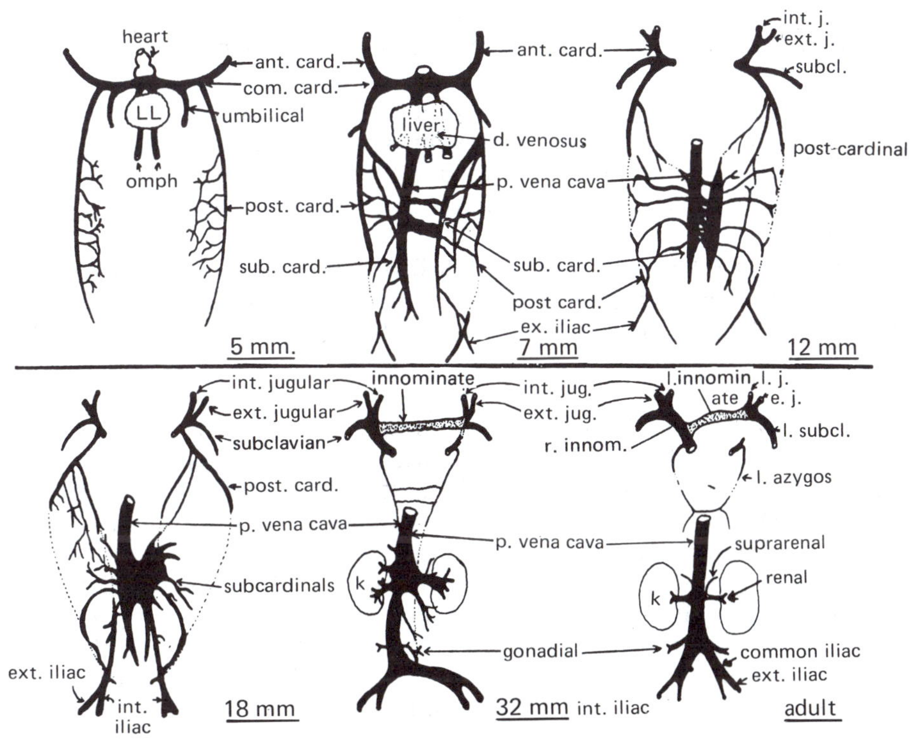

VENOUS SYSTEM OF PIG EMBRYO

Modified from Patten, 1948.

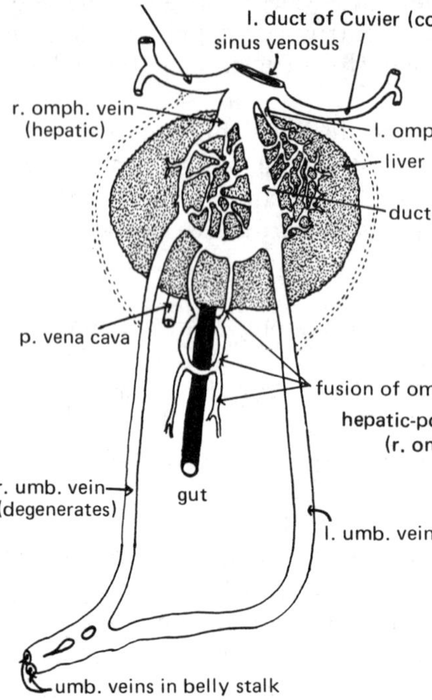

HEPATIC CIRCULATION
10-MM STAGE

ANTERIOR CARDINAL VEINS – these bring blood from the capillaries fed by the internal carotids which cover the brain and will become the internal jugulars.

EXTERNAL JUGULAR VEINS – new vessels from the mandibular region which joins the anterior cardinals.

SUBCLAVIAN VEINS – from the anterior limb buds.

COMMON CARDINAL VEINS – the fusion of all of the above vessels and the posterior cardinals as they join the sinus venosus and the right atrium. These vessels are also known as the ducts of Cuvier. The left common cardinal joins the postcaval before entering the sinus venosus.

LEFT INNOMINATE VEIN – vessel which forms an anastomosis from the left to the right so that blood in the left anterior cardinal system passes into the right common cardinal and a remnant of the left common cardinal becomes the coronary vein from the heart muscles.

POSTERIOR CARDINAL VEINS – drain blood from the dorsolateral regions of the posterior body wall and particularly the mesonephros. Midway in the body these vessels are interrupted by mesonephric sinusoids, and on the ventromesial surface of these embryonic kidneys appear the subcardinal veins. The cardinal and subcardinal vessels are joined through the mesonephros for a time. The right subcardinal is the larger and will later join the common hepatic vein to form the inferior vena cava.

HEPATIC PORTAL VEINS – derived from the paired embryonic vitelline veins lying lateral to the gut. These vitelline veins are broken up into sinusoids by the developing liver. Posterior to the liver the vitelline veins develop connections spiraling around the gut. Anterior to the liver a remnant of the right vitelline vein becomes the hepatic vein, which empties into the postcaval vein just as it enters the sinus venosus.

POSTCAVAL VEIN – (inferior vena cava) the vessel which, in the adult, conveys most of the venous blood to the heart. It develops from the median fusion of the paired subcardinals, particularly from the right subcardinal.

UMBILICAL VEINS – veins bringing blood from the allantois along the body stalk and into the liver sinusoids. The right umbilical degenerates, while the left enlarges to form the large short cut through the liver, known as the ductus venosus, which joins the postcaval vein before it enters the heart.

SUPERIOR MESENTERIC VEIN – new vessel which arises in the mesentery of the intestinal loop to join the left vitelline vein.

MESONEPHRIC VEINS – small veins arising within the mesonephros, which join the postcaval vein.

INTERSEGMENTAL VEINS – small veins which drain the dorsolateral body wall and empty into the posterior cardinals. (See Arey: 83-89, Patten (1948: 547-82, and Bodemer: 365-90.)

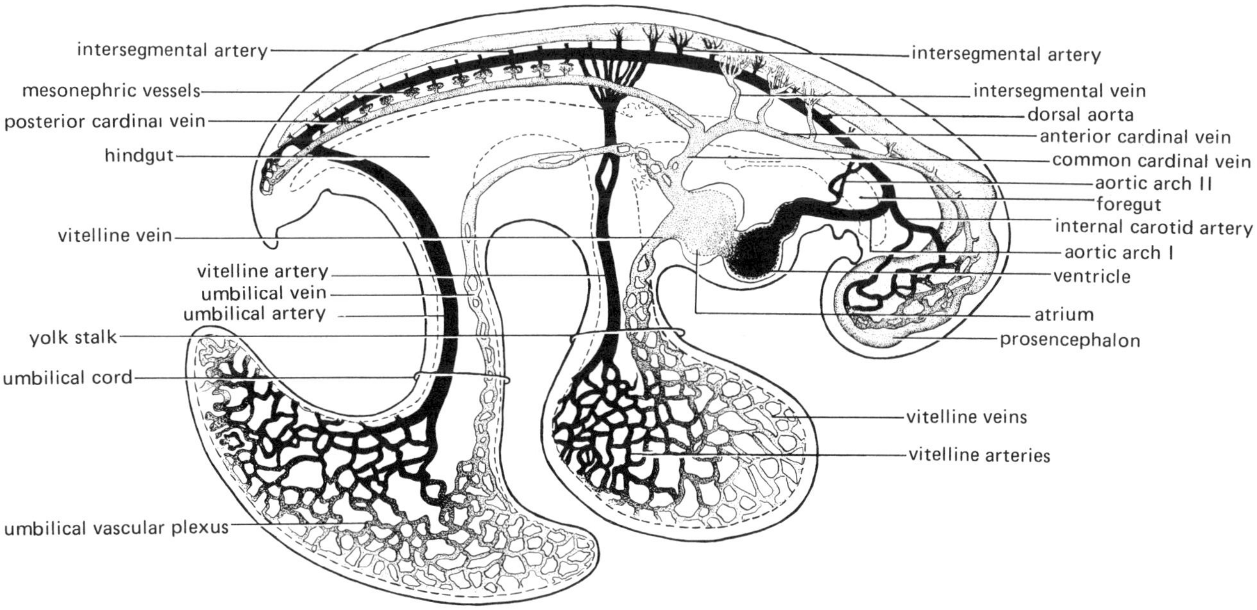

SCHEMATIZED DRAWING OF CIRCULATION OF EMBRYONIC PIG (RIGHT SIDE ONLY)

Drawing by R. H. Van Dyke.

"The higher organisms made themselves largely independent of their immediate environment. Their tissues are bathed from within by a fluid (blood) which they carry around with them, a sort of internal environment."

A. J. Lotka

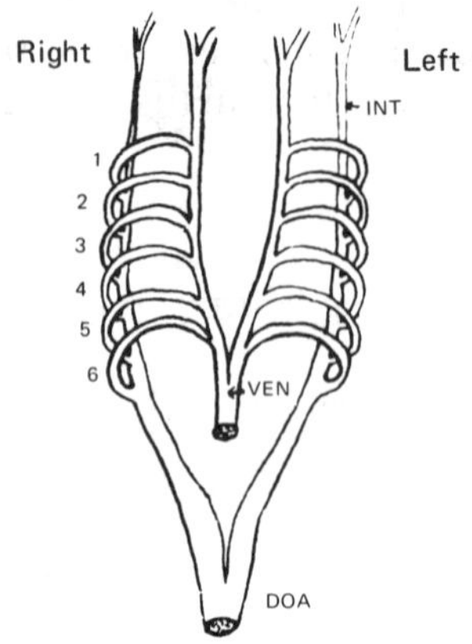

GENERALIZED (EMBRYONIC) PLAN
OF AORTIC ARCHES

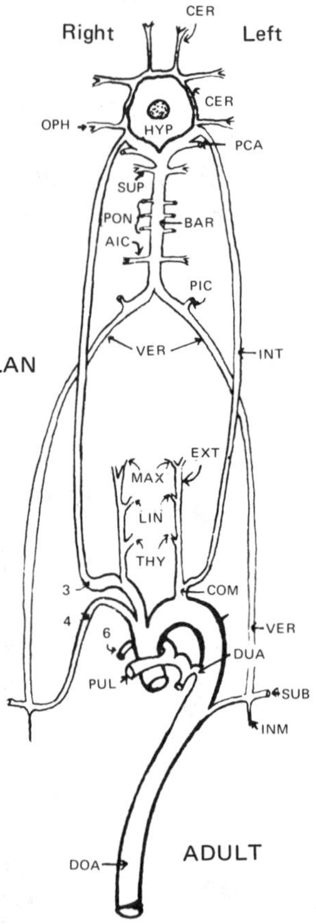

ADULT

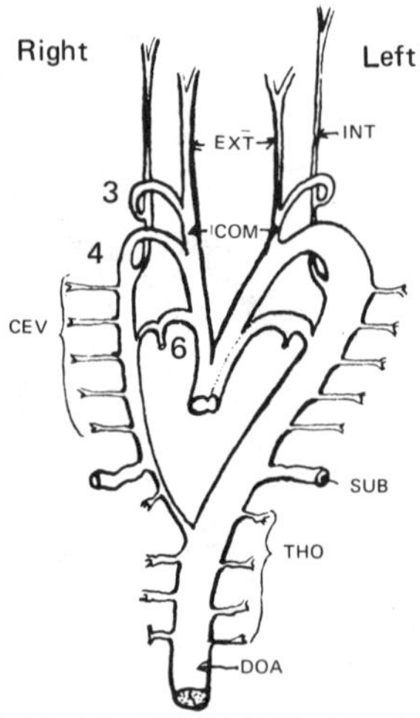

CHANGES IN EMBRYONIC CIRCULATION

THE PIG EMBRYO

Redrawn from Patten, 1948.

ABBREVIATIONS FOR DRAWINGS OF CIRCULATORY CHANGES

AIC – anterior inferior cerebral artery
BAR – basilar artery
CER – cerebral artery
CEV – cervical artery
COM – common carotid artery
DOA – dorsal aorta
DUA – ductus arteriosus
EXT – external carotid
HYP – hypophysis
INM – internal mammary artery
INT – internal carotid artery
LIN – lingual artery
MAX – maxillary artery
OPH – ophthalmic artery
PCA – postcerebral artery
PIC – posterior inferior cerebral artery
PON – pontine artery
PUL – pulmonary artery
SUB – subclavian artery
SUP – superior cerebral artery
THO – thoracic artery
THY – thyroid artery
VEN – ventral aortic root
VER – vertebral artery

See Patten (1948) and Patten (1957): 247-536.

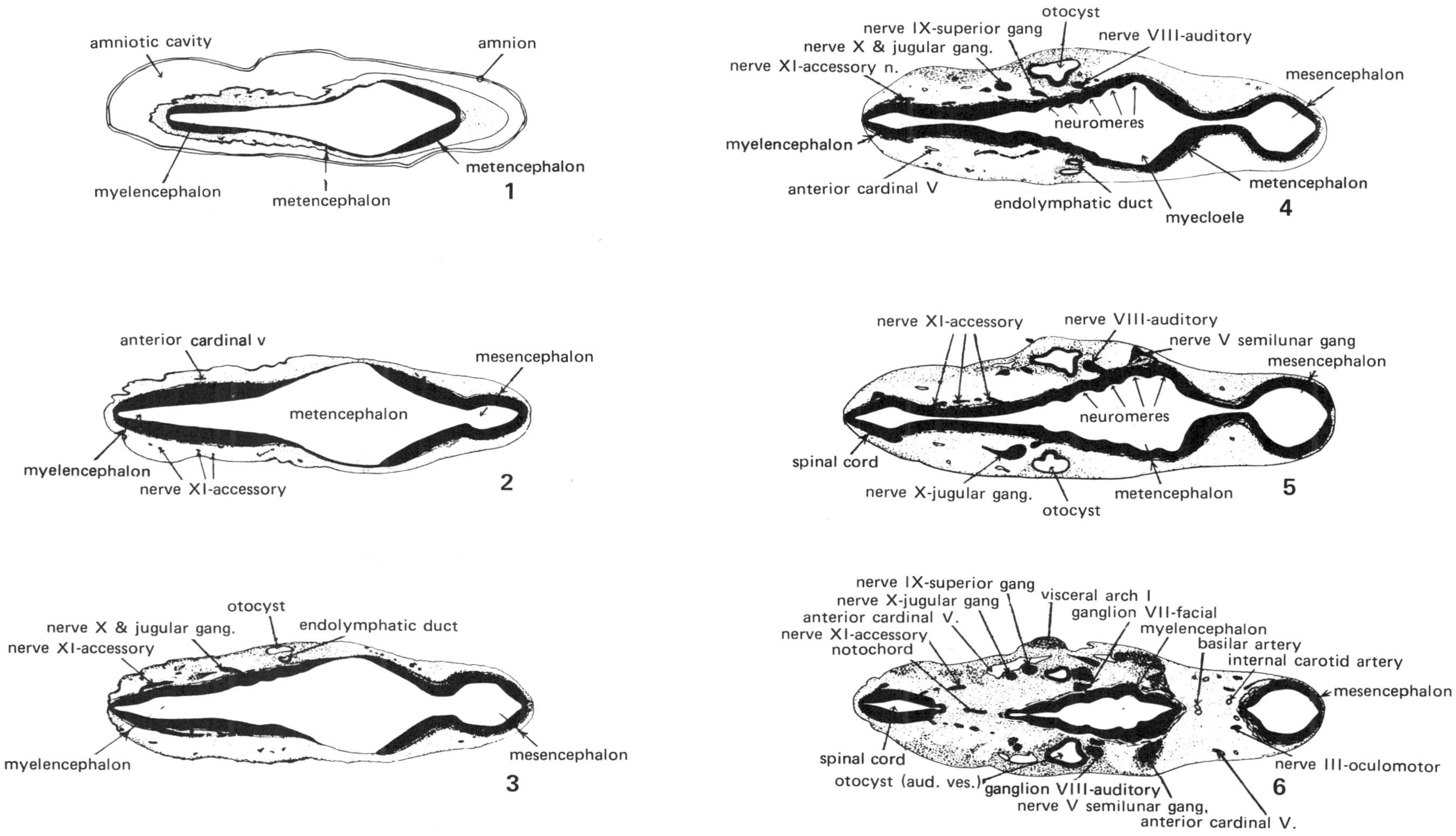

TRANSVERSE SECTIONS OF
10-mm PIG EMBRYO

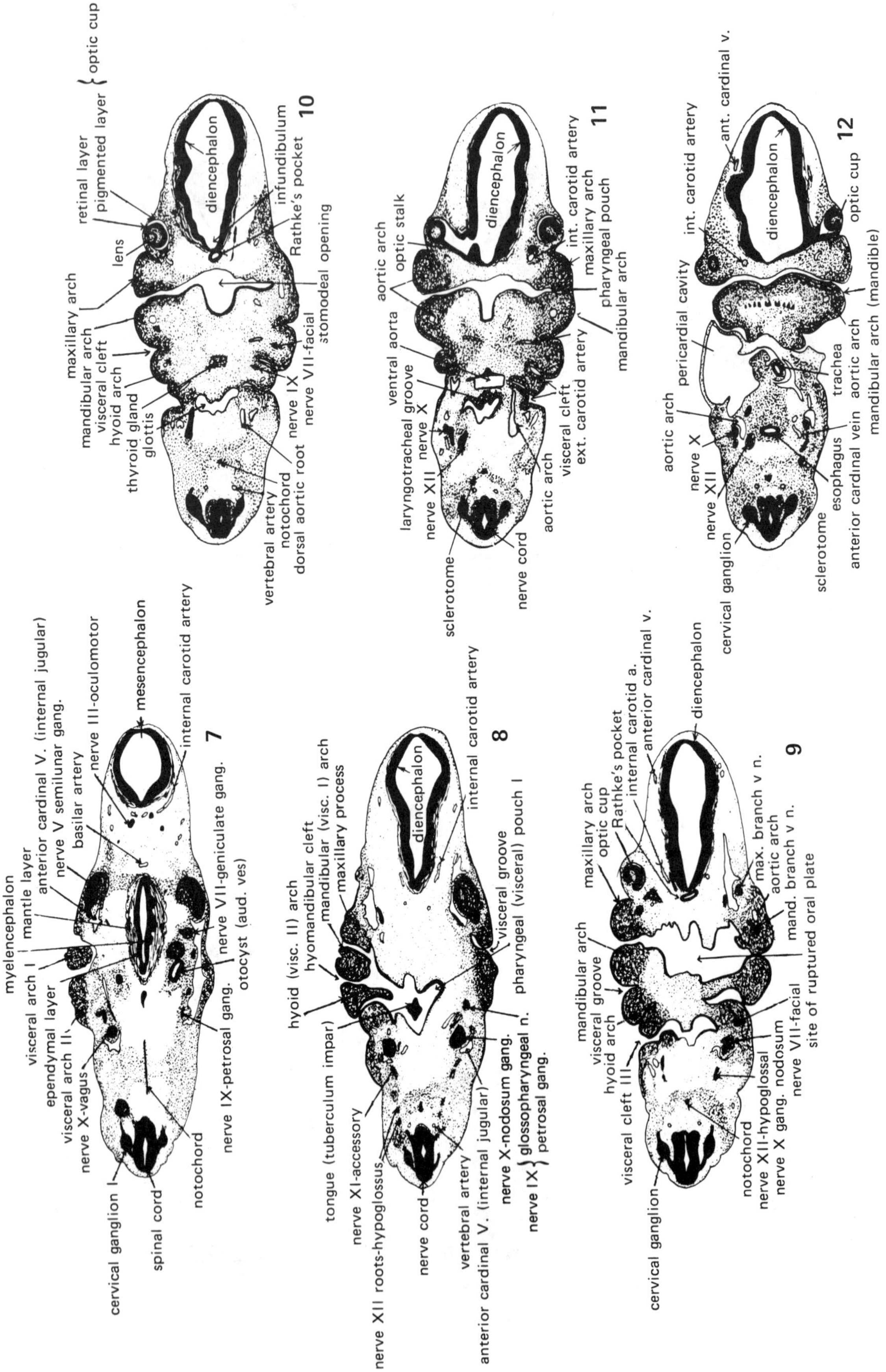
7
myelencephalon
visceral arch I
mantle layer
ependymal layer
anterior cardinal V. (internal jugular)
nerve V semilunar gang.
basilar artery
nerve III-oculomotor
mesencephalon
internal carotid artery
visceral arch II
nerve X-vagus
cervical ganglion I
spinal cord
notochord
nerve IX-petrosal gang.
otocyst (aud. ves)
nerve VII-geniculate gang.
8
hyoid (visc. II) arch
hyomandibular cleft
mandibular (visc. I) arch
maxillary process
diencephalon
internal carotid artery
tongue (tuberculum impar)
nerve XI-accessory
nerve XII roots-hypoglossus
nerve cord
vertebral artery
anterior cardinal V. (internal jugular)
nerve X-nodosum gang.
nerve IX glossopharyngeal n.
petrosal gang.
visceral groove
pharyngeal (visceral) pouch I
9
mandibular arch
visceral groove
hyoid arch
visceral cleft III
cervical ganglion
maxillary arch
optic cup
Rathke's pocket
internal carotid a.
anterior cardinal v.
diencephalon
notochord
nerve XII-hypoglossal
nerve X gang. nodosum
nerve VII-facial
site of ruptured oral plate
mand. branch v n.
aortic arch
max. branch v n.
10
maxillary arch
mandibular arch
visceral cleft
hyoid arch
thyroid gland
glottis
lens
retinal layer
pigmented layer
optic cup
diencephalon
infundibulum
Rathke's pocket
stomodeal opening
nerve VII-facial
nerve IX
dorsal aortic root
notochord
vertebral artery
11
laryngotracheal groove
ventral aorta
aortic arch
optic stalk
diencephalon
nerve X
nerve XII
sclerotome
nerve cord
aortic arch
visceral cleft
ext. carotid artery
int. carotid artery
maxillary arch
pharyngeal pouch
mandibular arch
12
aortic arch
nerve X
nerve XII
cervical ganglion
pericardial cavity
int. carotid artery
ant. cardinal v.
diencephalon
optic cup
sclerotome
anterior cardinal vein
esophagus
trachea
aortic arch
mandibular arch (mandible)

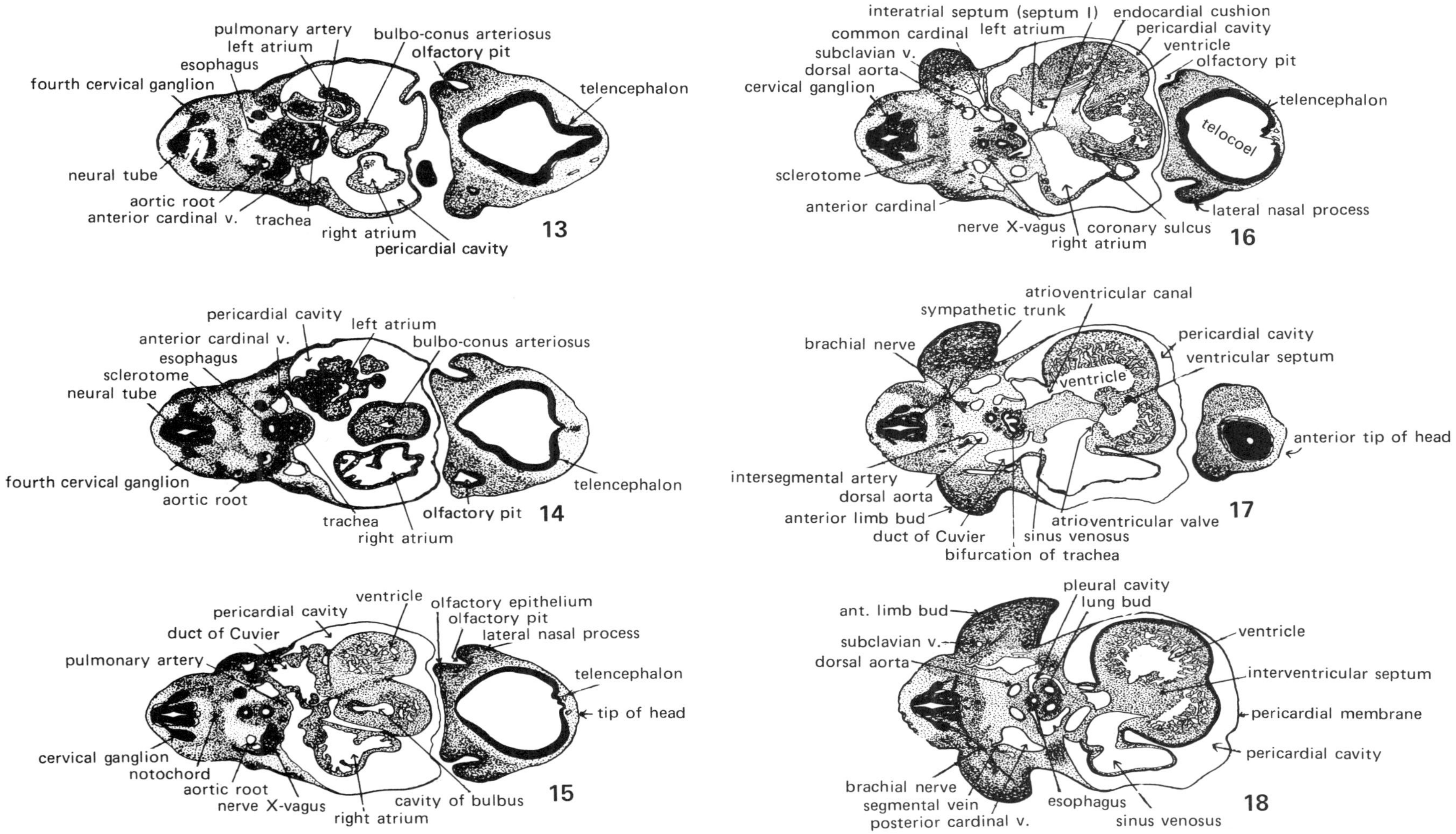

TRANSVERSE SECTIONS OF 10-mm PIG EMBRYO

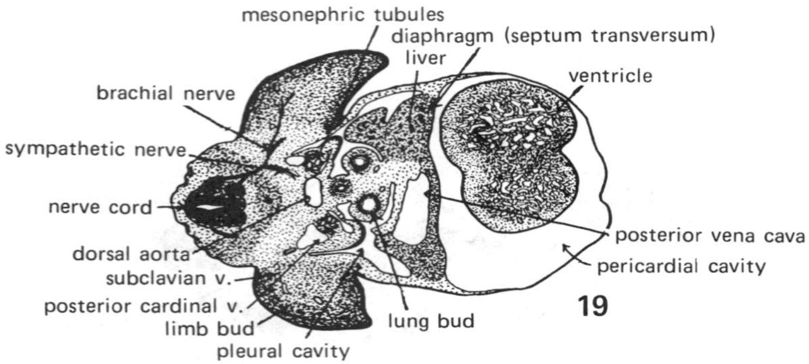

19

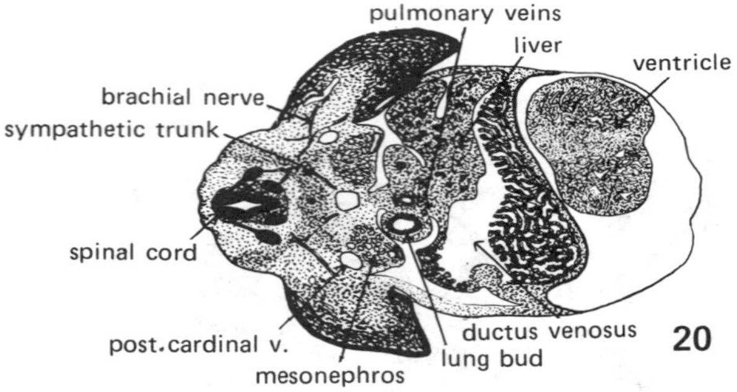

20

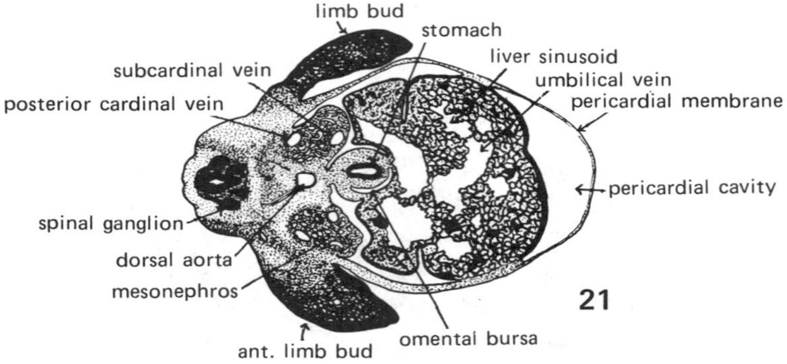

21

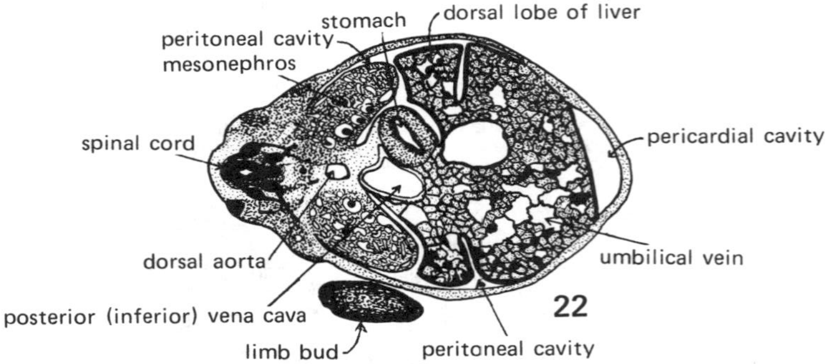

22

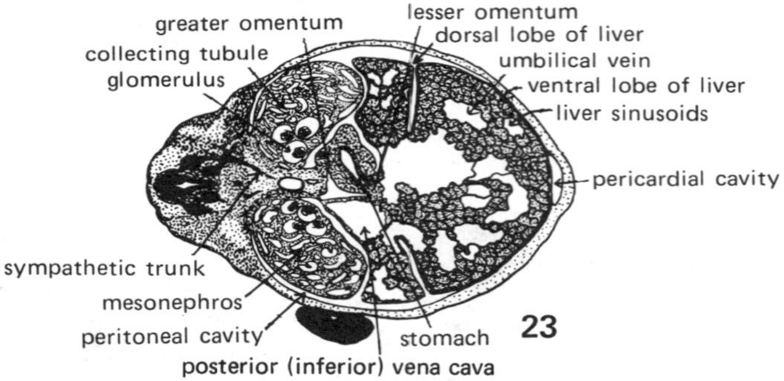

23

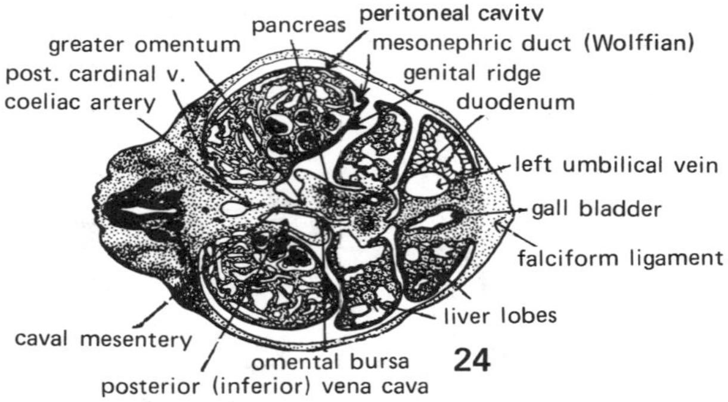

24

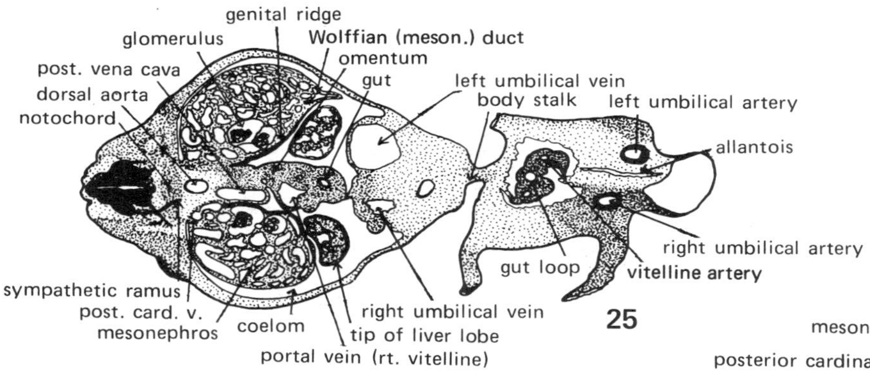

25

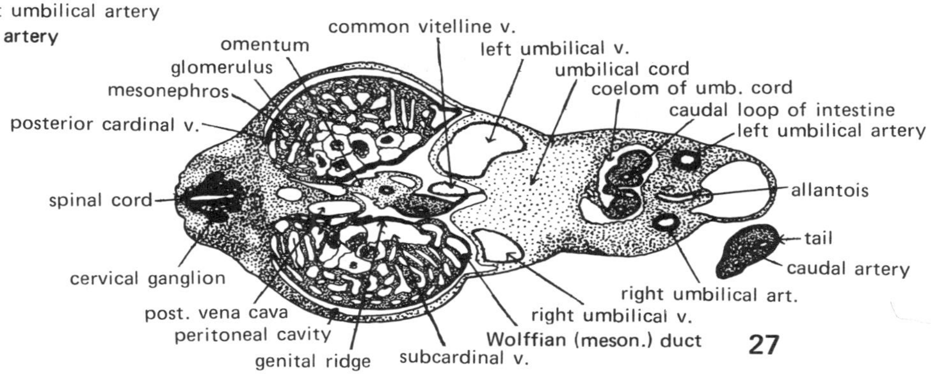

27

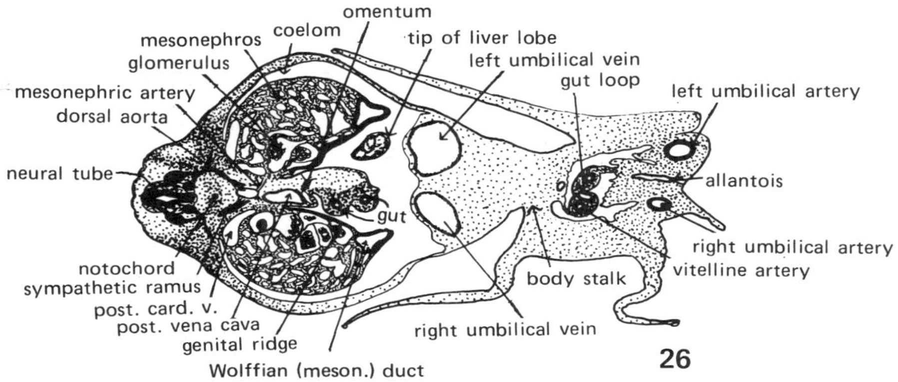

26

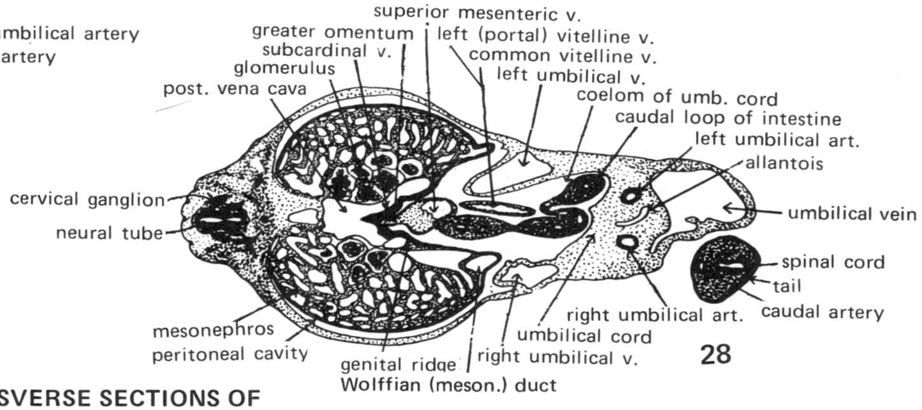

28

TRANSVERSE SECTIONS OF 10-mm PIG EMBRYO

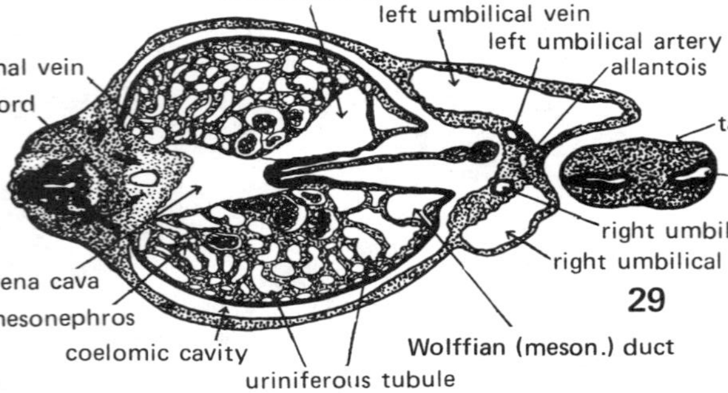

29

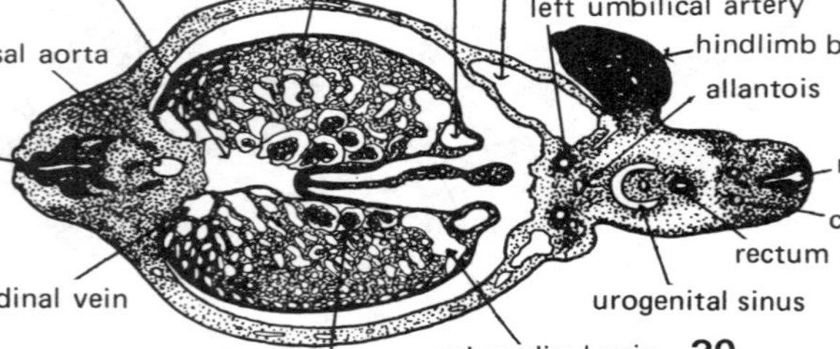

30

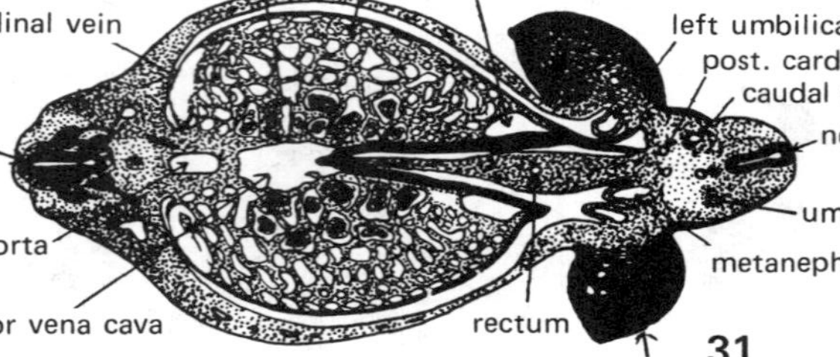

31

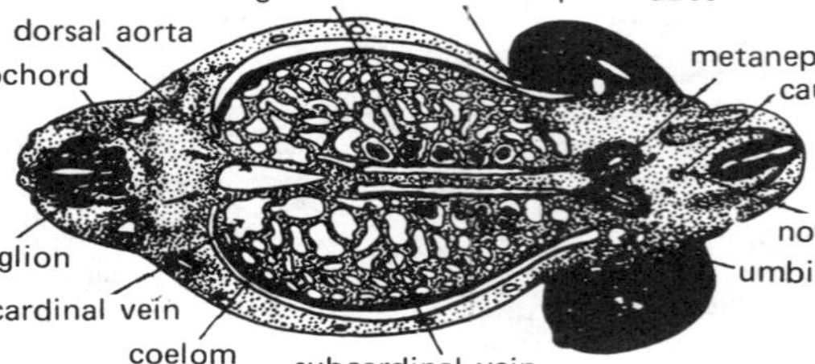

32

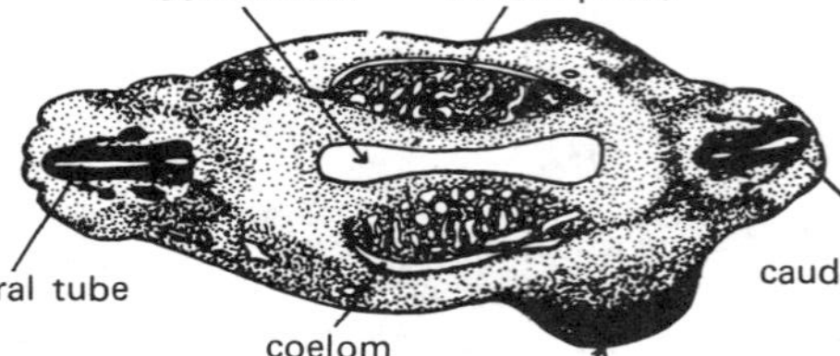

33

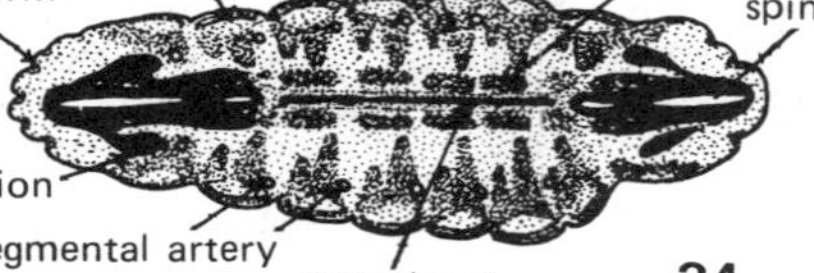

34

TRANSVERSE SECTIONS OF 10-mm PIG EMBRYO

DISSECTION OF THE PREGNANT UTERUS OF THE PIG

If available, a pregnant sow's uterus should be provided for each six students for group dissection and study.

The sow's ovaries lie in the pelvic portion of the abdominal cavity, each almost completely surrounded by the dilated ostium tubae or fimbriated ends of the oviducts (Fallopian tube of the human). Each oviduct enlarges into a uterine horn (cornu uteri) which unites mesially with the other uterine horn (bicornuate uterus) to form the body (corpus) of the uterus. The cavities of the oviducts continue through the uteri and the common cervix uteri into the vagina. The vagina, therefore, functions both as a birth canal and an organ of copulation. The broad ligament attaches the uterus to the body wall.

Examine closely the ovary of the sow. Oöcytes may be indicated by their Graafian follicles, which appear as clear vesicles on the surface of the ovary. Larger and slightly yellow protrusions are the corpora lutea, each of which represents a previous ovulation and now functions as an organ of internal secretion. By dissection it will be seen that the corpus luteum is relatively solid. The number of implantations in the uterus should correspond to the number of large corpora lutea present.

While in many mammals the chorion and allantois of the embryo fuse with the maternal uterine mucosa to form a placenta, the relationship in the pig is not so intimate, and the two layers (embryonic and maternal) can be peeled apart. The epithelial surface of the rather simple pig chorion follows the folds in the surface of the uterine mucosa so that there is contact throughout; the tissues are separated by a thin film of uterine secretion.

Make a small incision in the wall of the uterus, avoiding injury to the embryo or invasion of the uterine cavity, and gently pull the uterine wall away from the embryonic tissues. Identify the muscular and mucosal layers of the uterus and the relationship of the blood vessels.

Now, cut through a chorionic vesicle (really fused chorion and allantois), allow the fluid to escape from the enlarged allantois, and identify the embryo and its attachment to the maternal tissue by the umbilical cord. The allantois has expanded to such an extent that it has filled the chorionic cavity. Locate the amnion close to the embryo and note that it is covered by one wall of the allantoic membrane. Actually (as in the chick) one mesodermal membrane (allantoic) is fused with another mesodermal membrane (amniotic). In the region of fusion of the amnion and chorion, locate the reduced yolk sac. Examine further the remaining chorionic vesicles with contained embryos and determine the exact means of attachment to the maternal tissues. Since there is no localized zone of placentation, the sow has a *diffuse type* of placenta.

Nutritional and other substances which pass from sow to embryo must traverse a series of membranes by diffusion. These include the endothelial lining of the maternal blood vessels, connective tissue of the uterus, uterine epithelium, space between uterus and chorion, chorionic epithelium, connective tissue of the chorion, and, finally, the endothelial lining of the embryonic circulation. (See Patten, 1948, Chapter VI).

DISSECTION OF THE FETAL PIG

The student who understands chick, mouse, and pig embryology from serial transverse and sagittal sections is ready for a dissection study of the pig fetus of 15 cm. or longer. Fetal specimens of varying sizes may be secured from slaughterhouses, and the umbilical arteries can be injected immediately with a coloring mixture as follows:

Distilled water	1,000 cc
Glycerine	200 cc
Formaldehyde	200 cc
Cornstarch	2 lb

Add cornstarch to other ingredients, mix well to saturate, and let stand a day or so. Stir frequently. If color is desired, use ordinary, good-grade, water-base paint (the best is canary yellow), adding desired quantity directly to mixture above.

Upon reaching the laboratory, the brain cavity (through fontanelle) and the peritoneal cavity should be injected with 20% formalin. Specimens are best when immersed in 20% formalin for several months.

EXTERNAL FEATURES OF THE FETAL PIG

Even the early pig fetus shows the characteristic features of the mammal, having hair (at first anteriorly) and *milk line* or mammary rudiments. In later stages the long, stiff, tactile whiskers known as *vibrissae* may be found lateral to the snout and near the eyes.

The general shape of the body is due in part to the confinement within the uterus, but one can distinguish the head, neck, trunk, and tail. Locate the following:

HEAD:

1. Transverse *mouth* bounded by hairy lips.
2. Triangle-shaped *nostrils* dorsal to the end of the snout.
3. Lateral *eyes,* each protected by three eyelids. (The third lid or nictitating membrane can be pulled across the eye from the anterior angle, known as the canthus.)
4. External *ears* (pinnae), each of which surrounds an auditory aperture.

NECK: Note stout muscular nature of neck.

TRUNK:

1. FORELIMBS – generally covered by skin. Feel for shoulder (pectoral) girdle cartilage and the digits.
2. HINDLIMBS – like forelimbs, except that they are generally larger.
3. THORAX – identified by cartilaginous ribs.
4. ABDOMEN – nonrigid portion of trunk containing the viscera and from which extends the cylindrical umbilical cord.
5. TEATS or mammae, arranged in pairs on thorax and abdomen.
6. UMBILICAL CORD – cut a cross section of the cord and identify two arteries, one vein, the allantoic duct, and Wharton's jelly.
7. In male specimens (older) find *urogenital opening* posterior to the umbilical cord and *scrotal sacs* below the anus.
8. SACRUM – rigid because of the pelvic girdle and hindlimb attachments.
9. ANUS – small median opening beneath the base of the tail.
10. External FEMALE ORIFICE (vaginal opening) near anus of older specimens.

TAIL: Short and curled, without significant structure.

ORGANS OF SPECIAL SENSE:

Carefully dissect the eye, eyeball, eye muscles, and ear to locate all morphological parts in the fetus.

FETAL CIRCULATION

This will be described briefly because the student should have in mind the changes from fetal to adult circulation and should identify the strictly fetal blood vessels as they are exposed. (See page 248.)

The umbilical vein carries nutrient and oxygenated blood from the placenta to the fetus. This vein carries blood through the liver where it joins the postcaval vein by means of the ductus venosus, but there is a short-cut connection with the portal vein in the region of the liver. Since the portal blood enters the liver sinuses, later to emerge in the hepatic veins, it also eventually joins the postcaval vein and then goes into the sinus venosus. Placental (oxygenated) blood is mixed with venous blood from both the portal and postcaval systems. Postcaval blood of the fetus passes from the right to the left auricle by way of the temporary foramen ovale, guided by movements of the Eustachian valve. There is further mixture later, in the left auricle, as a small amount of venous blood is brought to the heart from the pulmonary veins. The left ventricle and the aorta, therefore, carry mixed blood, most of which is placental, and much of this blood passes directly to the head and anterior extremities, the remainder passing posteriorly through the descending aorta. Venous blood from the head enters the right auricle by way of the precavals and goes to the right ventricle and then to the pulmonary arteries. Most of the blood in these arteries passes through the ductus arteriosus into the corresponding descending aorta. The abdominal viscera are supplied by the dorsal aorta, but most of this mixed blood passes out again to the placenta by way of the allantoic arteries.

Since the fetus has no way of aerating its own blood, no part of the body receives absolutely pure blood. At parturition (birth), the following changes occur:

1. The cutting of the umbilical circulation increases the CO_2 content of the blood, which in turn stimulates the respiratory center of the brain to cause the first inspiration of air.
2. The entire pulmonary circulation is set up so that the ductus arteriosus becomes nonfunctional. The ductus closes to become the cordlike ligamentum Botalli in a few days.
3. The foramen ovale is closed by increased pulmonary flow against its flap valves, so that the postcaval blood now passes from the right auricle into the right ventricle.
4. The ductus venosus and umbilical vein degenerate.
5. The allantoic vessels degenerate, except for those portions which supply the bladder.

It is suggested that the student dissect carefully the fetal heart.

INTERNAL FEATURES OF THE FETAL PIG

Fasten each foot of the fetal pig to the corners of the dissecting board with the specimen on its back. With scalpel make a median incision through the skin from the sternum to the umbilical cord, around both sides of the cord, and posteriorly to a point near the anus. If the specimen is a male, avoid injury to the partially covered penis. Now, make lateral cuts below the diaphragm and above the symphysis, laying open the abdominal cavity. Note the following:

PERITONEUM: lining of the abdominal cavity; a shiny membrane.

GREAT OMENTUM: folds of the peritoneum containing fat in older specimens.

LIVER: red-brown, lobulated organ just posterior to the diaphragm.

STOMACH: found beneath the liver as a white muscular sac.

SPLEEN: blood-red body lying along the greater curvature of the stomach.

SMALL INTESTINE: much convoluted tube which fills the right posterior quadrant of the abdominal cavity, beginning at the stomach.

LARGE INTESTINE: larger, less convoluted portion of alimentary canal, made up of colon and caecum.

ALLANTOIC BLADDER: saclike enlargement of the allantoic duct, accompanied by two large allantoic arteries and passing to the umbilicus.

Tie off the posterior vena cava anterior and posterior to the liver, and then remove all of the visceral organs listed above without injuring the urinogenital system. Find the following:

KIDNEYS: dark-red bodies embedded in fat along the dorsal body wall. (Remove and slit open one kidney to note internal development.)

ADRENAL BODIES: long, slender, yellowish glands lying anteriomesially on the face of the kidneys.

URETERS: small, tough tubes extending posteriorly from each kidney along the dorsal abdominal wall and entering the bladder. (Note differences in the two sexes.)

BLADDER: thick-walled enlargement of allantoic duct which empties into the rather undeveloped urethra.

REPRODUCTIVE SYSTEM:

MALE: (remove skin from region of symphysis pubis, then make a median incision separating the two halves).

SCROTAL SAC: tough-walled, saclike evaginations of abdominal cavity, ventral to the pelvis.

TESTES: bluish red, elliptical bodies found somewhere between the kidneys and the scrotal sac, depending upon state of descent.

EPIDIDYMIS: mass of coiled tubes along mesial edge of the testis.

VAS DEFERENS: whitish tube which extends from the epididymis through the inguinal ring and into the abdominal cavity.

URETHRA: posterior extension of the allantoic bladder, ending in the penis.

PENIS: elongated, muscular organ carrying the urethra and extending anteriorly to a point just posterior to the umbilicus.

FEMALE: remove skin from the symphysis pubis and tease away surrounding tissue, but avoid the genital tract.

OVARIES: pale yellow, oval bodies suspended by mesovarium just posterior to the kidneys.

FALLOPIAN TUBES: slender, white tubes extending from the ovaries to the uterus.

UTERUS: paired horns leading from Fallopian tubes to junction in midline to form the body (corpus) of the uterus. This leads by way of a long neck or cervix into the vaginal cavity.

VAGINA: small cavity within the urinogenital sinus, containing the cervix.

VULVA: slitlike external opening of the vagina (urinogenital sinus).

CLITORIS: rounded, pimplelike structure within urinogenital sinus, homologous to the penis of the male.

Remove the diaphragm and then cut out a triangular area (including skin, ribs, and muscles) to expose the thoracic cavity. Avoid injury to the pleura.

PLEURA: membrane lining the paired cavities and surrounding the lungs.

MEDIASTINUM: large cavity between the median walls of the pleural sacs.

THYMUS: large, bilobed gland which covers anterior third of the heart and frequently extends into the neck region; largely fetal structure.

LUNGS: paired, semisolid, pinkish-white organs lying within the pleural cavity of the thorax. Circulation is poor in the fetus, and lungs are not expanded.

PERICARDIUM: tough, membranous sac surrounding the heart. (Remove to see the heart.)

HEART: cone-shaped, muscular organ with apex pointing posteriorly and to the left. Fully developed in 15-cm fetus.

The vascular system of the fetus will have been exposed in the above dissections, and it is assumed that the student who has had comparative anatomy and this course in embryology will be able to identify all of the blood vessels encountered.

NERVOUS SYSTEM: There are 33 pairs of *spinal nerves.* This includes eight cervicals, fourteen thoracics, seven lumbars, and four sacrals. Some of these spinal nerves may be seen emerging from the cord along the dorsal body wall. Locate also the *brachial plexus* (last three cervicals and first thoracics) and the *lumbosacral plexus* (last three lumbar and the first sacral nerve). The sciatic is the largest nerve in the body and arises largely from the sixth lumbar and the first sacral nerves, passing into the muscles of the hindleg.

The SYMPATHETIC NERVOUS SYSTEM consists of a chain of ganglia extending along each side of the vertebral column receiving communicating branches from spinal nerves and frequently forming plexuses.

SOLAR PLEXUS: between the stomach, adrenals, aorta, and dorsal portion of the diaphragm, receiving branches from the (tenth) vagus and supplying the abdominal viscera.

CARDIAC PLEXUS: between branch of the vagus and sympathetic nerves, lies at the base of the heart and around the pulmonary artery and the aorta, supplying the heart.

(For those who have the time, it is suggested that a more complete study of the fetal pig might be made with the guidance of Baumgartner's *Laboratory Manual of the Foetal Pig.* [New York: The Macmillan Company, 1924.])

"The stabilization of our institutions rests ultimately upon our ability to know and to test assumptions, and upon willingness to revise them without partizanship, or bitterness, or distress."

Simpson, 1922, ***Am. Math. Month***

"Man is the only animal who in any considerable measure bequeathed to his descendants the accumulated wisdom of past generations."

A. J. Lotka, 1925

"Truth comes out of error more readily than out of confusion."

Bacon

Exercise 25

EMBRYOLOGY OF THE MOUSE

INTRODUCTION

The study of mammalian embryology is hampered by two factors—first, the availability of the material and, second, the usual large size of the pig or other embryos that have been used. The mouse eliminates both of these objections or complications by being readily available for any laboratory and being relatively small throughout development.

Any species or strain of mouse may be used. The description here will be based upon the pure strain of white albino mouse which is widely used for research. It is not a hybrid and thereby avoids the wider variations in development attendant upon great heterozygosity. The stages described are essentially the same for any strain, the difference being in rate of development as it is presumed that the hybrid may grow slightly more rapidly than the pure strain. The gestation period of the form described here is 21 days. This is almost the same as that of the rat, so the development of these divergent strains may be compared.

SECURING OF THE MATERIAL

The mouse has a 5 or 5½ day estrous cycle and will accept the male only at the height of estrus, which generally occurs between 10 P.M. and 1 A.M. The most accurate method of timing conception is to observe the mating, but short of this it is possible to estimate the conception time rather closely, within several hours. Except for the early fertilization stages, such accuracy is not necessary.

Generally a box of five or six females is provided with a sexually mature male at 5 p.m., and the next morning at 9 A.M. each female is examined for the presence of a vaginal plug, a certain sign of successful copulation. The plug consists of secretions from the vesicular coagulating glands of the male. The vesicular glands are known as the seminal vesicles. They do not contain spermatozoa but do produce a secretion. The coagulating glands are the prostates whose secretion, when mixed with that of the vesicular glands, forms the coagulum that blocks the vaginal orifice. A vaginal smear, made by inserting a flamed pipette (filled with 0.5 cc of physiological saline) into the vagina and sucking out some of the contents and examining with dark field illumination under the microscope for living spermatozoa, is more tedious but another method of selecting those females that have mated.

Since mating could occur between 5 P.M. and 9 A.M. but is more likely to occur between 10 P.M. and 1 A.M., there is some variation in the early stages. This could be as much as 16 hours. The first cleavage occurs 24 hours after fertilization, so that one can easily obtain the precleavage stages of fertilization.*

Statistically one should obtain about 10 percent copulations by this method. If the cycle takes 5+ days, some 20 percent should be in estrus during any 24-hour period. If mating is allowed during only 12-16 hours (one night) this cuts the probability in half, so that one could expect a maximum of 10 percent matings in a large group of females properly exposed to mature males. It is possible for one male to mate more than once during a night, but it is statistically unlikely that more than

*Most accurately time pregnancies are obtained by exposing the estrus females to mature males between 8 - 10 a.m. or even 8 - 9 a.m., thus limiting the time of conception to 1 or 2 hours. This is our current practice.

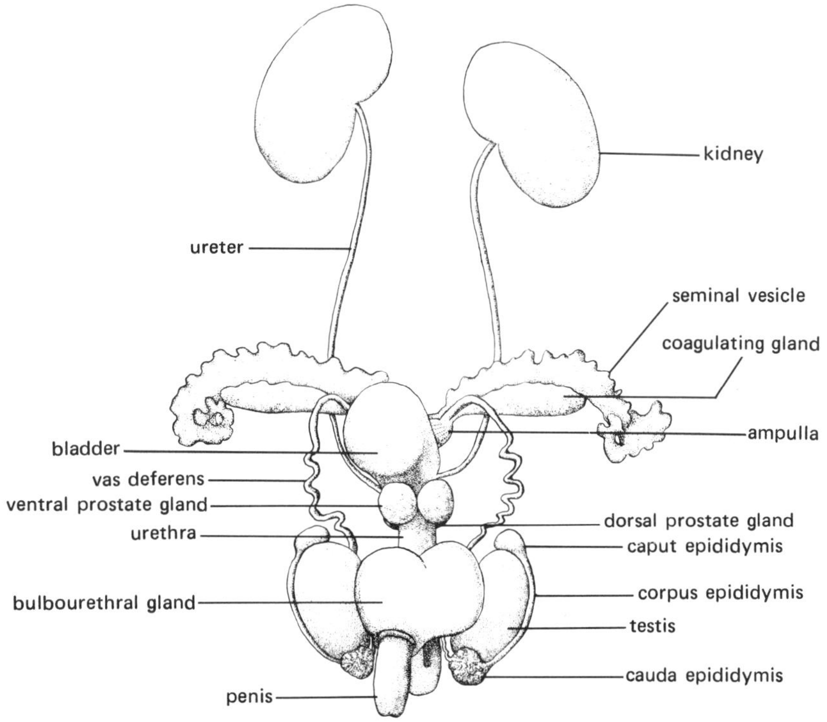

VENTRAL VIEW OF UROGENITAL SYSTEM OF MALE MOUSE

From R. Rugh. *Vertebrate Embryology,* 1964. Harcourt, Brace and World, New York.

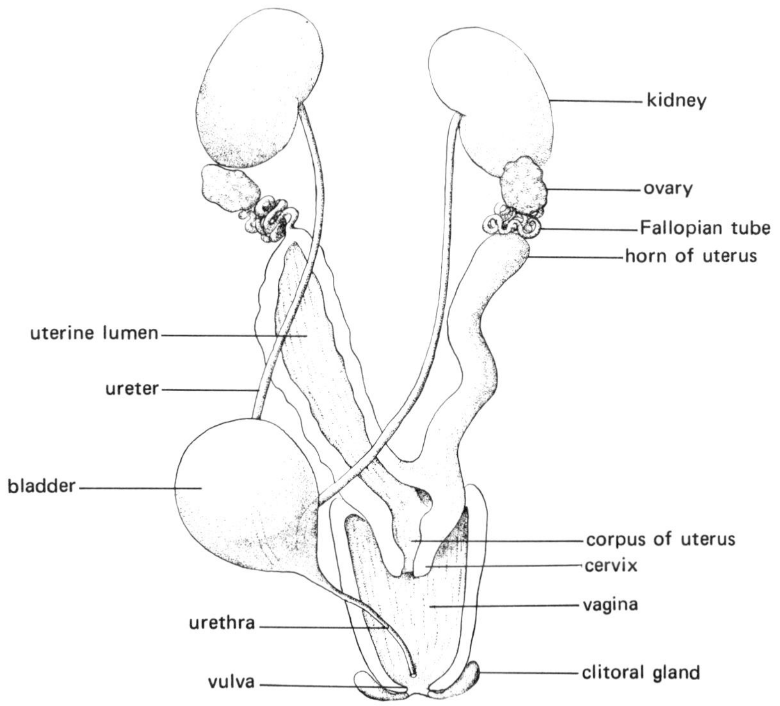

VENTRAL VIEW OF UROGENITAL SYSTEM OF FEMALE MOUSE

From R. Rugh. *Vertebrate Embryology.* 1964. Harcourt, Brace and World, New York.

DRAWINGS OF MOUSE GONADS: GROSS AND MICROSCOPIC

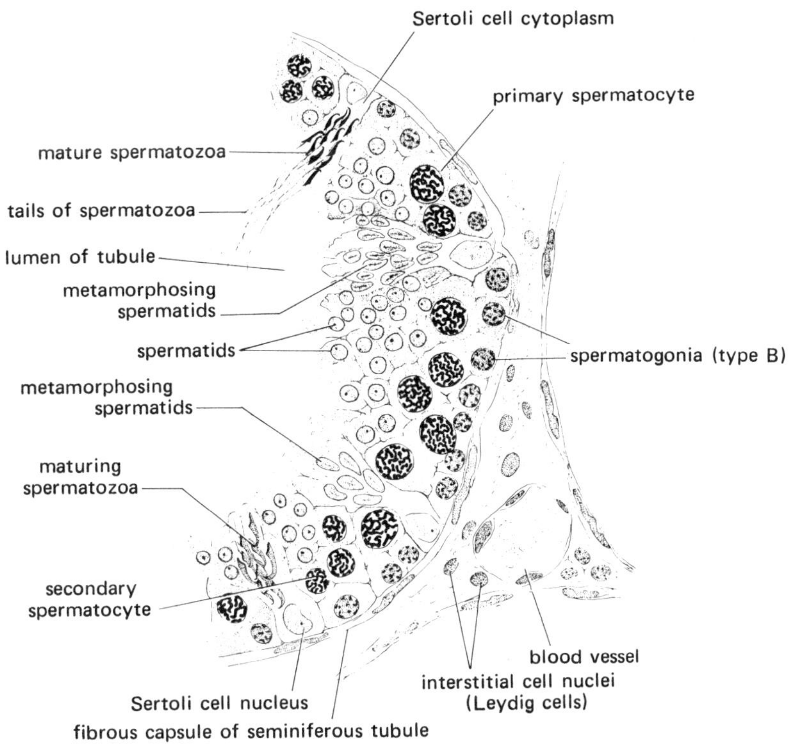

MATURATION OF SPERMATOZOA IN THE MOUSE

From R. Rugh. *The Mouse.* 1968. Burgess, Minneapolis.

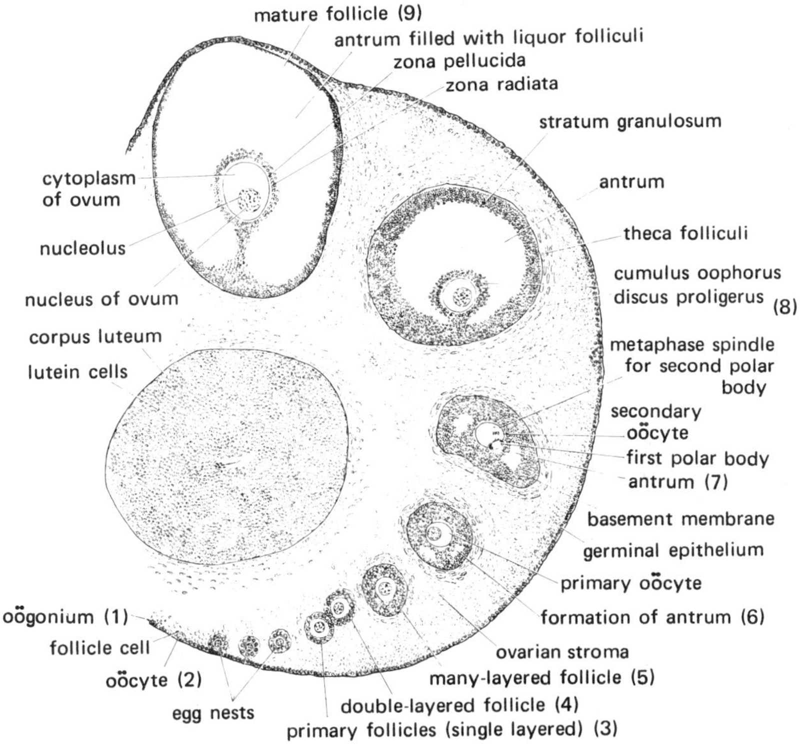

DEVELOPMENT OF THE MOUSE OVUM AND OVARIAN FOLLICLE

(Parenthetical numbers refer to maturation stages)

From R. Rugh. *The Mouse.* 1968. Burgess, Minneapolis.

one female in five will be receptive at one time. Thus, if it is necessary to obtain 20 pregnancies at any one time, one should expose some 200 or more mature females to 40 or more males for the mating period.

Successive exposures of the same females to males should be spaced at least one week apart in order to identify missed pregnancies. The vaginal plugs are sloughed off, or passed with urine, or eaten away by the female, so that successful matings may often be missed or undiagnosed. At the same time it is rare, but it does occur, that there may be a plug but no pregnancy.

At stated times after conception, embryonic material may be obtained as follows: Kill the female by cervical dislocation, and quickly dissect out the bicornuate uteri and attached ovaries. While still fresh, pin the double uterus outstretched onto a card with insect needles, and place in (alcoholic) Bouin's fluid. If the two uteri are placed close together for fixation, subsequent sectioning will be simpler. This material should be fixed for 24-48 hours and for later stages (7½ days or older) should be fixed 4-5 days. Embryos of 12½ or more days may not be properly fixed unless the uterus is opened or flooded internally with fixative. Such embryos may be temporarily fixed *in situ,* later to be dissected out and further fixed as free embryos. The later preparturition stages may be dissected out immediately and fixed while alive, free of all membranes and placenta.

During the first 4 days after conception the eggs and early embryos are located in the upper half of the oviduct and uterus and are generally clustered together. Sections including this part, plus the related ovary, are valuable teaching material. After 4 days all blastulae are evenly distributed throughout the uterus so that longitudinal sections of the entire duct should be made. Implantation is at 4½ days, completed by 6½ days, and organogenesis begins shortly thereafter.

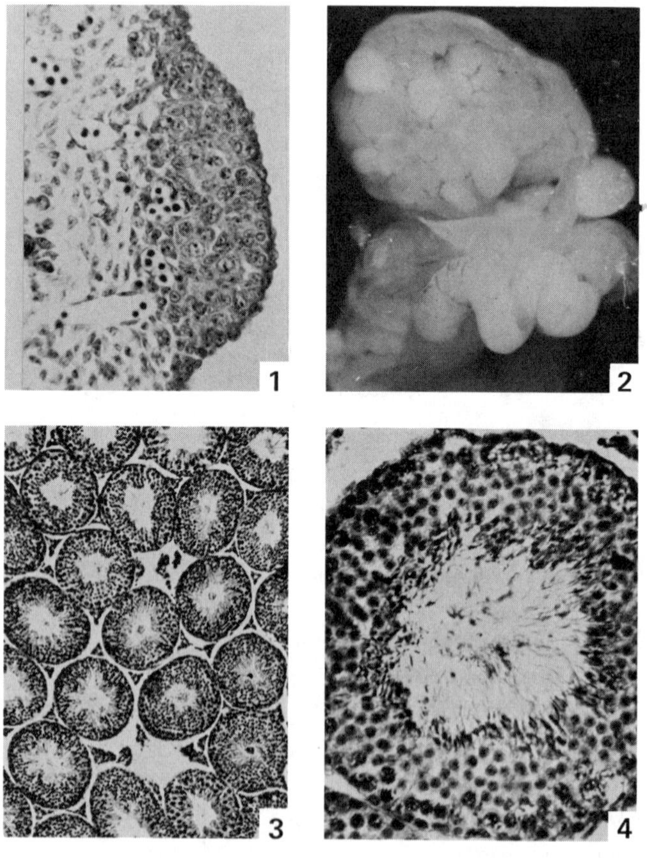

⇦ MOUSE GONADS

1 – Gonad primodium attached to mesonephros, at this stage undifferentiated.
2 – Mature ovary of mouse showing whitish areas which are corpora lutea.
3 – Testis of mature mouse, showing seminiferous tubules.
4 – Enlarged view of seminiferous tubule, showing the various stages of maturation, always present during period of reproductive potential.

EARLY CLEAVAGE IN THE MOUSE ⇨

1 – Typical living ovum, not showing polar body.
2 – Mature ovum, showing first polar body in perivitelline space.
3 – Fertilized ovum, showing female pronucleus.
4 – First cleavage, anaphase.
5 – First cleavage, telophase. Note cleavage furrow and polar body.
6 – 2-cell stage. Note dividing polar body. Second polar body not seen.
7 – 4-cell stage.
8 – From 4 to 8 cells, boundaries indistinct.
9 – 8-cell stage.
10 – 8 to 16 cells, usually about 12, all still within the zona pellucida.
11 – Succeeding stage when all cell boundaries are temporarily obscured. Leads to morula.

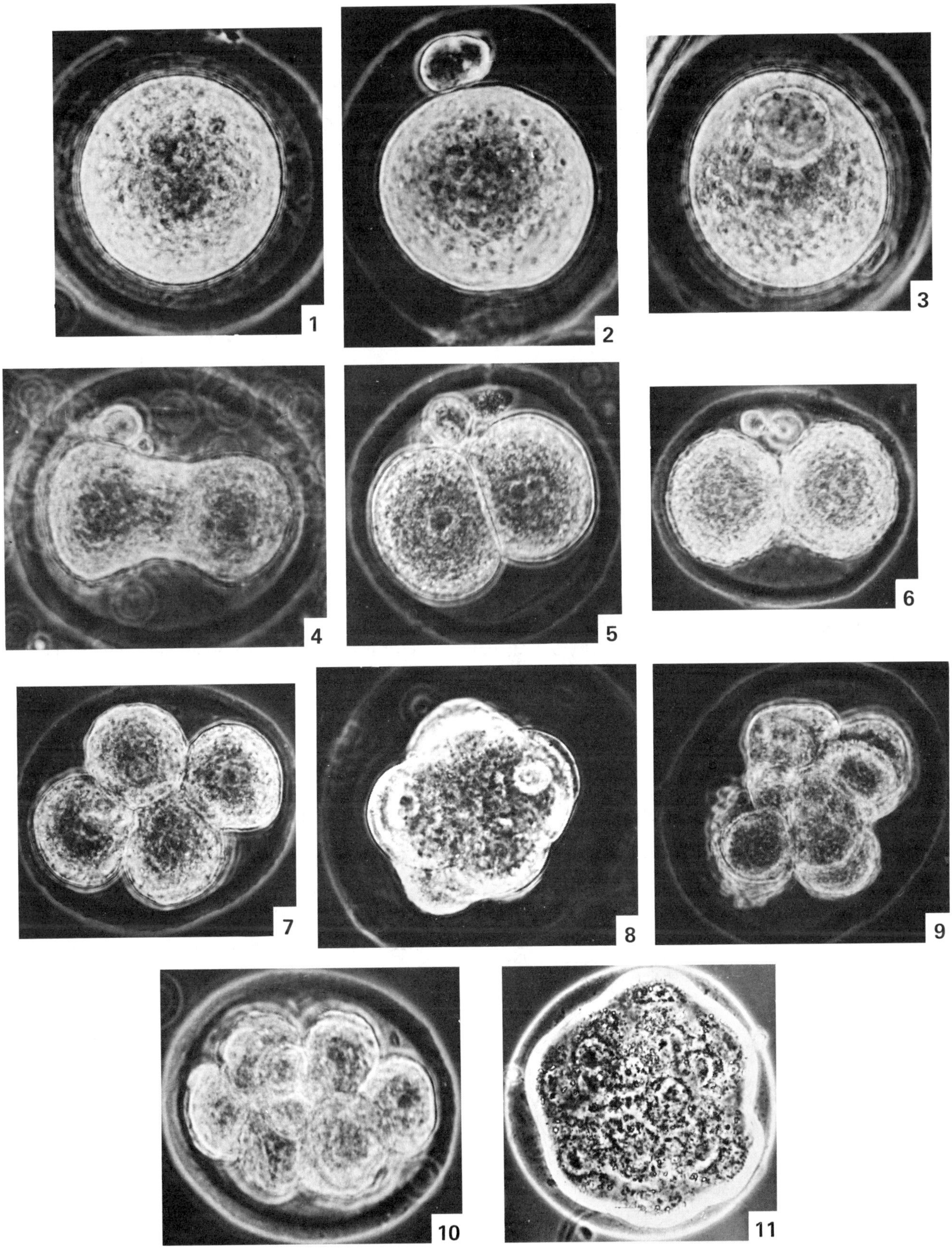
1
2
3
4
5
6
7
8
9
10
11

FERTILIZATION

The newly ovulated egg (80-85 μ diameter) of the mouse is surrounded by follicle cells, which are characteristically small, oval, and uniform, with a heavily staining nucleus. A nonliving zona pellucida may be seen as a clear, jellylike area surrounding the egg, adding 10 μ to the diameter. Fertilization occurs in the upper end of the oviduct, closely approximating the ovary. The sperm generally reach the region of the ovaries a short time before ovulation, both by their own movements and by the undulations of the uteri and oviducts. The period of heat (estrus), when the female will accept the male, begins several hours before actual rupture of the eggs from the ovary. With fertilization, the zona pellucida swells (as does the jelly of the frog's egg) so that the outside diameter may be as much as 115 μ, just visible to the naked eye.

Polyspermy is not known, so that a single sperm generally enters and prevents subsequent invasions by other sperm. This means that the response of the egg is instantaneous. The zona pellucida swells, and the egg may shrink a bit so that a perivitelline space develops between. Since the first polar body is formed while the egg is in its ovarian follicle, it may now be found within the perivitelline space.

The sperm head enters the vitellus but may often take with it the middle piece and the tail. Within 2-3 hours the second polar body is formed and everted into the perivitelline space, and the two pronuclei (male and female) appear and begin to migrate towards each other. The female nucleus may be identified either by its proximity to the newly everted polar body or by its larger size. These nuclei do not fuse before the first cleavage. After this first division their respective chromosomes are indistinguishable.

CLEAVAGE

The first two cleavages of the mouse egg occur while it is still within the oviduct, and generally one will find groups of five or six such cleavage stages close together in each oviduct. The first cleavage occurs about 24 hours after fertilization, but since this may vary in time, one estimates it to occur at about midnight so that by 9 A.M. development is considered to be already ½ day. Thus, on successive mornings, one estimates the ages as 1½, 2½ days, etc.*

Cleavage itself takes only about 10 minutes so that mitotic figures are rarely seen. The intermitotic time is many hours during the early cleavages. Each successive cleavage occurs more quickly, and by 2½ days after fertilization the embryo is a 16-cell morula (solid ball of 16 similar

---

THE MOUSE EMBRYO: ½ TO 2½ DAYS POST CONCEPTION

Fig. 1 – Female mouse at 9 A.M. after previous night of mating opportunity showing presence of vaginal plug.
Fig. 2 – Contents of bicornuate uterus after copulation showing viscous coagulum containing abundant spermatozoa.
Fig. 3 – Sperm entrance into the egg cortex, showing dissolution of the cortex at point of invasion. Note protruding sperm tail.
Fig. 4 – Contents of sperm head, without tail, being reorganized within the egg cytoplasm.
Fig. 5 – Early stage in polar body formation. Note chromosomes.
Fig. 6 – Polar body now separate from egg cytoplasm, with distinct nucleus and small amount of cytoplasm.
Fig. 7 – Extruded polar body quite distinct from the egg.
Fig. 8 – Large postmaturational female germinal vesicle (nucleus).
Fig. 9 – Same as Fig. 8 but stained heavily to show cell membranes and rather granular cytoplasm.
Fig. 10 – Two pronuclei as they are approaching each other, female (larger) on the left and male (smaller) on the right.
Fig. 11 – Approximation of the two pronuclei, female the larger.
Fig. 12 – Upper oviduct ½ day after conception, showing four eggs close together.
Fig. 13 – Fertilization complete in that polar body is extruded, and the two pronuclei are approaching each other. Male pronucleus appears to swell slightly.
Fig. 14 – Both polar bodies show, but pronuclei are unstained.
Fig. 15 – 2-cell stage at 1½ days after conception, also showing polar body.
Fig. 16 – Upper oviduct containing three embryos as seen also in Fig. 17.
Fig. 17 – Enlarged view, showing two 2-cell stages and a single cell with prominent polar body.
Fig. 18 – 4-cell stage.
Fig. 19 – 4 + cells, showing persistent polar body, at 2½ days.

*These estimates follow all-night matings. The shorter 1- or 2-hour matings in early morning delimit the time of fertilization which is designated as zero time, with 24 hours as 1-day development.

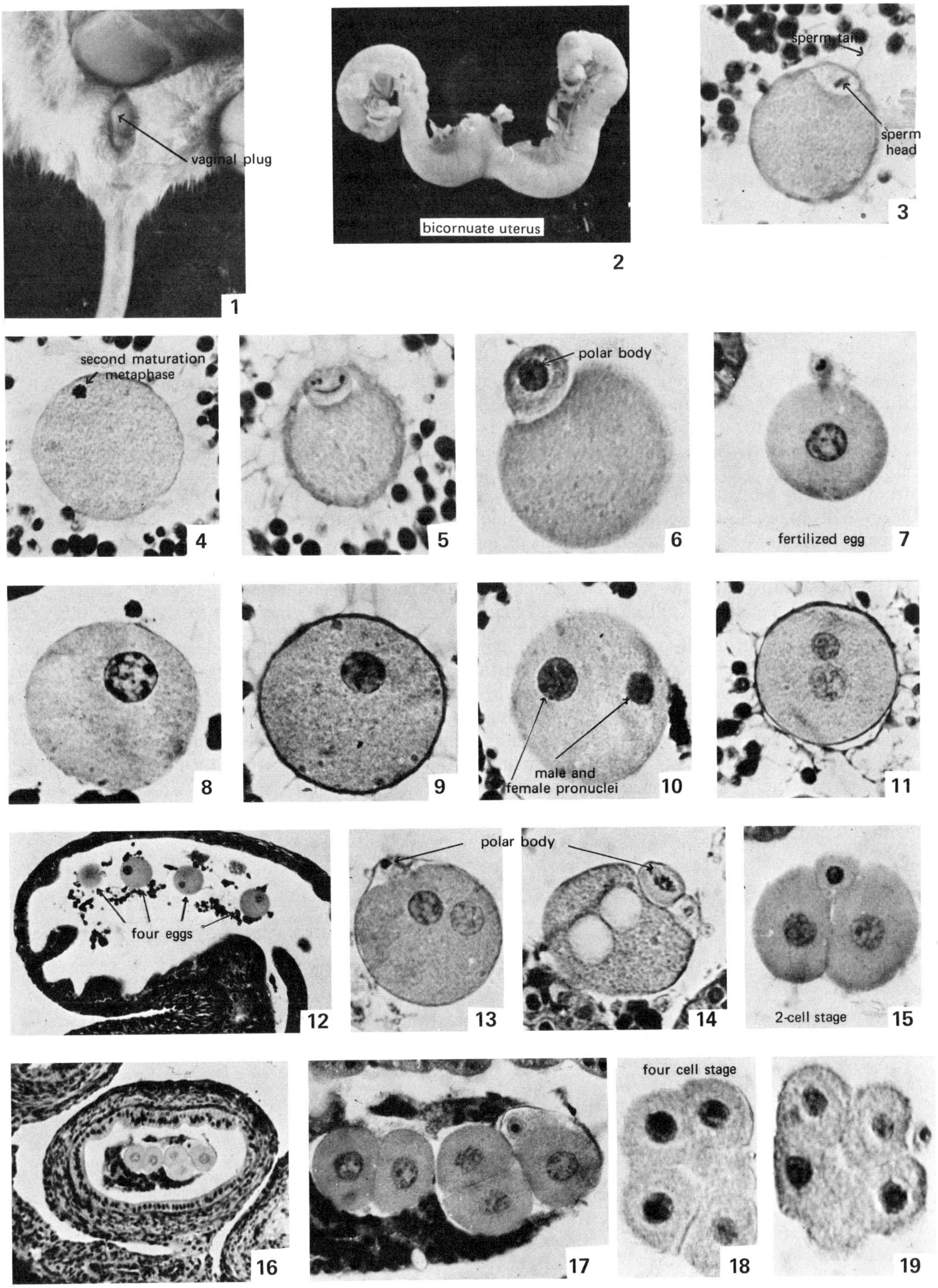
vaginal plug
1
bicornuate uterus
2
sperm tail
sperm head
3
second maturation metaphase
4
5
polar body
6
fertilized egg
7
8
9
male and female pronuclei
10
11
four eggs
12
polar body
13
14
2-cell stage
15
16
17
four cell stage
18
19

cells). Cleavage is not synchronous so that intermediate numbers of cells, between 4 and 16, for instance, may be found. By 3+ days the eggs descend from the oviduct into the uterus but may still be clustered. One of the unexplained situations then follows, namely the rather even distribution of the embryos for the length of the uterus, with spacing between each embryo for adequate implantation and growing room. Never does one find two closely implanted embryos.

THE BLASTULA STAGE

Shortly after reaching the 32-cell stage the morula begins to acquire an eccentric cavity, filled with a fluid, known as the *blastocoel* (segmentation cavity). This enlarges rather rapidly while the cells at one side appear to multiply and cluster to form the *inner cell mass.* The related single layer of cells which continue around the blastocoel will become the *trophectoderm.* While these changes are occurring, the embryos descend into the uterus and take up positions for implantations equidistant from each other.

IMPLANTATION OF THE EMBRYO

Starting from the innermost epithelial lining of the uterus outward, first comes the relatively thick mucosa, then the inner circular and outer longitudinal muscles, and finally the mesometrium, which attaches the uterus to the dorsal body wall. Both the circular and longitudinal muscles function to give the uterus undulating movements, which undoubtedly help in moving the embryo along through the lumen. Each blastula becomes located in an epithelial crypt, quite equidistant from adjacent blastulae. The nuclei in the vicinity of the uterine epithelium of the blastula then begin to disintegrate, and within 15-20 hours this epithelium is entirely sloughed off. This will be the implantation site for the particular blastula examined. The trophectoderm cells of the blastula make the contact with the epithelium, and there begins a response in the nearby mucosa which results in sufficient swelling by 5 days which may be noted with the naked eye. This swollen mucosa is known as the decidua; it will be an almost immediate source of nutriment for the growing embryo.

The zona pellucida disappears from the egg during the fourth and fifth cleavages even though there appear to be no actual increases with blastulation, and, as soon as implantation is achieved, the embryo appears to accelerate its growth processes. The blastocoelic fluid distends the blastula so that occasionally one can find a blastula making contact on two sides of the uterine epithelial crypt. At the same time the inner cell mass, cells on one side of the blastula, increase in number, giving it a multilayered appearance as compared with the trophectoderm. The blastocoel is now known as the *yolk cavity,* the inner cell mass as the *egg cylinder,* and the embryo as a *blastocyst.* It is the *trophectoderm* which makes the invasion into the uterine mucosa, establishing nutritional contact with the maternal tissues.

THE MOUSE EMBRYO: 3½ to 5½ DAYS POSTCONCEPTION ⇨

Fig. 20 – Longitudinal section of gravid uterus at 3½ days after conception, showing three blastulae in suspension within the uterine cavity.
Fig. 21 – Enlarged view of portion of Fig. 20.
Fig. 22 – Slightly later stage when the blastula is "trapped" within a uterine crypt, prior to implantation.
Fig. 23 – Embryo at 3½ days, showing cell structure. Preimplanted blastula.
Fig. 24 – Inner cell mass, trophectoderm, and blastocoel easily identified.
Fig. 25 – Blastula at about 4 days, showing embryonic and extraembryonic ectoderm, proximal and distal endoderm, and blastocoel (yolk cavity).
Fig. 26 – Blastula at the moment of adhesion to the uterine epithelium, low power.
Fig. 27 – Same as Fig. 26 but at high magnification.
Fig. 28 – Active implantation processes have begun, at 4½ days.
Fig. 29 – Embryo at 5½ days, showing little change in the surrounding uterus.
Fig. 30 – Enlargement of the embryo to distinguish between the inner solid embryonic ectoderm and the outer thin layer of proximal endoderm.
Fig. 31 – Separation of the first two primary germ layers.
Fig. 32 – Separation of distal and proximal endodermal layers, with a central core of solid ectoderm. Note early dissolution of contiguous uterine epithelium.

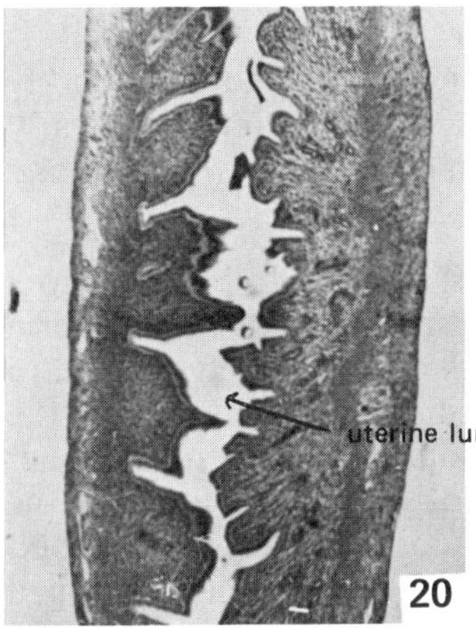

20

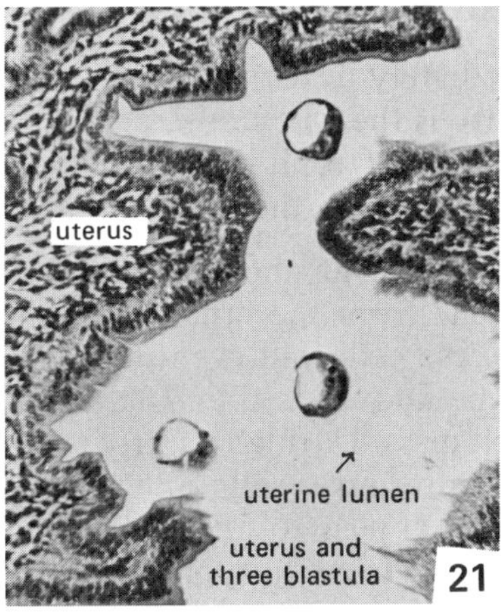

21

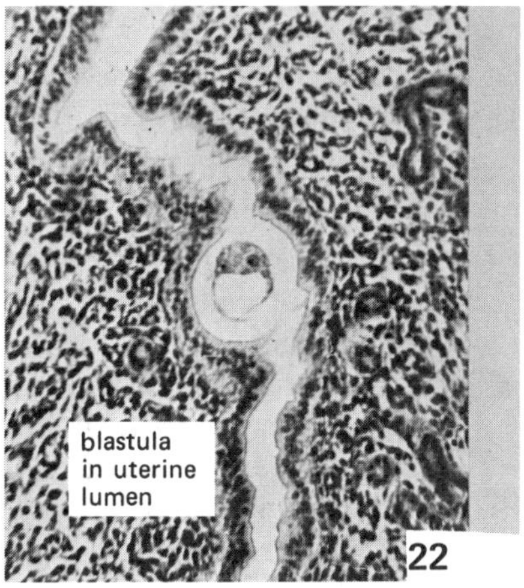

22

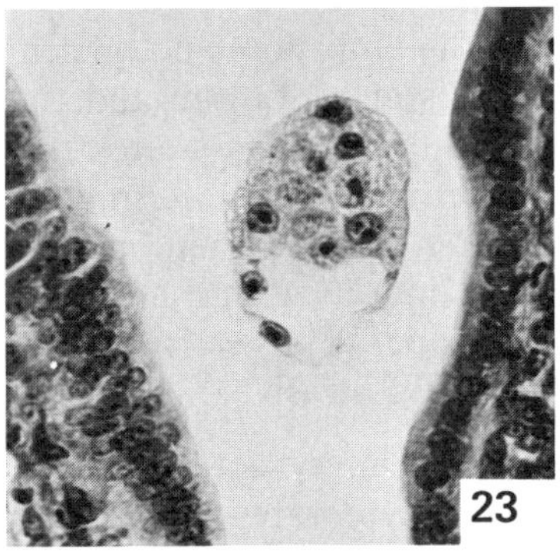
23

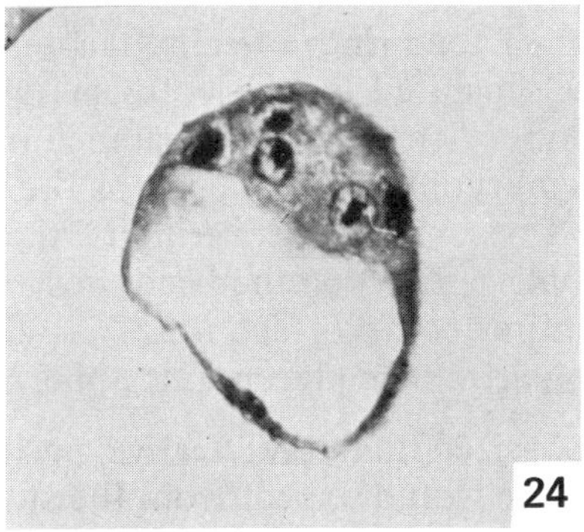
24

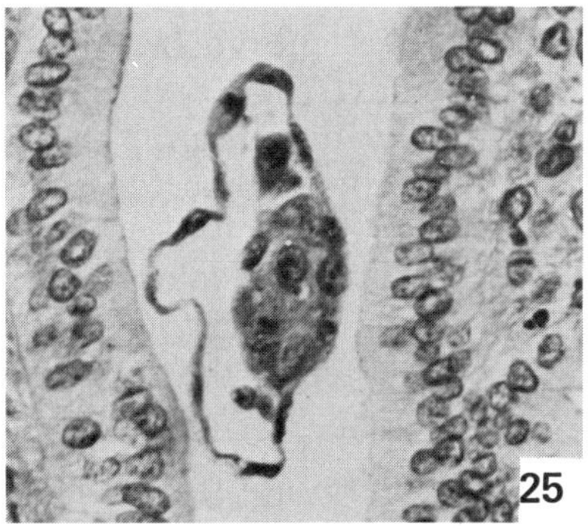
25

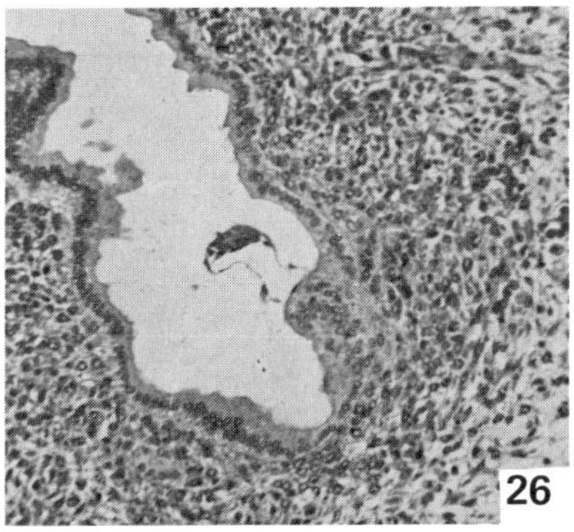
26

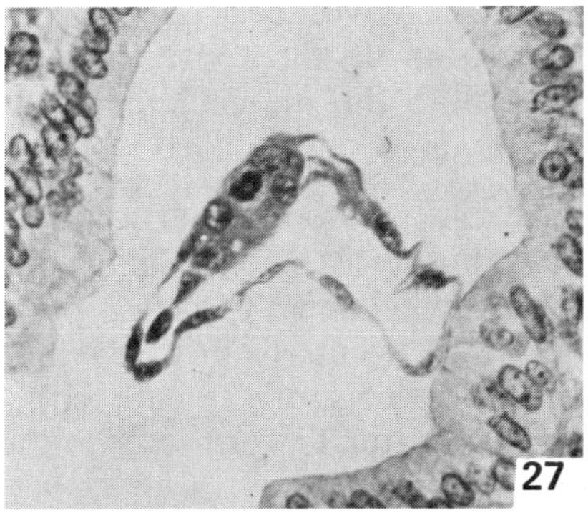
27

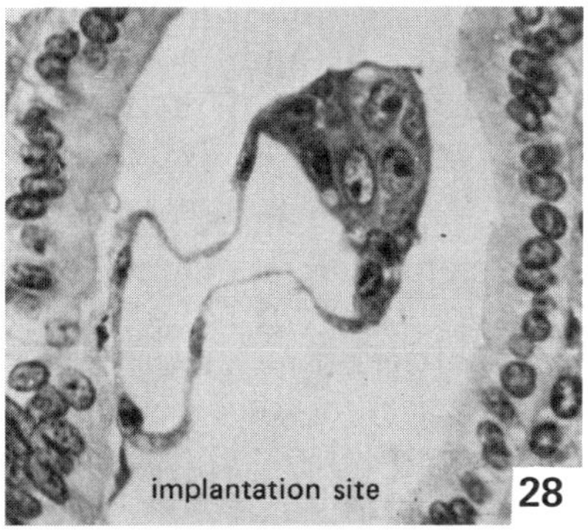

28

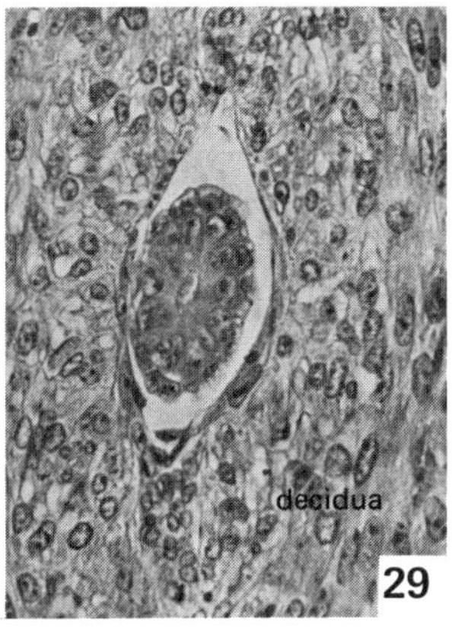

29

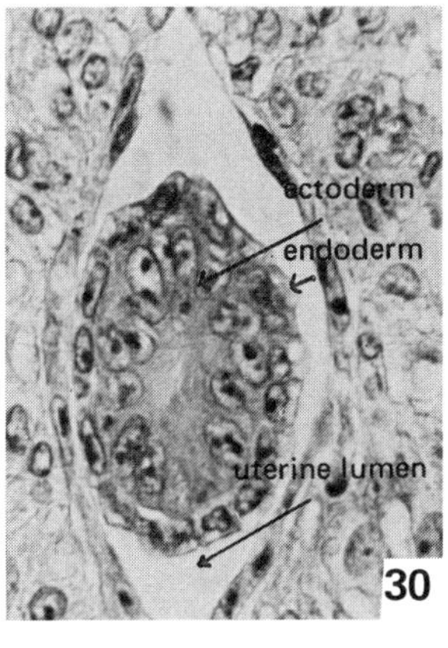

30

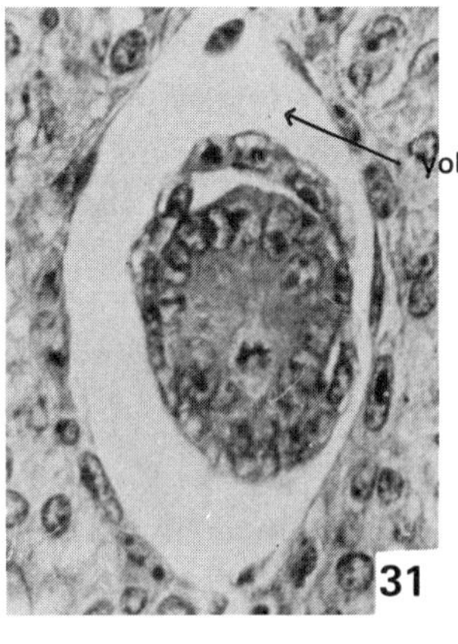

31

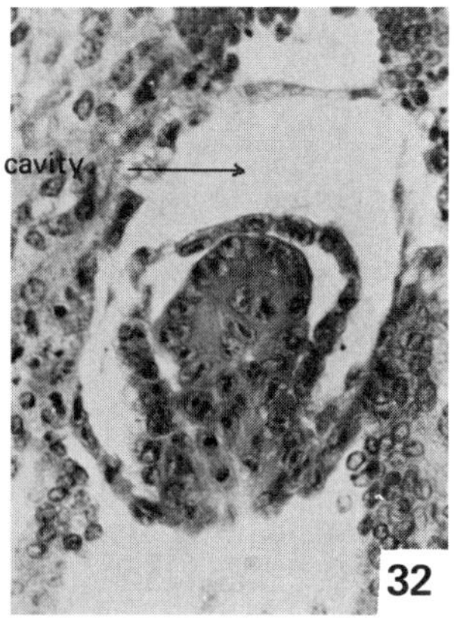
32

THE PRIMARY GERM LAYERS

The inner cell mass almost immediately develops two types of cells: the thinner and more darkly staining layer next to the yolk cavity is the *endoderm,* and all other cells of the blastocyst are early *embryonic ectoderm.* The trophectoderm (troph = Greek for nourishment) does not give rise to any embryonic structures. The embryo arises from the inner cell mass.

The endoderm of the inner cell mass, at about 4½ to 5 days, gives rise by proliferation to a column of cells which appear to come from the junction of the endoderm and the trophectoderm and project into the yolk cavity. These cells will eventually line the yolk cavity, closely applied to the outer trophectoderm, and will be known as the *distal endoderm* to distinguish them from the *proximal endoderm* of the inner cell mass. During this process the embryonic ectoderm of the inner cell mass has thickened, and the outermost cells appear to form another layer known as the *extraembryonic* ectoderm. These cells appear to invade the maternal mucosa and destroy its blood vessels, forming the *ectoplacental cone.* The embryonic ectoderm has, in the meantime, developed its own cavity known as the *proamniotic cavity.*

The embryo at 6 days consists of four rings of cells. The innermost is the embryonic ectoderm, the next the proximal endoderm, separated by the yolk cavity from the distal endoderm and the rapidly disappearing trophectoderm. With this disappearance one finds many invading giant cells. While the trophectoderm is extraembryonic, it disappears as the placenta forms. The ectoderm gives rise to the ectoplacental cone and becomes the covering of the *egg cylinder.* The extraembryonic ectoderm not comprising the ectoplacental cone then acquires a lumen, continuous with that of the proamniotic cavity, and the entire embryo acquires an elongated shape. The cells of the ectoplacental cone will give rise to part of the placenta. It appears to be actively invasive.

At this point it must be emphasized that the relative positions of the primary ectoderm and endoderm are the reverse of what one would expect from the study of other vertebrates (except the rat, rabbit, and guinea pig). The embryonic ectoderm is the innermost layer of cells, and by 6½ days the *mesenchyme* (embryonic mesoderm) will begin to appear intermediate between this inner ectoderm and the outer endoderm. After all, the confusion (if any) arises from human prejudice in the naming of presumptive layers, while the mouse allocates its cells and layers with dependable regularity and always comes out "a mouse."

As in most forms, the embryonic mesoderm arises by proliferation from prior existing embryonic ectoderm, near the junction with the extraembryonic ectoderm. This is the site of the *primitive streak.* At first the mesenchyme consists of loosely spaced cells which work their way between the ectoderm and endoderm, particularly in the vicinity of the junction of embryonic and extraembryonic ectoderm (the future anterior of the embryo). Some mesenchyme also invades the region of the extraembryonic ectoderm later to form the double-layered extraembryonic *yolk sac,* which will entirely envelop the embryo.

TOPOGRAPHICAL CONSIDERATIONS

The original site of the mesenchyme identifies the position of the thickened ectoderm known as the primitive streak. This also is found toward the posterior of the egg cylinder but obviously to one side. Since the original position of the inner cell mass is later marked by the ectoplacental cone, the embryo develops in a direction opposite to that of the inner cell mass but within the yolk cavity. The embryo always develops with its anterior-posterior axis perpendicularly to the mesometrium and the dorsal-ventral axis parallel to the mesometrium. Thus, if one cuts the gravid uterus lengthwise, he is apt to cut the early embryo either transversely or frontally, and a transverse cut of the gravid uterus should provide a sagittal section of the embryo. One must remember, however, that by the time the embryo is shaped like a capital "S," transverse cuts may be misleading. Eventually, by an inversion of the foregut and hindgut, the embryo assumes the capital "C" shape so characteristic of the higher vertebrates.

MEMBRANES AND CAVITIES

The proamniotic cavity has already appeared. It is surrounded by embryonic ectoderm and is internal. The mesoderm, appearing first between the ectoderm and endoderm at the junction of the embryonic and extraembryonic regions, causes a bulging of cells into the proamniotic cavity. As this bulge acquires a cavity the combined mesoderm and ectoderm form the *posterior amniotic fold.* In sagittal sections one cannot see it, but this mesoderm is continuous around the egg cylinder to form the *lateral* and finally the *anterior amniotic folds.* This mesodermal sheet, as it thickens, tends to constrict the egg cylinder at about its midpoint. The cavity or cavities which develop, largely in the posterior mesoderm, coalesce to form the *exocoelom* (extraembryonic coelom), lined with mesoderm. When the constriction of the encircling mesoderm is completed (7 days) the embryonic structures are found toward one pole of the egg cylinder (ventral) and the extraembryonic structures at the other pole (dorsal).

The original proamniotic cavity is now subdivided into the ventral *amniotic cavity* (ectodermal), the *exocoelom* (mesodermal), and the *ectoplacental cavity* (ectodermal), of which the latter cavity will ultimately disappear. The double (ectodermal and mesodermal) membrane which separates the amniotic cavity from the exocoelom is the *amnion,* and the similarly constructed *chorion* separates the exocoelom from the extraplacental cavity.

THE NEURAL AXIS

The *head process* develops at 7 days as an anterior prolongation of the cells from the dorsal *primitive streak,* probably largely ectoderm although originally in the position where one would expect mesoderm, between the ectoderm and endoderm. It is continuous around the ventral side of the egg cylinder as squamous endoderm. This gradually becomes columnar endoderm as one views the sheet of cells around toward the ventral side of the embryo. The primitive streak then extends from the most anterior tip of the egg cylinder to the level of the amnion, or about two-fifths of the length of the cylinder. In transverse sections through the head process one can see that its cells are loose and appear to come from the inner, thicker ectoderm. They extend about three-quarters of the way around the embryo but are much thicker at the dorsal side. They are contiguous with mesenchyme and can be distinguished only by virtue of their position and destination. It is noted, however, that the cells of the head process remain in contact with the outer endoderm while the mesenchyme adheres to the inner ectoderm. By 7½ days the mesoderm has almost completely separated the ectoderm from the endoderm, including all but the mid-dorsal region of the embryo where the notochord and neural axis will form.

The notochord arises from the outer thin endoderm, first by a slight thickening and then an invagination to form a groove. The inverted groove of endoderm is then the primitive archenteron, with the cells pinching off above it to form the notochord. At this level there are no mesenchyme cells between the ectodermal and endodermal layers.

The *neural groove* appears directly above the region of the notochord as an inverted depression or trough in thick ectoderm which extends forward in the midsagittal line as far as there is embryonic ectoderm. By 7½ days it is easily identifiable as a deepened (inverted) groove with thickened sides, and by 8½ days it is closed as a *neural tube,* continuous with the embryonic brain.

The *notochord* develops during these neural changes, with its originally oval nuclei becoming reoriented in a plane perpendicular to the ectoderm of the neural groove. The notochord retains its connections with the *archenteron* (gut endoderm) for a considerable time, but eventually (as in Amphioxus) the notochord becomes pinched off from the archenteron and forms the axis around which the embryo develops its central nervous and skeletal structures. Thus, the head process gives rise to the gut endoderm as well as the notochord.

The *allantois* of the mouse is not associated with the endoderm, as in so many vertebrates, but is purely mesodermal, arising in association with the original exocoelom into which it grows, after which it acquires its cavity. By 7½ days it appears as a globular mass, apparently destined to fill the exocoelom. Its function, as in all vertebrates, is nutritional and respiratory. The *chorion*, composed of ectoderm and mesoderm, is thickened and tends to compress and all but obliterate the ectoplacental cavity. By 8 days the allantois becomes adherent to the chorion, thereby establishing continuity between the embryo (primitive streak) and the region of maternal contact (ectoplacental cone). Thus, the groundwork is laid for close circulatory, nutritional, and respiratory connections with the maternal tissues.

The *foregut* primordium appears first as a notch in the thick columnar endoderm toward the anterior limits of the egg cylinder at about 7 days of development. The notch enlarges into the direction of the amniotic cavity, taking with it the overlying ectoderm, thus forming a distinctive *foregut pouch*. The greatly thickened ectoderm anterior to this foregut is the *head fold ectoderm*. During the next day the foregut enlarges rapidly to form a deep pocket and then a pouch (which becomes constricted laterally and ventrally by the appearance of mesoderm), with the anterior intestinal portal being continually shifted posteriorly (caudally). Eventually, the mesoderm will come to encircle the foregut as it is formed. Up to this stage, the germ layers have appeared to be in a position opposite that of other vertebrates studied, so that this inversion of the endoderm brings it and the surrounding mesoderm and ectoderm into the relationship one expects of the adult.

At the posterior margin of the primitive streak, beginning shortly after the invagination of the foregut, will appear the invagination of the hindgut. It will form in the same manner as the foregut, developing a posterior intestinal portal, and eventually it will become continuous with the foregut at the level of the last-to-form midgut. The *stomodeum* (invagination of head ectoderm) and *proctodeum* (invagination of caudal ectoderm) will eventually rupture through to the gut to form the mouth and the anus, which first appear at about 8 days. One may even see an *oral* and *anal plate*, points of contact of ectoderm and endoderm prior to rupture.

The head fold is accentuated by the developing foregut as the ectoderm thickens and is pushed forward into the amniotic cavity. With neurogenesis this head fold and process then become the aggressively active differentiating center of the embryo.

By 7½ days of development the embryonic mesenchyme has invaded almost all parts of the embryo and extraembryonic structures, between the ectoderm and endoderm, and also constitutes the lining of the exocoelom. The allantois, invading the exocoelom, is entirely mesodermal. About the only region the mesenchyme cannot invade is where the endodermal notochord is being

---

THE MOUSE EMBRYO: 6½ TO 14½ DAYS POST-CONCEPTION

Fig. 33 – Mouse embryo at 6½ days, showing relation to surrounding uterine epithelium. Note breakdown of adjacent uterine tissue.

Fig. 34 – Section distinguishing between inner solid ectoderm and outer flocculent-type endoderm.

Fig. 35 – Thickening of inner ectoderm of the head process, at 6½ days.

Fig. 36 – Section through uterus at level of 7½-day embryo, showing extent of uterine involvement. Ectoplacental cone is solid mass toward center of section.

Fig. 37 – Transverse section through more posterior level of embryo, showing exocoelom and two portions of proamniotic cavity.

Fig. 38 – Longitudinal section at about 7 days, showing first appearance of all three germ layers.

Fig. 39 – Longitundinal (sagittal) section at 7½ days, showing all structures from ectoplacental cone (above) to head process (below).

Fig. 40 – Mouse embryo at 8½ days. Note earliest neural invagination to form the central nervous system in inner thick ectoderm. Section through entire uterus at this level.

Fig. 41 – Enlarged view of 8½ day embryo, showing earliest neural differentiation.

Fig. 42 – Embryo at 8½ days, showing amniotic membrane across center of embryo, saccular allantois above and to the left, ectoplacental cavity slitlike and above, large amniotic cavity (below) with early differentiation of the primary neural axis (thickening below.)

Fig. 43 – Transverse section through early head structures at 9½ days, showing two dorsal and two ventral aortae and enclosed lateral coelomic spaces. Archenteron is the central cavity above which is the notochord and the closed neural tube. Easily compared with frog, chick, or any other mammal.

Fig. 44 – Section, at 10½ days, of the head process level with brain ectoderm still unclosed but with notochord and archenteron distinct.

Fig. 45 – Section of eye at 12½-days development, at level of origin of the optic nerve. Note state of development of the lens, cornea, and retina.

Fig. 46 – Eye of the mouse embryo at 14½ days, showing considerable further development beyond that of Fig. 45.

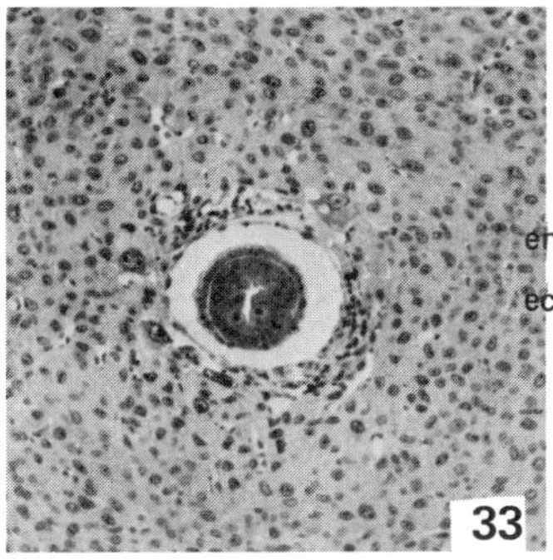
33

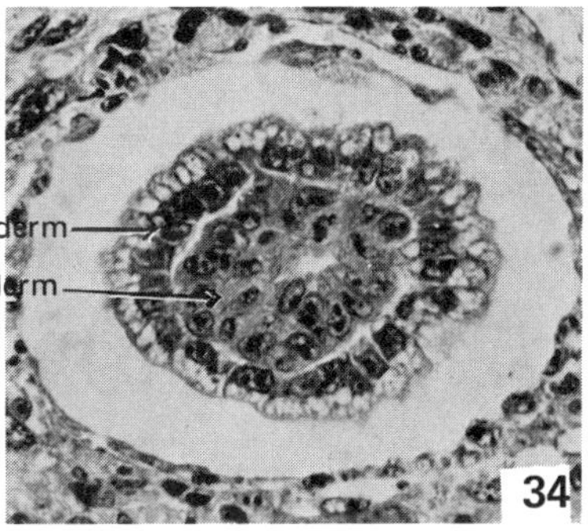
endoderm
ectoderm
34

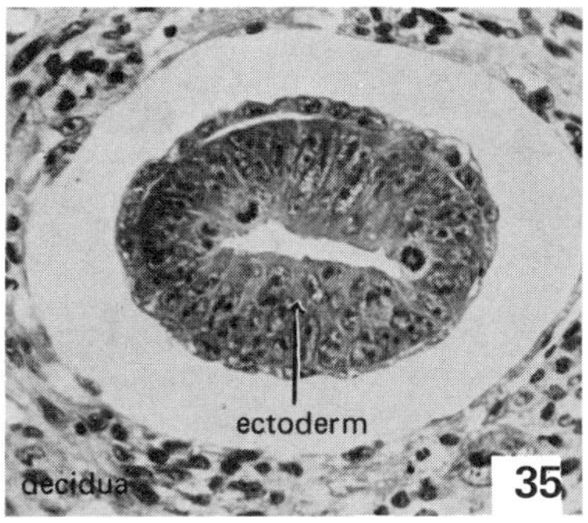
ectoderm
decidua
35

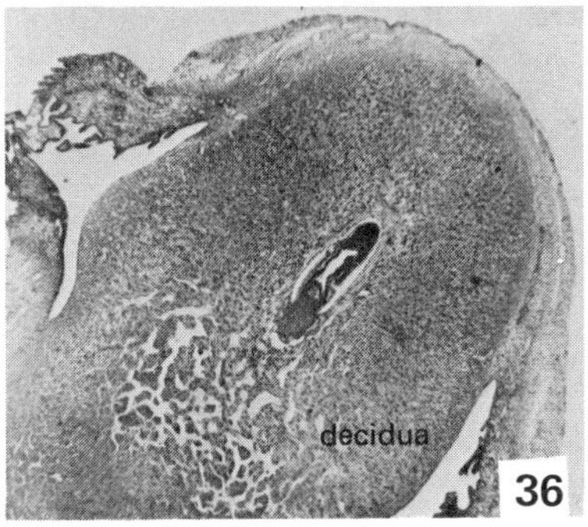
decidua
36

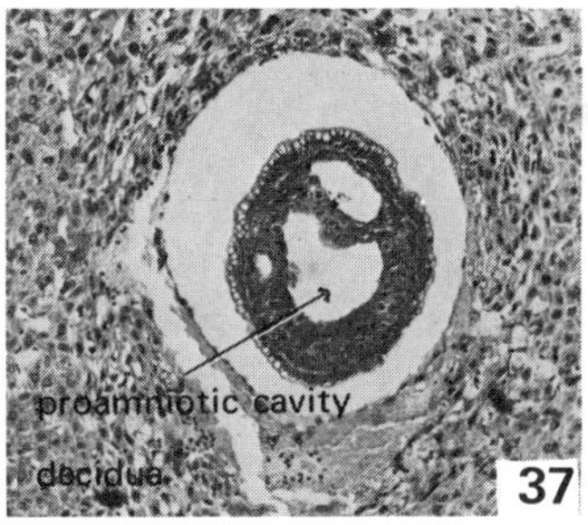
proamniotic cavity
decidua
37

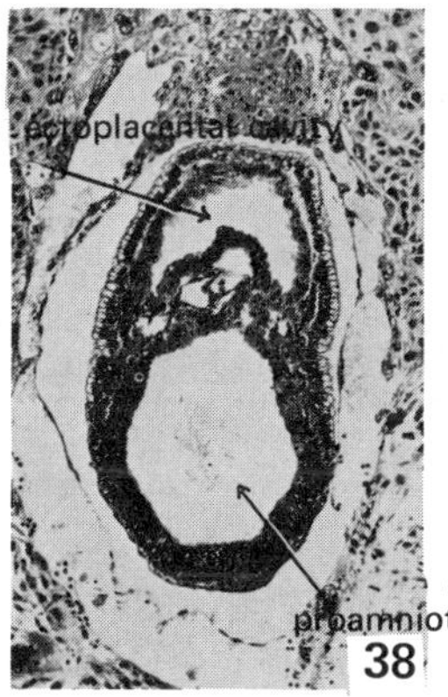
ectoplacental cavity
proamniotic cavity
38

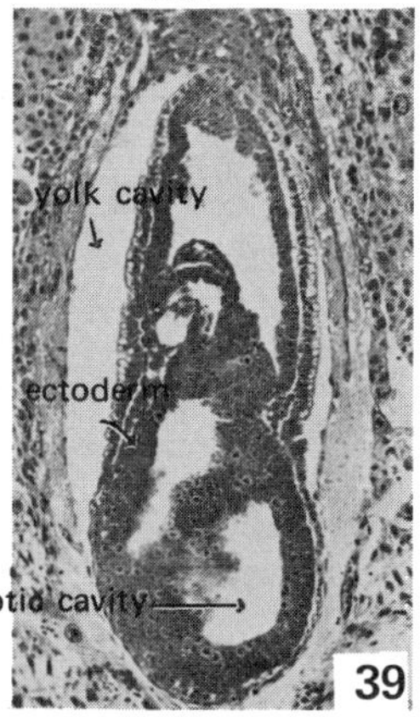
yolk cavity
ectoderm
proamniotic cavity
39

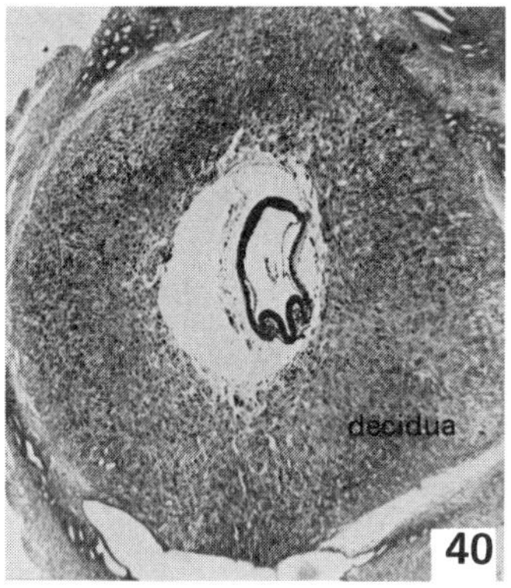
decidua
40

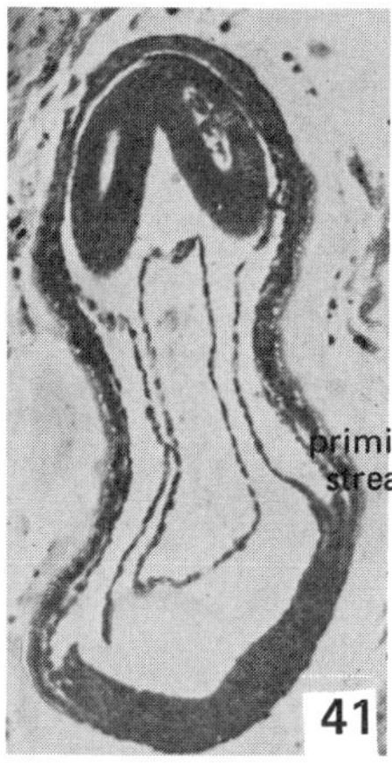
41

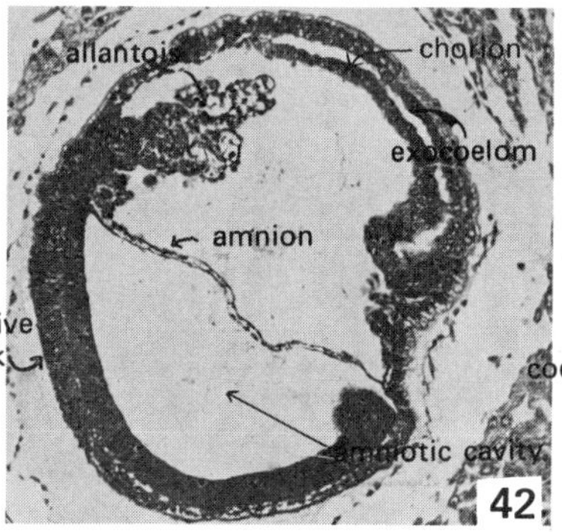
allantois
chorion
exocoelom
amnion
primitive streak
amniotic cavity
42

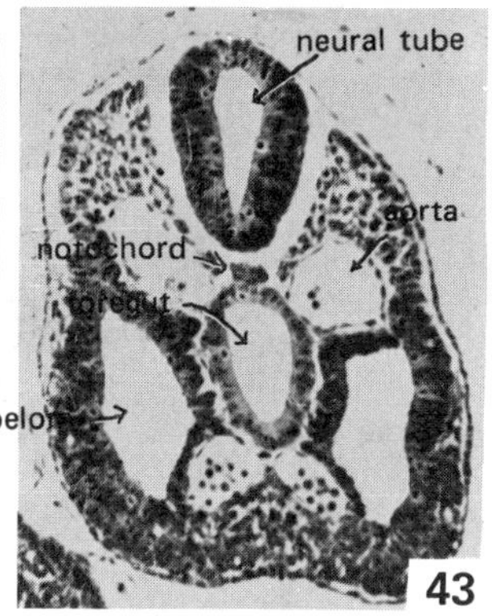
neural tube
aorta
notochord
foregut
coelom
43

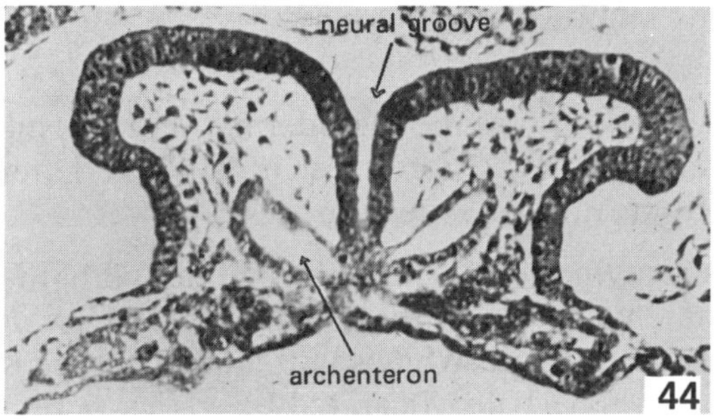
neural groove
archenteron
44

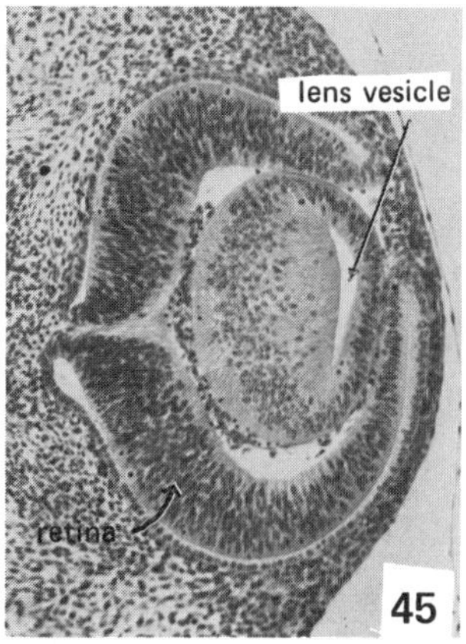
lens vesicle
retina
45

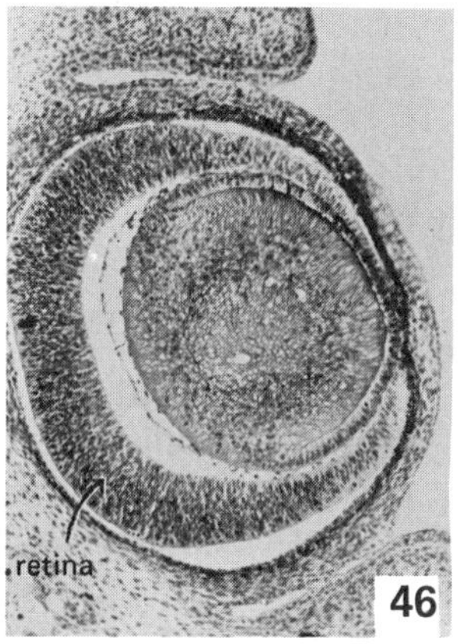
retina
46

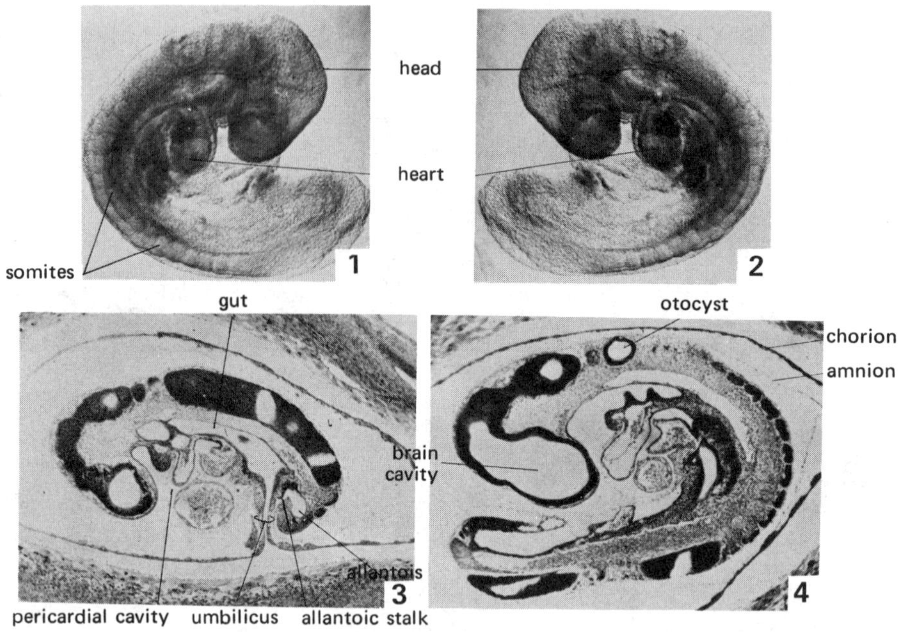

EMBRYO MOUSE AT 9 DAYS

1 & 2 – Living whole embryos showing basic organization.
3 & 4 – Sagittal sections showing various structures labelled.

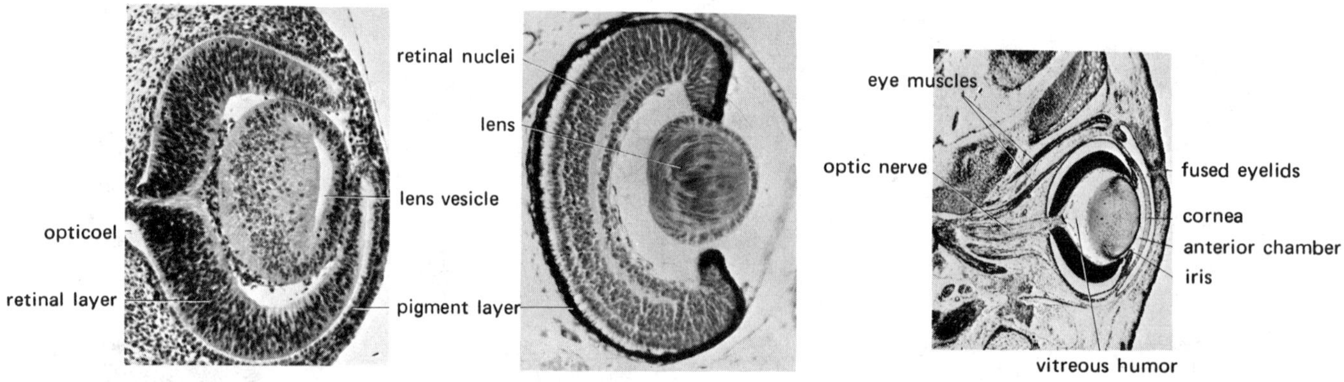

12 DAYS | 15 DAYS | POSTNATAL

MOUSE EYE DEVELOPMENT

developed, in the midsagittal plane. Beneath the head fold and process the mesodermal sheets come together to form the primitive rudiments of the heart, anterior to the foregut as it develops. Later the heart is ventral to the foregut due to the forward development of the head process.

On each side of the notochord the sheets of mesenchyme, which are continuous with the extraembryonic mesoderm, become organized so that by 8+ days there appear clusters of mesoderm, the primordia of the *somites* just lateral to the notochord. While somite mesenchyme is at first contiguous with the more lateral mesoderm, it becomes partially separated by a constriction which demarks the lateral plate mesoderm that acquires a cavity known as the *coelom*. The first definitive somite appears at about 7½ days, and by 8 days there may have developed as many as six or seven pairs. The first to develop is the most anterior, and it appears to be actually further anterior than the primitive streak, dorsal to the developing foregut. The second appears next just posterior to the first and becomes constricted off from it to form the characteristic somite block of cells seen in all developing vertebrates. Eventually there develop about 65 pairs of somites. It is likely that the first one, or several, lose their identity with the development of the head structures. During this development one can study the various stages in somite formation simply by examining the mesenchyme lateral to the notochord from the most anterior (most developed) to the most posterior (least developed) for several days of gestation from 8 days onward.

As in all typical vertebrates the primitive streak is the organizing center, forward from which develops the embryo. Actually, it gives rise to all three germ layers and is in this respect comparable to the dorsal lip of the blastopore of other forms. The allantois, usually arising from the floor of the hindgut, arises here from the posterior extremity of the primitive streak, still close to the hindgut. The embryo arises anteriorly to the streak and at its expense, but as the embryo develops it, too, shows cephalic activity which provides the cephalic or *head fold.* But differentiation begins at the most anterior extremity of the embryo so that the more posterior primitive streak remains the most retarded or undifferentiated region even though it gives rise to the germ layer prerequisites of the embryo.

The *coelom* is developed, as in other vertebrates, by a split in the lateral mesoderm to form the intervening cavity. The amnion is a continuation of the *somatopleure* (ectoderm and mesoderm). The *splanchnopleure* (mesoderm and endoderm) continues as the membrane laterally forming the floor of the exocoelom. Ventral to the forming foregut the mesodermal cavity (coelom) becomes the *pericardial cavity.* The relationship of these precardial mesodermal parts can be seen best at about 8 days of development in serial sections.

Reichert's membrane has no counterpart in other forms, is peculiar to rodent development, is noncellular, elastic, tough, and lies between the endoderm of the developing yolk cavity and the outer trophectoderm. It may arise from the endoderm and is the outermost, all-inclusive membrane of the embryo. But it is nonliving. The *amnion* is structurally and functionally similar to that of other vertebrates, but the *chorion* has lost its protective function to the Reichert's membrane and is insignificant. As the embryo becomes involved with maternal tissues, the Reichert's membrane becomes obscured and probably disappears. It may best be seen on day 6½. By 8 days, the amnion extends over the entire dorsal surface of the developing embryo and later envelops the embryo so that it is free to float and develop within.

Since nutrition for the mammalian embryo is derived through the maternal tissues, there is no need for a yolk sac, but in rodents it retains some vestige of its ancestral function by establishing a nutritional relationship between mother and embryo through at least a good portion of development. It is said that the yolk cavity arises from the original blastocoel but comes to lie between Reichert's membrane and the egg cylinder and that its membrane is that middle portion which comes from the extraembryonic splanchnopleure. As the embryo grows and its membranes and cavities expand, this yolk sac, by 8½ days, envelops the very large exocoelom and all but the eventual midgut region of the embryo. When the midgut finally closes, the entire mass is enclosed by the yolk sac (amnion, embryo, and all). At about 7½-8 days the inner mesodermal layer of the yolk sac acquires blood islands, the precursors of the embryonic and extraembryonic blood vascular systems, which are continuous.

The presence of a "parasite" in the uterine cavity, in the form of a developing embryo, causes responses on the part of the uterus which can be observed with the naked eye in the form of swellings as soon as 5½ days after fertilization. Implantation begins at about 4½ days so that this reaction appears quite soon thereafter. This is known as the decidual swelling, because it is due entirely to changes in the uterus itself.

The 4½-day embryos are suspended within the uterine cavity, and, for reasons not understood, they come to be spaced equidistantly throughout this lumen. Each embryo then is associated with the uterine epithelium, becoming attached often by the trophectoderm. A crypt develops around the embryo and soon envelops it, cutting it off from the uterine lumen at 7½ days. While there are some variations, generally the uterine tissue which finally closes the embryo off from the uterine cavity will give rise to much of the placenta, being located dorsal to the ectoplacental cone in most cases. By 8 days the embryo is completely cut off from the uterine cavity, and for a while it obstructs the lumen completely at this level, the only remaining lumen being between implantation sites. By 9 days the lumen on each side of the embryo has reestablished connections, but ventral to the embryo, representing a new spacial connection. Thus, the old uterine lumen is obliterated by the *decidua basalis* and eventual *placenta,* while the embryo will come to lie freely within a new uterine lumen, located ventrally to the embryo outside of the *decidua capsularis.*

Changes in the uterine mucosa are first seen at 5½ days when *giant cells* appear ventral to the embryo in association with degenerating uterine epithelium. At the opposite pole, and probably derived from the ectoplacental cone, there will then appear outside of the trophectoderm, rather numerous giant cells, characterized by extending protoplasmic strands. These cells move to lie between Reichert's membrane and the decidua and may help to establish and maintain a close relation between these two. A third group of less dense giant cells appears at about 7 days; they may be binucleate or multinucleate and are found within the decidua. They are distinguished by their hyperchromatic nuclei and are known collectively as the *symplasia.*

The decidua is entirely maternal, at least at the beginning. Dorsal to the embryo, toward the ectoplacental cone, is the decidua basalis, which is not easily distinguishable from the uterine mucosa. Between the decidua basalis and the implanted embryo there will appear the richly vascular

---

THE MOUSE EMBRYO: 6½ to 18½ DAYS POSTCONCEPTION

Fig. 47 – Paired uteri of mouse at 6½ days, showing 11 implantation sites. Note ovaries at upper end and coiled oviducts.
Fig. 48 – Genital system of pregnant mouse at 7½ days with 11 implantations.
Fig. 49 – Same at 8½ days. Note enlarging embryonic masses.
Fig. 50 – Same at 9½ days. Note only two implantations on left.
Fig. 51 – Same at 10½ days with embryos on left exposed; 14 implantations.

Fig. 52 – Enlarged view of embryos *in situ* at 10½ days. Note very vascular amnion covering lowermost embryo.
Fig. 53 – Exposed mouse embryos at 11½ days. Note appendage development.
Fig. 54 – Implantation sites of left uterus at 12½ days. Right side not shown.
Fig. 55 – Uteri at 13½ days with four embryos partially exposed. Next to lowest implantation on right is probably a resorption site.
Fig. 56 – Entire litter of six (only) at 16½ days liberated from their uterine sites but still attached by umbilical cords to placentae. Development such that they should be regarded as fetuses.

Fig. 57 – Sagittal section of mouse fetus at 15½ days gestation. Note particularly the hepatic vein to the sinus venosus; the vertebrae; and the development of the cerebral vesicles.
Fig. 58 – Sagittal section of entire fetus at 16½ days gestation, showing brain and cord, aortic arch from ventricle, and prominent tongue muscles.
Fig. 59 – Sagittal section of mouse fetus at 17½ days. Note well differentiation of the brain and spinal cord, heart, and viscera.
Fig. 60 – Sagittal section of mouse embryo at 18½ days.
Fig. 61 – Same as Fig. 60 but at different level.

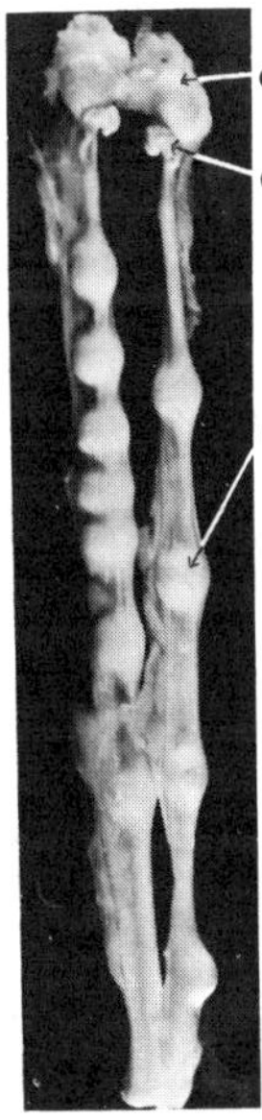

47

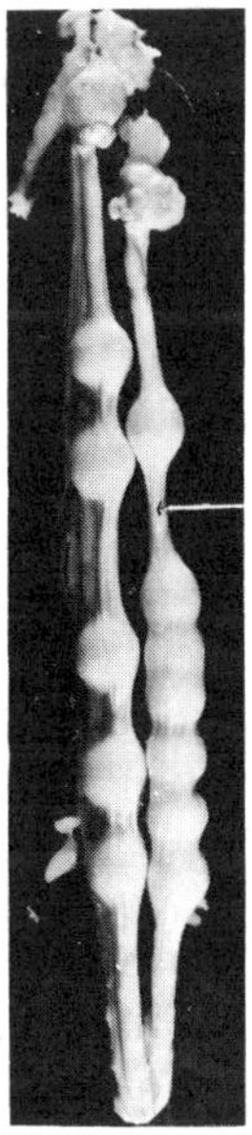

48

49

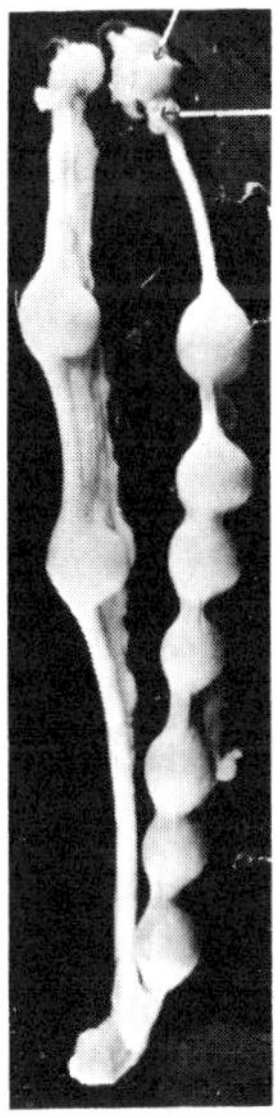

50

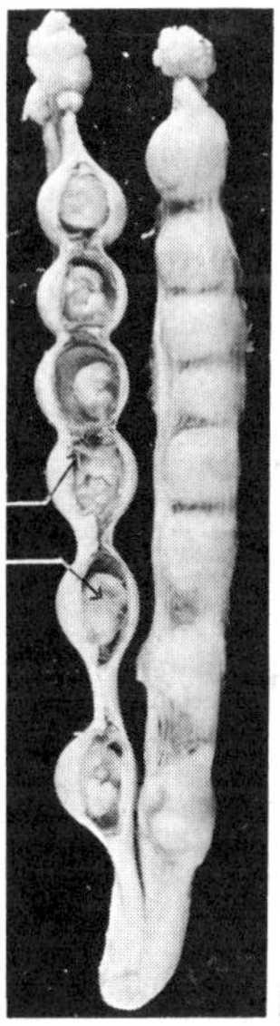

51

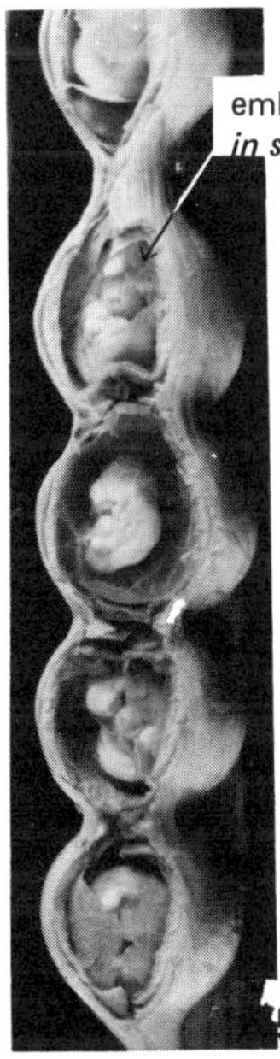

52

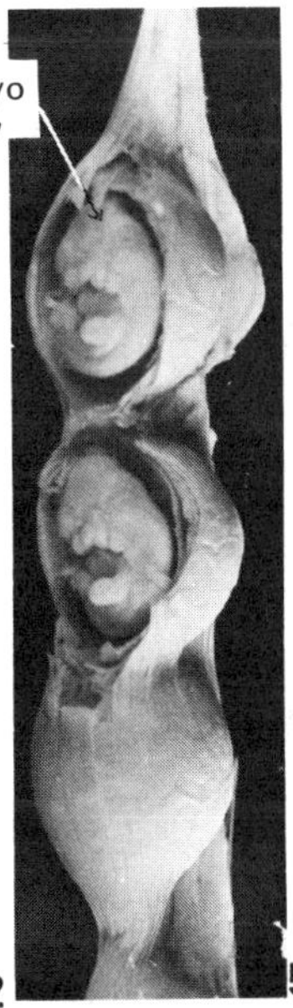

53

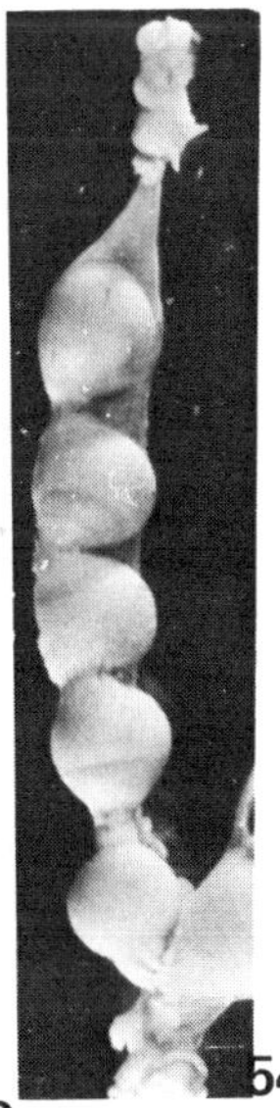

54

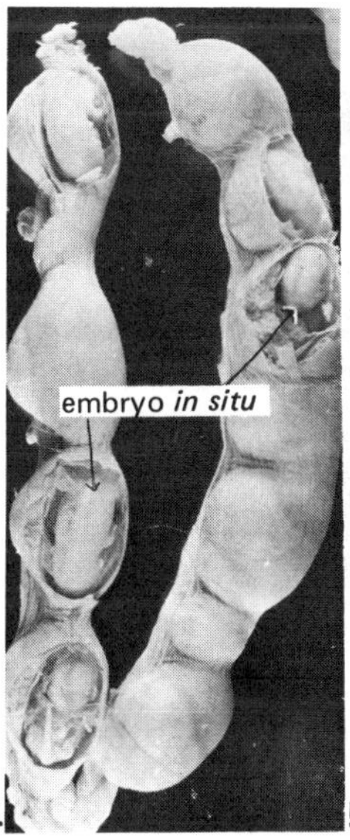

55

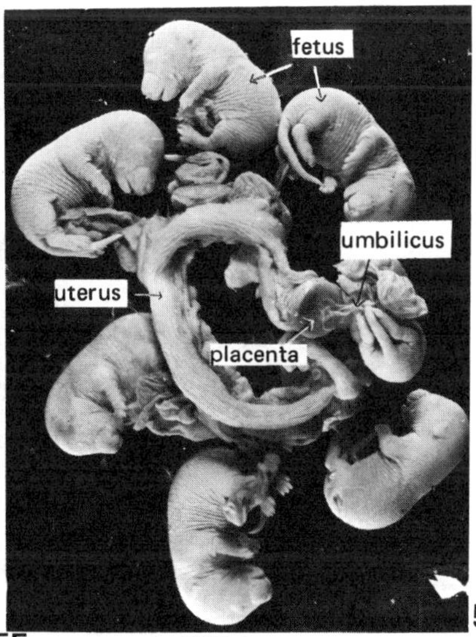

56

16.5 days

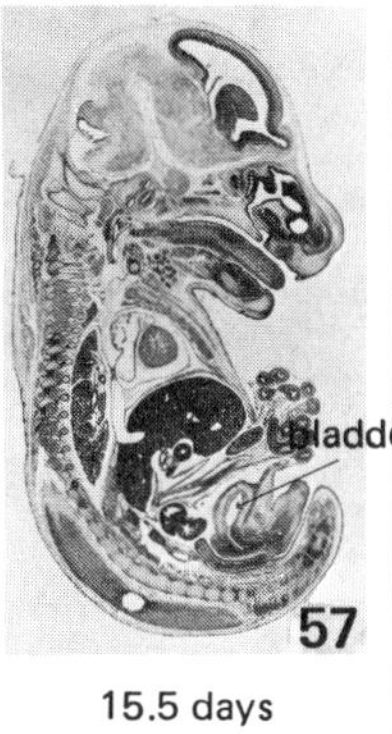

57

15.5 days

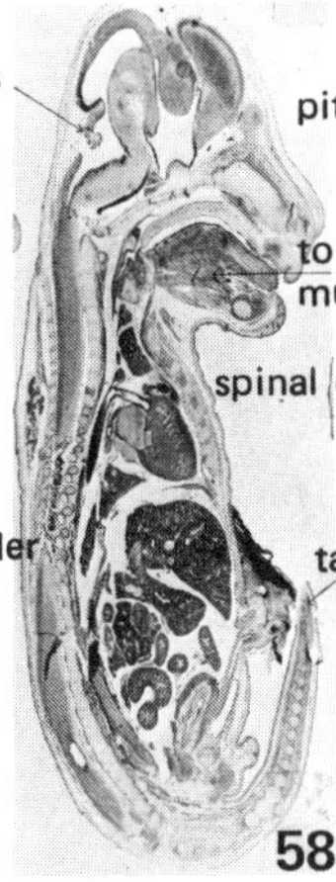

58

16.5 days

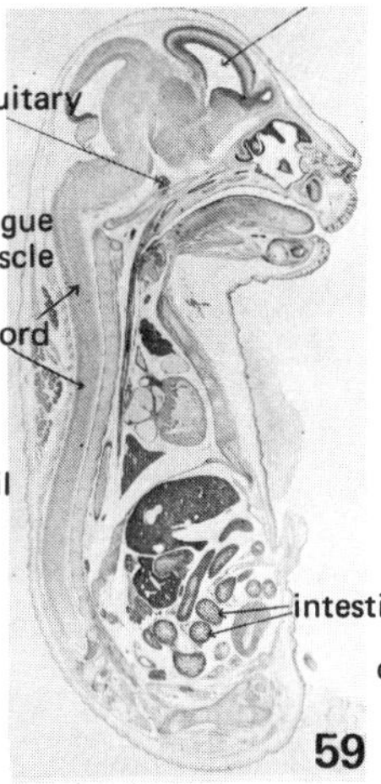

59

17.5 days

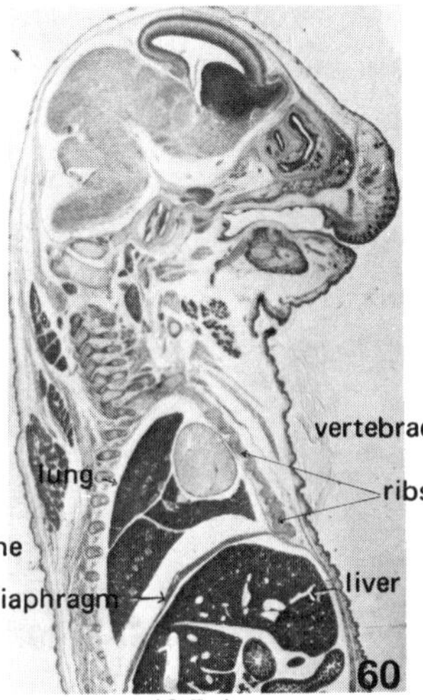

60

18.5 days

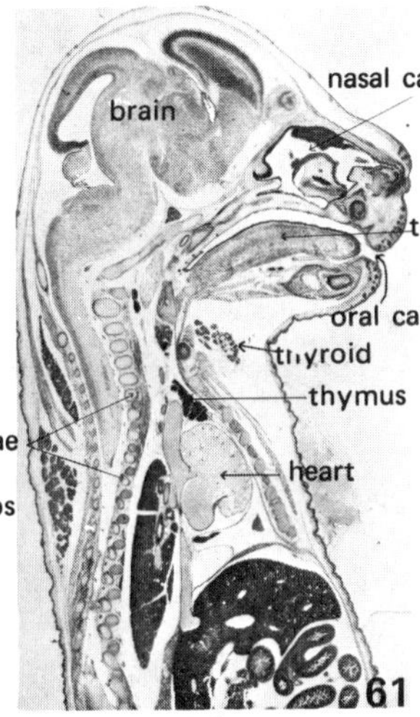

61

18.5 days

zone where the blood vessels appear actually to be ruptured. They appear more as sinuses than as formed blood vessels. Close around the embryo lie the giant cells, and ventral to it is the decidua capsularis, which is characterized by binucleate and possibly multinucleate giant cells in abundance. The decidua capsularis lies toward the newly establishing uterine lumen, the side of the implantation which will come to lie freely within the newly established uterine cavity. It is directly opposite to the placental connections of the decidua basalis.

The nutrition must reach the embryo by way of the maternal blood. This vascular zone between the embryo and the decidua basalis consists of ruptured maternal vessels which bathe the entire embryo in blood, changing continuously. This may be seen first at about 5½ days. The blood cells of the mother never actually penetrate the circulatory system of the embryo or fetus, being separated by Reichert's membrane, which is rather tough, and by the yolk cavity and the yolk sac. Thus, while Reichert's membrane is nonliving, it functions as a semipermeable membrane allowing soluble nutrients to pass through to the embryonic membranes and its blood vessels, particularly those of the so-called yolk sac. As the embryonic circulation develops, these nutrients are then circulated throughout to the developing tissues. The vascular zone expands through the decidual zone to the uterine lumen so that there may be some rupture of maternal blood into this lumen, beginning at about 7½ days and continuing for several days. The ultimate *placenta* is then made up of the decidua basalis and the remains of the ectoplacental cone, at the chorion and parts of the allantois forming a most efficient channel for nutrition from mother to embryo.

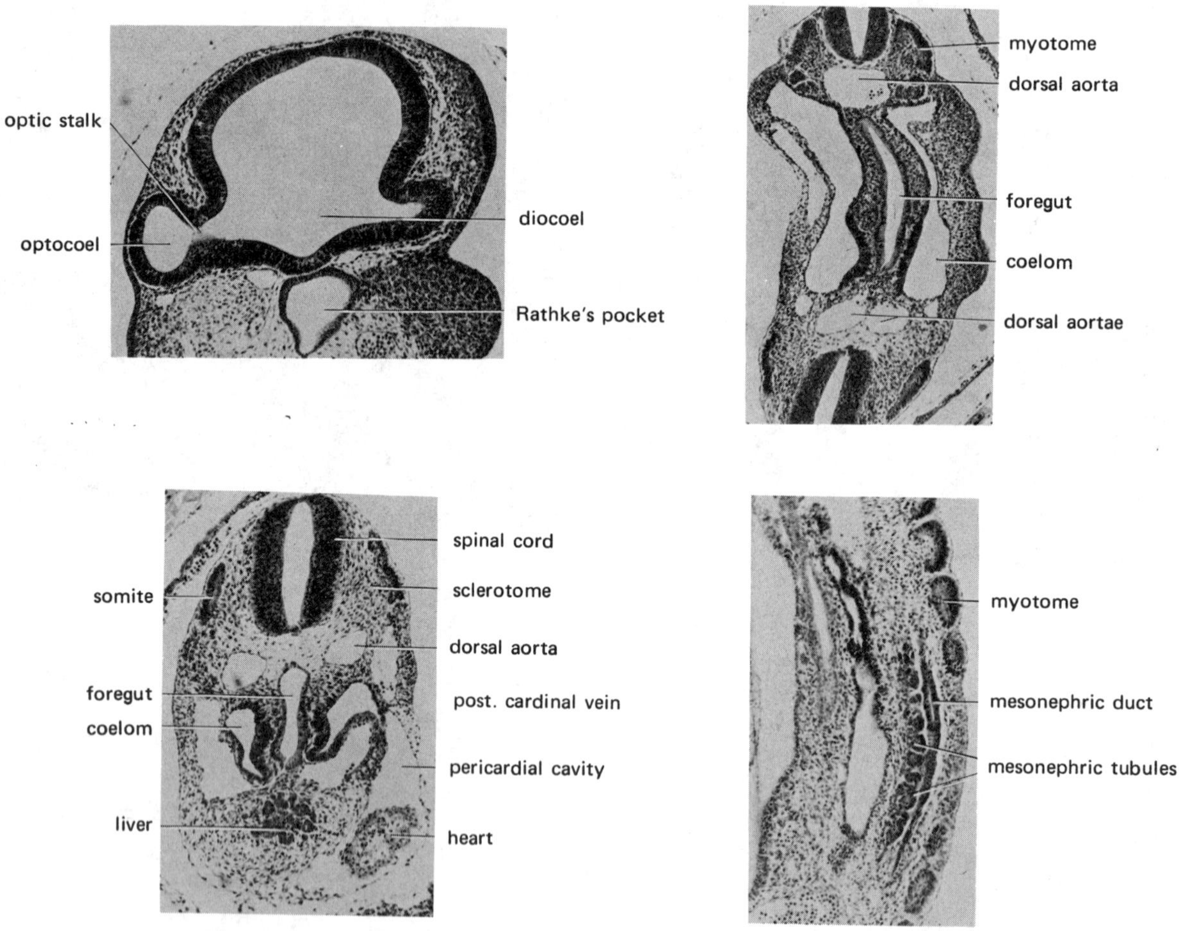

MOUSE EMBRYO AT 9 DAYS

The development of the embryo from this stage onward is sufficiently like that of other vertebrates not to require detailed description. Its general shape at 8 days is not the typical figure "C" but rather a reversed capital "S" when observed from the left side or in sagittal section. The ectoderm is on the inside with the neural groove open dorsally. Eventually, the germ layers will be seen to be in their correct relationship to each other by a sort of reversed folding of the parts accompanied by marked cephalization. The amnion, along with the chorion and yolk sac, form a covering of the embryo marking its dorsal aspect. The "S" contour of the 8-day embryo gradually changes, and by 9 days the embryo has turned so as to present the capital "C" shape so characteristic of the higher vertebrates. The turning begins at both the cephalic and caudal ends, and finally the midgut is involved at a time quite simultaneous with the closing of the neural folds and the appearance of the notochord. Naturally the rotation of either end must be in the same direction and is always toward the left side of the embryo. In sections at 8-8½ days, this may be confusing because within the same field may be sections of the cephalic and of the caudal ends, each apparently twisting in opposite directions. However, when the embryonic axis is considered, it is obvious that the entire embryo is turning in the same direction. The midgut of the mouse is ill-defined, largely because in its early development the archenteron is an inverted groove. With the turning of the embryo and the growth of the ends, the closure of the foregut and hindgut, the lateral folds (splanchnopleure) connecting the two eventually come together to form a continuous tube by 9 days. The embryo comes to lie on its left side close to the yolk sac, and its right side is turned toward the opposite placenta.

The mouse *heart* develops from the splanchnopleuric mesoderm, after the initial appearance of the cardiac coelom (pericardial cavity) as a split between the double-layered splanchnopleure and somatopleure. The mesodermal portion of the splanchnopleure will give rise to both the covering and the muscles of the heart (*epicardium* and *myocardium*), originally known as the epimyocardium. Other mesoderm, lying between the epimyocardium and the endoderm, will give rise by vacuolization to the pericardial cavity (originally part of the coelom). Unlike the development of the chick heart, where it is formed from two approaching masses of mesoderm from the sides, the

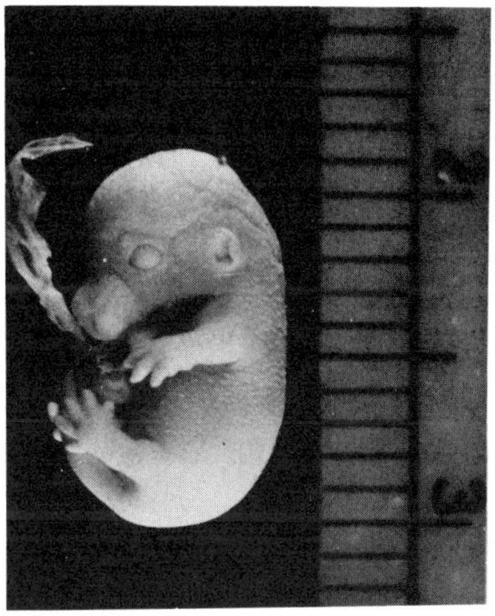

AT 14 DAYS

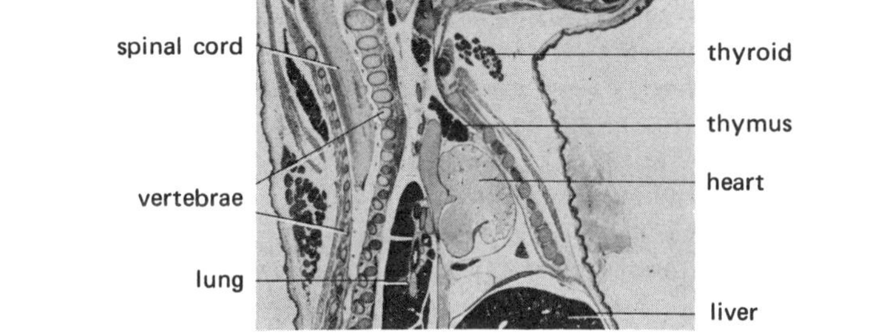

AT BIRTH

(Midsagittal section)

MOUSE FETUS

mouse heart parts have a capital "U" shape with the base of the formation appearing first just anterior to the anterior intestinal portal and ventral to the foregut and head fold. As the gut closes, the parts of the "U" come together to form the single tubular heart. Considerable cardiac differentiation may be seen by 9 days. Simultaneously with the development of the heart, the major arteries appear, the *dorsal aortae* extending the length of the body from the paired aortic arches. Posteriorly they fuse to form the single *omphalomesenteric artery* which passes out into the yolk sac, where, at 8½-9 days, it may disappear as a formed vessel and be seen as *blood islands.* But shortly the *omphalomesenteric veins,* as in the chick, appear and return the yolk sac blood to the embryo by a more ventral route. While the yolk itself is so rudimentary that it cannot supply nutrition for long, the yolk sac is a means of close association with the maternal tissues (vascular decidua) so that by this early circulatory relationship the embryo shortly begins to acquire nutriment from the mother.

Beginning at 9½ days one can see differentiated tissue areas, and as development proceeds the mouse embryo is the more difficult to distinguish the rat, pig, human, and other mammalian embryos. Its size and availability make it of increasing value as a teaching form for the understanding of mammalian embryology.

REFERENCES

Rugh: *Mouse.*

Rugh: *Vert. Emb.* 232-303

"We respond to fame's trumpeting of an individual scientist, but are often deaf to her orchestrations; yet of all human activities the occupation of science is more like a symphony than a solo."

W. Grey Walter

"About the fourth day the egg beginneth to step from the life of a plant to that of an animal. From that to the tenth it enjoys a sensitive and moving soul as Animals do, and after that it is completed by degrees and adorned with Plume, Bill, Clawes, and other furniture it hastens to get out. -- For all animals resemble one or other of those above mentioned (fowl, goose, duck, pigeon, frog, serpent, fish, crustacean, silkworms, sheep, goats, dogs, cats, deer, oxen, man) and agree with them either generally or specifically, and are procreated in the same manner, or the mode of their generation at least is referrable by analogy to that of one or other of them. -- Before man attains maturity, he was a body, an infant, an embryo. And then it is indispensable to inquire further as to what he was in his mother's womb before he was an embryo or fetus. -- Nature, by steps which are the same in the formation of any animal whatsoever, goes through the forms of all animals, as I might say egg, worm, embryo, and acquires perfection with each step."

William Harvey, 1578-1657

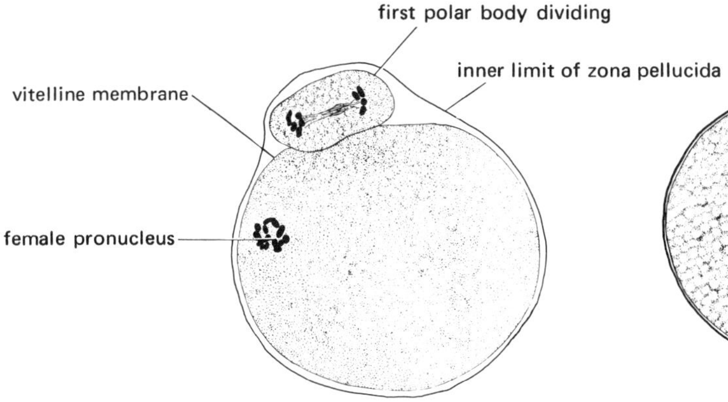

PREFERTILIZATION – MATURE OVUM

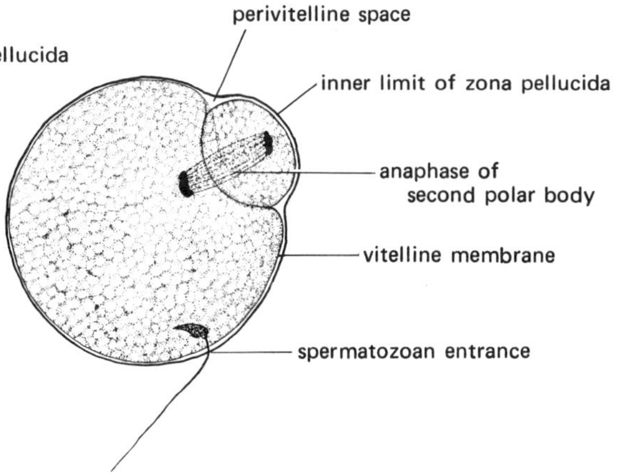

FERTILIZATION – ½ HOUR AFTER MATING

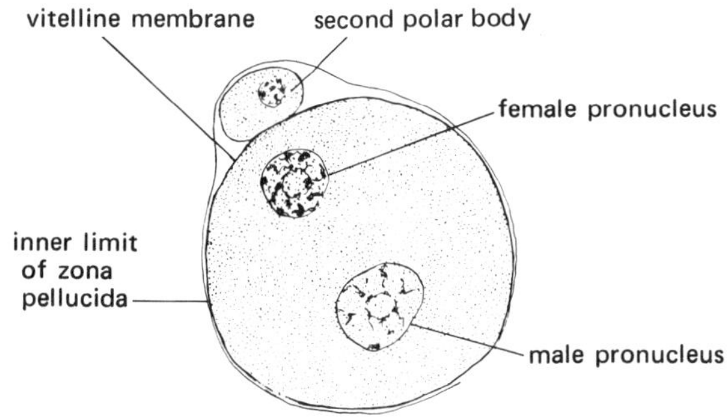

FERTILIZED OVUM
(LIVING MATERIAL)

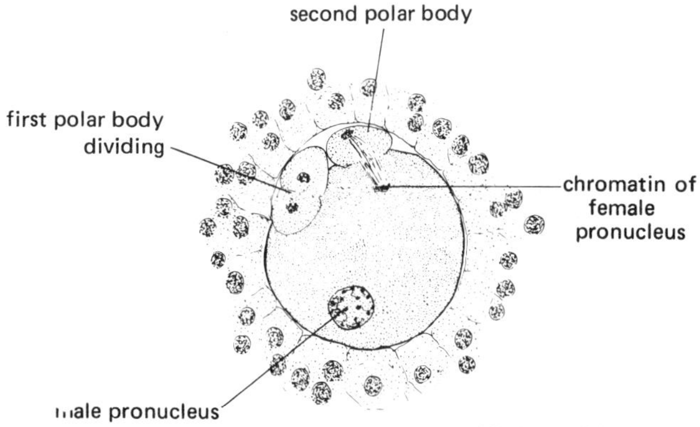

MOUSE OVUM
(FIXED MATERIAL)

2-CELL STAGE
(FIXED MATERIAL)

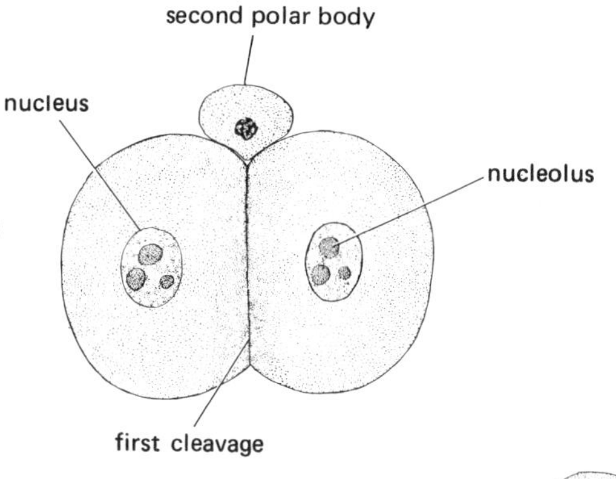

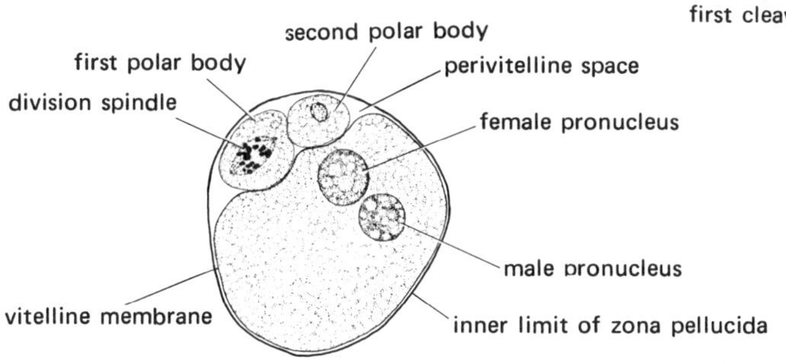

COMPLETION OF POLAR BODY FORMATION

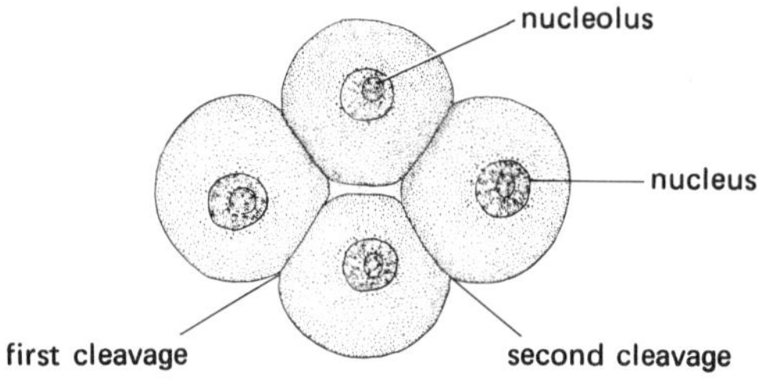

4-CELL STAGE
(FIXED MATERIAL)

Figures on this page are from R. Rugh. *The Mouse.* 1968. Burgess, Minneapolis.

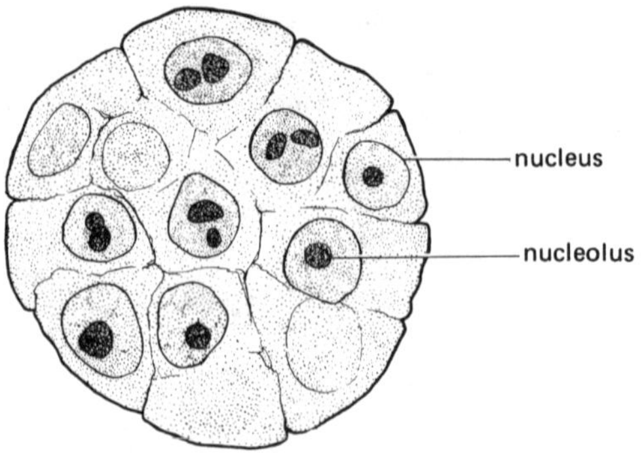

MORULA
3 GESTATION DAYS

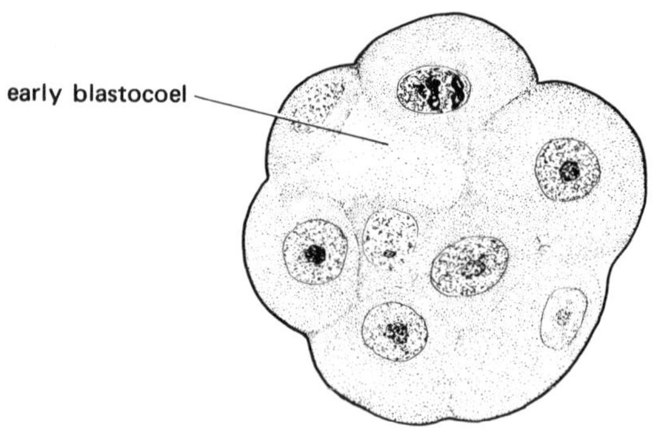

EARLY BLASTULA
(FIXED MATERIAL)

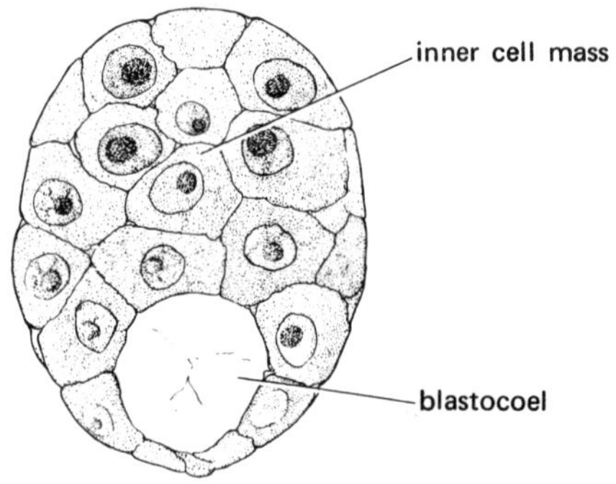

LATE BLASTULA

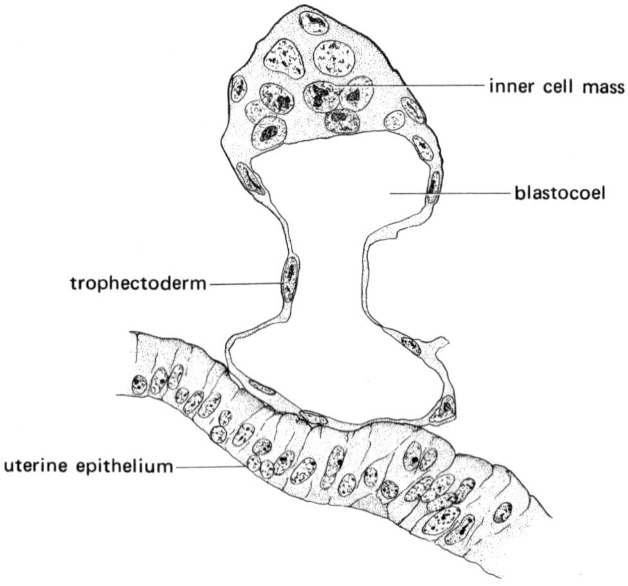

LATE BLASTULA
(SITE OF IMPLANTATION)

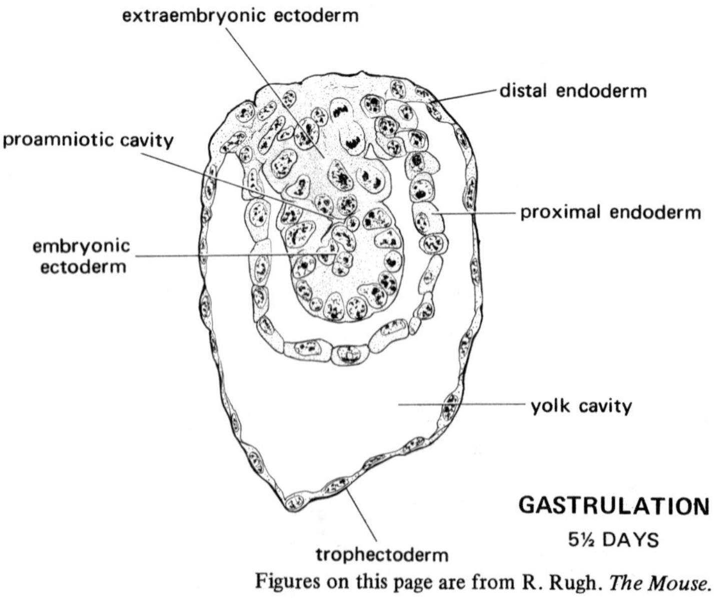

GASTRULATION
5½ DAYS

Figures on this page are from R. Rugh. *The Mouse.* 1968. Burgess, Minneapolis.

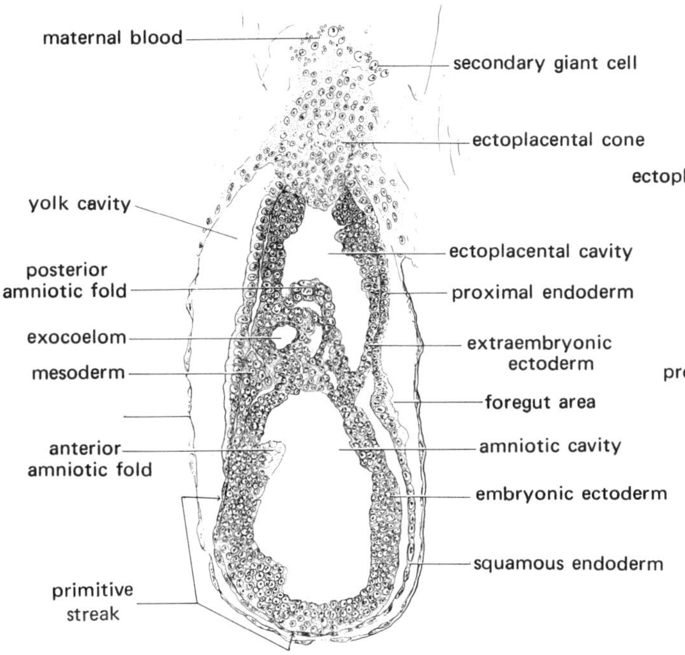

MOUSE EMBRYO, 7½ DAYS
'SAGITTAL SECTION

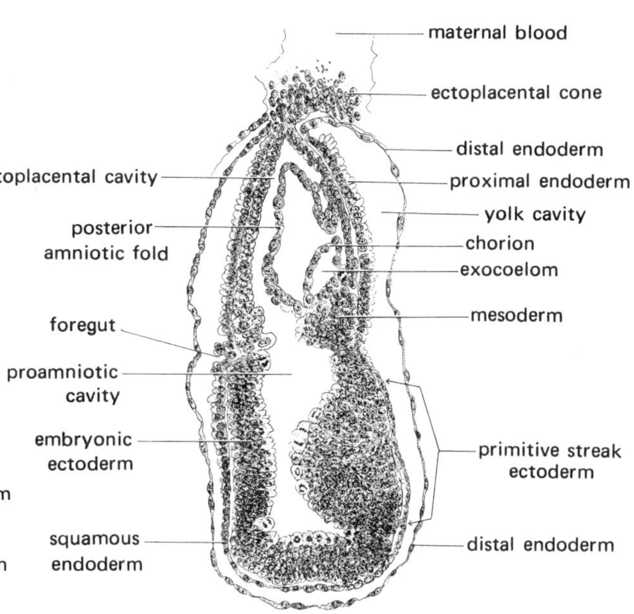

MOUSE EMBRYO, 7 DAYS
SAGITTAL SECTION

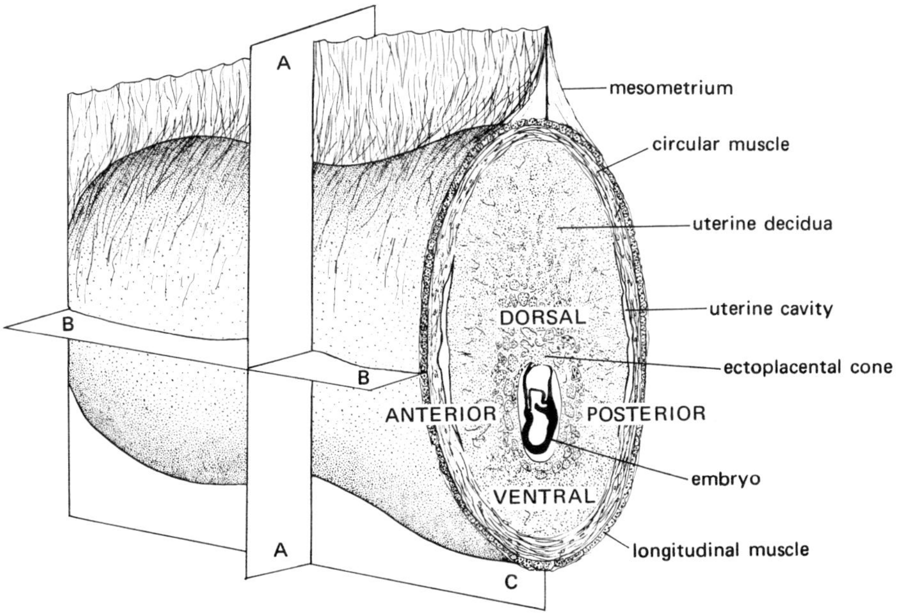

SCHEMATIC REPRESENTATION OF THE 7½-DAY MOUSE EMBRYO IN THE UTERUS

A. Sagittal to embryo
Transverse to uterus

B. Transverse to embryo
Horizontal to uterus

C. Frontal to embryo
Sagittal to uterus

Figures on this page are from R. Rugh. *The Mouse.* 1968. Burgess, Minneapolis.

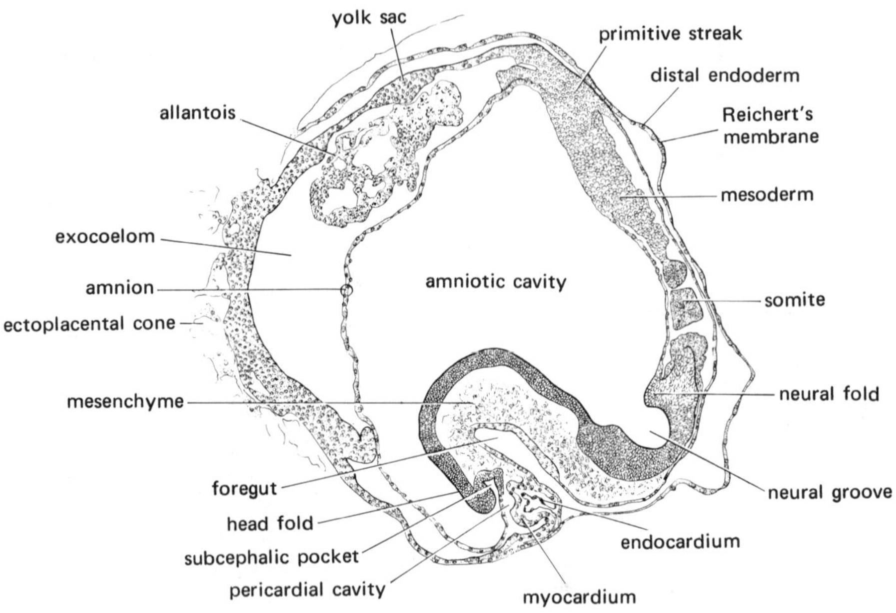

8-DAY MOUSE EMBRYO—HEAD PROCESS
SAGITTAL SECTION

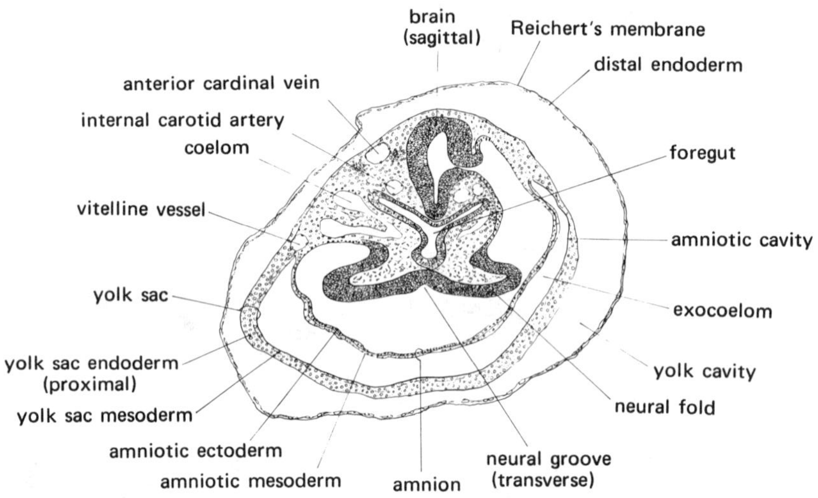

8-DAY MOUSE EMBRYO
(Twisted body results in parts being sectioned in two planes)

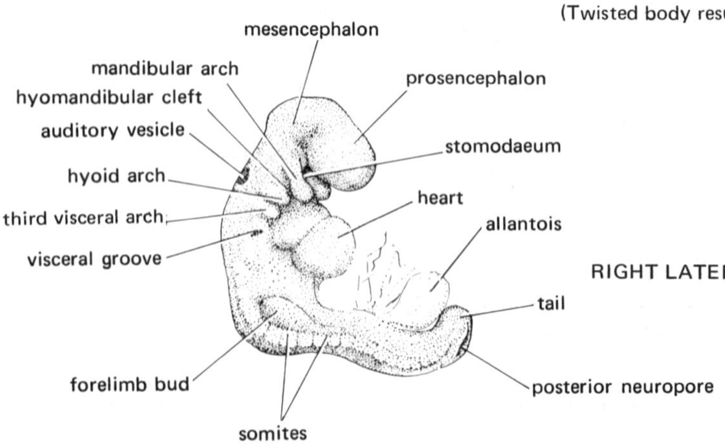

RIGHT LATERAL VIEW

MOUSE EMBRYO – 9 g.d.

Figures on this page are from R. Rugh. *The Mouse.* 1968. Burgess, Minneapolis.

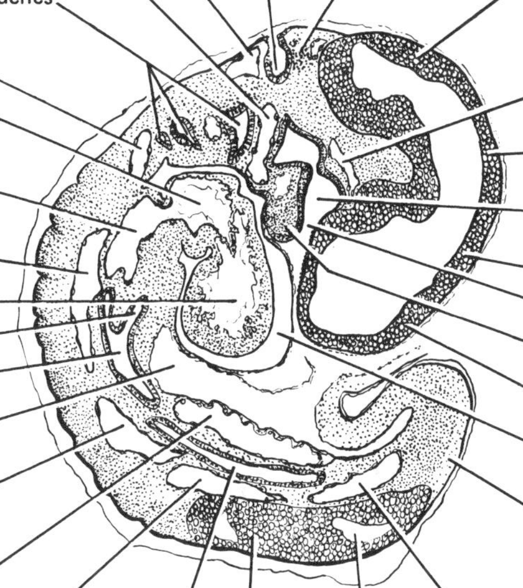

MOUSE EMBRYO – 9 DAYS

MIDSAGITTAL

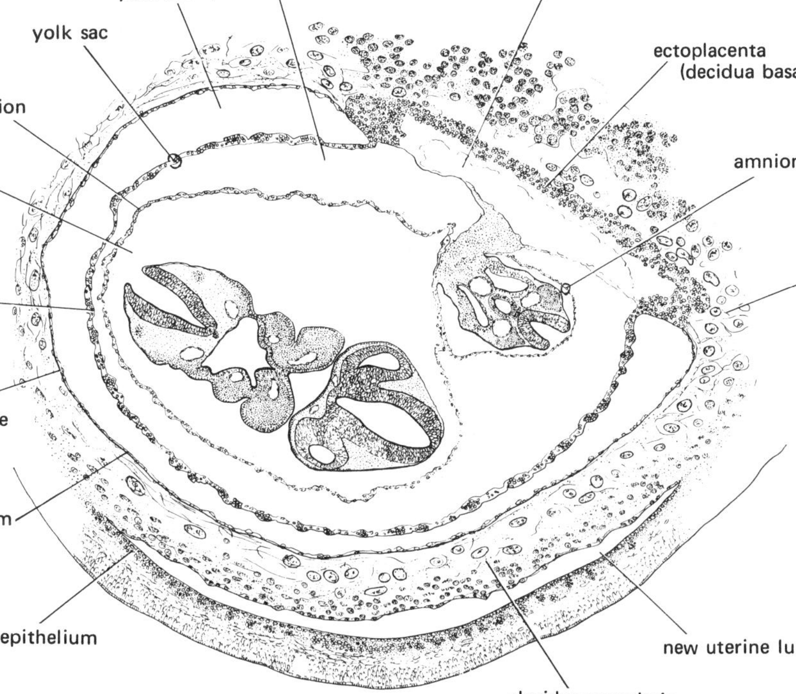

MOUSE EMBRYO – 9 DAYS

(SHOWING MEMBRANES)

Figures on this page are from R. Rugh. *The Mouse.* 1968. Burgess, Minneapolis.

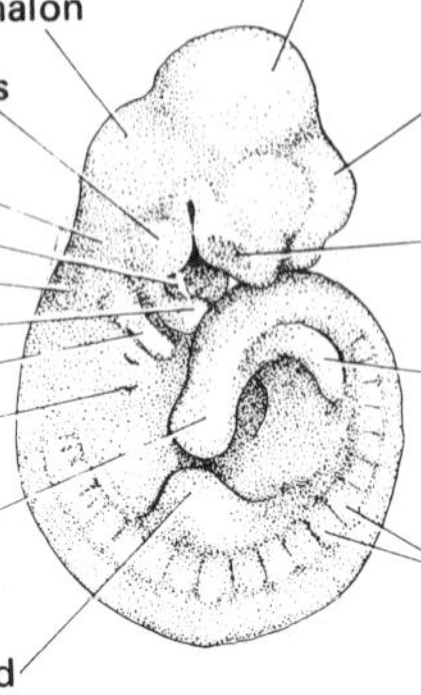

RIGHT LATERAL VIEW

MOUSE EMBRYO — 10 g.d.

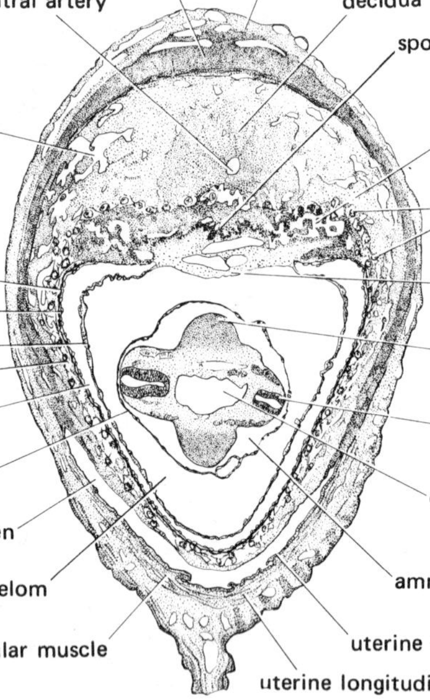

PLACENTAL RELATIONS OF THE 10-DAY MOUSE EMBRYO

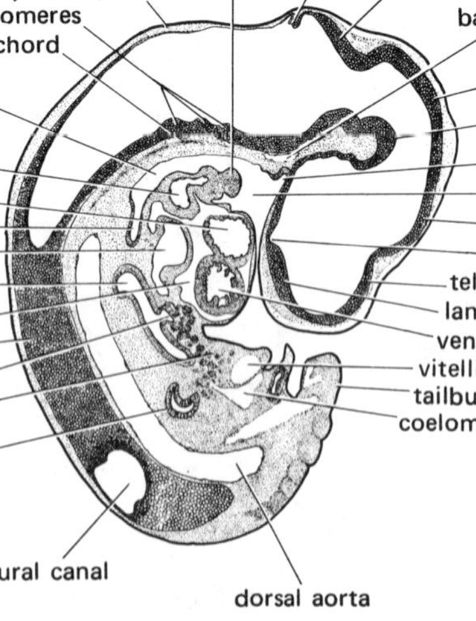

MOUSE EMBRYO — 10 DAYS

MIDSAGITTAL

Figures on this page are from R. Rugh. *The Mouse.* 1968. Burgess, Minneapolis.

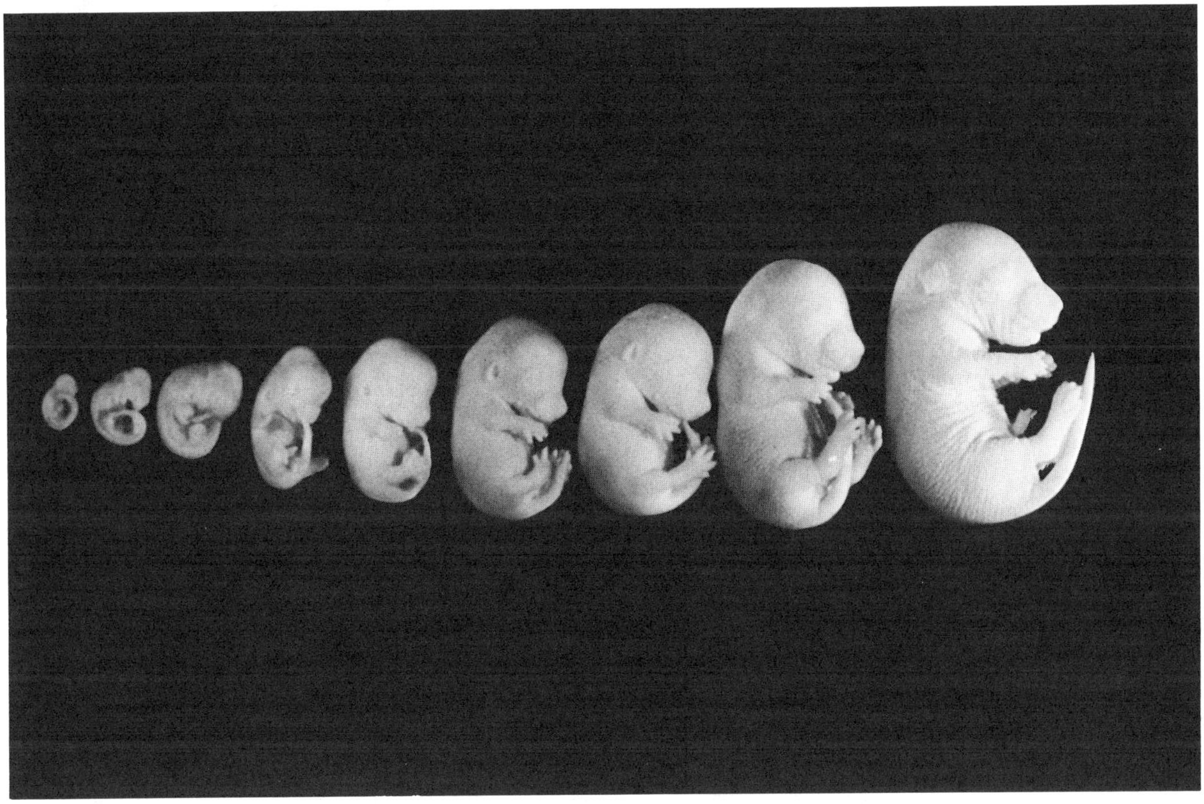

MOUSE EMBRYO: 10-18 DAYS GESTATION

From R. Rugh. *The Mouse.* 1968. Burgess, Minneapolis.

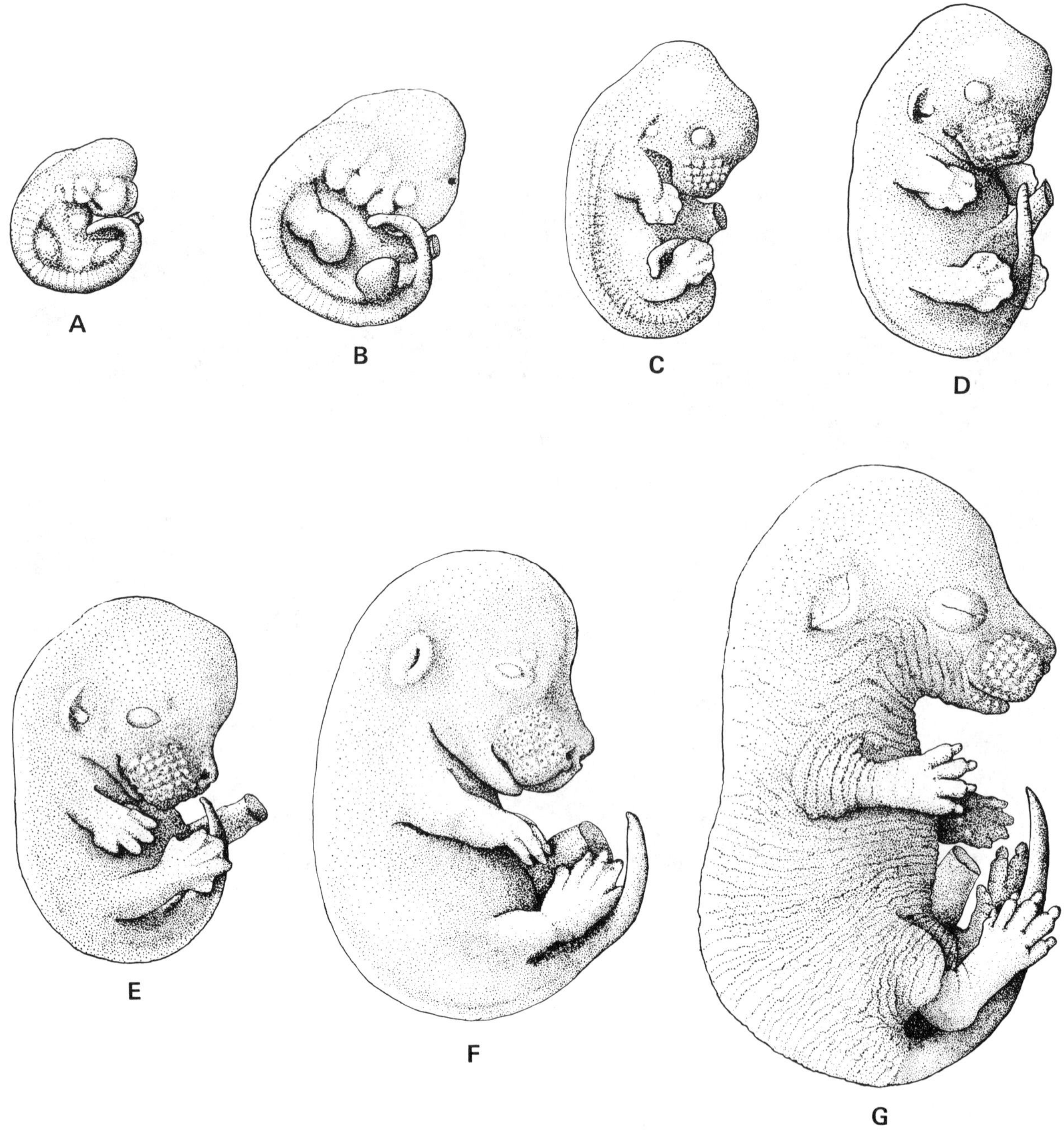

DIAGRAMS SHOWING DEVELOPMENT OF MOUSE EMBRYO FROM 10½ TO 16½ DAYS

From R. Rugh. *Vertebrate Embryology.* 1964. Harcourt, Brace and World, New York.

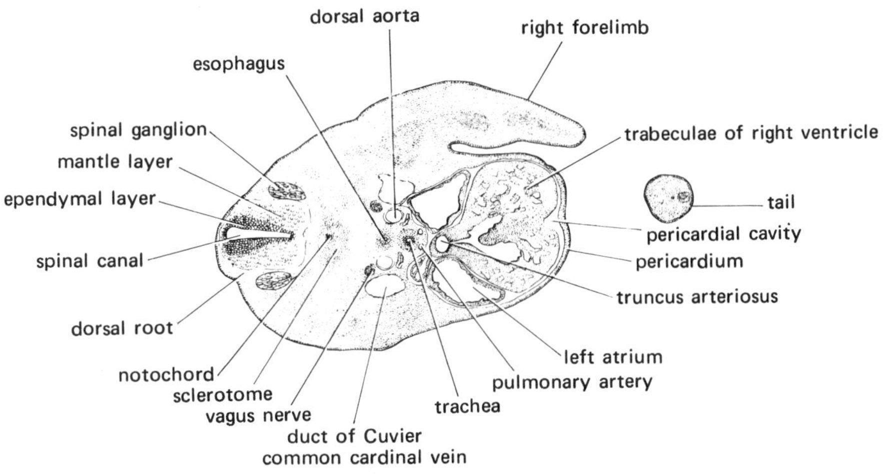

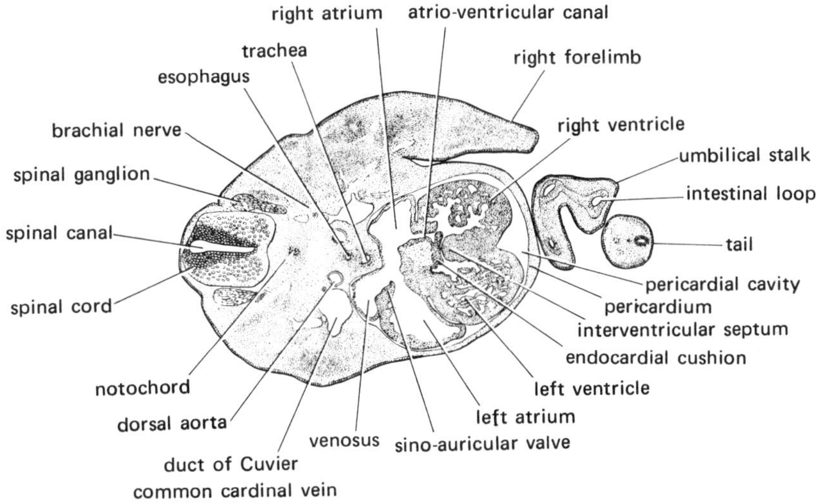

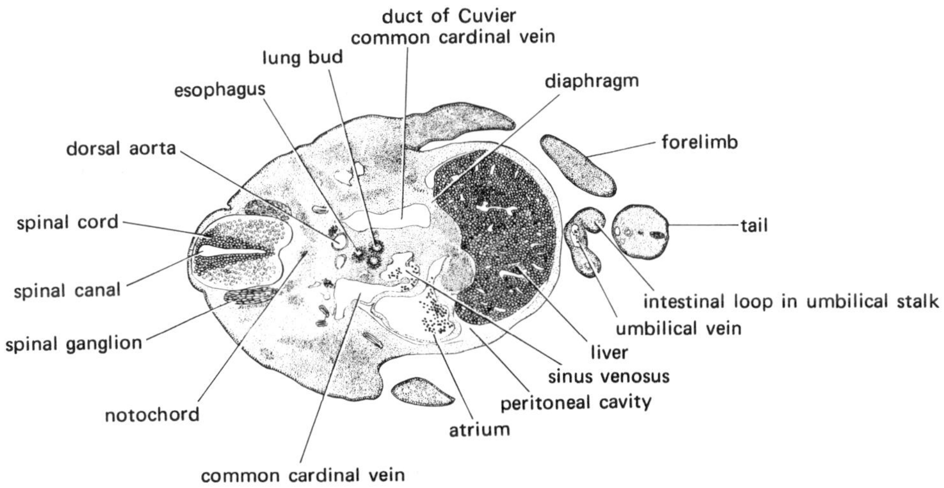

MOUSE EMBRYO – 12 g.d.

(TRANSVERSE SECTIONS)

Figures on this page are from R. Rugh. *The Mouse.* 1968. Burgess, Minneapolis.

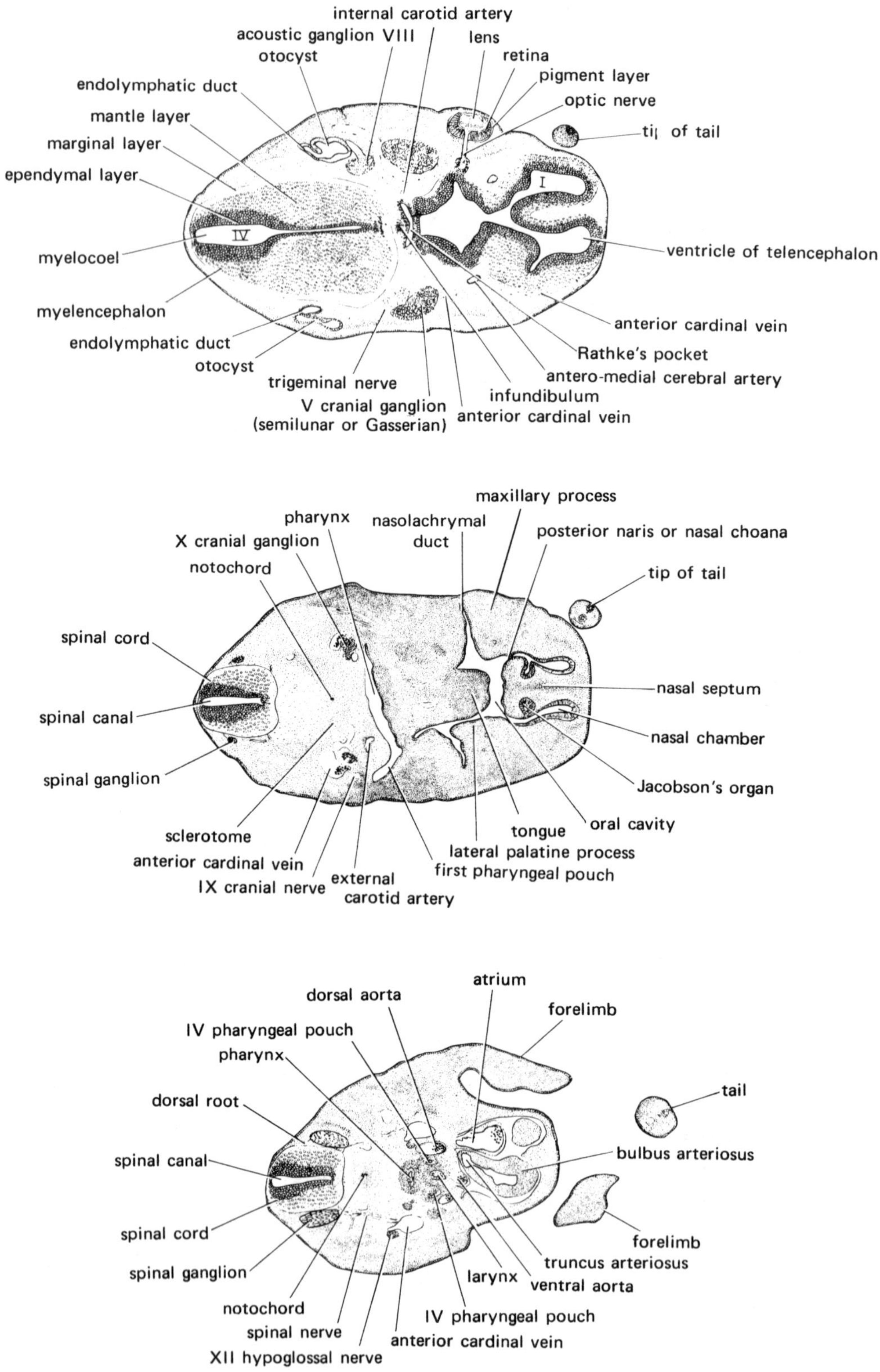

MOUSE EMBRYO — 12 g.d.
TRANSVERSE SECTIONS

Figures on this page are from R. Rugh. *The Mouse.* 1968. Burgess, Minneapolis.

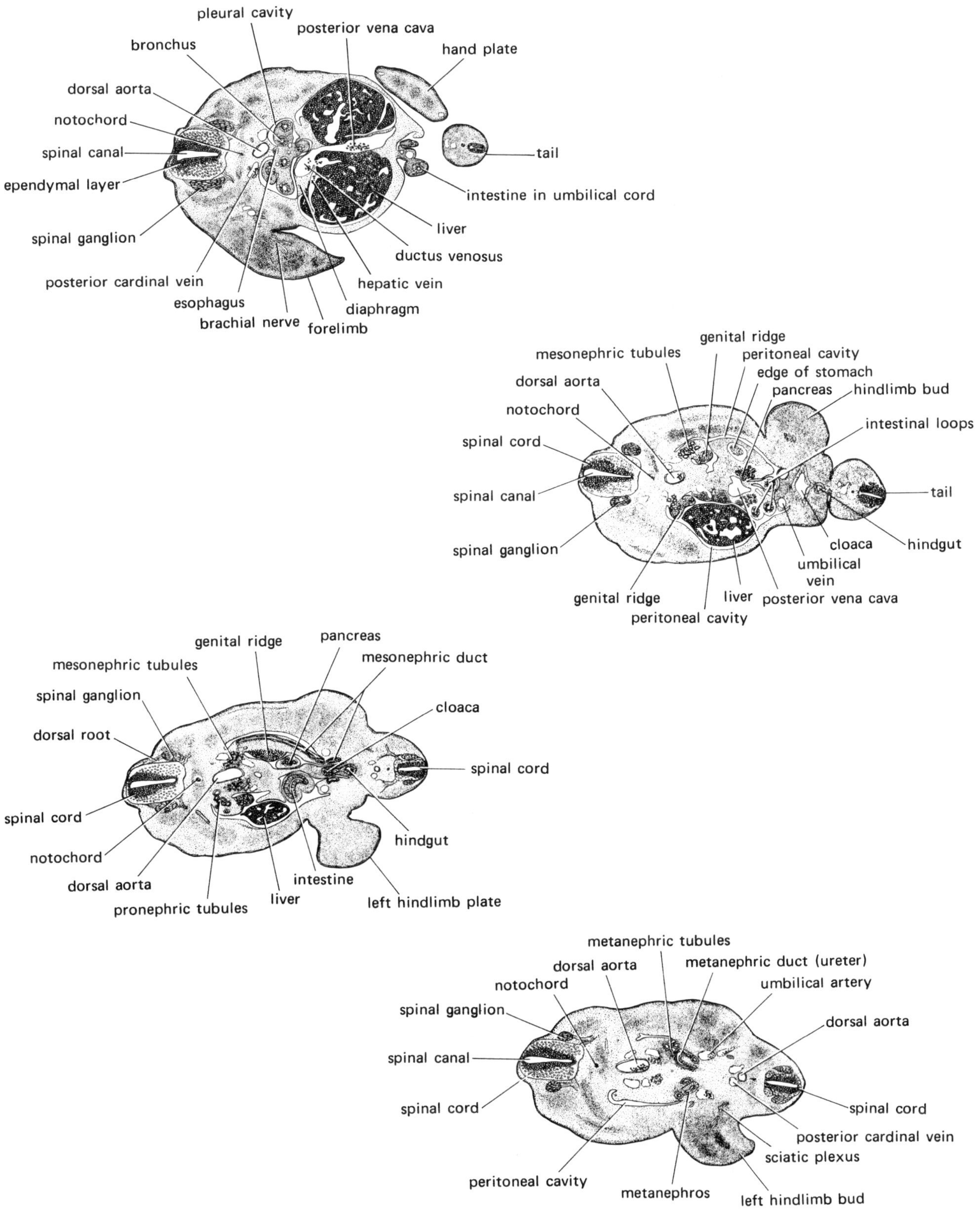

MOUSE EMBRYO — 12 g.d.

(TRANSVERSE SECTIONS)

Figures on this page are from R. Rugh. *The Mouse.* 1968. Burgess, Minneapolis.

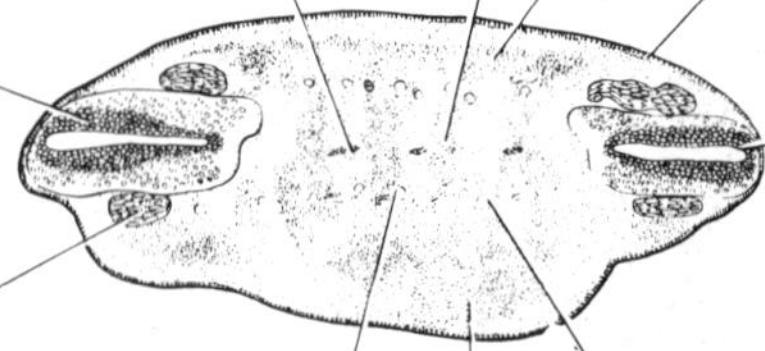

MOUSE EMBRYO – 12 g.d.

(TRANSVERSE SECTION)

From R. Rugh. *The Mouse.* 1968. Burgess, Minneapolis.

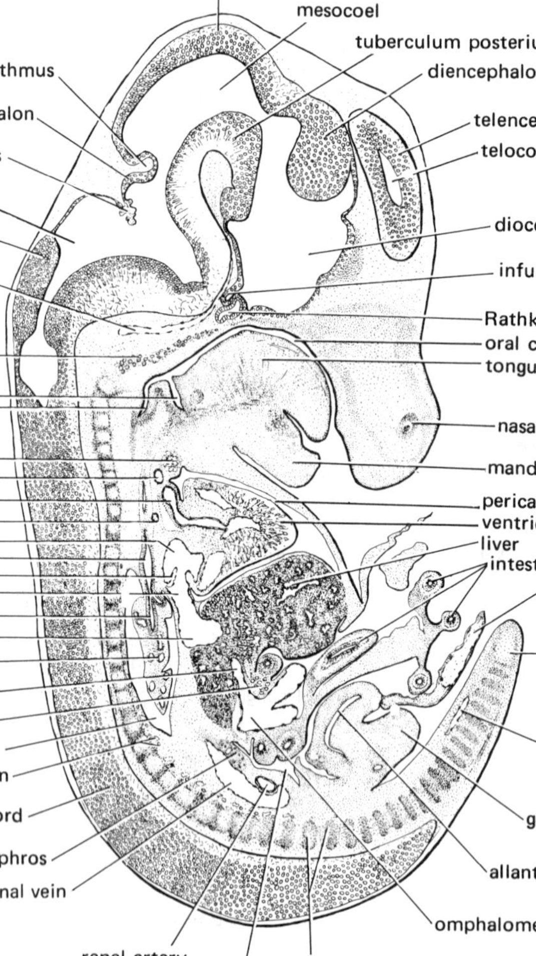

SAGITTAL SECTION OF MOUSE EMBRYO

(13½ GESTATION DAYS)

From R. Rugh. *The Mouse.* 1968. Burgess, Minneapolis.

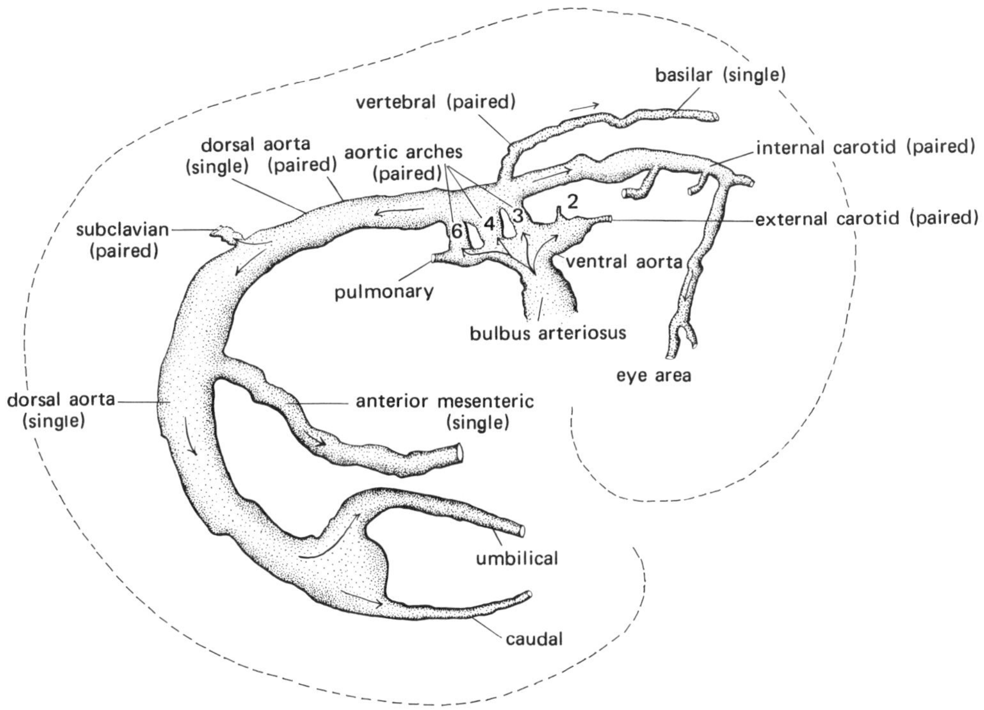

ARTERIES OF 11-DAY MOUSE EMBRYO OF RIGHT SIDE

(GRAPHIC RECONSTRUCTION)

From R. Rugh. *The Mouse.* 1968. Burgess, Minneapolis.

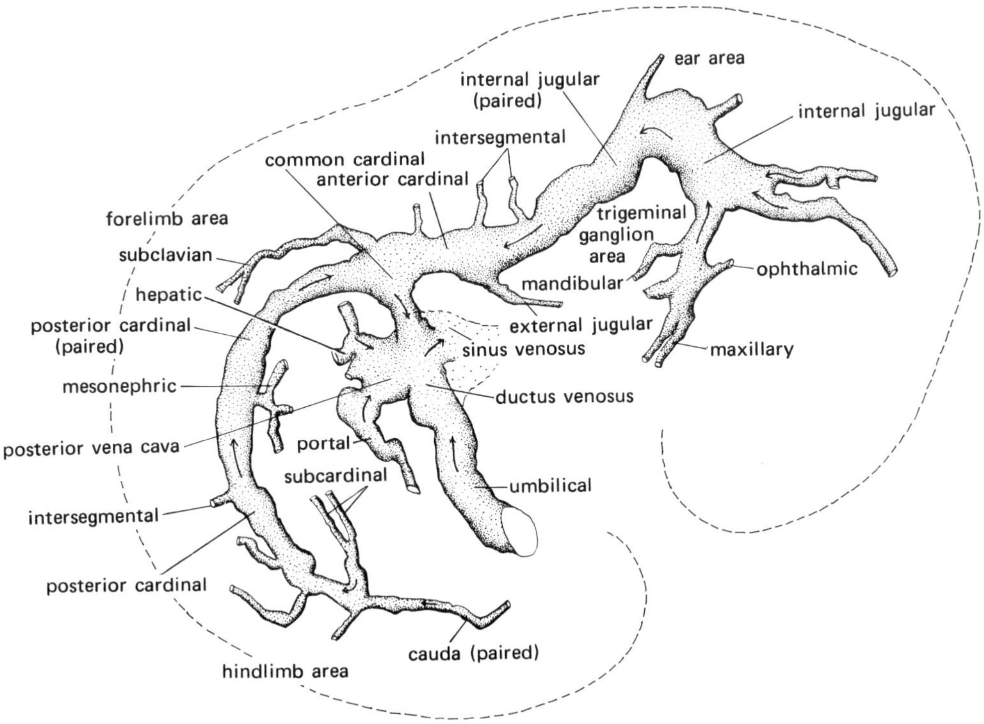

VEINS OF 11-DAY MOUSE EMBRYO OF RIGHT SIDE

(GRAPHIC RECONSTRUCTION)

From R. Rugh. *The Mouse.* 1968. Burgess, Minneapolis.

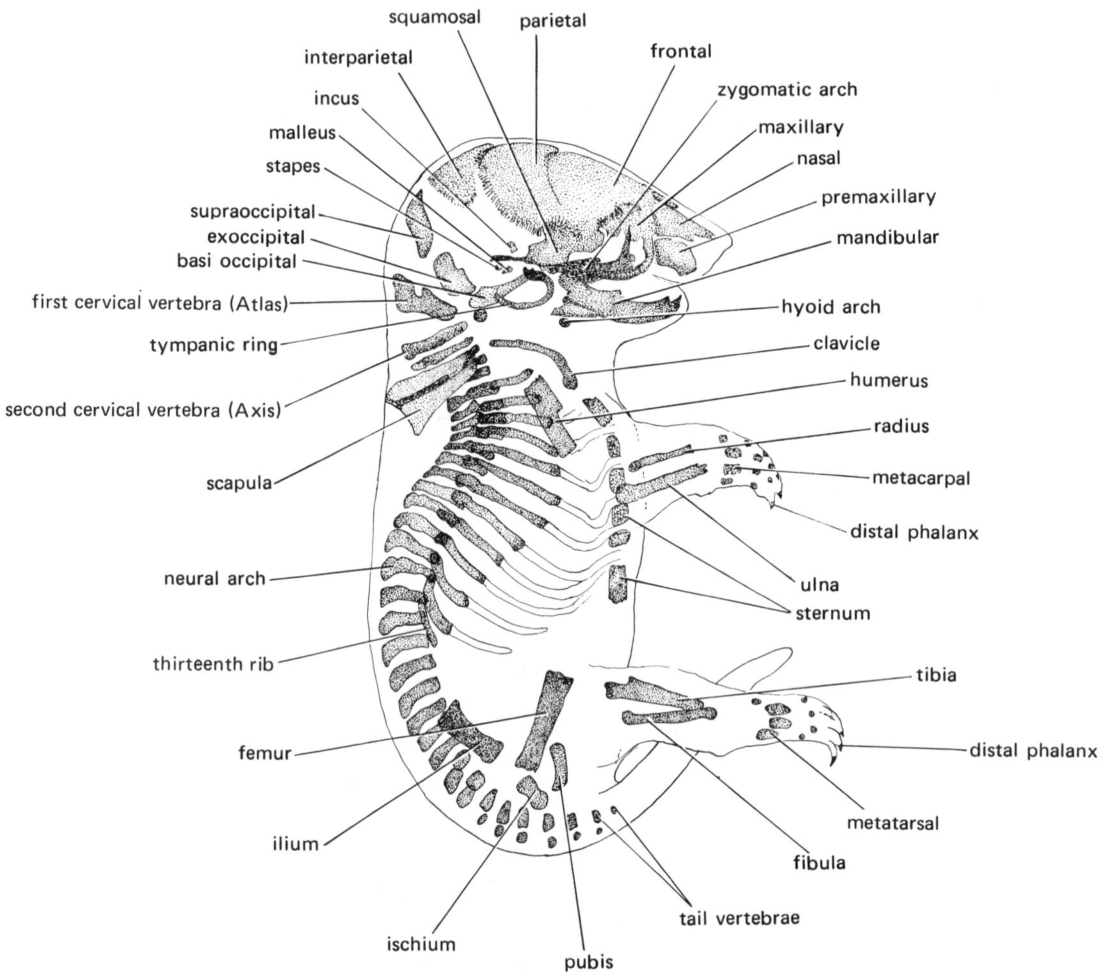

SKELETAL SYSTEM OF MOUSE AT 18 g.d.

From R. Rugh. *The Mouse.* 1968. Burgess, Minneapolis.

TYPES OF PLACENTAE AND GESTATION PERIODS OF VARIOUS MAMMALS

Animal	Placenta	Gestation Period, Days
opossum	none	13
mouse	discoid	20-21
rat	discoid	21-25
mole		30
squirrel		30-40
rabbit	discoid	30-43
hedgehog		31
woodchuck	discoid	35-42
kangaroo	none	39
ferret	zonary	42
mink	zonary	42-76
red fox	zonary	52-63
cat	zonary	60-63
dog	zonary	60-65
lynx	zonary	63
skunk		63
wolf	zonary	63
guinea pig	discoid	68-71
beaver	discoid	94-100
lion	zonary	106
tiger	zonary	106
pig	diffuse	115-120
goat	cotyledonary	140-160
sheep	cotyledonary	144-160
macaque monkey	discoid	147-160
armadillo		150
bear	zonary	210
mountain goat	cotyledonary	210
chimpanzee	discoid	250
elk	cotyledonary	250
marten	zonary	267-280
man	discoid	270-295
cattle	cotyledonary	282
deer	cotyledonary	300
zebra	diffuse	300-345
horse	diffuse	330-380
whale		334-365
seal	zonary	340-350
camel	cotyledonary	389-410
giraffe	cotyledonary	400-480
rhinoceros		488-540
elephant		510-730

Underlined mammals are considered in this *Guide*.

DRAWINGS: THE MOUSE EMBRYO

SOME EXPERIMENTAL PROCEDURES WITH THE MOUSE EMBRYO AND FETUS

Rarely does one have the opportunity of varying the environment of a mammalian embryo or otherwise subjecting it to controlled experimental variables. Now that the normal development of the mouse is quite thoroughly understood and pregnant mice are obtainable at any time of the year and in any part of the world, mouse embryos are being used for investigations to an ever-increasing extent.

1. NORMAL DEVELOPMENTAL SERIES: By placing mature males and females together for 2 hours (8-10 a.m.), one can expect that as many as 12 percent of the females will acquire vaginal plugs, which are the guarantee of successful mating. Of these, about 90 percent will produce embryos, fetuses, and offspring. Since the estrous cycle is 4½-5 days, such a short exposure would not be expected to produce a high percentage of conceptions, but by this method precisely timed embryos and fetuses may be obtained due to the limitation of the time for mating. Even so there may be a spread of several hours in the developmental stage even within a single litter, which may number from one to sixteen or more. By dissecting pregnant mice at daily intervals from days 9 to 19, the student can visualize directly the rate of growth, the development of the placenta, the presence of resorptions, and some dead and occasionally abnormal fetuses. Such fetuses, both normal and abnormal, if deprived of their membranes and properly preserved, will provide an excellent series for demonstration and study.

2. SUPEROVULATION: If a normal female mouse of 3-8 weeks of age is injected intraperitoneally with 5-15 international units of pregnant mare serum gonadotropin (Equinex, Ayerst) in 0.9% NaCl and 35-40 hours later with 10-15 units of human serum gonadotropin (Pregnyl, Organon), within the next 12-16 hours, she will actively ovulate many more ova than usual, sometimes as many as a hundred. If she is exposed to a sexually mature male at this time, many of these ova will be fertilized and begin development. A day and a half later if her oviducts are dissected into an appropriate culture medium, many 2-cell ova with their polar bodies will be found, and some can be cultured in groups of 20-30 under mineral oil in plastic dishes. The ideal culture medium consists of:

	Grams/Liter
NaCl	6.97
KCl	0.356
$CaCl_2$	0.189
KH_3PO_4	0.162
$MgSO_4$	0.294

	Grams/Liter
NaHCO3	2.106
Ca Lactate	1.147
Streptomycin	50μg/ml
Penicillin	100μ/ml
Crystalline bovine serum albumin	1.000 (Brinster '65)

With abundant fertilized ova available it is possible to culture them for a while, treat them in various ways, and then transplant them to a female of either the same or a different strain, even of a different hair color. One very important precaution must be emphasized. For proper implantation of a fertilized ovum, it must have achieved a certain state of activity and development and the receiving uterus must be properly prepared for its reception. The host or receiving female must receive the ova about days 5-6 after the time of her own ovulation. This can be determined by mating her with a vasectomized male who will provide a vaginal plug without sperm, thus indicating that she is receptive and hence ovulating. Five days later, the transplants may be made. It is suggested that an excess number of embryos be transplanted at first in order to increase the probability of at least one successful implantation.

3. CONGENITAL ANOMALIES: Some strains of mice have a high incidence of normally occurring anomalies in their offspring (e.g., C57 bl/6), while others seem to be relatively clean (CF1-S). In order to determine the incidence of congenitally induced anomalies, one must first establish the normal incidence of such anomalies since the two types can be confused. The treatment of pregnant females with Vitamin A excess or insufficiency, with trypan blue, or with x-rays will

produce congenital anomalies of great variety, especially when the insulting agent is given to the pregnant mouse between days 7-12 of gestation. Obviously, then, it is the reacting fetal system rather than the traumatic agent which is the common denominator, and there are certainly many other agents which should produce such nonheritable congenital anomalies. Reference is made to strychnine, arsenic, alkylating agents, radioiodine, Thalidomide, nicotine, LSD and other hallucinogens, excesses of heat (but apparently not cold since the mouse has a very efficient temperature regulatory system), and, most dramatically, ionizing radiations either from an x-ray source, cesium, or certain isotopes. A simple and safe experiment would be to subject pregnant mice at 8 days gestation to 150 R penetrating x-rays and sacrifice them at 19 days to determine the variety and frequency of induced congenital anomalies. If one allows abnormal fetuses to be delivered, the mother usually kills and devours them so that there is no record of their incidence. For this reason, examination by laparotomy at 19 days is routine in such studies.

4. THE FETAL SKELETON: One technique that is quite graphic and results in excellent demonstrations is to clear fetuses from 12 to 18 days gestation and stain them so as to determine the extent to which the skeleton has developed. The skeleton is sensitive to trauma and will show congenital effects from certain deficiencies and excesses, as well as radiation; these will show clearly if the directions below are followed closely:

   Technique for the Clearing of Fetal Skeletons: The following procedure will produce transparent specimens which will show the cartilage staining blue and the bone staining a brilliant red. It is excellent for the study of congenital anomalies which involve the skeleton. It is a modification of Spalteholz and Dawson, based upon experience in our laboratory.

   a. Fix mouse fetuses from 12 days to birth in 95% alcohol for 3 days or more. Specimens will shrink, but the skeletons will not be affected.
   b. If postnatal or large specimens are used or if there is an excess of fat, there should be a 3-day exposure to acetone to dissolve the fat, followed by washing in 95% alcohol for 24 hours. (Omit this step if mouse rather than large rat fetuses are used.)
   c. Place in Toluidin blue solution made up as follows:
      Toluidin blue – – – – – – –0.25 grams
      Alcohol 70% – – – – – –100.00 cc
      HCl – – 0.5% – – – – – –2.0 cc
      This solution should be made up in advance, filtered after standing 24 hours, and stored in tightly stoppered bottles. Leave in Toluidin solution for 3-4 days.
   d. Wash well in several changes of 95% alcohol to decolorize to the correct point. Observe regularly to stop at the right stage.
   e. Place in potassium hydroxide for 1-7 days. For small specimens it should be a 1% solution, for the largest specimens about 3% KOH. The bones should become clearly visible. If it is necessary to leave specimens overnight, in order to avoid complete maceration change to the weaker concentration or water. During the day when specimens can be observed, quicker treatment in the higher concentrations can be observed directly and terminated quickly in water when maceration has progressed far enough.
   f. Place in Alizarin red-S solution, made as follows: 1% in distilled water to which 1% KOH is added. When the formed bones have reached an appropriate red stain (1-7 days), staining is completed. If the staining takes more than 2 days, change solution. Fetuses stain rather quickly, possibly within 24 hours. (Human fetuses take weeks.)
   g. Clear in: KOH – – – – 1 gram
      Glycerin – – 20 cc
      Distilled Water – 100 cc
   h. Change this clearing solution every 24-48 hours until no more stain comes out.
   i. At 4-day intervals gradually change specimens from 50% to 70% to 90% glycerin. (All times should be extended for very large specimens.)
   j. Transfer to pure glycerin, making three changes each after a few days. Such specimens should be clear and skeletons beautiful for an indefinite period.

The mammalian fetus, because it is both warm-blooded and a parasite in the uterus, is rather inaccessible for surgical experimentation. However, there are many things which are instructive in experimental procedures that can be done to individual implantations, such as direct injections and typing off blood capillaries.* It will soon be possible to operate upon such fetuses to determine to what extent the solid foundation of experimental data from the frog and chick apply to the mammal in its development.

REFERENCES

Rugh: *The Mouse.*

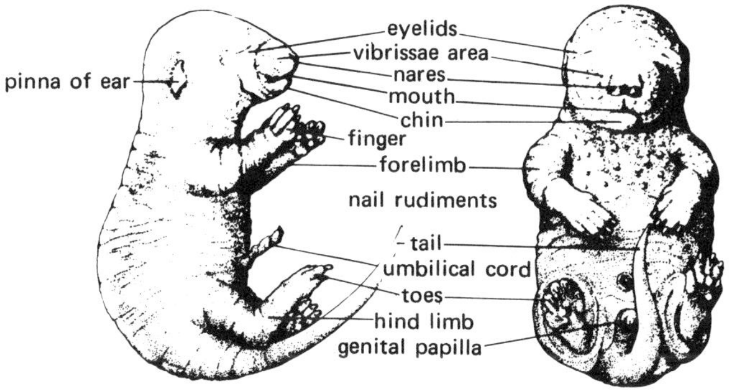

NORMAL

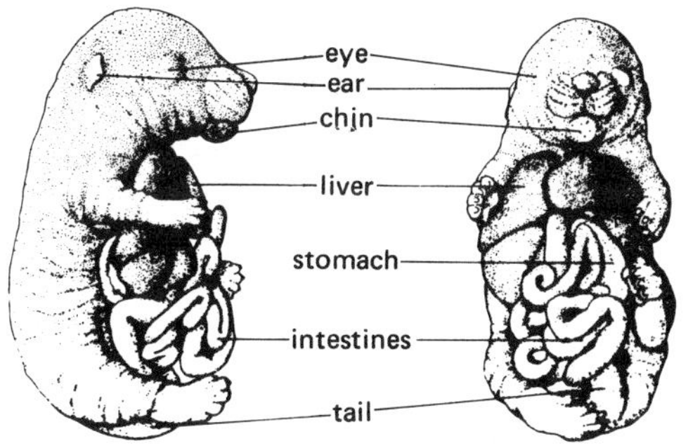

VISCERAL DYSRHAPHIE

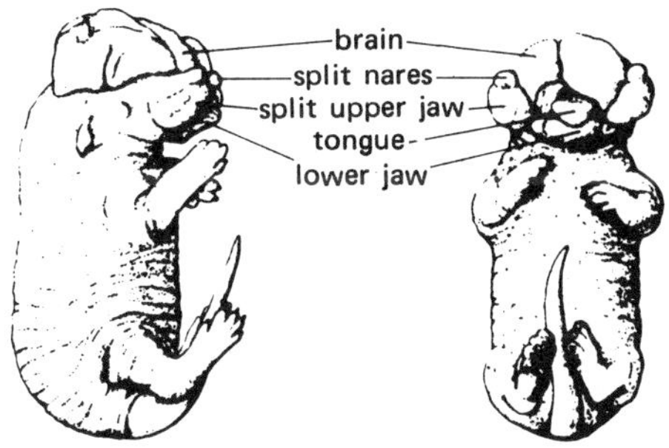

CRANIOFACIAL DYSPLASIA

Lateral View ____________ Ventral View

MOUSE FETUSES – NORMAL AND ABNORMAL

18 GESTATION DAYS

*For those interested in experimenting with mouse embryos and fetuses, examine the final exercise in the Guide for general suggestions. The author would welcome correspondence from any qualified student about proposed rodent experiments, since they merit special consideration.

Exercise 26 HUMAN DEVELOPMENT

It is a rare biology department or even medical school which has good slides and specimens of human embryos and fetuses for the student to study. However, it is important for the student to realize that development in the human is not unique and that from fertilization to delivery the human follows developmental patterns which are characteristic of all mammals, deviating only in details. Those details are indeed important and are predestined by the chromosome make-up of the human zygote at the instant of fertilization. That is the reason that the cover of this *Guide* was included to show that, superficially, at comparable stages, one could easily confuse a human with a mouse, a pig, or a chick and even with a turtle. If one understands basic embryology, one will understand and appreciate any available human material.

The major difference between mammalian development and that of any other vertebrate is that a placenta is formed which is the life line between mother and embryo (or fetus), so that the human is for a few months like an efficient parasite in the mother's body, getting from her all the nourishment and oxygen it needs and eliminating through her all its waste products and carbon

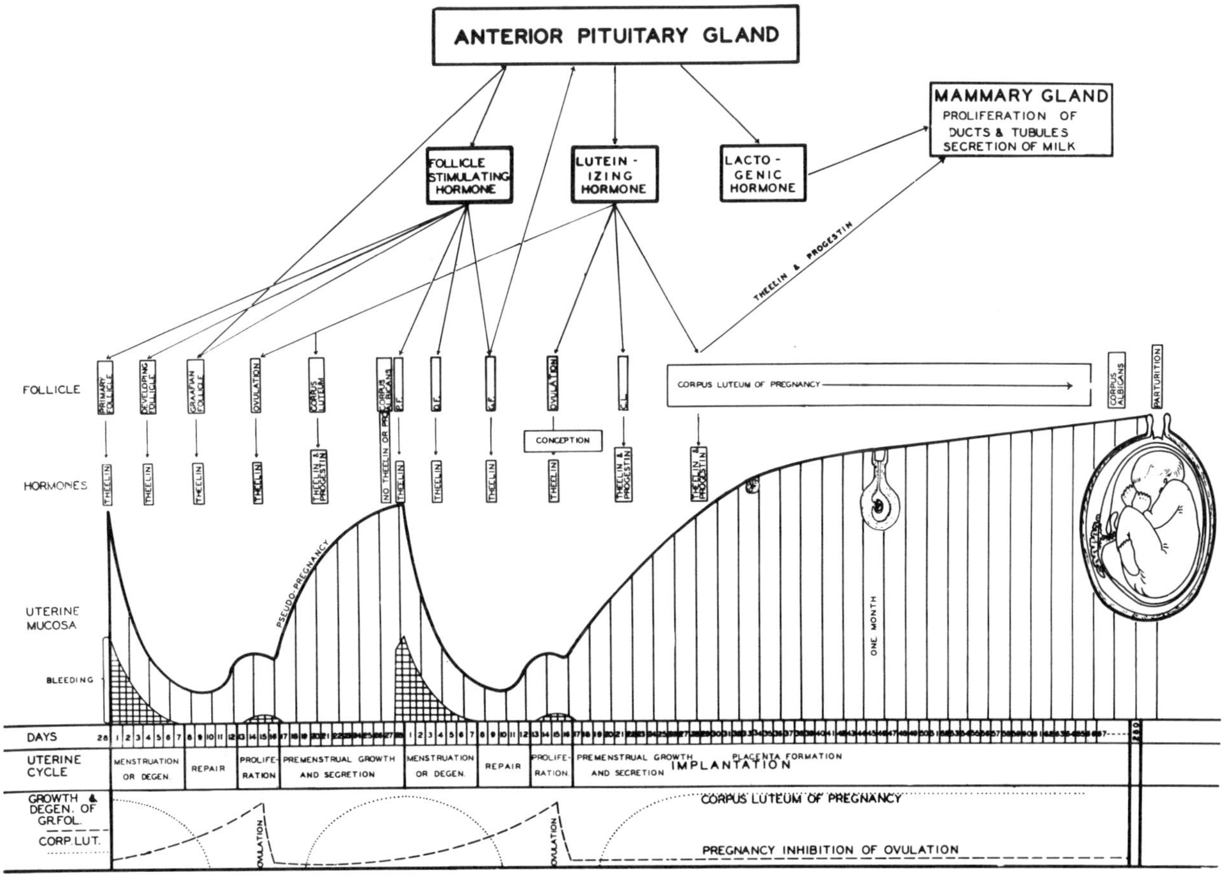

SEX CYCLE OF HUMAN FEMALE HORMONAL RELATIONS

dioxide. Not until it is a 5 months fetus or older could it possibly live outside of the uterus with extensive artificial help and without a placental association with the mother.

We will briefly survey here the chronology of early development of the human and illustrate it with photographs and drawings of actual specimens, most of them living, so as to convince the student that the human embryo and fetus conforms to that pattern of development found through the centuries to be the most efficient in the production of human beings. There must be some overriding reasons why all development follows the same pattern: fertilization, cleavage, cell proliferation, organogenesis, tissue differentiation—all integrated toward the goal of "the organism as a whole." all within a range of variation we call "normal." Deviations do occur, due sometimes to heredity and sometimes to congenitally acquired anomalies, but the remarkable fact is that the vast majority develop within a certain pattern that is recognized as "human." The forces which control this have thus far evaded scientific investigation, but the vast collection of scientific facts is leading us to a greater understanding and appreciation of what life really is. (Human fetuses at various ages are photographed in Plate 9, pages 388-89, and in Plate 10, page 390.)

From puberty to menopause, every normal woman will produce about 400 fertilizable ova, on the average of one each 28 days (generally, about 14 days after the onset of menstruation). Every male from puberty through old age, will produce uncountable trillions of viable sperm, of which some 350 million will be available at any time to fertilize a mature ovum.* This seems a terrible waste of sperm until one finds that to reach the mature ovum, in the vicinity of the infundibulum, the sperm must swim from the vagina through the cervix into the fallopian tube and then for about 9 inches more and that the majority of sperm are either lost enroute or devoured by the thousands of scavenging phagocytes marshalled by the tissues of the female to prevent any invasion. Probably not more than a few thousand sperm reach the ovum and only one of these, the first to make contact, will merge with the ovum in the process we know as fertilization. At that instant, all the potentialities of the individual embryo are predetermined: its heredity, its sex, its possibilities and limitations. Since the male- and female-producing sperm are formed in equal numbers, the differential in statistical production of males to females (106/100) must be explained on the basis of the minute difference in the chromosome make-up of the two kinds of sperm, one with a very small y chromosome and the other with the much larger X chromosome. Therefore, more males are produced. It appears that the sperm with the lighter chromosome weight get up to the egg a little more frequently than do the heavier X-carrying sperm. This can be somewhat controlled by timing coitus in relation to ovulation.

The ovum measures 1/175 of an inch in diameter, while the sperm measures 1/500 inch in length. Together, they can produce, with the help of the mother through the placenta, a baby that could weigh about 7 lbs. and have a sitting height of 20 inches in 266 days after conception. The first cleavage occurs about 36 hours after fertilization, but succeeding cleavages are accelerated, so that, by 3½ days, there may be as many as 64 cells aggregated into a mulberry-type structure, known as a morula. By 5 days, the embryo is a hollow blastula, and, by 9 days, it has begun to invade the lining membranes of the uterus, to reside there for the duration of development. It is still less than 1/100 of an inch in length, but it is now dependent upon an external source for nutrition, respiration, and elimination of wastes.

The first discernible primordium is the neural groove, which depresses to form a tube that is the forerunner of the brain and spinal cord. As early as 20 days, the foundation of the nervous system has been established. But even before this, a yolk sac is formed from which will be derived the primordial germ cells (of both ova and sperm) which are the guarantee of the possibility that another generation will be produced. This is a sort of security measure, a savings account to insure propogation of the line—really the first thing the embryo does. Next comes the heart, first to appear as a single tubular structure and pulsating at 24 days even before it is structurally complete and a circulatory system with blood cells is available.

During the first month after conception, the human embryo has acquired a completely closed

*Of course, there are variations among individuals, and some (20 percent of couples) are sterile.

neural tube, otic pits, optic vesicles, and lens primordia, olfactory placodes, cerebral vesicles, and mesencephalic and cervical flexures of the nervous system; fused endothelial heart tube, pharyngeal arch arteries, vitelline, umbilical and cardinal veins, and differentiation of atria, ventricle and bulbus cordis of the vascular system; head fold, liver primordium, and thyroid of the alimentary system; pronephros, followed immediately by mesonephros and ureteric bud, and gonad primordia of the U.G. system. Thus, most of the basic organ systems are represented, though in primitive state, by 30 days after conception.

During the second month, the lenses are separated, vitreous appears, the retina differentiates, prosencephalic and pontine flexures form, semicircular canals and cochlear duct form, eyelids and cornea may be seen, and eyelids fuse, anterior and hippocampal commissures develop. The lens cavity is obliterated, for the CNS; the septum primum forms in the heart, interventricular septum forms, atrial cushions fuse, interventricular foramen closes, and septum secundum appears in heart; tongue and tubo-tympanic recess form, lung buds and stomach dilation develop, thymus and parathyroids are distinguishable, and the palate begins to close for alimentary canal; mesonephros degenerates, male and female diverge in differentiation, tests show histological differentiation in U.G. system. Such anomalies as are congenital are caused by this time, prior to completion of basic organogenesis.

During the third month, there is evidence that the nervous system is functioning sufficiently to integrate the various organ systems that have been developed. Early breathing movements may be seen, drawing into the lungs some amniotic fluid. Reflex responses are evident but are at first whole body rather than of specific regions, but, by the end of the month touching of the lips causes sucking reaction and touching the eyelids causes the eyes to squint. This is the initial month of neuro-muscular activity. The genito-urinary systems are further separated in both sexes, and parts of the embryonic urinary system become essential parts of the genital system, with the male differentiating ahead of the prospective female. Urine is formed by the metanephric kidneys and conveyed to the bladder from whence it is eliminated into the amniotic fluid to be picked up by the mother and excreted through her urinary system. The eyes are well formed but lateral on the head, with optic nerve connections. The fetus now appears more human than previously.

The fetus measures only 1½ inches in sitting height and weighs 1/7 of an ounce. Its back is now straightened out, fingernails, toenails and hair follicles are forming, the eyes are sealed shut for the next three months, teeth are beginning to form, the fetal heartbeat can be detected with the stethescope through the mother's abdomen, and the skeleton is well formed. By the end of this third month, the fetus measures 3 inches in body length and weighs ½ ounce, its thumb can be opposed to its forefinger (a feat distinguishing it as a primate), the brain is fully formed but is not yet refined in structure (and will not be until some weeks after birth), tactile sense is well developed throughout the skin, blood is beginning to form in the marrow (previously in blood islands, spleen, and liver), secondary sex characters appear, and thyroid and pancreas are complete. Vocal chords form, but the fetus cannot emit sounds. The liver forms bile, and the digestive tract acquires some 20 million glands to anticipate its ultimate functions. This is the end of the first trimester. The second 3 months (second trimester) follow, during which all of the organ systems are further refined.

The second trimester (3 to 6 months) is the period of rapid growth and activity. Since most of the organ systems are well developed, they now begin to function. During the fourth month, the fetus measures 6 inches in sitting position and weighs 4 ounces, and, by the end of this trimester, it measures 10 to 12 inches and weighs about 1½ pounds. The skin begins to form sweat and sebaceous glands and hair follicles, all to protect it against the shocks of the ultimate extrauterine environment. The lungs are fully formed, but at 5 months the fetus could not breath without help. The fetus is now very active, but the mother is accustomed to this and can regulate her own and the fetus' activity and sleep so that both can be rested.

The third trimester (the last 3 months) shows the emergence of intelligence and personality. The fetus stands a fair (10 percent) chance of survival if delivered, and such chance is further improved with every day it remains in the uterus until about 266 days have elapsed from conception.

The mechanisms for controlling breathing, swallowing, and regulation of body temperature are not fully differentiated at 7 months, and delivered fetuses must be given immediate and special care for some time. During the eighth month, the fetus will measure about 13 inches in sitting height and weigh from 4 to 5 pounds. Its probability of survival is now 70%. The temperature control of the fetus ranks as the most seriously malfunctioning system at this time, with respiration second because the alveoli of the lungs have never been fully expanded with air. The third system is the digestive, and, when these three systems are fully able to support the fetus, it is ready to emerge into the air as an independent being. During the ninth month, the placenta gradually becomes nonfunctional so that the fetus must prepare to emerge into the outer world, with all of its capacities in order. It should measure about 15 inches in sitting height and weigh from 6 to 8 pounds.

The process of expelling the fetus is divided into labor, delivery, and elimination of the afterbirth or placenta. Labor means the contractions of the uterus and the stretching of the cervix until its opening is at least 10 cm in diameter and the baby can pass through into the vaginal canal. The passage of child through the vaginal canal to the outside normally occurs rather rapidly, with the flow of amniotic fluid lubricating the path. The placenta and membranes may follow immediately or be aided in their extrusion by the obstetrician.

Labor can last from 7 to 18 hours or even longer. The first uterine contractions last only 15 to 25 seconds and occur 10 to 15 minutes apart. Some women are hardly aware of them, and others experience severe labor pains. When the contractions become 3 to 5 minutes apart, they are more vigorous and birth (delivery) is imminent. The birth stage may last from 30 minutes to 2 hours without harm to the baby or mother, although the quicker delivery is better for both. The mother can help by bearing down on her abdominal muscles, hence some women prefer to have no anesthetic. In 95 percent of the cases, the baby emerges with the head in a downward position. The baby is usually handed to a pediatrician or midwife who tests it by the Apgar formula (breathing, pulse, muscle tone, reflex irritability, and color) and, if it scores 7 or above (on a 0, 1, 2 rating basis), its chances are very good, while a score of below 3 requires resuscitative measures.

The author has often said that in all biology there are two events that continue to awe him, no matter how many times he views them. The first is the initial cleavage of an ovum which has been fertilized and is on its way to developing into a fully viable organism. It does not matter whether it is a sea urchin, starfish, frog, salamander, turtle, mouse, or other mammalian ovum – the forces that initiate the initial cleavage of an ovum are most mysterious. The second event is the birth of a human child, because, within a few moments, the child must adjust its circulatory, respiratory, digestive, and heart regulatory systems to a new environment so different from its watery, warm, and completely adequate environment in the mother's uterus that it is nothing short of miraculous that the adjustment is usually made without incident. It occurs at the rate of 3¾ times every second around the world, and most of the children are "normal."

RECOMMENDED READING:

Developmental Anatomy, Arey (Saunders).
Essentials of Human Embryology, Dodds (Wiley).
Biography of the Unborn, Gilbert (Hafner).
Medical Embryology, Langman (Williams & Wilkins).
Human Embryology, Patten (Blakiston).
Vertebrate Embryology, Rugh (Harcourt, Brace & World).
From Conception to Birth, Rugh & Shettles (Harper & Row).

REFERENCES

Arey: 68, 609-47
Nelsen: 546-51
Patten: *Pig.*
Rugh: *Vert. Emb.* 304-77
Rugh and Shettles: *From Conception to Birth*

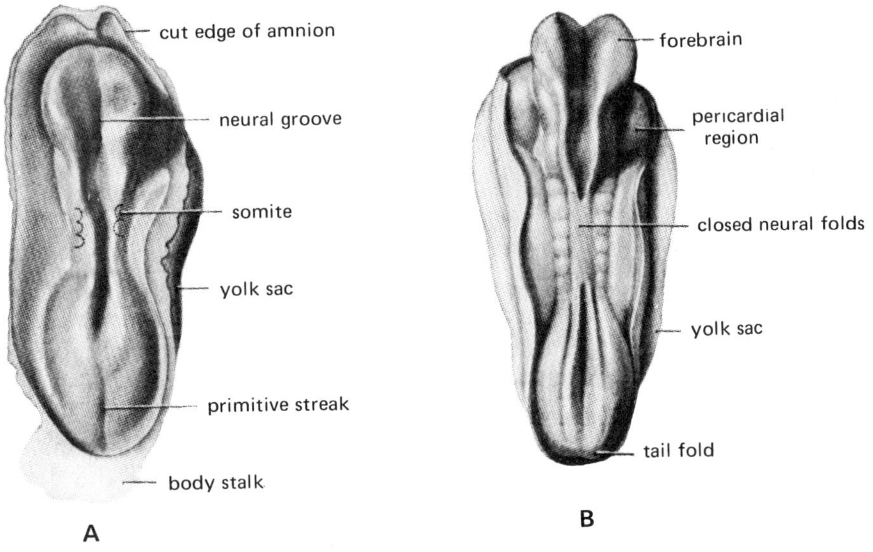

Human embryos of 22 days, in dorsal view (Streeter). A, Ingalls embryo of 1.4 mm., with three somites (X42). B, Payne embryo of 2.2 mm., with seven somites (X27).
(X27).

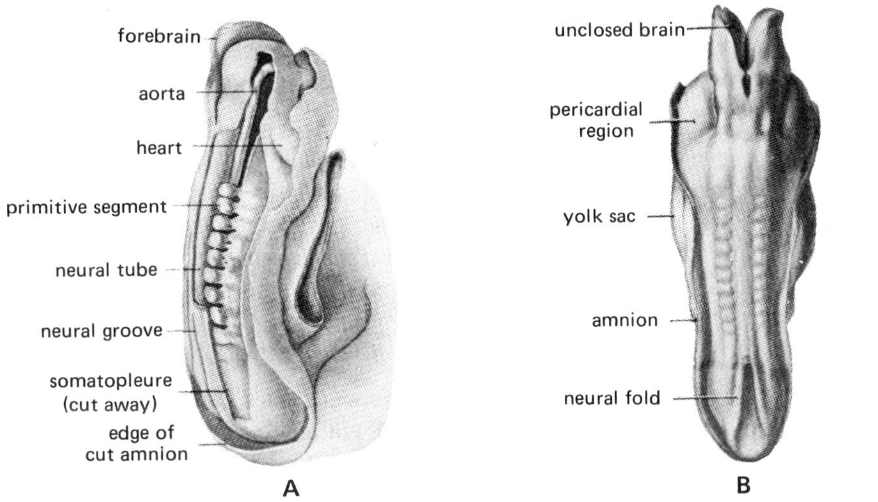

Human embryos of 23 days. A, Veti-Esch embryo of 2.3 mm. with nine somites, partially dissected and viewed from the right side (X25). B, Corner embryo of 1.7 mm., with 10 somites, in dorsal view (Streeter: X34).

From L. B. Arey. *Developmental Anatomy.* 4th ed. 1965. W. B. Saunders, Philadelphia.

See Patten (1964): 209-536

EXPERIMENTAL EMBRYOLOGY

Exercise 27

During the ontogeny of any organism, there is never any more sensitive stage with respect to environmental variables than that of the embryo. This is because there are rapid, constant, dynamic changes that bring the fertilized ovum to the fully formed and independent organism. By interrupting those changes by transplanting pieces of an embryo from their proper places to other sites, by subjecting the embryo to variables of temperature, salinity, pressure, etc., one can produce bizarre organisms and, at the same time, learn about the interdependence of the developing parts with respect to the organism as a whole. By producing freak organisms (with extra appendages, stunted growth, eyes developing on the tail, etc.), one is not creating animals that can reproduce themselves, but one is learning why the vast majority of organisms do develop within the range of normality. Isn't it remarkable that, while there are minor variations in the anatomy of human beings, they generally fall within a limited range of pattern variation so that books of anatomy can be used in the study of adult organisms, including the human? Much that we know about the adult organism is learned through knowledge of its early development and the reactions of the embryo and fetus to environmental variables. For this reason, students are encouraged to environmental variables. For this reason, students are encouraged to experiment with available vertebrate embryos, such as the frog, salamander, and chick, and even with a mammal, such as the mouse.

EQUIPMENT AND PROCEDURE IN EXPERIMENTAL EMBRYOLOGY

General Introduction

Biologically clean, these two words should be the most frequently emphasized in any laboratory of experimental embryology, even as they are in any tissue culture laboratory. Glassware, instruments, solutions, and hands must be "biologically clean" before any experimental results may be considered valid. The term means that, barring any experimental conditions imposed which might alter the situation, there is no possible contamination of the living material either by chemical substances, by living parasites, or by harmful organisms, such as bacteria or viruses. The experimental conditions should be such that any embryo, introduced into that environment, would be expected to survive. The following precautions, in the interest of biological cleanliness are suggested.

1. *Glassware:* Regardless of the source of the glassware used, it should be thoroughly scrubbed with hot soap and water, rinsed in running tap water for at least 2 hours, rinsed in distilled water, and air dried either by or under the direct supervision of the person planning to use it. If the glassware is cleaned with the usual cleaning fluid (potassium dichromate saturated in 10 percent sulfuric acid), it must be thorougly washed and rinsed for a longer period in order to remove the tenacious chemicals. Properly cleaned glassware may be wrapped in clean paper towelling and heat sterilized for ½ hour at 100°C. As long thereafter as the glassware remains wrapped, it may be considered as sterile.
2. *Hands:* A surgeon usually spends as much time scrubbing his hands as operating, and such cleanliness in experimental embryology will result in more dependable and reproducible results.

Note: The following pages are taken largely from the author's *Experimental Embryology,* which was designed for the candidate for a graduate degree in embryology. With a pair of watchmaker's forceps, sharp scissors, and a good dissecting microscope, any student who has completed a course in basic embryology should be able to plan and carry out some simple experiments with vertebrate embryos. The experience would be invaluable.

Formaldehyde, osmic or hydrochloric acid fumes, adherent to the hands, will contaminate instruments and ultimately the embryos. Thin, white surgeon's rubber gloves are advised in tissue culture experiments.

3. *Instruments:* If the instruments have never before been used, they may be thoroughly washed rinsed, and sterilized in the autoclave at 15 pounds pressure for 30 minutes, cooled, and immersed in 95 percent alcohol until used. Dissecting instruments from other laboratories should never be "trusted." Each student should provide himself with a new set of stainless steel instruments (never chromium plated) and should keep them in a plastic tube or in a cotton-filled box to be reserved exclusively for operational procedures with living embryos.

   It is often necessary to preserve eggs or embryos, but the operating instruments should never come into contact with any fixatives. The student may use an unsterile or even contaminated set of instruments for the handling of such material to be preserved. Often eggs, embryos, or even tissues may be introduced into a fixative with the flat end of a toothpick, thus avoiding the contamination of instruments. Once an instrument has come into contact with a fixative, it should thereafter be regarded as contaminated and not biologically clean.

4. *Embryos:* Dead or dying embryos are probably the most common source of contamination of cultures, because they are infested with bacteria and necro-toxins. Ailing embryos should be isolated, and crowding should be avoided. The culture medium should, at all times, appear to be clear. Healthy embryos may be passed through several changes of sterile medium in order to free them of adherent bacteria. Some stages of development, particularly of aquatic forms, may tolerate brief immersion in hypertonic salt solutions or dilute potassium permanganate, either of which appears to be bacteriocidal. Generally, such treatment is unnecessary and should be avoided if other precautions are taken. Also, if the embryos are kept at the lowest tolerable temperatures and the dishes are covered, there is less likelihood of infections killing them.

The sooner the student insures personal and operative biological cleanliness, the sooner will he or she obtain reproducible experimental data and enjoy the experiences of experimental embryology. It is a complete waste of time and talent to allow contamination to invalidate experimental data.

The Research Protocol

Purpose: Here the major purpose of the investigations should be succinctly stated. For example, it could be: "To determine the effect of temperature on the growth rate of the tadpole of the frog, *Rana pipiens*."

Such a statement gives immediately the one major variable, *temperature*, and the test object or situation, *growth rate,* and most important, the form used, *Rana pipiens tadpoles.* There can be no doubt by anyone that if the purpose is achieved, there will be established some correlation between temperature and growth rate.

But, one cannot always anticipate that results will be positive, so the purpose might be stated: "To investigate the possible relation between temperature and growth rate of *Rana pipiens tadpoles.*" There can be no doubt by anyone that if the purpose is achieved, there will be established some correlation between temperature and growth rate.

EXPERIMENTAL PROCEDURE: In this section, one must outline exactly what he intends to do.

a. The controls.
b. The number, age, sex stage of development, or other conditions of the animals or embryos.
c. The single experimental variable imposed.
d. The method of collecting data.

THE CONTROLS: Every experiment and every research project must include a "control" or "controls." It is difficult to conceive of a situation in which a control is not possible, and most experiments are not valid without adequate controls.

The control is the standard, the normal, the untreated organism, or the situation with which experimentally induced results are to be compared. One simply cannot evaluate experimental data without controls, without organisms or embryos exactly identical but without the imposed experimental variable. Only by direct comparison of an experimental situation with a control situation can we evaluate any deviations caused by the experimental variable.

The ideal control, in any animal experiment, is a genetically identical individual, if such is obtainable. Identical twins, derived from the same zygote by cell separation at the 2-cell stage, would therefore be ideal in the sense that one could be regarded as the exact (genetic) biological duplicate of the other. One could be kept protected from the single variable of interest, while the other could be subjected to it to determine the effect. Isolation cultures among the protozoa or bacteria, resulting in clones which may be separated and followed, all derived from a single parent, would constitute ideal experimental material. This is because these unicellular forms have a sexual reproduction, the progeny resulting from binary fission and therefore being identical with each other as well as with the single parent.

Among multicellular forms, and all embryos, this ideal situation is generally impractical so that we usually satisfy the condition of a control with litter mates, derived from the same parents and at the same time. Certainly, these are not genetically identical, but, nonetheless, they are more closely identical than any other possible combination of paired individuals.

Thus, while one's interest may be directed toward the effect of some experimental variable, one cannot even recognize the results if one does not have proper and adequate controls with which to compare. The CONTROL is absolutely essential to every experiment.

The novice might ask about controls for extirpation or transplantation experiments. Obviously, if an extirpation results in a specific organ loss, the control would be an extirpation from another region to determine whether these results a similar organ loss. Likewise, in transplantations, there is the donor tissue and the location in the host field, either condition which must have a control to specifically circumscribe potentialities. In these qualitative experiments, large numbers of controls are not necessary, but similarily placed transplants of other donor material, or varied locations within the host area of the same anlage, would constitute adequate control. The function of the control is to prove positively that the results obtained could not have been obtained with any other set of circumstances.

THE NUMBER, AGE, SEX, ETC., OF ANIMALS USED: The results of many experiments are the more convincing the greater the numbers of experimentals and controls used, within certain limits. If every experimental animal shows a particular response never seen among the controls, then 10-20 controls and experimentals should suffice. However, if the differences between experimentals and controls are (calculated to be) small, then, on a purely statistical basis, larger numbers are necessary. In the latter instance, the results are primarily quantitative, even though they may be concerned with qualitative differences. Thus, it is important to plan on a sufficient number of animals all used at one time. There are, for instance, seasonal differences even within a species, and also diurnal differences (cycles) in certain rhythmic functions, so that in order to obtain large numbers, it is not safe to accumulate data from small numbers over an extended period.

The purpose of any experiment is generally to study the effect of a *single variable,* e.g. temperature. It is important, therefore, that all other physical and biological variables be eliminated between the experimental and controls, or at least be reduced and recognized. One variable among animals that is not always recognized is sex, though the differences are usually obvious! There are few biological experiments in which sex does not play a part, though sometimes a negative one. In general, the female is the more hardy, the more resistant to trauma, etc. Another variable that must be recognized is age. Certainly, the embryos of different ages are very different, so that one cannot put together data from temperature observations on the blastula, gastrula, and neurula stages of the frog tadpole, for instance. Another area of variation is in physiological activity, which is more difficult to recognize. Thus, one must try to obtain a large number of identical animals or embryos, half of which may then be subject to the one experimental variable in order to determine the valid biological responses.

THE SINGLE EXPERIMENTAL VARIABLE IMPOSED: The above statements emphasize the necessity of preparing the animals or embryos so that the results obtained are clearly due to the single variable imposed. If one is experimenting on temperature effects, one must be absolutely certain that no other variable is involved. Aside from a few biological variables above mentioned, among the physical variables would be light, agitation, salinity, number of organisms per unit of space (concentration), etc., any one of which could easily becloud the results of the intended variable of temperature. Both the biological and the physical and chemical variables must be eliminated or so controlled that there remains but a single variable between the control and the experimental animals.

METHOD OF COLLECTING DATA: This will depend entirely upon the material and the nature of the experiment. In any case, it should be systematic and complete, so that anyone else could step in and complete the experiment at any point. Dates should appear by every set of data collected, and every measurable or detectable bit of information should be recorded in a book which never leaves the laboratory. The practice of collecting data on scraps of paper, later to be transposed to the record book, is to be discouraged.

Sketches, accompanied by photographs, are very useful in qualitative experiments. But, in every case, the illustration must be *fully labeled.* No one can remember accurately such minute data for any length of time. When the data are all available, it is well to reduce it to a table and then, if possible, to a graph.

Brief mention must be made here of the necessity in all scientific research for absolute honesty, to the minutest detail. One false entry or record, one faked operation, or even incomplete data will ruin forever the reliability of all data collected by that individual. Science must be a body of knowledge based upon fact, as truly as the human mind can perceive it. One quickly discovers that facts in science are more exciting than any concoction of the human imagination.

DISCUSSION: In this section will appear all references to related or pertinent experiments and objective discussion of their meaning with respect to the findings of the data presented. The discussion should never be exhaustive, should never be a means to "padding" the paper, and only those papers of recent date should be included. It is presumed that intelligent readers will be familiar with the earlier works in the field, and, through references included, other and earlier references become available. The author should freely admit disparities and point toward further work that would be instructive.

SUMMARY AND CONCLUSIONS: The summary may be a part of the discussion, although some authors offer a single paragraph summary as the first of the items in the conclusion. In a proper discussion, the points of a summary would naturally appear. The conclusions, if no summary is offered, should start with a very succinct statement of the project, animals used, and the general findings. Succeeding conclusion statements would then relate the newly proven facts. It should be remembered that often a reader will read the introduction and the conclusions, and possibly look at the illustrations, in order to decide whether the material is of particular interest to him. This lends importance to these sections. However, it is a cardinal error toinclude in any summary or conclusions any statements not born out by the data of the paper.

REFERENCES: There are many ways of presenting this information, and one should consult the journal to which he or she intends to send the manuscript. In any case, a complete reference should include the names and initials of all authors, the date of publication, the complete title, and the detailed reference, including the volume of the journal and the inclusive pages. This is a very important section of any scientific report.

The general outline, therefore, for a manuscript reporting research would be as follows:

Statement of purpose
Materials and methods
Experimental data
Discussion
Summary and conclusions
References

EQUIPMENT AND INSTRUMENTS NEEDED BY EACH STUDENT IN EXPERIMENTAL EMBRYOLOGY*

Optical Equipment

Standard microscope with usual lenses; oil immersion optional.
Dissecting microscope (binocular):
 Objectives 1.7X and 3.5X most useful, with 10X oculars.
 The U-shaped base should be removable for indirect lighting.
Microscope lamp: Spencer diaphram mody udrgul.
Heat-absorbing flask: A 250 cc, round-bottom Pyrex flask filled with distilled water and supported on vertical stand by screw clamp. This device will absorb the infrared heat rays and also provide a means of concentrating the cold light.

Glassware and Culture Dishes

2 enamel pans 3.5 x 12. 24 inches for culturing tadpoles
2 battery jars with weighted covers, for living frogs
2 crystalizing dishes with covers, 12 inch size, for early stages of amphibian development
12 Syracuse dishes
12 # 1 stenders with covers
12 # 2 stenders with covers
24 regulation fingerbowls
6 petri dishes with covers
6 shell depression slides
½ gross microscopic slides
1 oz. # 1 coverslips, square or round
2 ft. soft glass tubing, 3-mm diameter
2 ft. soft glass rod, 4-mm diameter
4 ft. soft glass rod, 2-mm diameter
2 hypodermic syringes, 2-cc capacity, with # 18 needles
2 glass plates, convenient size 5 x 7 inches

Metal Instruments and Equipment

2 watchmaker's forceps, stainless steel # 5
2 adjustable needle holders
3 scissors: fine pointed to heavy and blunt
2 fine, sharp scalpels, one a lancet type
2 rigid section lifters, two sizes
2 pairs of forceps, regulation but fine and heavy
1 triangle file
1 Bunsen burner with wire gauze and rubber tubing

Miscellaneous Equipment

½ lb. Permoplast, cream colored, to be used for operating depressions
¼ lb. beeswax, white, clear
¼ lb. soft (low melting point) paraffin
2 China marking pencils, one blue and the other red
1 diamond point or carborundum pencil
2 needle holders, to hold operating needles of various sizes
1 plastic cover for needle holder
2 ft. rubber tubing, 11-12-mm outside diameter
2 ft. rubber tubing, 8-9-mm outside diameter

*This is a complete reference list of equipment which would facilitate experiments with aquatic vertebrate embryos. However, excellent research can be accomplished with a minimum of equipment, accompanied by perserverance and imagination.

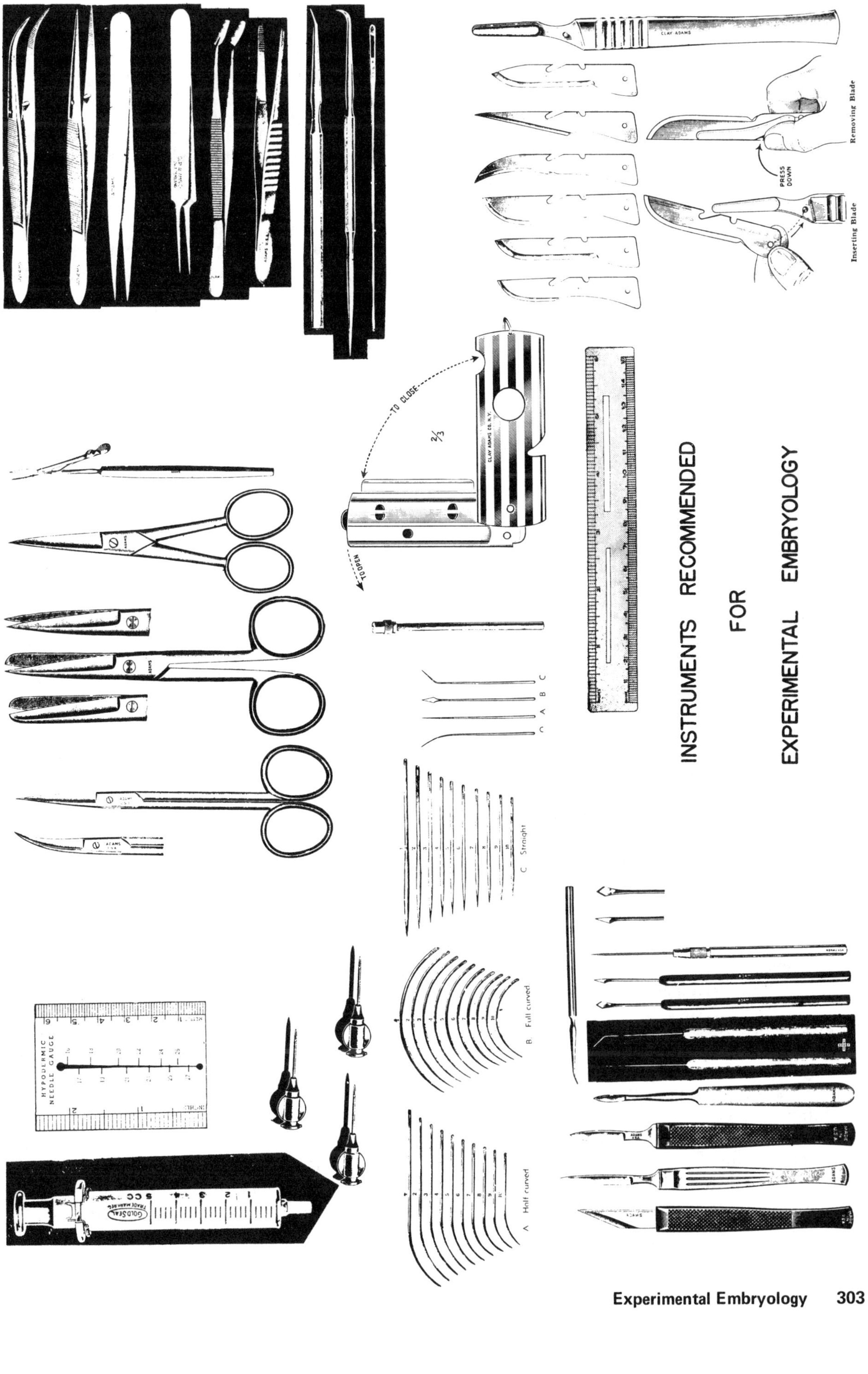

INSTRUMENTS RECOMMENDED FOR EXPERIMENTAL EMBRYOLOGY

The Preparation of Instruments

1. *The Micro burner:* This may be set up for class use. However, its preparation may be described since individuals may prefer to have accessible their own burner. Its function is to give a small but intense flame.

   Secure a 5-inch piece of 7-mm, soft glass tubing, and, in a hot flame, reduce the diameter of one end to about 1 mm. Then, place a right angle in the tube at its center (with a flame tip), and, when it is cool, pass the flamed end through a hole in a large cork. Mount the cork in a burette clamp attached to a ring stand so that the microburner is turned upward. To the other end of the tube attach a rubber tubing joined to the gas outlet. Apply a screw clamp to the rubber tubing. By regulating the screw clamp a microflame of intense heat may be secured. The usual type of microburner, where the tube is drawn out to a thin tip of small bore, is not so likely to remain uniform when in use. Hypodermic needles may be used as microflame tips. (See illustrations following).

2. *Glass needles:* There are two types of handles used for glass needles. The standard type consists of a 7-mm diameter glass rod cut into 10-cm lengths by flaming longer pieces in the center and drawing them out. In this manner one end of the rod is tapered. The tapering end should be brought into the flame so that it retracts to knob for attachment of the needle, made separately. See that the knob and rod are perfectly straight. Prepare 10 such needle handles. The second type of needle holder consists of the regulation steel needle holder with adjustable screw into which various needles may be inserted and fastened. This type is convenient and entirely satisfactory.

   The needles may be made on the electric needle puller, but this is not necessary. Secure some 4-5-mm diameter, soft glass rod, and in a flame tip draw it out perfectly straight until a thickness of about 1 mm is secured. Break this into about 7-8-cm lengths, and in the microflame put a hook or a bend toward the end of each piece. Hang this "hook" over any support (metal rod) attached to the ring stand in such a manner that it is directly above a 20 x 60 mm (or larger) glass vial. In the bottom of the vial place a small amount of cotton. At about 2 cm from the base of the hanging glass rod apply the microflame from the side. It will take practice to apply the correct amount of heat. When the glass at the point of heat application is melted, the weight of the hanging rod will drop it into the cotton in the glass vial, providing a needle point of microscopic dimensions. With practice, it may be possible to do this in two steps, the first heating will lengthen and thin out the rod and the second will draw an even finer point. In any case, the tapering micropoint should not be long and flexible and will probably have to be trimmed with sharp scissors. Draw out many such needles at one time and mount them temporarily in plasticene, i.e., until ready to attach them to holders or to mount them in a needle board. These needle points may be attached to the glass handles (described above) by bringing them together in a small flame. The needles may be attached at a slight angle which will facilitate operations. (See illustrations following.)

3. *Steel needles:* The ordinary steel needles are much too coarse for work with small embryos. However, the finest insect needles may be secured, cut short, and mounted in wooden handles and will be extremely useful.

4. *Hair loops:* This device is for handling embryos or isolated tissues. Draw out the end of some 5-6-mm, soft glass tubing so that the total length is about 10 cm and the smaller end has a diameter of about 1 mm. Close the larger end in a flame. The smaller end should be cut off with a diamond pencil or flamed to make it smooth. Secure some blond hair from a newborn infant and cut it into lengths. With forceps insert one end of the hair into the capillary opening and then the other end into the same opening. Regulate the insertion of the hair so that a relatively small loop protrudes. Melt a small amount of soft paraffin on a glass slide and dip the hair loop into the paraffin, whereupon some of the paraffin will run up into the tube, and harden upon cooling. This will hold the hair loop in place. To remove any paraffin adherent to the hair loop itself, warm a slide in a flame, place on it a small piece of filter paper, and

gently touch the hair loop to the filter paper. Avoid melting the paraffin within the capillary tube.

5. *Glass ball tip:* This instrument is also used for moving embryos about without injury to them. Using a 6-7-mm, soft rod about 18 cm long, draw out the center (with a flame tip) and break off the excess thread down to within 2-3 cm of the widening part of the rod. Hold the pointed end of the rod in a flame and a glass ball will form on the end. It can be kept symmetrical by constant spinning of the rod while heating. If it is planned to put a gentle curve in such a rod, this should be done when the rod is drawn out rather than later. Such ball tips are used also for making depressions in Permoplast, suitable for embryos. It is well, therefore, to have a small assortment of sizes of such ball tips.
6. *Glass pipettes:* Three types of pipettes will be used: wide mouthed, micropipette, and micropipette with lateral control (see diagrams).

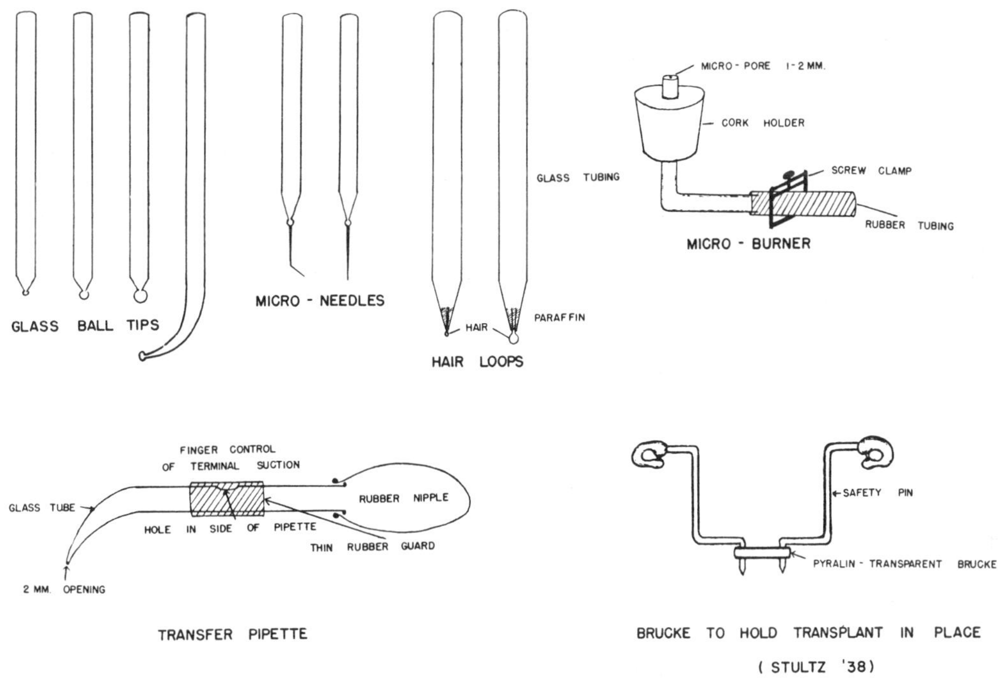

The wide-mouthed pipette is used for transferring embryos. It is simply a wide bore pipette (6-7-mm diameter) with a curved and smooth-edged tip. This can be made by cutting off the tip of a medicine dropper and should be available at all times.

The micropipette is made of soft glass with 7-8 mm outer diameter pulled out to microscopic dimensions in a manner similar to the making of microneedles (above). The tip end of such pipettes are often closed but may be trimmed off. These pipettes may be used with ordinary rubber nipple or may be used with rubber tubing and mouth suction.

The micropipette with lateral control has special use in the transfer of isolated pieces of embryos to regions where transplants are desired. It is made in the following manner. Pull out some 7-8 mm soft glass tubing so that the handle portion has a length of about 10 cm. It is better if the capillary end has a slight curve and tapers rapidly. Close the capillary end by melting it in the Bunsen flame and attach a piece of rubber tubing to the large open end of the pipette. Bring the side of the pipette, at about 3-4 cm above the capillary end, into a small hot flame. If the capillary end is curved, the point of heating should be on the opposite side of the curve. When this point is soft, blow gently into the rubber tubing and the melted glass will bulge outward. The size of the heated area will determine the size of the bulge. Break off the glass bubble down to the wall of the pipette and smooth off the edges in a microburner. With diamond point cut off the closed tip of the capillary end so that the aperture will be about 1mm in diameter. Cut a short piece of thin-walled rubber tubing and slip it (from larger end)

SPEMANN MICROPIPETTE FOR TRANSPLANTATIONS

over the pipette to the point where it covers the lateral hole. Heat the broad end of the pipette until very soft, and press down on metal surface to give a ridge to hold the rubber nipple. Add rubber nipple to this end, and the pipette is ready for use. (See illustration above.)

This latter pipette is of special value in the transfer of small pieces of tissue under solution. With nipple, suck in solution until the capillary end is filled almost to the lateral hole. Remove fingers from nipple and place a forefinger over the rubber-covered lateral hole. With the capillary point under the solution, force out a small amount of the contents of the pipette by gently pressing on the lateral rubber cover with the forefinger. Then, by releasing the gentle pressure, a small amount of fluid, or tissue, may be drawn up into the pipette and held there. This tissue may be oriented at any place by slight pressure on this lateral membrane. The capillary end must hold solution (not drip) even when held vertically with the point suspended. Practice the use of this pipette with small objects.

7. *Glass bridges:* There are small pieces of cover glass, thinnest grade, used to hold transplanted tissues in place for 15+ minutes while they "take" or heal onto the host. Using a diamond point pencil and ruler, cut thin cover glasses into strips about 2-3 mm wide and 5-10 mm long. With forceps run the edges of these glass strips through a microflame to make them smooth. To put a slight curve in some of the pieces of cover glass, grasp the edge of a piece with forceps and bring the center of it over a microburner with a 1-mm flame. The weight of the cold end will bend the coverslip slightly when the center is heated. The height of the bridge should be determined by the size of the embryo to be used as host. Frequently, flat coverslips will prove to be adequate, especially when the host is held in a depression. The bridges should be sterilized in 70 percent alcohol over cotton and kept thus in a covered stender dish.

There is a refinement of this type of bridge (or Brüke) described by Schultze (1938). The cover itself consists of transparent Pyralin of less than 2-mm thickness cut into rectangular blocks of 1/8 inch by 3/8 inch each. Holes are drilled through each end of the block with # 70 or # 71 wire gauge drill. The edges of the block are smoothed off with fine file and then dipped in very weak balsam to increase transparency. Take care that balsam does not fill the holes. Small safety pins (preferably gold plated to prevent rusting) are straightened out and then given three right angle bends as shown in the diagram on page 336. The pins are then forced through the holes, to the extent of 5-6 mm. Such a graft cover may be pressed down onto the graft exactly as desired, the pins anchored in the Permoplast and the Pyralin, thus allowing constant observation of the graft. In most transplantations, such an elaborate Brücke is not necessary.

Operating Procedures

Most operations on aquatic forms will be carried out in Syracuse dishes, salt cellars, or stenders with covers. Since embryonic tissues often adhere to glass, the base of the Syracuse dish is lined with one or another plastic substance. The following are satisfactory:

a. Permoplast – American Art Clay Co., Indianapolis, Indiana.
b. Beeswax to which has been added lampblack, to take glare off background.
c. Rainbow Wax – American Art Clay Co., Indianapolis, Indiana.
d. White refined beeswax.
e. Pieces of cellophane or Pyralin.
f. Agar, 2 percent or more concentrated.

The Permoplast is probably best for delicate operative procedures, as it is easily molded without heating. However, it tends to fragment under solution. It may be necessary to compensate for this by melting it with about 20 percent low-melting -point paraffin.

For frog work, melted beeswax to which enough lampblack has been added to give it a dark gray appearance will prove to be satisfactory. Occasionally, after the dishes are prepared, the wax base will, under water, float off the bottom. To prevent this, place a few pieces of 1-mm glass rod in the bottom of each dish before adding the melted beeswax. These will add sufficient weight to hold the wax in place. Have at least 10 operating dishes available at all times. Grooves to hold embryos may be made with ball tips before the beeswax hardens.

Operating Solutions

The operating medium will vary with the age, the condition of the embryo (or tissue), and the species used. Operations are generally performed in more concentrated solutions, and, after healing, specimens are returned to weaker salt solutions.

For operations on urodeles, the urodele operating medium is used, and the embryos will heal normally if left in these solutions. After the operation wound has healed, the urodele embryo is transferred to urodele growing solution. For the anura, double strength and then normal standard solutions are used. Brief boiling of the solutions to sterilize them may be necessary. If convenient, large volumes could be autoclaved, filtered, and stored as stock solutions. Controlled salt solutions are more satisfactory than spring water or conditioned tap water because there is greater uniformity.

In all cases remember that surface rather than volume is important. The embryos need be just covered in the solution and no more, providing evaporation is reduced to a minimum.

Glassware

Ovulating aquatic (fish, amphibians) animals should be kept in fish bowls or small battery jars, properly covered. If eggs are to be layed in these containers, it will be necessary to provide them with appropriate solutions. The urodeles generally attach their eggs to vegetation.

For fertilization of frog's eggs, fingerbowls or petri dishes are used, and all eggs from a single female may be inseminated in a single container providing they are flooded in 15 minutes and separated to lots of about 25 per fingerbowl before the first cleavage. Development will continue to hatching in fingerbowls, but beyond hatching a different procedure will be necessary.

Operations are perfomed in dishes described above. Following the operation, when the transplant has "taken," the embryo may be transferred to a covered stender, the bottom of which has a thin layer of 4-5 percent agar. This provides a softer base than glass so that there is less danger of injury. Such operated animals should be kept at constant temperature toward the lower levels of tolerance for the particular species under investigation.

Asepsis

The amphibian embryo is relatively immune to infection, or bacterial contamination. Operated embryos are, of course, exposed to such conditions, but the wounds heal so rapidly that simple aseptic precautions are generally sufficient to protect the embryo.

Detwiler, Copenhaven, and Robinson (1947) have demonstrated that sodium sulfadiazine

(0.5%) in any of the various operating or culture media is perfectly harmless to the embryo and will reduce the operative casualties on the central nervous system from 95 percent to 5 percent. The sulfadiazine may partially crystallize out, but these crystals are harmless except as they may mechanically pierce the ectoderm of the embryo.

At later larval stages, or just after metamorphosis, amphibia may become susceptible to fungus or other skin infections. These can be treated locally with dilute mercurochrome, neo-silvol, or very concentrated salt solutions painted onto the affected area. Generally, a copper penny in the same tank will keep down all fungi.

Membranes

The eggs of all amphibians are surrounded by secondary (jelly) membranes secreted by the oviduct in addition to the vitelline membrane. These membranes presumably protect the embryo from bacteria and injury during development. Any operations on such embryos require that such membranes be removed. It must be remembered, therefore, that the unprotected embryo is rather easily injured and that it is more susceptible to physical and chemical changes in the environment, as well as to bacterial infection.

1. URODELA
   The urodele jelly is rather tough and may be peeled off of the eggs with sharp-pointed watchmaker's forceps. The vitelline membrane will be resistant to puncturing so that it may be necessary to use glass microneedles to make the initial break. Urodele egg capsules should be opened in urodele growing solution.
2. ANURA
   The anuran egg is surrounded by loose jelly which is adherent to the vitelline membrane. Attempts have been made to remove this jelly by ultraviolet light or by dilute KCN, but there is no reliable chemical method of jelly removal which allows the embryo to survive. Should someone discover a method of fertilizing body cavity eggs (devoid of jelly) this would be a boon to experimental embryology. The anuran egg jelly can be removed to some extent by rolling the egg on filter paper or paper toweling, pushing it along with the flat side of the scalpel. The danger in this method is, of course, excessive mechanical injury to the egg so that the correct rate and amount of absorption (by filter paper) and the exact amount of rolling will have to be determined empirically. With practice, it is possible to pierce the jelly with one prong of the forceps, slide a prong of a second pair of forceps along the first, and cut through the jelly with an outward movement. Then, the split jelly capsule can be shelled off of the egg. The enzyme hyaluronidase, so effective in denuding the mammalian egg, may prove of similar value with the amphibian egg.

STANDARD LABORATORY EQUIPMENT IN EXPERIMENTAL EMBRYOLOGY

Major Equipment

1. Water table—wood or stone table provided with current of tap water to depth of 1 inch to be used for holding fingerbowls, etc., at fairly constant temperatures.
2. Refrigeration—electric refrigerators (or constant temperature rooms) with heating units installed so that the temperatures may be regulated. Best temperatures are 4° C, 10° C, and 20° C.
3. Incubators—thermostatically controlled boxes which can be regulated at temperatures above the laboratory temperatures. Best temperatures are 25° C, 29° C, 32° C and 35° C. Chick incubators regulated at 103° F for few eggs (Oaks, Chicago) or for several hundred eggs (Buffalo) should be on hand.
4. Centrifuge—heavy electric centrifuge with large capacity as well has smaller electric and hand centrifuges.

5. Animal cage—a wooden frame with slatted or no floor, screened sides and top, should be made to fit into the water table if the table is large enough. The cage may be divided into compartments with the doors opening on top. If the cage is to sit in water, the wood should be treated with some waterproofing. Frogs and salamanders may be kept in such a cage on the water table in healthy condition if they are provided with cool running water.

6. Aquaria—the sizes and shapes of aquaria depend upon the particular animals concerned. For Urodele and Anuran tadpoles (larvae), the low, flat tanks with considerable surface are best. Ideal dimensions are the 12 x 12 x 4-inch enamel (restaurant) pans.

   For fish, four sizes are desirable:
   —for fry — ½-gallon battery jars or dimestore aquaria.
   —for tropicals and breeding pairs — 16¼ x 8 1/5 x 8½ inches wide is best (5 gallons).
   —for larger fish, or large groups of nonbelligerent fish — 20¼ x 10½ x 12¼ inches high (about 10-gallon capacity).
   —for zebra fish which are very active — 24½ x 6½ x 6 inches wide.

   All fish tanks should have slate bottoms and glass covers. One corner of the glass cover may be cut off to facilitate removal for feedings.

   The larger tanks may be subdivided by cutting a piece of glass to fit and then covering the edges of the glass with split rubber tubing and inserting the glass into the tank. The rubber tubing will hold the plate in position and at the same time block the passage of small fry from one compartment to the other.

7. Microtome—Spencer rotary probably the best.

8. Embedding ovens—Columbia probably the best.

9. Slide warmer—Chicago Apparatus Company, thermostatically controlled electric warmer.

10. Balances—Coarse (200 + gram capacity) and sensitive types.

Glassware

1. Battery jars or aquaria for individual use of students; each with weighted wire screen cover.
2. Crystalizing dishes or 10-inch finger bowls, with covers.
3. Fingerbowls—regulation size, made to fit into each other.
4. Petri dishes (10 cm) with covers.
5. Stender dishes with covers, both #1 and #2.
6. Syracuse dishes—regulation size.
7. Coplin staining jars, homeopathic and shell vials.
8. Salt cellars for embedding oven.
9. Erlenmeyer flasks—500-cc capacity for storing sterile media.
10. Lantern slide covers—used as glass base for operating and for protecting binocular and compound microscope stages when using wet mounts.
11. Graduates—10 cc and 100 cc.
12. Beakers—100 cc and 600 cc.
13. Slides: regulation microscope slides.
    depression slides, cell type.
14. Coverslips—glass. Best size, 7/8-inch square and #1.
15. Glass rods—diameters from 4 to 7 mm. Soft glass.
16. Glass tubing—diameters from 4 to 7 mm. Soft glass.

Solutions and Reagents

1. Distilled water—glass distilled preferred. Supply in large carboys with siphon and pinch clamp.

2. Spring water—if possible supply in carboys with siphon and pinch clamp. Great Bear Spring Water (New York City) is excellent.

3. Conditioned tap water—to be stored in carboys after conditioning. This will be necessary where the city water supply is treated so that embryos cannot survive in it. This may be tested with sperm or early embryos. Conditioning is achieved by running tap water into large tank in which there will be maximum of surface exposed and in which plant material is abundant. Artificial aeration will help to eliminate chlorine. Passing tap water through fine gravel and charcoal is rarely necessary but would aid in conditioning highly toxic water. Three or four days of such conditioning should be sufficient.

4. Standard (Holtfreter's) solution—this should be available to the students in several concentrations and large volumes. For the convenience of the instructor, the dry salts may be made up in appropriate concentrations and stored in vials to be added to carboys of glass-distilled water when needed. The concentrations needed are 200 percent, 100 percent, and 20 percent. The formula for Holtfreter's solution is:

NaCl	0.35 g
KCl	0.005 g
$CaCl_2$	0.01 g
$NaHCO_3$	0.02 g
Water.	100.00 cc (glass distilled preferred)

5. Amphibian Ringer's—should be available in concentrated form:

NaCl	0.66 g
KCl	0.015 g
$CaCl_2$	0.015 g
$NaHCO_3$	to pH 7.8
Water.	100.0 cc

6. Urodele stock solution:

Great Bear Spring Water	10 l
NaCl	70 g
KCl	1 g
$CaCl_2$	2 g

Urodele operating medium: (hypertonic)

Stock solution.	2 parts
Great Bear Spring Water	1 part

Urodele growing medium: (isotonic)

Stock solution.	2 parts
Great Bear Spring Water	4 parts

Steinbert medium:—solution made up complete and autoclaved. Essential to use glass -distilled water and can be used for raising whole embryos. (pH 7.4)

20 cc of 17 percent NaCl
10 cc of 0.5 percent HCl
10 cc of 0.8 percent $Ca(NO_3).4H_2O$
10 cc of 2.05 percent $MgSO_4.7H_2O$
4 cc of 1 percent N. HCl
560 mg of Tris*
946 cc glass-distilled water

* Buffer available from Sigma Chemical Company, St. Louis.

The Niu-Twitty solution—Three solutions made separately in hot water, cooled, and then mixed. Good for urodele embryos and isolates.

Solution A	Solution B	Solution C
3,400 mg NaCl	110 mg Na_2HPO_4	200 mg $NaHCO_3$
50 mg KCl	20 mg KH_2PO_4	250 cc distilled water
80 mg $Ca(NO_3).H_2O$	250 cc distilled water	
100 mg $MgSO_4$		
500 cc distilled water		

7. Locke's solution—for chick embryos:

NaCl	0.9	g
KCl	0.04	g
$CaCl_2$	0.024	g (anhydrous)
$NaHCO_3$	0.02	g
Water	100.0	cc

8. Special variations of the above solutions:

a. Calcium-free standard (Holtfreter's) solution.
b. Sodium-free standard (Holtfreter's) solution.
c. Potassium-free standard (Holtfreter's) solution.
d. Buffer-free standard (Holtfreter's) solution.
e. Nuclear medium for germinal vesicle studies (calcium-free Ringer's).

NaCl	0.6	g
KCl	0.01	g
Glass-distilled water	100.00	cc

9. Anesthetics:

In general, anesthetics are not necessary when eggs or embryos are to be fixed immediately. Occasionally, it is desirable to fix an embryo in a certain position or to reduce body movements during fixation and then anesthesia is in order.

a. MS - 222: This imported poison is the most satisfactory anesthetic available, used in 1/3000 concentration either in standard (Holtfreter's) solution, spring water, or Locke's solution (for chick embryos). The embryos are immobilized in about 1 minute and, after return to normal medium, recover in about 10 minutes without ill effects. Must be used fresh. (Sandoz Chemical Company, 63 Van Dam Street, New York, N. Y.

b. Chloretone: Generally, 0.5 percent concentration in whatever medium the embryo is accustomed will give slow but quite satisfactory anesthesia.

c. Magnesium sulphate (epsom salts): Simply drop a few crystals into small volume of water containing the embryo and await immobilization.

d. Cyanide: KCN 1/1000 in salt solutions acts as an anesthetic. Must avoid anoxia.

e. Ether and chloroform: These volatile anesthetics are for air-breathing forms and so are of little use with embryonic material.

f. Chilling: Embryos are rapidly retarded in all of their activities by adding to their media some cracked ice. Such embryos may be operated upon and will recover upon return to normal temperatures, without ill effects. It may take 10-15 minutes to adequately stupefy the organisms.

10. Killing and fixing solutions:

a. Smith's fixative—best for yolk-laden amphibian eggs. Two solutions to be mixed just before use. Not to be used if discolored.

Solution A:	K Bichromate .	0.5	g
	Water.	87.5	cc
Solution B:	Formalin	10	cc
	Acetic (glacial).	2.5	cc

b. Bouin's fixative—safest of all fixatives, particularly for late stages.

Saturated (aqueous) picric acid .	75	cc
Formalin (commerical)	25	cc
Glacial acetic acid	5	cc

(Can use 95 percent instead of water for more rapid fixation.)

c. Bouin - Dioxan—excellent for yolk embryos and rapid technique. Use Bouin's and Dioxan in equal parts (Caution: Dioxan is volatile and toxic).

d. Michaelis' fluid—for yolk embryos.

Conc. $HgCl_2$, aqueous	20	parts
Conc. picric, aqueous	20	parts
Glacial acetic acid	1	part
Water distilled	40	parts

e. Kleinberg's picro-sulphuric—for polar bodies and spindles (Mc Clung).

Water.	200	volumes
Conc. H_2SO_4.	2	volumes
Picric acid—to saturation		

f. Chrom-acetic fixative—excellent for cytological studies of amphibian egg and embryos.

Chromic acid 10 percent . .	25	parts
Glacial acetic acid	10	parts
Saturated aqueons picric acid	100	parts

g. Gilson's fixative:

Nitric acid—80 percent . . .	15	cc
Glacial acetic acid	4	cc
Corrosive sublimate.	20	g
Alcohol 80 percent	100	cc

h. Gatenby's fluid:

K. bichromate—2 percent aqueous	100	cc
chromic acid 1 percent	100	cc
Nitric acid	6	cc

i. Stockard's solution—for fish embryos.

Formalin	5	parts
Clacial acetic	4	parts
Glycerine.	6	parts
Water.	85	parts

j. Acetic alcohol:

Absolute alcohol	90	cc
Glacial acetic.	10	cc

k. Corrosive acetic:

$HgCl_2$	5 g
Glacial acetic	10 cc
Water	90 cc

l. Formalin—two concentrations should be available, 10 percent and 4 percent.

11. Bleaching Agents:

a. Javelle water—potassium hypochlorite.
b. Mayer's chlorine—to be made up just before use.
   (1) Place few crystals potassium chlorate in vial.
   (2) Add 2-3 drops HCl.
   (3) When green chlorine fumes evolve, add 2-10 cc of 70 percent alcohol.
   (4) Transfer specimens from pure 70 percent alcohol to this mixture until bleached.
c. Ammoniated alcohol—2 percent NH_4OH in 70 percent alcohol to decolorize picric acid stain.

Cytoplasmic Stains

a. Eosin—0.5 percent in 95 percent alcohol.
b. Light green (Grubler's)—0.5 percent in 95 percent alcohol.
c. Fast green (National Aniline Company, N. G. f. 3) 0.5 percent in 95 percent alcohol.
d. Safranin O (National Aniline Company, N. S. – 10) 1 percent in aniline water.
e. Orange G—salt water in clove oil.
f. Masson stains (A, B, C.)—Excellent for cell types (pituitary).
g. Alizarine S—(used in Spaltoholz technique).

Vital Dyes*

a. Nile blue sulphate—1/200,000.
b. Methylene blue—0.5 percent.
c. Neutral red—1 percent.
d. Bismark brown—1 percent.
e. Janus green—1 percent

Nuclear Stains

a. Iron haematoxylin.
b. Feulgen, nucleal reaction.
c. Harris acid haemalum.

12. Miscellaneous reagents:

a. Iodine—saturated iodine in 70 percent (to follow Zenker's fixation).
b. Ammoniated alcohol—5 percent NH_4OH in 70 percent alcohol for removal of picric.
c. Glacial acetic acid—and normal acetic (60 cc glacial/liter).
d. Hydrochloric acid—concentrated, 1 percent and Normal.
e. Sodium hydroxide—concentrated, 1 percent Normal and 0.003N.
f. Ammonium hydroxide—concentrated.
g. Potassium hydroxide—1 percent.
h. Hydrogen peroxide—concentrated (can be used as bleaching agent).
i. Acetone.
j. Glycerine.
k. Cleaning fluid (K. bichromate saturated in sulphuric acid).
l. Clearing agents:
   (1) Xylol (toluene, toluol, benzene (benzol).

* These are concentrated stock solutions.

(2) Cedar oil.
(3) Aniline oil.
(4) Oil of wintergreen (methyl salicylate).
(5) Oil of cloves.
(6) Dioxan (volatile and the fumes are toxic).
(7) Chloroform.

m. Mounting media:
(1) Canada balsam—dissolved in xylol.
(2) Gum damar.
(3) Clarite.
(4) Egg albumen—best as albumen water (3 percent).

n. Embedding media:
(1) Paraffin: M. P. range from 45°C to 58°C.
(2) Beeswax: pure white.
(3) Bayberry wax. (Candle Factory, Falmouth, Massachusetts).
(4) Rubber—white rubber to be added to paraffin.

o. Vaseline.

p. pH indicators (LaMotte sets for entire range).

Miscellaneous Equipment

a. Operating base
(1) Permoplast—American Art Clay Company, Indianapolis, Indiana.
(2) Rainbow wax—American Art Clay Company, Indianapolis, Indiana.
(3) Beeswax with lampblack to give proper background shade.

b. Lampblack.

c. Agar.

d. Cellophane, pliofilm, or pyralin.

Removal of Jelly Capsules (Fish and Amphibians)

It is easier and allows better fixation if the jelly membranes are removed from eggs and embryos prior to the killing process.

The urodele egg is provided with a distinct jelly capsule which may be punctured with needles or sharp watchmaker's forceps, and pulled off of the egg. If a tear is made by means of a pair of forceps, the embryo will usually "shell out." If the capsule is placed on a piece of paper towelling, filter, or blotting paper, to which it will adhere, this operation may be facilitated.

The anuran egg generally has looser but more adherent jelly capsules. This jelly may be removed by cutting single eggs away from the mass and placing them on coarse paper and rolling them along with the flat side of a scalpel until the bulk of the jelly rolls off onto the paper. A better method is to pierce the jelly with one prong of the #5 watchmaker's forceps, to slide a prong of a second pair of forceps along the first, and, with an outward motion, to cut through the jelly. It may then be peeled off.

Remove frog egg jelly with 10 percent chlorox bleach or by brief trypsinization.

It has been reported that the ultraviolet light will dissolve off the jelly capsules of eggs, but it is very likely that the same irradiation will damage the egg or embryo. Chemical removal of the jelly, after fixation, may be achieved by placing the embryo and capsule into 10 percent sodium hypochlorite or chlorox diluted with 5 or 6 volumes of water. The jelly can be shaken off within a few minutes. Javelle water (potassium hypochlorite) diluted three or four times may also be used. Following fixation in Gilson's fluid, the jelly hardens sufficiently so that it may be picked off the embryo which is subsequently hardened in alcohol.

GLOSSARY

ACROSOME – apical organ at tip of mature spermatozoon, derived from the idiosome, and presumably functional in aiding penetration of the egg cortex by the spermatozoon during fertilization. Syn., perforatorium.

ACTIVATION – process of initiating development in the egg, normally achieved by the spermatozoon of the same species but also accomplished artificially (parthenogenesis); term also used to refer to stimulation of spermatozoon to accelerated activity by chemical (fertilizin) means.

ADNEXA – extraembryonic structures (e.g., yolk sac) discarded before the adult condition is attained.

AESTIVATION – reduced activity of some animals during the heat of the summer, opposed to term "hibernation."

AFTERBIRTH – extraembryonic membranes which are delivered after the emergence of the mammalian fetus. It consists of placenta, chorion, amnion and yolk sac.

AGGLUTINATION – cluster formation; a spontaneously reversible reaction of spermatozoa to certain chemical situations (e.g., the fertilizin in egg water).

AGGREGATION – the coming together of cells, such as spermatozoa, without sticking; a nonreversible response comparable to chemotropism.

ALAR PLATE – dorsolateral wall of the myelencephalon, separated from the basal plate by the sulcus limitans.

ALBUGINEA OF TESTIS – the stroma of the primitive testis which forms a layer between the germinal epithelium and the seminiferous tubules.

ALBUMEN – a protein substance secreted by the walls of the oviducts around the egg of reptiles and birds.

ALBUMEN SAC – a two-layered ectodermal sac which encloses the albumen of the chick egg during early development of the embryo, separated for a time from the yolk by the vitelline membrane; later to release some of its contents into the amniotic cavity through the ruptured seroamniotic connection.

ALECHITHAL EGG – (see EGG).

ALLANTOIN – a nitrogenous component of the allantoic fluid.

ALLANTOIS – saclike organ of respiration and excretion found in embryonic reptiles, birds, and mammals, consisting of splanchnic mesoderm and evaginated hindgut endoderm.

ALLELOMORPH – one of a pair of genes, at the same chromosome locus, the two expressing dominance or recessiveness with respect to a particular character.

AMELOBLASTS – cells which secrete the enamel cap of the (mammalian) tooth.

AMITOSIS – direct nuclear division without chromosomal rearrangements; generally thought to be a sign of decadence or of high specialization, if it occurs at all.

AMNIOCARDIAC VESICLES – paired primordia of the coelomic cavity which appear in the mesoblast lateral to the head fold of the 24-hour chick embryo and grow beneath the foregut to give rise to the pericardial cavity. Named because of embryonic relation to both amnion and pericardium.

AMNION – thin double membrane enclosing the embryos of some invertebrates and of reptiles, birds, and mammals. In the vertebrate groups, it is derived from the somatopleure. It is filled with a fluid which functions as a protection to the fetus.

AMNIOTIC RAPHE – region of junction of the amniotic folds as they encircle the embryo. Syn., seroamniotic or chorio-amniotic junction of raphe; amniotic umbilicus.

AMPHIBLASTULA – double-structured blastula as in the porifera (sponges).

AMPHIMIXIS – mixing of germinal substances accomplished during fertilization.

AMPHITENE – one end of the chromosome is thick, one end thin, moving toward full pachytene during maturation.

AMPLEXUS – the sexual embrace of the female amphibian by the male, a process which may (frogs and toads) or may not (urodeles) occur at the time of oviposition.

ANALOGY – similarity of parts in respect to function rather than to structure.

ANAL PLATE – a thickening and invagination of midventral ectoderm just posterior to the primitive streak at the 14-hour stage (chick) which meets evaginating endoderm of the hindgut, later to be perforated as the anus. Syn., cloacal membrane.

ANAMNIOTA – forms which never develop an amnion, e.g., cyclostomes, fish, amphibia.

ANAPHASE – phase of mitosis when the paired chromosomes separate at the equatorial plate and begin to move toward the ends of the spindle.

ANASTOMOSIS – joining together as of blood vessels and nerves, generally forming a network.

ANDROGENESIS – development of an egg with paternal (sperm) chromosomes only, accomplished by removing or destroying the egg nucleus after fertilization but before syngamy.

ANESTRUS – the quiescent period which follows estrus in the reproductive cycle of the female mammal.

ANGIOBLAST – the migratory mesenchyme cell associated with formation of vascular endothelium. The primordial mesenchymal tissue from which embryonic blood vessels and vascular epithelium are differentiated.

ANIMAL POLE – the region of the egg where the polar bodies are formed; region of telolecithal egg containing the nucleus and bulk of cytoplasm. Gives rise largely to ectodermal derivatives. Syn., apical pole or hemisphere.

ANLAGE – German term for primordium, now in disuse (see PRIMORDIUM.)

ANTERIOR – toward the head; head end. Syn., cephalic, cranial, rostral.

ANTERIOR INTESTINAL PORTAL – the entrance from the yolk sac into the foregut.

ANURA – tailless amphibia (e.g., frogs and toads). Syn., salientia.

ANUS – posterior opening of the digestive tract, region of original proctodeum.

AORTIC ARCH – blood vessel which connects the dorsal and ventral aortae by way of the visceral arch.

AORTIC BULB – embryonic bulge of truncus arteriosus (pig) where it turns toward the midline to give rise to the aortic arches.

AQUEDUCT OF SYLVIUS – ventricle of the mesencephalon (mesocoel) becomes the aqueduct of Sylvius, connecting third and fourth ventricles. Syn., iter, cerebral aqueduct.

AQUEOUS HUMOR – fluid which fills the anterior and posterior chambers of the eye, probably derived from mesoderm.

ARCHENCEPHALON – prechordal brain, e.g., forebrain. Brain anterior to the anterior end of the notochord.

ARCHENTERIC POUCH – (see ENTEROCOEL).

ARCHENTERON – primitive gut found in the gastrula and communicating with the outside by the blastopore; precursor of the embryonic gut. Syn., gastrocoel, enteron.

ARCUALIA – small blocks of sclerotomal connective tissue involved in the formation of vertebral arches.

AREA OPACA – marginal opaque ring of the chick blastoderm which surrounds the area pellucida; less transparent because of the underlying adherent yolk.

AREA PELLUCIDA – the central portion of the chick blastoderm, originally represented by the central cells, where the blastoderm is separated by a cavity from the underlying yolk and, therefore, appears to be more transparent.

AREA VASCULOSA – the three-layered portion of the area opaca in which blood islands develop, appearing first posterior to the embryo and progressing forward, opposite the sides of the embryo, following the expansion of the mesoblast.

AREA VITELLINA – two-layered peripheral portion of the area opaca, itself subdivided into area vitellina interna and area vitellina externa, the latter including the zone of junction.

ASEXUAL REPRODUCTION – reproduction without union of gametes; generally no maturation divisions.
ASTER – aggregate of lines radiating from centrosome during animal mitosis.
ASTRAL RAYS – lines which make up the aster.
ASTROCYTES – stellate-shaped cells arising from the spongioblasts of the mantle layer of the brain classified under the more general term of neuroglia.
ATTACHMENT POINT – point of chromosome to which the spindle fiber is attached and, therefore, the portion of the chromosome nearest the centrosome in anaphase. Syn., centromere, heterogenetic chromatin mass, kinetochore.
ATTRACTION SPHERE – (see CENTROSPHERE).
AUTOGAMY – self-fertilization.
AUTOSOME – any chromosome except the so-called sex (X or Y) chromosomes.
AUXOCYTE – premeiotic germ cell. Syn., primary cyte, meiocyte.
AXIAL FILAMENT – the central fiber in the tail of a spermatozoon.
AXIAL MESODERM – that portion of the epimeric mesoderm nearest the notochord. Syn., vertebral plate.
AXIS – central or median (imaginary) line, generally correlated with a gradient.
AXIS OF THE CELL – a line (imaginary) passing through the centrosome and the nucleus of a cell, generally through the geometrical center of the cell also. In an egg, such an axis generally is also the gradient axis of materials, such as cytoplasm, yolk, pigment, etc.
AXIS OF THE EMBRYO – line (imaginary) representing the anteroposterior axis of the future embryo.

BALANCERS – cylindrical and paired projections of ectoderm with mesenchymatous cores, used as tactile and balancing organs in the place of (anuran) suckers. Found in many urodele amphibia.
BALFOUR'S LAW – "The velocity of segmentation in any part of the ovum is, roughly speaking, proportional to the concentration of the protoplasm there, and the size of the segments is inversely proportional to the concentration of the protoplasm."
BASAL PLATE – ventrolateral wall of myelencephalon, separated from dorsolateral alar plate by the sulcus limitans.
BIDDER'S ORGAN – anterior portion of the anuran progonad, somewhat ovarian in character, developing from part of the gonad rudiment consisting wholly of cortex; its development indicates failure of the medullary substance to diffuse to the anterior extremity of the gonad rudiment.
BIOGENETIC LAW – Individuals in their embryonic development pass through stages similar in general structural plan to the stages their species passed through in their evolution; i.e., ontogeny is an abbreviated recapitulation of phylogeny (also called the recapitulation theory).
BLASTEMA – during regeneration, an indifferent group of cells about to be organized into definite tissue, kept together by intercellular matrix of the constituent cells; primitive embryonic, relatively undifferentiated regenerating cell masses.
BLASTOCOEL – the cavity of the blastula. Syn., segmentation or subgerminal cavity.
BLASTOCYST – mammalian blastula, containing large blastocoel. Syn., blastodermic vesicle.
BLASTODERM – the living portion of the egg from which both the embryo and all of its membranes are derived. The cellular blastodisc.
BLASTODERMIC VESICLE – (see BLASTOCYST).
BLASTODISC – the disc of protoplasm within which cleavage occurs in eggs of fishes, reptiles, or birds. Found in telolecithal eggs having meroblastic or discoidal cleavage. Syn., germinal disc.
BLASTOMERE – a cellular unit of the developing egg or early embryo prior to the time of gastrulation. Smaller blastomeres are micromeres; intermediate are mesomeres; larger are macromeres, where there is great disparity in size.
BLASTOPORE – the opening of the archenteron (gastrocoel) to the exterior, occluded by the yolk plug in amphibian embryos; consisting of a slitlike space between the elevated margin of the blastoderm and the underlying yolk of the chick egg; represented in the amniota as the primitive streak.

BLASTOPORE, DORSAL LIP OF – the region of first involution of cells in the amphibian gastrula; general area of the "organizer"; original gray crescent area; the cells which turn in beneath the potential central nervous system (Amphioxus) and form the roof of the archenteron. Syn., germ ring.

BLASTOPORE, VENTRAL LIP OF – region of blastopore opposite the dorsal lip; region which gives rise to ectoderm overlying the central nervous system of Amphioxus or to the peristomial mesoderm of the frog. Syn., germ ring.

BLASTULA – the stage in embryonic development between the appearance of distinct blastomeres and the end of cleavage (i.e., the beginning of gastrulation); a stage generally possessing a primary embryonic cavity or blastocoel; invariably one cell layer (see specific types under specific names).

BLOOD ISLANDS – prevascular groups of mesodermal cells found in the splanchnopleure, from which will arise the blood vessels and corpuscles.

BOWMAN'S CAPSULE – double-walled glomerular cup associated with uriniferous tubule.

BRANCHIAL – having to do with aquatic respiration. Syn., gill.

BRANCHIAL ARCHES – the visceral arches, beginning with the third pair, containing blood vessels which (phylogenetically) have respiratory function during embryonic development. The mesodermal components which support those blood vessels are the branchial arches. Syn., gill arch (see VISCERAL ARCHES).

BRANCHIAL ARTERY – the blood vessel which actually passes through the gills (external or internal) of the frog embryo. Syn., gill artery.

BRANCHIAL CHAMBER – a closed chamber (except for a single spiracular opening on the left side) which encloses the internal gills of the frog embryo. Syn., opercular or gill chamber.

BRANCHIAL CLEFT – an opening between branchial arches formed by invaginating head ectoderm and evaginating pharyngeal endoderm (pouch) through which water passes from the pharynx to the outside of the frog. Syn., gill cleft or slit, some visceral clefts.

BRANCHIAL GROOVE – ectodermal invagination anterior or posterior to visceral arch which joins branchial pouch to form branchial cleft, in most instances.

BRANCHIOMERY – a type of metamerism exemplified in the visceral arches.

BUD – an undeveloped branch, generally a primordium of the appendage (e.g., limb or wing bud).

BUDDING – reproductive process by which a small secondary part is produced from the parent organism and gradually grows to independence.

BULBUS ARTERIOSUS – the most anterior division of the early tubular embryonic heart which leads from the ventricle to the truncus arteriosus.

BURSA FABRICII – endodermal cavity derived from the posterior portion of the embryonic cloaca of the chick and communicating with the dorsal part of the proctodeum at the level of the urodele membrane.

CANALS OF GARTNER – remnants of mesonephric ducts in the broad ligament close to the uterus and vagina.

CARDINAL VEINS – anterior, posterior, common, and subcardinal veins; anterior veins receive blood from the head, including the first three segmental veins; posterior receive blood from all 29 pairs of trunk segmental veins of the chick and from veins of the Wolffian bodies; paired subcardinals enlarge, fuse, left half degenerates, and the balance fuses with the developing inferior (posterior) vena cava.

CARUNCLE – raised area in the uterus which determines the position of the maternal cotyledons.

CAVAL FOLD – the inferior vena cava develops in the lateral of the two plicae mesogastrica (at 96 hour in the chick), which is known as the caval fold.

CELL LINEAGE – the study of the origin and fate of specific blastomeres in embryonic development. Syn., cytogeny.

CELL THEORY – the body of any living organism is composed of structural and functional units, the primary agents of organization called cells. Each cell consists of a nucleus and its sphere of influence, consisting of cytoplasm, generally circumscribed by a membrane. "Omnis cellula e cellula."

CELLULATION – development of cytoplasmic areas around normal (syncytial) nuclei or by nuclei migrating from living blastomeres as in the chick blastoderm.

CEMENTOBLASTS – cement forming cells of (pig) tooth.
CENTRAL CANAL – (see NEUROCOEL).
CENTRAL CELLS – cells of the chick blastoderm which are bounded on all sides in surface view. May be connected with each other below by contiguous syncytial protoplasm. Surrounded by marginal periblast.
CENTRAL PERIBLAST – the protoplasmic layer beneath the segmentation cavity of the early fish blastula.
CENTRIOLE – the granular core of the centrosome.
CENTROSOME – the granule (centriole) and the surrounding sphere of rays (centrosphere) which function as kinetic centers in mitosis. The center of the aster which does not disappear when the astral rays disappear.
CENTROSPHERE – the rayed portion of the centrosome; the structure in the spermatid which gives rise to the acrosome. Syn., spermatosphere, idiozome; also attraction sphere.
CEPHALIC FLEXURES – ventral bending of the embryonic head at the levels of the midbrain and hindbrain.
CEREBRAL FLEXURES – flexures of the head region, including the cranial, pontine, and the cervical flexures.
CEREBRAL PEDUNCLES – longitudinal tracts in the floor of the mesencephalon.
CERVICAL FLEXURE – extensive transverse flexure of the chick embryo occurring at the level of the junction of the head and trunk about the end of the second day.
CERVICAL SINUS – depression between the third and fourth bronchial clefts (pig) and the body.
CHALAZA – twisted opalescent chord of heavy albumen immediately surrounding the vitelline membrane of the chick egg which helps to hold the yolk in position.
CHIMERA – compound embryo generally derived by grafting major portion of two embryos, usually of different species; may be derived by abnormal chromosomal distributions in cleavage after normal fertilization; exchange of parts too great to be called a transplant.
CHOANA – the opening of the internal nares into the oral cavity, primitively in the anterior part of the mouth but secondarily moved posteriorly by growth of the palatine process. Two choanae of chick never fuse.
CHONDRIFICATION – the process of forming cartilage by the secretion of an homogeneous matrix between the more primitive mesoderm cells.
CHONDRIN – a chemical substance in cartilage matrix which makes it increasingly susceptible to basic stains.
CHONDROCRANIUM – that portion of the skull which is originally cartilaginous.
CHORDA DORSALIS – (see NOTOCHORD).
CHORDA-MESODERM – region of the late (amphibian) blastula, arising from the grey crescent area, which will give rise to the notochord and mesoderm and will, if transplanted, induce the formation of secondary medullary folds.
CHORIO-ALLANTOIC GRAFT – the transplantation of parts of embryos to the rich vascular field of the chorio-allantois.
CHORIO-ALLANTOIS – a common membrane formed by the fusion of the inner wall of the chorion and the outer wall of the allantois (both mesoderm) in the chick embryo.
CHORION – an extraembryonic membrane of the chick embryo which develops from the somatopleure as a corollary of the amnion and which encloses both the amnion and the allantois; consists of inner mesoderm and outer ectoderm. Syn., serosa, false amnion. The term also used to designate the tough membrane deposited by cells of the ovarian follicle on fish eggs.
CHORION FRONDOSUM – that portion of the mammalian chorion which forms the placenta and adheres to the decidua basalis. Its villi are long, branched, and profuse.
CHORION LAEVE – portion of the mammalian chorion, except the chorion frondosum.
CHOROID COAT – a mesenchymatous and sometimes pigmented coat within the sclerotic coat of the eye but surrounding the (pigmented layer of the) eye in vertebrate embryos.
CHOROID FISSURE – a ventral groove in the optic cup and stalk the lips of which later close around blood vessels and nerves that enter the eyeball.

CHOROID KNOT – a thickened region of the fused lips of the choroid fissure, near the pupil, from which arise the choroid cells of the iris.
CHOROID PLEXUS – vascular folds of the thin roof of the telencephalon, diencephalon, and myelencephalon into their respective cavities.
CHROMAFFIN TISSUE – tissue of the developing medullary part of the adrenal gland which exhibits characteristic reactions with chromic acid salts.
CHROMATID – one of the parts of a tetrad; really a longitudinal half of a chromosome.
CHROMATIN – deeply staining substance of the nuclear network and the chromosomes, consisting of nucleoprotein; gives Feulgen reaction for DNA and stains with basic dyes.
CHROMATOPHORE – pigment-bearing cell frequently capable of changing size, shape, and color; cells responsible for superficial color changes in animals; behavior under control of the sympathetic nervous system and/or the neurohumors.
CHROMIDIA – granules within the cytoplasm which stain like chromatin and which may actually be extruded chromatin granules.
CHROMOMERE – unit of chromosome recognized as a chromatin granule.
CHROMONEMA – slender thread of chromatin which is the core of the chromosome during mitosis, a purely descriptive term without functional implications.
CHROMOPHOBE – cells whose constituents are nonstainable (have no affinity for dyes).
CHROMOSOMES – the chromatic or deeply staining bodies probably derived from the nuclear network and containing a matrix and one or more chromonemata during the process of mitosis; bodies found in all somatic cells of the normal organism in a number characteristic of the species; bearers of the genes.
CICATRIX – the thin nonvascular side of the ovarian follicle where rupture will occur to allow the extrusion of the egg into the body cavity in vertebrates. Syn., stigma.
CILIARY PROCESS – supporting and contractile elements of the iris which originate from the lenticular zone on the eighth day in the chick embryo.
CIRCLE OF WILLIS – an arterial circle formed by anastomoses between the internal carotids and the basilar artery; surrounds the pituitary gland.
CIRCULATORY ARCS – intraembryonic; vitelline; allantoic circulatory channels, each involving afferent and efferent tracts and interpolated capillary bed.
CLEAVAGE – the mitotic division of the egg resulting in blastomeres. Syn., segmentation.
CLEAVAGE, ACCESSORY – cleavage in peripheral or deeper portions of the (chick) germinal disc caused by supernumerary sperm-nuclei following (normal) polyspermy.
CLEAVAGE, ASYMMETRICAL – extremely unequal divisions of the egg as in the Ctenophora.
CLEAVAGE, BILATERAL – cleavage in which the egg substances are distributed symmetrically with respect to the median plane of the future embryo.
CLEAVAGE, DETERMINATE – cleavage in which certain parts of the future embryo may be circumscribed in certain specific (early) blastomeres; cleavage which produces blastomeres that are not qualitatively equipotential; i.e., when such blastomeres are isolated, they will not give rise to entire embryos. Syn., mosaic development.
CLEAVAGE, DEXIOTROPIC – cleavage resulting in a right-handed production of daughter blastomere(s), as in spiral cleavage.
CLEAVAGE, DISCOIDAL – (see CLEAVAGE, MEROBLASTIC).
CLEAVAGE, EQUATORIAL – cleavage at right angles to the egg axis, opposed to vertical or meridional; often the typical third cleavage plane. Syn., latitudinal or horizontal.
CLEAVAGE, HOLOBLASTIC – complete division of the egg into blastomeres, generally equal in size, although not necessarily so (e.g., Amphioxus). Syn., total cleavage.
CLEAVAGE, HORIZONTAL – (see CLEAVAGE, EQUATORIAL).
CLEAVAGE, INDETERMINATE – cleavage resulting in qualitatively equipotential blastomeres in the early stages of development; when such blastomeres are isolated from each other, they give rise to complete embryos. Opposed to mosaic development. Syn., regulatory development.
CLEAVAGE, LATITUDINAL – (see CLEAVAGE, EQUATORIAL).
CLEAVAGE LAWS – (see specific laws under names of Balfour, Hertwig, and Sachs).

CLEAVAGE, LEVOTROPIC – cleavage resulting in left-handed or counterclockwise production of daughter blastomere(s), as in some cases of spiral cleavage.

CLEAVAGE, MERIDIONAL – cleavage along the egg axis, opposed to equatorial; generally the first two cleavages of any egg. Syn., vertical.

CLEAVAGE, MEROBLASTIC – cleavage restricted to peripherally located protoplasm, as in the chick egg. Syn., discoidal cleavage.

CLEAVAGE NUCLEUS – the nucleus which controls cleavage. This may be the syngamic nucleus of normal fertilization; the egg nucleus of parthenogenetic or gynogenetic eggs; or the sperm nucleus of androgenetic eggs.

CLEAVAGE PATH – path taken by the syngamic nuclei to the position awaiting the first division.

CLEAVAGE, RADIAL – holoblastic cleavage which results in tiers of cells.

CLEAVAGE, SPIRAL – cleavage at an oblique angle with respect to the egg axis so that the resulting blastomeres (generally micromeres) lie in an interlocking fashion within the furrows of the original blastomeres, due to intrinsic genetic factors (e.g., mollusca).

CLEAVAGE, SUPERFICIAL – cleavage around the periphery of centrolecithal eggs. Syn., peripheral cleavage.

CLOACAL MEMBRANE – the endoderm of the large terminal cavity of the hindgut which fuses with the superficial ectoderm at the base of the tail to form a membrane that closes the cloaca to the outside, later to rupture as the anus (see ANAL PLATE). Syn., urodeal membrane, anal membrane, proctodeal membrane.

COCHLEA – portion of the original otic vesicle associated with sense of hearing; supplied by vestibular ganglion of eighth cranial nerve, having to do with equilibration.

COELOBLASTULA – spherical ball of blastomeres with a central cavity (e.g., echinoderms).

COELOM – mesodermal cavity from the walls of which develop the gonads; cavity subdivided in higher forms into pericardial, pleural, and peritoneal cavities (see also EXTRAEMBRYONIC and EXOCOEL).

COITUS – copulation of male and female, term generally used in connection with mammals.

COLLECTING TUBULE – portion of nephric tubule system leading to the nephric duct (Wolffian, etc.); term also used to refer to tubules which conduct spermatozoa from the seminiferous tubule to the vasa efferentia, within the testis.

COLLICULI, INFERIOR – posterior pair of thickenings of dorsolateral walls of mesencephalon (two of corpora quadrigemina) containing synaptic centers for auditory reflexes (mammal).

COLLICULI, SUPERIOR – anterior pair of thickenings of dorsolateral walls of mesencephalon (two of corpora quadrigemina) containing synaptic centers for visual reflexes (mammal). Syn., optic lobes.

COLUMELLA – a bone in the tubotympanic cavity of frog and chick which aids in auditory sensations. Syn., plectrum, malleus.

COMMISSURE, ANTERIOR – axis cylinders of spinal cord neuroblasts which originate in the mantle layer, grow ventrally, and cross over to the opposite side of the cord. Refers also to the fibers connecting the cerebral hemispheres, developing in the torus transversus of the lamina terminalis.

COMMISSURE, INFERIOR – floor of the diencephalon between the mammillary tubercles.

COMMISSURE, POSTERIOR – roof of the brain between the anterior limit of the mesencephalon and the synencephalon of the forebrain.

COMMISSURE, TROCHLEARIS – region of the dorsal isthmus, located in the roof between the metencephalon and the mesencephalon.

COMPETENCE – ability of an embryonic area to react to a stimulus (e.g., evocator).

CONCRESCENCE – the coming together of previously separate parts (cells) of the embryo, generally resulting in a piling up of parts; one of the corollaries of gastrulation in many forms (see also CONFLUENCE).

CONES OF GROWTH – the end enlargement of an outgrowing axone.

CONFLUENCE – similar to concrescence, except that this term refers specifically to the "flow" of cells (or areas) together, whether or not they are piled up.

COPRODEUM – a slight dilation of the occluded hindgut, temporarily separated from the cloaca in chick embryos from about the seventh day to hatching.

COPULATION PATH – the second portion of sperm migration path through the egg toward the egg nucleus when there is any deviation from the entrance or penetration path; the path of the spermatozoon which results in syngamy.

CORNEA – a transparent layer composed of head ectoderm acid underlying mesoderm which lies directly in front of the eye.

CORONA RADIATA – layer of follicle cells which immediately surround the egg of the bird or mammal, the cells which elongate and acquire intercellular canals radiating outward from the egg to the surrounding theca. Syn., follicular epithelium.

CORPORA QUADRIGEMINA – two pairs of rounded elevations arising from the dorsolateral thickened walls of the mesencephalon (mammal), associated with centers of hearing and vision.

CORPUS ALBICANS – a retrogressed corpus luteum of the mammal following its absorption.

CORPUS HAEMORRHAGICUM – blood clot found in recently ruptured ovarian follicle of the mammal.

CORPUS LUTEUM – a yellow granular tissue derived from the zona granulosa which fills in the follicular vesicle after ovulation in mammals; performs endocrine function relative to gestation and ovulation. Produces a substance called lutein.

CORPUS STRIATUM – center of coordination of certain complex muscular activities, derived from the mantle layer of the ventrolateral walls of the mammalian telencephalon.

CORPUS VITREUM – (see VITREOUS HUMOR).

COTYLEDONOUS PLACENTA – a chorion, as in the cow and other ruminants, which is made up of brushlike groups of villi (opposed to diffuse and intermediate placentae).

COWPER'S GLANDS – derivatives of the urethral epithelium, adjacent to the prostate, providing fluid for spermatozoa. Syn., bulbourethral glands.

CRANIAL – relative to the head; "craniad" means toward the head. Syn., rostral, cephalad.

CRANIAL FLEXURE – bending of the forebrain toward the underlying yolk of the chick embryo, with the angle of the bend occurring transversely at the level of the midbrain, at the 14-somite stage.

CRESCENT, GRAY – crescentic area between the original animal and vegetal pole regions on the surface of the frog's egg, gray in color because of the migration of pigment away from the area and toward the sperm entrance point; region of presumptive chorda-mesoderm and the future blastopore and anus.

CRESCENT, MESODERMAL – region of the Amphioxus egg seen in the 4-cell stage, due to pigment migration, which will give rise to the embryonic mesoderm.

CRESCENT, YELLOW – crescentic area on the surface of the Ascidian egg, yellow in color, which will give rise to the embryonic mesoderm.

CREST SEGMENT – the original neural crests become divided into segments from which develop the spinal, and possibly the cranial, ganglia.

CROP – a spindle-shaped dilatation of the esophagus of the chick appearing first on the eighth day of development.

CROSS-FERTILIZATION – union of gametes produced by different individuals which, if they are of different species, may produce hybrids.

CROSSING OVER – mutual exchange of portions of allelomorphic pairs of chromosomes during the process of synapsis in maturation.

CRYPTS – depressions found in the uterine wall for the reception of chorionic cotyledons.

CUMULUS OÖPHORUS – the aggregation of cells which immediately surround the mammalian egg within its Graafian follicle.

CUSHION SEPTUM – two endothelial thickenings arise within the atrioventricular canal of the heart and grow together to form a partition between right and left atrioventricular canals.

CUTICLE OF SHELL – the outermost, thinnest layer of the three layers of the chick egg shell; porous but otherwise structureless.

CUTIS PLATE – (see DERMATOME).

CYANOSIS – mixing of arterial and venous blood in the newborn due to the failure of the atrophy of the ductus Botalli and the incompletely formed interauricular canal, resulting in "blue babies" of high mortality. Bluing of the skin as a result of a circulatory defect.

CYCLE, SEXUAL – cyclic breeding activity, most evident in the female mammal, associated with definite changes in the genital and endocrine systems.
CYCLOPIA – median fusion of eyes which may be due to suppression of the rostral block of tissue which ordinarily separates the eyes.
CYST – tubular portion of the testis within which aggregations of germ cells mature, often (e.g., Rhomaleum) containing cells all in the same stage of maturation.
CYSTIC DUCT – narrow, proximal portion of embryonic bile duct leading from the gall bladder to the common bile duct.
CYTASTERS – asters arising apart from the nucleus in the cytoplasm.
-CYTE – a suffix meaning cell (e.g., osteocyte for bone cell; oöcyte for egg cell). See specific definitions.
CYTOLOGY – the study of cells.
CYTOLYSIS – the breakdown of the cell indicated by a dispersal of formed components.
CYTOPLASM – the material of the cell exclusive of the nucleus; protoplasm apart from nucleoplasm.

DECIDUA – the portion of the uterine wall cast off at the time of parturition.
DECIDUA BASALIS – that portion of the uterine wall to which the placenta is attached. Syn., decidua serotina.
DECIDUA CAPSULARIS – the portion of the uterine mucosa and epithelium which covers the mammalian blastocyst opposite the placenta. Syn., decidua reflexa.
DECIDUATE – the type of placentation which is characterized by the destruction of maternal (uterine) tissue and hemorrhage at parturition. True placenta.
DECIDUA VERA – the portion of the uterine wall aside from that associated with either the decidua basalis or the decidua capsularis but which will eventually be in contact with the decidua capsularis with the expansion of the embryo.
DELAMINATION – a separation (of cell layers) by splitting, a process in mesoderm formation.
DENTAL PAPILLA – the mesenchymal portion of the tooth primordium.
DENTINE – the main portion of the tooth derived from mesoderm.
DERMAL BONES – bony plates which originate in the dermis and cover the cartilaginous skull.
DERMATOME – the outer unthickened wall of the somite which gives rise to the dermis. Syn., cutis plate.
DERMIS – the deeper layers of the skin entirely derived from mesoderm (dermatome).
DERMOCRANIUM – the portion of the skull which does not go through an intermediate cartilaginous stage in development. Syn., membranocranium.
DETERMINATION – process of development indicated when a tissue, whether treated as an isolate or a transplant, still develops in the originally predicted manner.
DETERMINATION OF SEX – method by which the sex of the unborn is revealed. Not to be confused with sex determination which is generally achieved at the time of fertilization.
DEUTOPLASM – yolk or secondary food substances of the egg; nonliving.
DIAKINESIS – a stage in maturation when the double chromosomal threads fuse to form the haploid number, generally in curious shapes, and then line up for the first of two maturational divisions. The nuclear membrane is still present.
DIAPHRAGM – (see SEPTUM TRANSVERSUM).
DIENCEPHALON – the portion of the forebrain posterior to the telencephalon, including the second and third neuromeres. Embryonic source of the optic cups.
DIESTRUS – short period of quiescence immediately following estrus in the mammalian sexual cycle.
DIFFERENTIATION – increase in the visible and invisible complexities of the organism; progressive diversification leading to histogenesis; the process which changes a simple organism into one of complexity.
DIFFUSE PLACENTA – placenta in which the individual villi are scattered over the whole inner surface of the uterus or intercotyledonary areas (e.g., in the giraffe).
DIOCOEL – the cavity of the diencephalon, the ultimate third ventricle.

DIPLOID – the normal number of chromosomes in somatic and primordial germ cells; twice the haploid number characteristic of mature gametes.

DIPLOTENE – the stage in maturation following the pachytene when the chromosomes again appear double and do not converge toward the centrosome. Sometimes refers to split individual chromosomes.

DISCOBLASTULA – a disc-shaped blastula found in cases of discoidal (meroblastic) cleavage (e.g., cephalopoda and the chick).

DISCOIDAL CLEAVAGE – (see CLEAVAGE, DISCOIDAL).

DISCOIDAL PLACENTA – placenta developing on only one side of the blastocyst (e.g., rodents) in the general shape of a button or disc. Opposed to zonary placenta.

DISCUS PROLIGERUS – the oöcyte of the mammal is attached to the inner wall of the follicle by a neck of cells which, along with the cells surrounding the ovum, are together known as the discus proligerus.

DISTAL – farther from any point of reference, away from the main body mass.

DIVERTICULUM – the blind outpocketing of a tubular structure (e.g., liver or thyroid primordium).

DIVISION, POSTREDUCTION – refers to the second maturation division being meiotic, following the first, which is mitotic.

DIVISION, PREREDUCTION – refers to the first maturation division being meiotic (reductional) and the second mitotic.

DOMINANCE – parts of a system which have greater growth momentum and which also gather strength from the rest, such as the dorsal lip of the blastopore. Refers also to subjection of expression of one allelomorphic gene by another, usually at the same time.

DORSAL ROOT GANGLION – the aggregation of neuroblasts which are derived from the neural crests and which send their processes into the dorsal horns of the spinal cord.

DORSAL THICKENING – the roof of the mesencephalon which gives rise to the optic lobes.

D-QUADRANT – one of the four early blastomeres of the annelid embryo which has the prospective function of giving rise chiefly to mesoderm.

DUCT – (see ducts under specific names).

DUCTUS ARTERIOSUS – (see DUCTUS BOTALLI).

DUCTUS BOTALLI – the dorsal portion of the sixth pair of aortic arches which normally becomes occluded after birth, the remainder of the arch giving rise to the pulmonary arteries. Syn., ductus arteriosus.

DUCTUS CHOLEDOCHUS – common chamber associated with the duodenum into which the three pancreatic ducts, the hepatic duct, and the cystic duct (from the gall bladder) all empty prior to their developing separate openings into the gut.

DUCTUS COCHLEARIS – the connection between the lagena and the utricle of the ear.

DUCTUS CUVIERI – union of somatic veins which empty directly into the heart, specifically the vein which unites the common cardinals and the sinus venosus. (Sometimes regarded as synonymous with common cardinal.)

DUCTUS CYSTO-ENTERICUS – the original caudal liver duct which connects with the right lobe of the liver.

DUCTUS ENDOLYMPHATICUS – the dorsal portion of the original otic vesicle which has lost all connection with the epidermis and which is partially constricted from the region which will form the semicircular canals.

DUCTUS HEPATO-ENTERICUS – the original cephalic duct which connects with the left lobe of the liver.

DUCTUS VENOSUS – the anastomising sinusoids of the umbilical blood stream forming a major channel through the substance of the mammalian liver to receive blood from the left and right hepatic (omphalomesenteric) veins as it leaves the liver to join the posterior vena cava and then enter the right auricle.

DUODENUM – portion of the embryonic gut associated with the outgrowths of the pancreas and the liver (bile) ducts.

DYADS – aggregations of chromosomes consisting of two rather than four (tetrad) parts, a term used to describe a condition during maturation process.

ECHINOCHROME – red pigment of echinoderm eggs which probably has respiratory function.
ECTAMNION – the thickening of the ectoderm at the anterior boundary of the proamnion of the chick embryo occurring at about the eight-somite stage, in anticipation of the amniotic fold.
ECTOBLAST – (see EPIBLAST).
ECTODERM – primary germ layer from which are derived the epidermis (skin) and all of its derivatives, the nervous system, and part of the mouth and anus; the outermost layer of the didermic gastrula.
ECTOPLASM – external layer of protoplasm of the (egg) cell; the layer immediately beneath the cell membrane. Syn., egg cortex.
EDEMA – general puffiness or swelling of parts due to the accumulation of fluid in the tissues.
EFFERENT DUCTULES – some mesonephric tubules adjacent to the testes (mammal) which, together with a portion of the mesonephric duct, become the epididymis.
EGG, ALECITHAL – eggs with little or no yolk. Literally means "without yolk."
EGG, CLEIDOIC – eggs which are covered by a protective shell such as those of reptiles, birds, and oviparous mammals.
EGG, ECTOLECITHAL – an egg having the yolk around the formative protoplasm; opposed to centrolecithal.
EGG ENVELOPE – material enveloping the egg but not necessarily a part of the egg, such as vitelline membrane, chorion, jelly, albumen.
EGG, GIANT – abnormal polyploid condition where chromosome complexes are multiplied, resulting in giant cells and embryos.
EGG, HOMOLECITHAL – egg (e.g., mammal) with little yolk which is scattered throughout the cytoplasm.
EGG, ISOLECITHAL – eggs with homogeneous distribution of yolk, may be isolecithal, alecithal, or homolecithal.
EGG JELLY – the mucin covering deposited on the amphibian egg as it passes through the oviduct.
EGG, MACROLECITHAL – egg with large amount of yolk, generally telolecithal.
EGG MEMBRANES – includes all egg coverings such as vitelline membrane, chorion, and the tertiary coverings.
EGG, MICROLECITHAL – egg with small amount of yolk. Syn., meiolecithal, oligolecithal.
EGG, TELOLECITHAL – egg with large amount of yolk concentrated at one pole.
EGG TOOTH – a mammiform hard structure with pointed nipple found on the dorsal side of the tip of the upper jaw at the time of hatching of the chick embryo, with the sole function of aiding the chick in the hatching process.
EGG WATER – watery extract of materials diffusing from living eggs, presumably the "Fertilizin" of Lillie. Syn., egg water extract.
EJACULATION – the forcible emission of mature spermatozoa from the body of the male.
EJACULATORY DUCT – the short portion of the mesonephric duct (mammal) between the seminal vesicles and the urethra.
EMBOITEMENT – the preformationist theory of Bonnet and others based on the idea that the ovary of the first female (Eve?) contained the miniatures of all subsequently existing human beings. Syn., encasement theory.
EMBRYO – any stage in the ontogeny of the fertilized egg, generally limited to the period prior to independent food getting. The stage between the second week and the second month of the human embryo (see FETUS).
EMBRYONIC DISC – the portion of the early mammalian embryo where the ectoderm and endoderm are in close contact with each other.
EMBRYONIC FIELD – region of formative processes within the embryo, larger than the area of ultimate realization of structures concerned.
EMBRYONIC KNOB – the inner trophoblastic cell mass of the mammalian embryo (ectodermal).
EMBRYONIC MEMBRANES – refers to the amnion, chorion (serosa), allantois, and the yolk-sac splanchnopleure of the chick embryo.
EMBRYONIC SHIELD – the region of the fish blastoderm which will give rise to the embryo proper; the area of ectoderm and adherent endoderm out of which the mammalian embryo will develop.

EMBRYOTROPH – the materials obtained when the maternal tissues are broken down by the mammalian embryo prior to the establishment of the placental circulation.
EMBRYOTROPHY – the nourishment of the mammalian embryo.
ENAMEL ORGAN – the ectodermal portion of the tooth primordium.
ENDOCARDIAL CUSHION OF ATRIOVENTRICULAR CANAL – a median partition dividing the atrium (mammal) into right and left channels (see CUSHION SEPTUM).
ENDOCARDIUM – delicate endothelial tissue forming the lining of the heart.
ENDOCHONDRAL BONE – bone preformed in cartilage. Syn., cartilage bone.
ENDOLYMPHATIC DUCT – (see DUCTUS ENDOLYMPHATICUS).
ENDOLYMPHATIC SAC – (see SACCUS ENDOLYMPHATICUS).
ENDOMETRIUM – (see UTERINE MUCOSA).
ENDOPLASM – inner medullary substance of the (egg) cell which is generally granular, soft, watery, and less refractive than the ectoplasm.
ENTELECHY – Driesch's theory of an (intangible) agent controlling development. Syn., élan vital, vital force, psychism, perfecting principle, etc.
ENTEROCOEL – the cavity or pouch within the mesoderm just formed by evagination of the gut (enteron) endoderm as in Amphioxus. Syn., gut pouch, coelomic pouch, archenteric pouch.
ENTERON – the definitive gut of the embryo, always lined with endoderm.
ENTOBRONCHI – six dorsal outgrowths of the mesobronchus on the sixth day (chick) related to the respiratory system of the adult.
ENTOMESODERM – refers to the portion of the invaginating blastoporal lips which will induce the formation of medullary fields in the amphibian embryo.
ENTRANCE CONE – the temporary depression on the surface of the egg following the entrance of the spermatozoon.
ENTRANCE PATH – (see PATH, PENETRATION).
ENTYPY – a method of amnion formation in the mammal in which the trophoblast above the embryonic knob is never interrupted; a method of gastrulation (rodents) wherein the endoderm comes to lie externally to the amniotic ectoderm.
EPENDYMAL CELLS – narrow zone of nonnervous and ciliated cells which surround the central canal (neurocoel), from the outer ends of which branching processes extend to the periphery, such processes forming a framework for other cellular elements in the spinal cord and brain.
EPIBLAST – outermost layer of the early embryo from which the various germ layers may be derived.
EPIBOLY – growing, spreading, or flowing over; process by which the rapidly dividing animal pole cells or micromeres grow over and enclose the vegetal pole material.
EPIBRANCHIAL PLACODE – a placode (thickening) external to the gills related to the lateral line organs and the tenth cranial nerves. Syn., suprabranchial placode.
EPICARDIUM – outer thin layer covering the myocardium; originally part of the epimyocardium. Syn., visceral pericardium.
EPIDERMIS – the ectodermal portion of the skin including the cutaneous glands, hair, feathers, nails, hoofs, and some types of horns and scales, in various vertebrates.
EPIDIDYMIS – a tubular portion of the male genital system, derived from the anterior part of the Wolffian body, which conducts sperm from the testis to ductus deferens.
EPIGENESIS – development of systems starting with primitive, homogeneous, lowly organized condition and achieving great diversification. (Opposite of preformation.)
EPIMERE – the most dorsal mesoderm–that lying on either side of the nerve and notochord, which gives rise to the somites. Syn., axial mesoderm.
EPIPHYSIS – a dorsal evagination of the diencephalon of vertebrates which becomes separated from the brain as the pineal (endocrine) gland of the adult.
EPIPLOIC FORAMEN – opening from the peritoneal cavity (mammal) into the omental bursa, formed with change in position of the stomach. Syn., foramen of Winslow.
EPITHELIOID BODIES – endodermal masses arising from second and third visceral pouches of the amphibia and the third and fourth visceral pouches of the chick. In the adult, these are found near the lower pole of the thyroid (chick).
EPITHELIUM – a thin covering layer of cells; may be ectodermal, endodermal, or mesodermal.

EPOOPHORON – the rudimentary anterior portion of the mesonephros in the female; homologous with the epididymis in the male bird and mammal.

EQUATIONAL MATURATION DIVISION – the maturational division in which there is no (qualitative) reduction in the chromosomal complex; similar in results to mitosis.

EQUATORIAL PLATE – the chromosomes, lined up on the mitotic spindle, prior to any anaphase movement.

ESOPHAGUS – elongated portion of the foregut between the future glottis and the opening of the bile duct of the frog embryo; a narrow tube between the pharynx and the stomach primordium of the 36-somite chick embryo; temporarily occluded on the eighth day of the chick just behind the glottis but open again on the eleventh day; gives rise to the crop of the chick.

ESTRIN – an hormone found in the mammalian ovarian follicle.

ESTROGEN – a hormone secreted by the ovary which controls estrus and endometrial growth.

ESTROUS CYCLE – periodic series of changes in the endocrine, endometrial, and reproductive activities of the mammal, divided into phases of varying lengths depending upon the species: estrus, metestrus, diestrus, proestrus. Terms related to the preparation of the uterus for implantation of the ovum and to repair.

ESTRUS – period of the reproductive cycle of the mammal when the uterus is prepared for implantation of the ovum. Period of sexual receptivity in most animals.

EUSTACHIAN TUBE – vestige of the endodermal portion of the hyomandibular pouch connecting middle ear and pharyngeal cavity and lined with endoderm.

EVAGINATION – the growth from any surface outward.

EXOCOEL – the cavity within mesoderm beyond the limits of the (chick) embryo, continuous with the coelom. Syn., extraembryonic body cavity; seroamniotic cavity.

EXOGASTRULA – gastrulation modified experimentally by abnormal conditions so that invagination is partially or totally hindered and there remains some mesendoderm not enclosed by ectoderm.

EXTERNAL GILLS – outgrowths of (amphibian) branchial arches which function as temporary (anura) or permanent neotonic (urodela) respiratory organs.

EXTRAEMBRYONIC – refers to structures apart from the embryonic body, such as the membranes.

FALCIFORM LIGAMENT – portion of the original ventral mesentery between the liver and the ventral body wall, supporting the liver.

FALLOPIAN TUBES – the oviducts of the mammal.

FALSE AMNIOTIC CAVITY – a temporary cavity arising in the dorsal trophoblast of the mammalian embryo, having no connection with the true amniotic cavity.

FERTILIZATION – activation of the egg by sperm and syngamy of the pronuclei; union of male and female gamete nuclei.

FERTILIZATION CONE – a conical projection of cytoplasm from the surface of the egg to meet the spermatozoon which invades the egg, makes contact, and then draws the sperm into the egg.

FERTILIZATION MEMBRANE – a nonliving membrane seen to be distinct from the egg shortly after fertilization, very probably the vitelline membrane elevated from the egg.

FERTILIZIN – chemical substance in the cortex of the mature egg, apparently necessary for normal fertilization. Syn., sperm iso-agglutinin.

FETAL MEMBRANES – (see EXTRAEMBRYONIC).

FETUS – embryo after it has attained definite characteristics of the species; the embryo of the mammal when it has attained sufficient development to be recognized as belonging to a particular species. (Mouse = 15 days; human = 2 months.)

FEULGEN REACTION – chemical test for thymo-nucleic acid, used as a specific staining test for chromatin.

FIELD – mosaic of spatiotemporal activities within the developing organism

FIELD, MORPHOGENETIC – an embryonic area out of which will normally develop certain specific structures.

FLEXURE – refers to a bending such as the cranial, cervical, and pontine flexures. Also dorsal and lumbosacral flexures of the pig.

FOLLICLE, GRAAFIAN – the follicle of the mammalian ovary, including a double-layered capsule, the membrana granulosa, discus proligerus, corona radiata, and follicular fluid.

FOLLICLE, OVARIAN – a cellular sac within which the egg generally goes through the early maturation stages.

FORAMEN – (see under specific names as INTERATRIAL, EPIPLOIC, etc.).

FORAMEN OVALE – an opening between the embryonic auricular chambers of the heart.

FOREBRAIN – the most anterior of the first three primary brain vesicles, associated with the lateral opticoels, first distinct from midbrain at the six-somite stage in the chick. Syn., prosencephalon.

FOREGUT – the more anterior portion of the enteric canal, the first to appear in the chick, aided by the development of the head fold. Its margin is the anterior intestinal portal.

FRONTAL – a plane at right angles to both the transverse and sagittal, dividing the dorsal from the ventral. Syn., coronal.

GAMETE – a differentiated (matured) germ cell, capable of functioning in fertilization, e.g., sperm, or egg cell; germ cell.

GAMETOGENESIS – the process of developing and maturing germ cells.

GANGLION – an aggregation of neurons, generally derived from a neural crest, e.g., cranial and spinal ganglia.

GANGLION, ACOUSTIC – eighth (VIII) cranial ganglion from which the fibers of the eighth cranial nerve arise, purely sensory. Ganglion (in pig) later divides into vestibular and spiral ganglia.

GANGLION, ACUSTICO-FACIALIS – early undifferentiated association of the seventh and eighth cranial ganglia.

GANGLION, GASSERIAN – the fifth (V) cranial ganglion, carrying both sensory and motor fibers. Syn., trigeminal ganglion, semilunar ganglion (See TRIGEMINAL GANGLIA).

GANGLION, GENICULATE – the ganglion at the root of the facial (VII) cranial nerve, carrying both sensory and motor fibers.

GANGLION, NODOSAL – ganglion associated with the vagus (X) cranial nerve which carries afferent fibers from pharynx, larynx, trachea, esophagus, and thoracic and abdominal viscera.

GANGLION, PETROSAL – ganglion associated with the glossopharyngeal (IX) cranial nerve, more peripheral than the superior ganglion carrying sensory fibers from pharynx and root of tongue. From myelencephalon.

GARTNER'S CANAL – remains of the mesonephric duct in female mammals.

GASSERIAN GANGLION – the fifth cranial or trigeminal ganglion, attached to the hindbrain.

GASTRAEA THEORY – theory that, since all higher forms have gastrula stages, there may have existed a common ancestor built on the plan of a permanent gastrula, as are the recent coeloenterata.

GASTRAL MESODERM – mesoderm derived from the dorsolateral bands (enterocoelic) in Amphioxus or from the dorsal lip in the frog. Opposed to peristomial mesoderm.

GASTROCOEL – major cavity formed during the process of gastrulation. Syn., archenteron.

GASTROHEPATIC OMENTUM – that portion of the persisting ventral mesentery between the liver and the stomach. Syn., ventral mesogastrium.

GASTRULA – the didermic or double cell-layered embryo, possessing a newly formed cavity, the gastrocoel or archenteron.

GASTRULAR CLEAVAGE – the separation of ectoderm and endoderm, during gastrulation, by a slitlike crevice, actually the compressed blastocoel.

GASTRULATION – dynamic process involving cell movements which change the embryo from a monodermic to either a didermic or tridermic form. Generally involves inward movement of cells to form the enteric endoderm.

GENITAL – refers to the reproductive organs or processes, or both.

GENITAL DUCTS – any ducts which convey gametes from their point of origin to the region of insemination, e.g., collecting tubules, vas deferens, vas efferens, epididymis, seminal vesicle, oviduct (Fallopian tube), uterus, etc. Syn., gonoduct.

GENITAL RIDGE – initial elevation for the development of the external genitalia; paired mesodermal thickenings between the mesonephros and the dorsal mesentery of all vertebrates, which are the gonad primordia.

GENITAL TUBERCLE – the ridge at the base of the phallus, also the primordium of the mammalian labioscrotal swellings, primordium of either the penis or clitoris.

GERM – the egg throughout its development or at any stage.

GERM BANDS – distinguishable bands of material in the (molluscan) egg which will give rise respectively to the ectoderm and mesoderm of the embryo.

GERM CELL – a cell capable of sharing in the reproductive process, in contrast with a somatic cell, e.g., sperm or egg cell. Syn., gamete.

GERMINAL DISC – the small protoplasmic disc on the surface of the huge yolk mass of the (hen's) egg.

GERMINAL EPITHELIUM – the peritoneal epithelium out of which the reproductive cells of both the male and female develop. Syn., germinal ridges, gonadal ridges.

GERMINAL SPOT – synonym for nucleolus found in the ovum.

GERMINAL VESICLE – the prematuration nucleus of the egg.

GERM LAYER – a more or less artificial spatial and histogenic distinction of cell groups beginning in the gastrula stage, consisting of the ectoderm, endoderm, and mesodermal layers. No permanent and clear-cut distinction, as shown by transplantation experiments.

GERM PLASM – the hereditary material, generally referring specifically to the genotype. Opposed to somatoplasm.

GERM RING – ring of cells showing accelerated mitotic activity, generally a synonym for the lips of the blastopore. The rapidly advancing cells in epiboly.

GERM WALL – the syncytial periblast of the chick embryo which is being added to the central, well-defined, cellular blastoderm; the advancing peripheral boundary of the chick blastoderm.

GESTATION – period of carrying the young (mammal) within the uterus.

GILL – (see BRANCHIAL, including arches, chamber, cleft, and groove).

GILL PLATE – elevated and thickened areas of ectoderm posterior to the sense plate of the (amphibian) embryo where the visceral grooves will subsequently form.

GILL RAKERS – ectodermal fingerlike obstructions which sift the water as it passes from the oral cavity into the gill chambers of the frog tadpole.

GLIA CELLS – small, rounded, supporting cells of the central nervous system derived from the germinal cells of the neural ectoderm.

GLOMERULUS – an aggregation of capillaries associated with the branches of dorsal aorta but lying within the substance of the functional kidney; function is excretory.

GLOMUS – the vascular aggregations within the head kidney or pronephros, never to become a glomerulus.

GLOTTIS – the opening between the pharynx and the larynx.

GONAD – the organ within which germ cells are produced and generally matured (e.g., ovary or testis). Syn., sex or germ gland.

GONIUM – suffix referring to a stage in the maturation of a germ cell prior to any maturation divisions (e.g., spermatogonium, or oögonium).

GONOCOEL – cavity within the gonad, generally the ovary, into which (Amphioxus and some fish) the eggs may rupture.

GONODUCT – (see GENITAL DUCTS).

GRAAFIAN FOLLICLE – (see FOLLICLE, GRAAFIAN).

GRADIENT – gradual variation of developmental forces along an axis; scaled regions of preference (see AXIS).

GRANULOSA – the layer of follicle cells which surround the bird or mammalian ovum, so called because of their granular appearance when crowded.

GRAY CRESCENT – (see CRESCENT, GRAY).

GROWTH – a developmental increase in total mass of protoplasm at the expense of raw materials; an embryonic process generally following differentiation.

GYNANDROMORPH – condition where parts of an animal may be male and other parts female, not to be confused with hermaphroditism.

GYNOGENESIS – development of sperm-activated egg but without benefit of the sperm nucleus.

HAPLOID – having a single set of chromosomes not appearing in allelomorphic pairs, as in the mature gametes. Opposed to diploid (somatic) cells.
HARDERIAN GLAND – a solid ingrowth of ectodermal cells of the conjunctival sac appearing for the first time at the innermost angle of the nictitating membrane of the 8-day chick embryo.
HARMONIOUS-EQUIPOTENTIAL SYSTEM – an embryonic system in which all parts are equally ready to respond to the organism as a whole. Isolated blastomeres of such a system may give rise to complete embryos.
HATCHING – the beginning of the larval life of the amphibian, accomplished by temporarily secreted hatching enzymes which aid the embryo to escape its gelatinous capsule; the process of emergence of the chick embryo from its shell. Involves critical changes in the body, such as closing of the umbilicus, contraction of the ductus arteriosus, and drying up of the membranes. Hatching of chick aided by beak tooth.
HEAD FOLD – an anteriorly projecting semicircular fold of the medullary plate of the chick appearing first at about 21 hours. The fold involves both ectoderm and endoderm and is found also in mammals. Becomes head process.
HEAD FOLD AMNION – fold of the amnion which envelops the head of the chick embryo.
HEAD PROCESS – originally a faint line of cells anterior to Hensen's node and continuous with the primitive streak but slightly out of line with the streak, later to enlarge as an actual anterior projection of ectoderm and endoderm, the first indications of the embryonic axis.
HEAT, PERIOD OF – period of strong mating impulse in some female mammals. Syn., estrus.
HEMIPLACENTA – the chorion, the yolk sac, and generally the allantois, which together serve as an organ of nutritional supply to the uterine young of marsupials.
HENSEN'S KNOT – the anterior termination of the primitive streak, consisting of a knot of cells possibly homologous to the dorsal lip of the blastopore of the amphibia. Syn., Hensen's node, primitive knot or node.
HEPATIC PORTAL VEINS – remnants of the posterior portions of the left vitelline (amphibia) or left omphalomesenteric (chick) vein, supplied with blood mainly from the yolk sac but also from veins of the alimentary canal and connected with the hepatic veins only through sinuses within the liver. Function eventually assumed by the mesenteric vein.
HEPATIC SINUSOIDS – maze of dilated and irregular capillaries between the loosely packed framework of hepatic tubules.
HEPATIC VEINS – veins from the liver to the heart, originating as the anterior portions of the vitelline veins of the amphibia or the omphalomesenteric veins fo the chick embryo.
HERMAPHRODITE – an individual capable of producing both spermatozoa and ova.
HERMAPHRODITE, PROTANDROUS – male elements mature prior to female elements in hermaphrodite.
HERMAPHRODITE, PROTOGYNOUS – female elements mature prior to male elements in hermaphrodite.
HERTWIG'S LAW – the nucleus tends to place itself in the center of its sphere of activity; the longitudinal axis of the mitotic spindle tends to lie in the longitudinal axis of the yolk-free cytoplasm of the cell.
HETEROPLASTIC TRANSPLANT – a graft from one individual to another, generally of a different species.
HETEROTAXIS – exceptional case wherein the chick embryo rotates so as to lie on its right side, often associated with situs inversus viscerum; may be induced experimentally.
HETEROZYGOUS – condition where the zygote is composed of gametes bearing allelomorphic genes; opposed to homozygous.
HIBERNATION – spending the cold (winter) period in a state of reduced activity.
HINDBRAIN – the most posterior of the three original brain divisions, the first neuromere of which is larger than the succeeding neuromeres, there being a total of five at the 12-somite stage in the chick embryo. Syn., rhombencephalon.
HINDGUT – portion of embryonic gut just anterior to the neurenteric canal of amphibia and in the

chick embryo the archenteron posterior to the posterior intestinal portal. Level of origin of the rectum, cloaca, postanal gut, allantois, and caudal portions of the urogenital systems.

HISTOGENESIS – the development of the tissues; histological differentiation of tissues.

HOFFMAN'S NUCLEUS – group of neuroblasts at the external margin of the white matter just above the central roots in the spinal cord of the chick and extending the length of the cord.

HOMOIOTHERMAL – an adjective which refers to condition where the temperature of the body of the organism is under the control of an internal mechanism; body temperature regulated under any environmental conditions. Opposed to poikilothermal.

HOMOLOGY – similarity in structure based upon similar embryonic origin.

HOMOPLASTIC – refers to a graft to an organism of the same species or even to another position on the same individual. Syn., autoplastic.

HOMOZYGOUS – conditions where the zygote is composed of gametes bearing identical rather than allelomorphic genes.

HOMUNCULUS – imaginary human form presumed to exist within the egg or sperm; preformationism.

HORIZONTAL – an unsatisfactory term sometimes used synonymously with frontal, longitudinal, and even sagittal plane or section. Actually means across the lines of gravitational force.

HORMONE – a secretion of a ductless (endocrine) gland which can stimulate or inhibit the activity of distant parts of the biological system already formed.

HYALOPLASM – the viscous liquid regarded as the essential living protoplasm; substance of cell apart from contained nucleus.

HYBRID – a successful cross between different species, e.g., horse and ass give a (sterile) mule.

HYOID ARCH – the mesodermal mass between the hyomandibular and the first branchial cleft or between the first and second visceral pouches or clefts, which gives rise to the columella and parts of the hyoid apparatus. Syn., second visceral arch.

HYOMANDIBULA – refers to the pouch, cleft, or slit between the mandibular and the hyoid arches.

HYPERPLASIA – overgrowth; abnormal or unusual increase in elements composing a part.

HYPERTROPHY – increase in size usually due to increase in metabolic demands upon the part concerned.

HYPOCHORDAL ROD – a transitory string of cells constricted off between the dorsal wall of the midgut and the notochord of the amphibian embryo, between the level of the pancreas and the tail and disappearing before hatching time. Syn., subnotochordal rod.

HYPOMERE – the most ventral segment of mesoderm out of which develop the somatopleure, splanchnopleure, and coelom. Syn., lateral plate mesoderm.

HYPOPHYSIS – an ectodermally derived solid (amphibia) or tubular (chick) structure arising from the stomodeum and growing dorsally toward the infundibulum to give rise to the anterior and intermediate parts of the pituitary gland. Syn., in chick, Rathke's pocket, an embryonic structure.

HYPOPLASIA – undergrowth or deficiency in the elements composing a part.

HYPOTHALAMUS – ventral portion of lateral thickening of diencephalon, not clearly distinguishable from the main portion of the optic thalamus.

HYPOTHESIS – a presumption based on fragmentary but suggestive data.

IDIOSOME – the material out of which the acrosome is formed during the metamorphosis of spermatid to spermatozoon.

IMPLANTATION – the process of adding, superimposing, or placing a graft within a host without removal of any part of the host; the attachment of the mammalian blastocyst.

INCUBATION – the application of regulated heat to a developing egg.

INDECIDUATE PLACENTA – the type of placenta in which each villus simply fits into a crypt, as a plug fits into a socket, from which it is withdrawn at birth without serious hemorrhage or destruction of maternal tissues. Syn., nondeciduate.

INDUCTION – successive and purposeful influences which bring about morphogenetic changes within the embryo.

INFUNDIBULUM OF THE BRAIN – funnellike evagination of the floor of the diencephalon which, along with the hypophysis, will give rise to the pituitary gland of the adult.

INFUNDIBULUM OF THE OVIDUCT – (see OSTIUM TUBAE ABDOMINALE).

INNER CELL MASS – the spherical cells on the inner side of the mammalian blastocyst.

INSEMINATION – the process of impregnation; fertilization.

INTERATRIAL FORAMEN – (see FORAMEN OVALE).

INTERAURICULAR SEPTUM – a sheet of (mesodermal) tissue which grows ventrally from the roof of the auricular chamber to divide it into right and left halves.

INTERKINESIS – resting stage between mitotic divisions.

INTERMEDIATE CELL MASS – the narrow strip of mesoderm which, for a time, joins the dorsal epimere with the ventral hypomere, being made up of an outer layer continuous with the dermatome of the somite and the somatic mesoderm and an inner layer continuous with the ventromedial wall of the somite and the splanchnic mesoderm. Source of the excretory system. Syn., nephrotome or middle plate.

INTERNAL GILLS – filamentous outgrowths on the posterior side of the first three pairs of branchial arches and a single row on the anterior side of the fourth pair of branchial arches of the frog tadpole, which have a respiratory function concurrent with and following the absorption of the external gills. They are internal because they are enclosed by the operculum.

INTERNAL LIMITING MEMBRANE – a membrane which develops on the innermost surface of the inner wall of the optic cup during the fourth day of chick development.

INTERSEX – an individual without sufficient sexual differentiation to be diagnosed as either male or female.

INTERSTITIAL TISSUE OF TESTIS – cell aggregates between the seminiferous tubules of the testis, which elaborate a male sex hormone.

INTERVENTRICULAR SEPTUM – a partition growing anteriorly from the apex of the ventricle, which extends from the auricle to the bulbus arteriosus and divides the ventricle.

INTERVERTEBRAL FISSURE – a cleft between the caudal and cephalic divisions of the sclerotome.

INTERZONAL FIBERS – portion of the spindle fibers located between chromosome groups in the anaphase and telophase stages.

INTESTINAL PORTAL – an opening from the midgut into either the anterior or posterior levels of the formed gut in the chick embryo.

INVAGINATION – the folding or inpushing of a layer of cells into a preformed cavity, as in one of the processes of gastrulation. Do not confuse with involution.

INVOLUTION – the rolling inward or the turning in of cells over a rim, as in the gastrulation of the chick embryo.

IRIS – the narrow zone bounding the pupil of the eye in which two layers of the optic cup become blended so that the pigment from the outer layer invades the material of the inner layer, giving the eye a specific color by variable reflection.

ISOGAMY – similar gametes, without differentiations into spermatozoa and ova.

ISOLATION CULTURE – removal of a part of an organism and its maintenance in a suitable medium in the living condition.

ISTHMUS OF THE BRAIN – a depression in the dorsal wall of the embryonic brain which partially separates the mesencephalon from the metencephalon.

ISTHMUS OF THE OVIDUCT – the short tubular posterior end of the (chick) oviduct in which the fluid albumen and the shell membranes are applied to the egg.

ITER – (see AQUEDUCT OF SYLVIUS).

JACOBSON'S ORGAN – ventromedial evaginations from the olfactory pits (amphibia and reptiles) which later become the glandular and sensitive olfactory epithelia.

JELLY – mucin covering of the (amphibian) egg derived from the oviduct and applied outside the vitelline membrane.

JUGULAR VEINS – veins which bring blood from the head, the superior or internal jugular being the anterior cardinal veins, and the inferior jugular veins bringing blood from the lower jaw and mouth to the base of each ductus Cuvieri or common cardinal.

KARYOPLASM – protoplasm within the confines of the nucleus.

KUPFFER'S VESICLE – a cavity at the posterior end of the early fish embryo presumed to represent a rudiment of the neurenteric canal.

LABIOSCROTAL SWELLINGS – the primorida of the labia majora (female) or of part of the scrotum (male) in fetal mammals.

LACHRYMAL GROOVE – a shallow groove between the lateral nasal and the maxillary processes of the chick embryo of 5 days or older.

LACUNA – literally a "little lake," referring to the numerous gaps in the mammalian (mouse) trophoderm through which maternal blood flows when the vessels are destroyed.

LAGENA – a derivative of the pars inferior labyrinthii of the original endolymphatic sac, related to the ductus cochlearis.

LAMINA TERMINALIS – the point of suture of the anterior neural folds (i.e., the anterior neuropore) where they are finally separated from the head ectoderm. It consists of a median ventral thickening at the anterior limit of the telencephalon from the anterior side of the optic recess to the beginning of the velum transversum and includes the anterior commissure of the torus transversus.

LANUGO – the fine hairy covering of the fetal mammal.

LARVA – stage in development when the organism has emerged from its membranes and is able to lead an independent existence but may not have completed its development. Generally (except in cases of neotony or paedogenesis), larvae cannot reproduce.

LARYNGOTRACHEAL GROOVE – a transverse narrowing of the postbranchial region of the (chick) embryonic pharynx and the groove which leads posteriorly to the lung primordia; apparent at the 35-somite (72-hour) stage.

LARYNX – the anterior part of the original laryngotracheal groove which opens into the pharynx by way of the glottis.

LATEBRA – the bulbous portion of the flask-shaped mass of white yolk located toward the center of the chick egg and connected by a neck of similar material with the nucleus of Pander.

LATERAL – either the right (dextral) or left (sinistral) side; laterad means toward the side.

LATERAL AMNIOTIC FOLDS – folds of the amnion extending up over the sides of the chick embryo, developing as corollaries to or, in consequence of, the head and tail amniotic folds.

LATERAL LIMITING SULCUS – (see LIMITING SULCUS).

LATERAL LINE ORGANS (or SYSTEM) – a line of sensory structures along the side of the body of fish and amphibia, generally embedded in the skin and innervated by a branch from the vagus ganglion, presumably concerned with the recognition of low vibrations in water. Appears first at about 4-mm stage in the frog embryo. Syn., ramus lateralis.

LATERAL MESOCARDIUM – septum posterior to the heart extending from the base of each vitelline vein obliquely upward to the dorsolateral body wall, representing one of the three parts of the septum transversum.

LATERAL MESODERM – (see LATERAL PLATE MESODERM).

LATERAL NEURAL FOLDS – (see NEURAL FOLDS).

LATERAL PLATES OR LATERAL PLATE MESODERM – the lateral mesoblast within which the body cavity (coelom and exocoel) arises. Syn., lateral mesoderm.

LATERAL ROTATION – twisting which begins at the anterior end of the chick embryonic body and extends posteriorly until the embryo is lying on its left side.

LATERAL TONGUE FOLDS – folds to the right and the left of the tuberculum impar which give rise to the sides of the chick tongue.

LATERAL VENTRICLES OF THE BRAIN – the thick-walled and laterally compressed cavity of the prosencephalon which opens into the third ventricle by way of the foramina of Monro, the walls of which will become the cerebral hemispheres.

LECITHIN – organismic fat which is phosphorized in the form of phosphatides; abundant in eggs.

LENS – a thickening in the head ectoderm opposite the optic cup at about the time of hatching in the frog embryo and at the 26-somite stage in the chick, which becomes a placode. This invaginates to acquire a vesicle and then pinches off into the space of the optic cup as a lens. Inner surface convex; substance fibrous.

LENS PLACODE – the early thickened ectodermal primordium of the lens.

LENTICULAR ZONE – the portion around the rim of the optic cup adjacent to the pupil and separated from the retinal zone in later (chick eye) development by the ora serrata.

LEPTOTENE – a stage in maturation which follows the last gonial division and is prior to the synaptene stage. It is structurally similar to the resting cell nuclear chromosome stage. The chromatin material is in the form of a spireme. The term means thin, diffuse.

LESSER PERITONEAL CAVITY – the growth of the liver to the right mesonephros and finally to the portal vein cuts off a portion of the peritoneal cavity of the chick at about the 160-hour stage, giving rise later to the greater and lesser omental spaces on either side of the coeliac fold. Syn., bursa omenti (major and minor).

LIMITING SULCUS – the separation of the chick embryo from the blastoderm by an undercutting groove or folding of tissue between the embryonic and extraembryonic regions. Consists of anterior, posterior, and lateral sulci and is found only in telolecithal eggs.

LINGUAL GLANDS – solid ingrowths beneath the lateral margin of the chick embryonic tongue, following somewhat the lingual cartilage.

LIP GROOVES – grooves in the upper beak of the chick embryo presumably homologous to the lip grooves of other vertebrates.

LIPIDS – fats and fatty substances such as oil and yolk (lecithin) found in eggs, e.g., cholesterol, ergosterol.

LIPS OF THE BLASTOPORE – (see BLASTOPORE, LIPS OF).

LIQUOR FOLLICULI – the fluid of the mammalian Graafian follicle into which the mature ovum is freed when the discus proligerus is broken, finally to be liberated from the ovary at the time of ovulation. Contains hormones.

LOCULUS – a local enlargement of the mouse uterus which contains an embryo and its associated membranes.

LUMBOSACRAL FLEXURE – the most posterior of the four flexures of the pig embryo.

LUNAR PERIODICITY – maturation and oviposition during certain periods of the lunar cycle.

LUTEIN – the yellow-colored material contained in the cells which fill the empty mammalian Graafian follicles (see CORPUS LUTEUM).

MACROMERE – larger of blastomeres where there is a conspicuous size difference.

MALPIGHIAN BODY – a unit of the functional kidney including Bowman's capsule and the glomerulus. Syn., renal corpuscle, Malpighian corpuscle.

MAMMARY GLANDS – multiple milk glands of the typical mammal, derived from the milk ridges occasionally seen as early as the 10-mm stage in the pig embryo.

MAMMILLAE – the innermost of the three layers of the chick egg shell, consisting of minute calcareous particles fused together with their conical faces impinging on the shell membrane.

MANDIBULAR ARCH – the mesodermal rudiment of the lower jaw or mandible, anterior to the first or hyomandibular pouch. Gives rise (chick) to the palatoquadrate and to Meckel's cartilage.

MANDIBULAR GLANDS – series of solid ingrowths of the chick oral mucosa about the eighth day, extending on both sides of the base of the tongue to near the mandibular symphysis.

MANTLE FIBERS – those fibers of the mitotic spindle which attach the chromosomes to the centrosomes.

MANTLE LAYER OF THE CORD – layer of the developing spinal cord with densely packed nuclei, slightly peripheral to the germinal cells from which they are derived. Includes the elongated processes of the central ependymal cells.

MARGINAL BELT – ring of presumptive mesoderm of the amphibian blastula, essentially similar to the gray crescent of the undivided egg.

MARGINAL LAYER OF THE CORD – layer of the spinal cord peripheral to the mantle layer, practically devoid of nuclei, and consisting of bundles of axons.

MARGINAL NOTCH – occasionally the blastopore of the chick closes (primitive streak) so as to produce a notch at the margin of the blastoderm, or posterior limit of the streak.

MARGINAL VELUM – a narrow, nonnucleated margin in the lateral walls of the neural tube, occupied by the outer ends of epithelial cells in the 50-hour chick embryo.

MARGIN OF OVERGROWTH – (see ZONE OF OVERGROWTH).

MASSA INTERMEDIA – fusion of thick lateral walls of diencephalon across the third ventricle in the pig embryo.
MATERNAL PLACENTA – uterine mucosal portions of the typical mammalian placenta.
MATURATION – the process of transformation of a primordial germ cell (spermatogonium or oögonium) into a functionally mature germ cell, the process involving two special divisions, one of which is always meiotic. Divisions known as equational and reductional.
MEATUS VENOSUS – the junction of the primitive omphalomesenteric veins posterior to the sinus venosus, around which develops the substance of the liver. Later, it will also receive the left umbilical (allantoic) vein of the chick. Syn., ductus venosus.
MECKEL'S CARTILAGE – the core of the lower jaw derived from the ventral part of the cartilaginous mandibular arch.
MEDIAN PLANE – "middle" plane, as of an embryo. May be median sagittal or median frontal.
MEDULLA OBLONGATA – that portion of the adult brain derived from the myelencephalon.
MEDULLARY – (see terms under NEURAL, such as canal, fold, groove, plate, tube.)
MEDULLARY CORDS – that portion of the suprarenal glands of the chick which is derived from the sympathetic nervous system; central cords. Also that portion of the embryonic gonad presumably derived from migratory germ cells upon reaching the genital ridge.
MEGASPHERES – yolk masses which are cut off from the floor of the subgerminal cavity of the chick blastoderm and which lie within this cavity.
MEIOSIS – a process of nuclear division found in the maturation of germ cells, involving a separation of members of pairs of chromosomes. Syn., reductional division.
MELANOPHORE – cell with black or brown (melanin) pigment, derived from the neural crests and migrating throughout the body.
MEMBRANA GRANULOSA – the layers of follicle cells which bound the mammalian follicular cavity.
MEMBRANA REUNIENS – a membrane which extends dorsally from the neural arches around the upper part of the neural tube; the line of later chondrification.
MEMBRANE BONE – bone developed in regions occupied by connective tissue, not cartilage.
MEMBRANE, FERTILIZATION – (see FERTILIZATION MEMBRANE).
MEMBRANES – (see EGG MEMBRANES).
MEMBRANE, VITELLINE – (see VITELLINE MEMBRANE).
MEMBRANOUS LABYRINTH – the parts of the internal ear, lined with ectodermal epithelium and filled with endolymphatic fluid, including the ductus endolymphaticus, the pars superior labyrinth, and the pars inferior labyrinth.
MENSTRUATION – process in primates (and other vertebrates) caused by decrease in secretion of estrogen and progestrone, involving loss of endometrium and some hemorrhage.
MEROBLASTIC CLEAVAGE or OVA – (see under CLEAVAGE or EGG).
MESENCEPHALON – the section of the primary brain between the posterior level of the prosencephalon and an imaginary line drawn from the tuberculum posterius to a point just posterior to the dorsal thickening. Gives rise to the optic lobes, crura cerebri, and the aqueduct of Sylvius. Syn., midbrain.
MESENCHYME – the form of embryonic mesoderm or mesoblast in which migrating cells unite secondarily to form a syncytium or network having nuclei in thickened nodes between intercellular spaces filled with fluid; often derived from mesothelium.
MESENDODERM – newly formed layer of the (urodele) gastrula before there has been any separation of endoderm and mesoderm. Syn., mesentoderm.
MESENTERY – sheet of (mesoderm) tissue generally supporting organ systems, e.g., mesorchium, mesocardium.
MESIAL – (see MEDIAN, MEDIAL, MIDDLE).
MESOBLAST, GASTRAL – (see GASTRAL MESODERM).
MESOBLAST, PARAXIAL – (see AXIAL MESODERM).
MESOBLAST, PROSTOMIAL – mesoderm in the region of, and fused to, the primitive streak (chick).
MESOBRONCHUS – division of the primary (chick) lung, parts of which give rise to air sacs.

MESOCARDIUM – the mesentery of the heart; may be dorsal, ventral, or lateral (see under LATERAL MESOCARDIUM).

MESOCOLON – that portion of the embryonic dorsal mesentery which supports the colon.

MESODERM – the third primary germ layer developed in point of time; may be derived from endoderm in some forms and from ectoderm in others, and lies between ectoderm and endoderm (see other terms such as MESOBLAST, MESENCHYME, LATERAL PLATE MESODERM, EPIMERE, MESOMERE, HYPOMERE, GASTRAL, PERISTOMIAL, AXIAL, etc.).

MESOGASTRIUM – dorsal and ventral mesenteries which support the developing gut and its derivatives.

MESOMERE – cell of intermediate size where there are conspicuous size differences in an early embryo (also refers to INTERMEDIATE CELL MASS).

MESOMETRIUM – attachment of the uterus to the coelomic wall.

MESONEPHRIC DUCT – the duct which grows posteriorly from the mesonephros to the cloaca and later becomes vas deferens in male chicks. Syn., Wolffian duct.

MESONEPHRIC TUBULES – primary, secondary, and sometimes tertiary tubules developing in the Wolffian body, functioning in adult amphibia and embryonic chick and mammal.

MESONEPHROS – the Wolffian body, or intermediate kidney, functional as kidney in the adult fish and amphibian but only in the embryonic bird and mammal (see EPIDIDYMIS and VASA EFFERENTIA).

MESORCHIUM – mesentery (mesodermal) which surrounds and attaches the testis to the body wall.

MESOTHELIUM – epithelial layers or membranes of mesodermal origin.

MESOVARIUM – mesentery (mesodermal) which suspends the ovary from the dorsal body wall.

METAMERISM – serial segmentation, as seen in the nervous, muscular, and circulatory systems.

METAMORPHOSIS – the end of the larval period of amphibia when growth is temporarily suspended. There is autolysis and resorption of old tissues and organs, such as gills, and the development of new structures, such as eyelids and limbs; changes in structure correlated with changes in habitat from one that is aquatic to one that is terrestrial; change in structure without retention of original form, as in change from spermatid to spermatozoon.

METANEPHRIC DIVERTICULUM – budlike outgrowth of the mesonephric duct just anterior to its juncture with the cloaca.

METANEPHROS – the permanent kidney of birds and mammals derived from the nephrogenous tissue of the most posterior somite level (renal corpuscles and secreting tubules) and from a diverticulum of the posterior portion of the Wolffian duct (collecting tubules and definitive ureter).

METAPHASE – stage in mitosis when the paired chromosomes are lined up on the equatorial plate midway between the amphiasters, supported by the mitotic spindle, prior to any anaphase movement.

METATELA – thin roof of fourth ventricle of the brain (pig).

METENCEPHALON – the anterior part of the hindbrain (rhombencephalon) which gives rise to the cerebellum and the pons of the adult brain, separated from the mesencephalon by the isthmus and including the sixth neuromere.

METESTRUS – short period of regressive changes in the uterine mucosa in which the evidences of fruitless preparation for pregnancy disappear. Syn., postestrus.

MICROMERE – smaller of blastomere cells when there is a conspicuous difference in size, characteristic of annelids and molluscs (see section on Crepidula).

MICROPYLE – an aperture in the egg covering through which spermatozoa may enter; in such eggs the only possible point of insemination, e.g., many fish eggs.

MIDBRAIN – (see MESENCEPHALON).

MIDGUT – that portion of the archenteron which will give rise to the intestines and, in birds, to the yolk stalk; bounded in the early chick embryo by the anterior and the posterior intestinal portals.

MILK RIDGE – a band of tissue between the somites (dorsally) and the level of the heart, liver, and mesonephros (ventrally) in the pig embryo, which gives rise to the mammary glands; first seen at 12 mm and well developed by 15 mm.

MITOCHONDRIA – small permanent cytoplasmic granules or filaments which stain with Janus Green B and Janus Red; granules which have powers of growth and division; probably lipoid.

MITOSIS – the process of cell division in somatic cells, as distinct from germ cells, in which each resulting cell is provided with a set of chromosomes similar to the other and to that of the parent cell; the division consists of prophase, metaphase, anaphase, and telophase.

MITOTIC INDEX – the proportion of the dividing cells in any tissue and at any specified time.

MONESTROUS CYCLE – forms which have single breeding season during the year have a monestrous cycle (see estrous cycle).

MONOSPERMY – fertilization accomplished by only one sperm. Opposed to polyspermy.

MONRO, FORAMINA OF – tubular connections between the single third and the paired lateral ventricles of the forebrain.

MORULA – a spherical mass of cells, as yet without segmentation cavity.

MŬLLERIAN DUCTS – (see OVIDUCTS).

MUSCLE PLATE – (see MYOTOME).

MYELENCEPHALIC TELA – the thin roof of the myelencephalon. Syn., metatela.

MYELENCEPHALON – the posterior portion of the hindbrain (rhombencephalon) which has a thin roof that becomes the choroid plexus of the fourth ventricle and thick ventral and ventrolateral walls which give rise to the medulla oblongata. The cranial ganglia fifth to twelfth inclusive are associated with this portion of the brain.

MYELOCOEL – cavity of the myelencephalon; fourth ventricle.

MYOBLASTS – formative cells within the myotome or muscle plate which will give rise to the true striated muscles of the adult.

MYOCARDIUM – the muscular part of the heart arising from the splanchnic mesoblast.

MYOCOEL – the cavity within which the ovaries of Amphioxus develop; temporary cavities within the myotomes which may have been connected with the coelom; the comparable position in the chick myotome is filled with loose mesenchyme.

MYOTOME – the thickened primordium of the muscle found in each somite. Syn., muscle plate.

NARES, EXTERNAL – the external openings of the tubes which are connected with the olfactory organs.

NARES, INTERNAL – the openings of the tubular organ from the olfactory organs into the anterior part of the pharynx of the 12-mm frog tadpole; the openings are posterior in the chick pharynx.

NASAL CHOANAE – opening of the olfactory chambers into the mouth.

NASAL PIT – (see OLFACTORY PIT).

NASOFRONTAL PROCESS – a median projection overhanging the mouth and separating the olfactory pits, beginning on the fifth day of chick development.

NASOLACHRYMAL GROOVE – groove between junction of nasolateral and maxillary processes, extending to the mesial angle of the eye. Portion becomes tear (nasolachrymal) duct which drains fluid from conjunctival sac into the nose (pig).

NASOLATERAL PROCESS – lateral projection to nasal (olfactory) pit in pig.

NASOMEDIAL PROCESS – median elevation ventral to the nasal (olfactory) pit in pig.

NASOOPTIC FURROW – Syn., nasolachrymal groove.

NEBENKERN – cytological structure complex, Golgi apparatus and mitochondria, near the nucleus of the early spermatid.

NEOTONY – condition of many urodeles and of experimentally produced (thyroidless) anuran embryos where the larval period is extended or retained, i.e., the larvae fail to go through normal metamorphosis, e.g., Axolotl, Necturus.

NEPHROCOEL – the cavity, found in the nephrotome or intermediate cell mass, which temporarily joins the myocoel and the coelom.

NEPHROGENIC CORD – continuous band of intermediate mesoderm (mesomere) without apparent segmentation, prior to budding off of mesonephric tubules.

NEPHROGENIC TISSUE – the intermediate cell mass, mesomere, or nephrotome which will give rise to the excretory system.

NEPHROSTOME – the funnel-shaped opening of kidney tubules into the coelom; the outer tubules of the (amphibian) mesonephric kidney acquire ciliated nephrostomal openings from the coelom, and these funnels shift their connections to the renal sinus.

NEPHROTOME – the intermediate cell mass.

NEPHROTOMIC PLATE – Syn., intermediate mesoderm, mesomere.

NERVES, ABDUCENS – sixth (VI) cranial nerves arising from the basal plate of the myelencephalon which control the external rectus muscles of the eye.

NERVES, ACCESSORY – eleventh (XI) cranial nerves, motor, with fibers arising (pig) from posterior myelencephalon and first six segments of the spinal cord.

NERVES, AUDITORY – eighth (VIII) cranial nerves, purely sensory, arising from acoustic ganglia and associated with the geniculate ganglia of the seventh nerves.

NERVES, FACIAL – seventh (VII) cranial nerves, both sensory and motor, related to taste buds and facial muscles, arising from basal plate of myelencephalon.

NERVES, GLOSSOPHARYNGEAL – ninth (IX) cranial nerves, mixed association sensory and motor with the superior and petrosal ganglia.

NERVES, HYPOGLOSSAL – twelfth (XII) cranial nerves, somatic motor, arising from posterior myelencephalon and extending to the muscles of the tongue.

NERVES, OCULOMOTOR – the third cranial nerves which arise from neuroblasts in the ventral zone of the midbrain near the median line at about 60 hours in the chick, and just before hatching in the frog tadpole.

NERVES, OLFACTORY – first (I) cranial nerves, sensory, without ganglia and with nonmedullated fibers, which arise from the epithelial linings of the olfactory pits and have synaptic connections in olfactory bulbs.

NERVES, OPTIC – second (II) cranial nerves, sensory, arise from neuroblasts of sensory layer of retina, pass through choroid fissure to enter brain at diencephalic floor. In contrast with other cranial nerves, these intersect so that each eye has connections with both sides of the brain (optic chiasma).

NERVES, VAGUS – tenth (X) cranial nerves, mixed, arising from the myelencephalon and associated with jugular ganglia.

NERVES, CRANIAL, PERIPHERAL, SPINAL – designated purely with respect to morphological position.

NERVOUS LAYER – the innermost of two layers found in the roof of the segmentation cavity of the amphibian blastula, from which the bulk of the central nervous system is developed.

NEURAL ARCH – the ossified cartilages which extend dorsally from the centrum around the nerve cord, involving both the caudal and the cephalic sclerotomes (chick). The cephalic arch of one sclerotome fuses with the caudal of the next to form a single arch which corresponds to a vertebra. Syn., vertebral arch.

NEURAL CANAL – (see NEUROCOEL and NEURAL TUBE).

NEURAL CREST – a continuous cord of ectodermally derived cells lying on each side in the angle between the neural tube and the body ectoderm, separated from the ectoderm at the time of closure of the neural tube and extending from the extreme anterior to the posterior end of the embryo; material out of which the spinal and possibly some of the cranial ganglia develop, and related to the development of the sympathetic ganglia by cell migration.

NEURAL FOLD – elevation of ectoderm on either side of the thickened and depressing medullary plate; folds which close dorsally to form the neural tube. Syn., medullary folds.

NEURAL GROOVE – the sinking in of the center of the medullary plate to form a longitudinal groove, later to be incorporated within the neural tube (spinal cord). Syn., medullary groove.

NEURAL PLATE – thickened broad strip of ectoderm along the future dorsal side of all vertebrate embryos, later to give rise to the central nervous system. Syn., medullary plate.

NEURAL TUBE – the tube formed by the dorsal fusion of the neural folds, the rudiment of the nerve or spinal cord.

NEURENTERIC CANAL – the posterior neurocoel where it is connected with the closing blastopore and posterior enteron of the amphibian; the large common nervous and enteric chamber of Amphioxus; the Kupffer's vesicle of the fish embryo; possibly the primitive pit of the chick embryo (see NOTOCHORDAL CANAL and PRIMITIVE PIT).

NEUROBLASTS – primitive or formative nerve cells, probably derived (along with epithelial and glia cells) from the germinal cells of the 3-day chick embryo neural tube and neural crests.
NEUROCOEL – the cavity of the neural tube, formed simultaneously with the closure of the neural folds. Syn., central canal, neural canal.
NEUROCRANIUM – the dorsal portion of the skull associated with the brain and sense organs.
NEUROGLIA – (see GLIA CELLS).
NEUROMERE – apparent metamerism of the embryonic brain, the divisions being prosencephalon-3, mesencephalon-2, and rhombencephalon-6.
NEUROPORE – a temporary opening into the neural canal due to a lag in the fusion of the neural folds at the anterior extremity; permanent in Amphioxus and in the vicinity of the epiphysis of higher vertebrates. There may be a temporary posterior neuropore in some, even in the chick.
NONDECIDUOUS PLACENTA – (see INDECIDUATE PLACENTA).
NOTOCHORD – rod of vacuolated cells representing the axis of all vertebrates, found beneath the neural tube and dorsal to the archenteron. Origin variable or doubtful, in most cases thought to be derived from or simultaneously with the endoderm.
NOTOCHORDAL CANAL – an exaggerated primitive pit in some mammals which extends into the head process. May be homologous to neurenteric canal.
NOTOCHORDAL SHEATH – double mesodermal sheath around the notochord consisting of an outer elastic sheath developed from superficial chorda cells and an inner secondary or fibrous sheath from chorda epithelium.
NUCLEOLUS – the circular body within the nucleus which has no affinity for chromatin dyes but stains with acid or cytoplasmic dyes. It fades and disappears just before mitosis and reappears later. Function unknown. Syn., plasmosome.
NUCLEUS, DEITER'S – synaptic center at the boundary between the myelencephalon and the metencephalon, where sensory neurones of semicircular canals le d.
NUCLEUS OF PANDER – the plate of white yolk beneath the blastodisc of the chick egg.
NUCLEUS, RED – synaptic center of the midbrain which acts as a coordinating pathway for synergic type of muscular control.

ODONTOBLASTS – dentine-forming embryonic cells; column-shaped outer cells of mammalian dental papilla.
OEDEMA – (see EDEMA).
OLFACTORY CAPSULE – the extreme anterior ends of the skull trabeculae which form the cartilaginous capsules around the olfactory organs.
OLFACTORY LOBES – the partially constricted anterior extremities of the telencephalic cerebral lobes, associated with the first pair of cranial nerves.
OLFACTORY PIT – depressions within the olfactory placodes of the 6-mm frog or the 36-somite chick embryo which will become the olfactory organs (external nares).
OLFACTORY PLACODE – the thickened ectoderm lateral to the stomodeal region found in the 5-mm frog embryo and the 26-somite chick embryo, primordia of the olfactory pits.
OMENTAL BURSA – pouch formed in the dorsal mesogastrium as the embryonic stomach of the pig changes its position.
OMENTUM – syn., mesogastrium; ventral mesogaster. Dorsal membrane which supports the gut.
OMNIPOTENT – term used in connection with a cell which could, under various conditions, assume every cytological differentiation known to the species or which, by division, could give rise to such varied differentiations.
OMPHALOMESENTERIC ARTERIES – originally paired vessels originating from the dorsal aorta of the chick embryo, later to fuse as result of formation of the dorsal mesentery; function to carry blood out to the extraembryonic vitelline vessels of the chick embryo. Syn., vitelline artery.
OMPHALOMESENTERIC VEINS – the vitelline veins as they enter the body of the chick embryo at the level of the anterior intestinal portal; united posterior to the sinus venosus as the meatus venosus. Venous rings join these vessels dorsal to the gut on the third and then the fourth day to become a single vessel emptying through the liver into the meatus venosus. The vitelline vein is continuous with the omphalomesenterics but is found on the yolk and is therefore extra-embryonic.

ONTOGENY – developmental history of an organism.

OÖCYTE – the presumptive egg cell after the initiation of the growth phase of maturation. Syn., ovocyte.

OÖGENESIS – the process of maturation of the ovum; transformation of the oögonium to the mature ovum. Syn., ovogenesis.

OÖGONIA – cells in the multiplication (mitotic) stage prior to maturation of the presumptive egg cell (ovum), found most frequently in the peripheral germinal epithelium.

OPERCULAR CHAMBER – (see BRANCHIAL CHAMBER).

OPERCULUM – integumentary growth posteriorly from each of the hyoid arches of the frog embryo, which covers and encloses the gills.

OPTIC CHIASMA – thickening, eventually consisting of crossing optic nerve fibers, in the prosencephalon floor anterior to the infundibulum.

OPTIC CUP – invagination of the outer wall of the optic vesicle to form a cup made up of two layers, a thick internal or retinal layer continuous at the pupil with a thinner external layer which is pigmented. The cavity of the cup becomes the future posterior chamber of the eye.

OPTIC LOBES – the thickened, evaginated, dorsolateral walls of the mesencephalon (frog and chick). (See COLLICULI, SUPERIOR.) Syn., corpora bigemina.

OPTICOEL – the cavity of the primary optic vesicle.

OPTIC RECESS – a depression in the floor of the forebrain anterior to the optic chiasma which leads to the optic stalks.

OPTIC STALK – the attachment of the optic vesicle to the forebrain, at first a tubular connection between the optic vesicle and the diencephalon. The lumen is later (8-day chick) obliterated by the development of optic nerve fibers.

OPTIC THALAMI – the thickened lateral walls of the diencephalon.

OPTIC VESICLE – evaginations of forebrain ectoderm to form the primary optic vesicles which in turn invaginate to form the secondary optic cups of the eyes.

ORAL PLATE – fused stomodeal ectoderm and pharyngeal endoderm to form the oral membrane. Breaks through to form the mouth. Syn., pharyngeal membrane, oral membrane, stomodeal plate.

ORAL SUCKERS – elongated, pigmented depressions at the anteroventral ends of the mandibular arches of the frog embryo which give rise to mucous glands with adhesive function.

ORA SERRATA – the line of separation between the retinal and lenticular zones of the eye cup.

ORGANIZER – the chorda-mesodermal field of the amphibian embryo; a tissue area which has the power of organizing indifferent tissue into a neural axis; possibly comparable to Hensen's node of the chick embryo.

OSSEIN FIBERS – organic fibers in bone which give it strength and resilience.

OSTEOBLASTS – mesenchymal cells which actively secrete a calcareous material in the formation of bone; bone-forming cells.

OSTEOCLASTS – bone-destroying cells; cells which appear in and tend to destroy formed bone; constantly active, even in the embryo.

OSTIUM TUBAE ABDOMINALE – the most anterior, fimbriated end of the oviduct in female vertebrates; the point of entrance of the ovulated egg into the oviduct; double in amphibia and on the left side only in the female bird. Syn., infundibulum of the oviduct (see TUBAL RIDGES).

OSTIUM UROGENITALE – opening of the urogenital sinus (mammal) prior to the rupture of the cloacal membrane.

OTIC VESICLE – (see OTOCYST). Syn., auditory vesicle.

OTOCYST – the original auditory vesicle appearing at the level of the rhombencephalon in the amphibian embryo just before hatching and in the chick embryo at about the 16-somite stage, forming first as a placode. Syn., auditory vesicle.

OTOLITH – granular concretion found within the (embryonic) ear.

OVIDUCAL MEMBRANES OF OVUM – tertiary membranes applied over the egg as it passes through the oviduct, in both amphibia and chick.

OVIDUCTS – the paired Müllerian ducts in both males and females, which generally degenerate in the males (except in amphibia).

OVIGEROUS CORDS – columns or strands of tissue which divide the germinal epithelium of the primordium of the ovary, carrying primoridal germ cells with them and later breaking up into nests of cells each of which contains an oögonium. Syn., egg tubes or cords of Pflüger (mammal).

OVIPOSITION – the process of laying eggs.

OVOCYTE – (see OÖCYTE).

OVOGENESIS – (see OÖGENESIS).

OVOGONIA – (see OÖGONIA).

OVULATION – the release of the egg from the ovary, not necessarily from the body.

OVUM – Latin for egg, the haploid and fully matured female gamete.

PACHYTENE – stage in maturation when the allelomorphic pairs of chromosomes are fused (telosynapsis or parasynapsis) so as to appear haploid, during which process crossing over may occur; stage just prior to diplotene. Syn., pachynema. The term means "thick" or "condensed."

PAEDOGENESIS – reproduction during larval stage; precocious sex development.

PALATINE GLANDS – oral glands appearing after the eleventh day of the chick development, anterior, lateral, and posterior to the choanae.

PALATO-QUADRATE – true ossified bone developing from the dorsal parts of the first three visceral arches, a portion of which gives rise to the annulus tympanicus.

PALLIUM – outer, thickened walls of the telencephalic vesicles which will give rise to the cerebral hemispheres, dorsal and posterior to the rhinencephalon.

PANCREAS – digestive and endocrine glands arising from the gut as single dorsal and paired ventral primordia in the vicinity of the liver (72-hour chick).

PANDER'S NUCLEUS – (see NUCLEUS OF PANDER).

PAPILLARY MUSCLES – muscles which arise in the heart in conjuction with the tendinous cords and control the heart valves.

PARABIOSIS – lateral fusion of embryos, accomplished by injuring mirror surface and allowing them to grow together (see TELOBIOSIS).

PARADIDYMIS – the rudiment of the posterior Wolffian body which represents degenerate, nonsexual tubules of the mesonephros, found in the male bird.

PARAPHYSIS – a pouchlike evagination of the telencephalon median.

PARASYNAPSIS – lateral fusion, term applied to chromosomes of maturation stages in synaptene to experimental fusion of embryos (see TELOSYNAPSIS).

PARATHYROIDS – endocrine glands derived from endoderm of the third and fourth pairs of visceral pouches (mammal) and which control calcium metabolism.

PARENCEPHALON – the anterior portion of the diencephalon, separated from the synencephalon by the epiphysis in the chick of 7 days or older.

PARIETAL CAVITY – (see PERICARDIAL CAVITY).

PARIETAL RECESS – passage between the pericardial and peritoneal cavities of mammalian embryos.

PARTHENOGENESIS – development of an egg without benefit of spermatozoon.

PARTHENOGENESIS, ARTIFICIAL – initiation of development of an egg by artificial means.

PARTHENOGENESIS, NATURAL – maturation of eggs of some forms leads directly to development without the aid of spermatozoa.

PARTHENOGENETIC CLEAVAGE – fragmentation of protoplasm of old and unfertilized chick eggs, originally thought to be true cleavage.

PARTURITION – the process or act of giving birth.

PATH, COPULATION – (see COPULATION PATH).

PATH, PENETRATION – the initial direction of sperm entrance into the egg, often shifting toward the egg nucleus along a new copulation path. Syn., entrance path.

PECTEN – an aggregation of blood vessels embedded in mesenchyme, exclusively found in the bird eye about the fourth day of development, forming a ridge of tissue projecting into the optic cup through the proximal end of the choroid fissure. Becomes folded and fan shaped by the eighth day.

PENETRATION PATH – (see PATH, PENETRATION).

PENIS – elongated genital tubercle, enclosed in genital folds (prepuce), and associated with genital swellings (scrotal pouches) in male mammal; organ of transfer of genital products from the testes of the male to the vaginal cavity of the female. Syn., phallus.
PERFORATORIUM – (see ACROSOME).
PERIAXIAL CORDS – the primordia of the trigeminus and acustico-facialis ganglia which appear first at the 10-somite stage of the chick embryo and later mark the paths of the trigeminal and facial nerves. Distinguished as more deeply stained and concentrated masses than mesenchyme just posterior to the optic vesicles.
PERIBLAST – the portion of the chick blastoderm in which the cells are not completely separated by cell membranes; consisting of a few central periblast cells above the yolk and an ever-expanding periphery of marginal periblast cells. Central periblast is known as subgerminal syncytium to distinguish it from the peripheral (marginal) syncytium, or periblast.
PERICARDIAL CAVITY – the cavity or membranous sac which encloses the heart, representing a cephalic portion of the (amphibian) coelom or (in the chick) the original amniocardiac vesicles within the embryonic body, bounded anteriorly by the proamnion and posteriorly by the omphalomesenteric veins. Syn., parietal cavity.
PERICARDIUM – the thin mesodermal membrane which encloses the pericardial cavity and heart.
PERICHONDRIUM – mesenchymal layer immediately around forming cartilage.
PERICHORDAL SHEATH – a thin, mesodermal (sclerotomal), continuous sheet of tissue immediately around the notochord.
PERILYMPH – at first loose and spongy mesenchymatous tissue between the bony and membranous labyrinths (or the ear) in the chick, later to become more spacious and vascular.
PERIOSTEUM – mesenchymal layer, often originally perichondrium, which will be found immediately around forming bone.
PERISTOMIAL MESODERM – mesoderm of the amphibian gastrula derived from the (ventral) lips of the blastopore; in Amphioxus derived from the substance of the germ ring; in the chick derived from the primitive streak itself. Opposed to gastral mesoderm.
PERITONEAL CAVITY – the body cavity (coelom) separated from the pleural cavity by the pleuroperitoneal septum, including the septum transversum.
PERITONEUM – coelomic mesothelium of the abdominal region reinforced by connective tissue.
PERIVITELLINE MEMBRANE – (see VITELLINE MEMBRANE).
PERIVITELLINE SPACE – the space between the vitelline (fertilization) membrane and the contained egg, generally filled with a fluid. In the mammal, it is the space between the zona radiata and the egg.
PETER'S OVUM – one of the rare specimens of a very early human embryo, from which a great deal of information has been gathered.
PFLÜGER, CORDS OF – the ovigerous layer which grows into the stroma of the mammalian ovary as ovigerous cords, carrying primitive ova with them.
PHALLUS – (see PENIS).
PHENOTYPE – the outward appearance of an organism regardless of its genetic make-up, opposed to genotype.
PIGMENT LAYER OF OPTIC CUP – thin outer wall of the optic cup which never fuses with the rods and cones of the retina and becomes the pigment layer of the retina.
PINEAL – (see EPIPHYSIS).
PITUITARY – (see HYPOPHYSIS).
PLACENTA – an extraembryonic vascular structure of placental mammals which serves as an organ of nutritive and respiratory exchange between the fetus and the mother. (See different types under COTYLEDENOUS, DECIDUATE, DIFFUSE, DISCOIDAL, INDECIDUATE, and ZONARY PLACENTA. See also TRUE PLACENTA.)
PLACODE – platelike thickening of ectoderm from which arise sensory or nervous structures.
PLACULA – a flat blastula which may be without a cavity, as in Cynthia.
PLANE – an imaginary two-dimensional surface; may be frontal, sagittal, transverse, median, or lateral.
PLASMOSOME – a true nucleolus (see NUCLEOLUS).

PLASMOTROPHODERM – outer layer of syncytial cells of the trophoderm following implantation. Syn., syncytiotrophoderm.

PLECTRUM – (see COLUMELLA).

PLEURA – membrane enclosing the cavity surrounding the lungs, consisting of splanchnic mesoderm.

PLEURAL CAVITY – the portion of the (chick and mammalian) coelomic cavity separated ventrally by the septum transversum (pleuroperitoneal septum) diaphragm in mammals from the peritoneal cavity and one into which the primary lung buds grow.

PLEURO-PERITONEAL SEPTUM – membranous outgrowth from the sides of the esophagus which helps to separate the pleural and peritoneal cavities, in the chick.

PLEXUS, CHOROID – vascular folds in the roof of the prosencephalon, diencephalon, and rhombencephalon.

POIKILOTHERMAL – adjective meaning cold blooded; animals whose body temperatures are subject to environmental changes because they lack regulating mechanisms. Opposed to homoiothermal.

POLAR – pertaining, in most cases, to the animal pole, although may refer to the vegetal pole or both.

POLAR BODY – relatively minute, discarded nucleus of the maturing oöcyte (generally three). Syn., polocytes.

POLARITY – axial distribution of component parts; animal and vegetal poles.

POLE, ANIMAL – region of the egg where the polar bodies are eliminated; ectoderm forming portion of the precleaved egg. Syn., apical or animal hemisphere.

POLE, VEGETAL – region of the egg opposite the animal pole; region of lowest metabolic rate; pole with greater density of yolk in telolecithal eggs, generally the endoderm-forming region of the egg.

POLYEMBRYONY – production of several separate individuals from one egg by an early separation of its blastomeres, possible origin of some identical twins.

POLYESTROUS CYCLE – reproductive cycles which occur at least several times a year, e.g., pig.

POLYESTRUS – mammals which have more than a single estrus cycle in 1 year.

POLYINVAGINATION – theory of Pasteels that individual (endoderm) cells of the chick blastoderm migrate beneath the surface of the blastodisc and spread out to form the second of the primary germ layers.

POLYSPERMY – insemination of an egg with more than a single sperm, generally occurring in the chick egg although but a single sperm nucleus is functional, in syngamy.

PONS VAROLII – the thickened floor and ventrolateral zones of the metencephalon.

PONTINE FLEXURE – cephalic flexure of the chick indicated by a ventral bulge in the floor of the myelencephalon, appearing first at about 72-hours.

POSTANAL GUT – a posteriorly projecting blind pocket of the hindgut of the chick embryo appearing first at about 72 hours; that portion of the hindgut posterior to the anal plate or proctodeal plate. Syn., postcloacal gut, Bursa of Fabricus.

POSTBRANCHIAL BODIES – epithelial rudiments of the fifth pair of vestigial visceral pouches in the chick embryo, similar to the embryonic tissues of the thymus and epithelial bodies.

POSTERIOR INTESTINAL PORTAL – opening from the unformed midgut into the formed hindgut of the chick embryo.

POSTERIOR TUBERCLE – (see TUBERCULUM POSTERIUS).

POSTESTRUS – short period of regressive changes in the uterine mucosa following fruitless preparation for pregnancy.

POSTREDUCTION – maturation in which the equational and reduction divisions occur in that order.

POUCH, MARSUPIAL – an abdominal pouch of marsupials within which the prematurely born young are protected and provided with nourishment until independent of the mother.

PREFORMATION – old theory that the adult is represented in miniature within the egg or sperm and that development is simply enlargement.

PREGNANCY – condition of actually bearing an embryo or fetus within the uterus.

PREMIGRATORY GERM CELL – in chick, yolk-laden cells of splanchnopleuric origin which migrate by way of blood vessels to the gonad primordia. Believed to be the precursors of gonad stroma and/or functional germ cells.

PREORAL GUT – the extension of the pharynx anterior to the oral plate and behind the hypophysis, which flattens out and finally disappears. Syn., Seessel's pocket.

PREREDUCTION – maturation in which the reductional and equational divisions occur in that order.

PRIMARY OÖCYTE – the termination of the growth phase in the maturation of the ovum from the oögonial stage, prior to any maturational divisions.

PRIMARY SPERMATOCYTE – stage in spermatogenesis whose division results in secondary spermatocytes; stage beginning with growth of the spermatogonia.

PRIMITIVE FOLDS – the uplifted sides of the primitive streak of the chick possibly corresponding to the lips of the amphibian blastopore.

PRIMITIVE GROOVE – a groove through the center of the primitive streak, bounded by the primitive folds and terminated anteriorly by the primitive pit and posteriorly by the primitive plate.

PRIMITIVE GUT – the space between the endoderm and the yolk of the chick embryo, derived from part of the subgerminal cavity. Syn., archenteron or primitive intestine.

PRIMITIVE KNOT – (see HENSEN'S KNOT OR NODE).

PRIMITIVE OVA – oögonia within the ovary which multiply by mitosis without growth, in the mammalian ovary found along the cords of Pflüger.

PRIMITIVE PIT – the anterior end of the primitive groove, surrounded by Hensen's node, and probably representing the homologue in the chick of the amphibian neurenteric canal.

PRIMITIVE PLATE – the expanded posterior end of the primitive streak of the chick embryo.

PRIMITIVE STREAK – the first sign of embryonic differentiation in the chick blastoderm; the material out of which the notochord and mesoderm originate anteriorly; possibly the homologue of the fused lateral lips of the amphibian blastopore. The embryo develops anteriorly to the primitive streak. The streak first appears at about 18 hours in the chick.

PRIMORDIAL GERM CELLS – the diploid cells which are destined to become germ cells. Syn., primitive germ cells, e.g., oögonia and spermatogonia.

PRIMORDIUM – a rudiment; a group of cells which indicate a prospective development into a part or an organ. Syn., anlage.

PROAMNION – the area pellucida immediately anterior and lateral to the head process of the early chick embryo which is devoid of mesoderm for some time; consists only of ectoderm and endoderm.

PROCESSUS VAGINALIS – peritoneal recess into scrotal pouch.

PROCTODEUM – an ectodermal pit in the region of the future cloaca which invaginates to fuse with hindgut, evaginating endoderm to form the anal or proctodeal plate, later to rupture and form the anus.

PROESTRUS – period of active preparation of the uterine mucosa (endometrium) leading to estrus.

PROGESTERONE – endocrine secretion from the corpus luteum which causes the thickening of the endometrium.

PROGESTIN – a hormone from the corpus luteum found in all placentalia.

PRONEPHRIC CAPSULE – mesodermal connective tissue covering of the pronephric masses (amphibia) derived from adjacent myotomes and somatic mesoderm.

PRONEPHRIC CHAMBER – a portion of the amphibian coelomic cavity containing the glomus which is open anteriorly and posteriorly but closed ventrally by the development of the lungs.

PRONEPHRIC DUCT – the outer portion of the pronephric nephrotomes which develops a lumen connected posteriorly with the mesonephric or Wolffian duct. Syn., segmental duct.

PRONEPHRIC TUBULES – the lateral outgrowths of the most anterior nephrotomal masses which acquire cavities in the amphibia but not in the chick, connected with the pronephric duct. Possibly become infundibulum of oviduct.

PRONEPHROS – the embryonic kidney of all vertebrates, extending from the fifth to the sixteenth somites of the chick embryo and consisting of as many primitive tubules as somites concerned; completely lost in all adult vertebrates except a few fish and cyclostomes. Syn., head kidney.

PRONUCLEUS – the egg nucleus after polar body formation and the sperm nucleus after entrance of the spermatozoon into the egg.

PROPHASE – the first stage in the mitotic cycle when the spireme is broken up into definite chromosomes, prior to lining up on the metaphase (equatorial) plate.

PROSENCEPHALON – (see FORERAIN).

PROSOCOEL – cavity of the prosencephalon.

PROSTATE GLAND – secondary sex gland derived from the urethral epithelium surrounding the urethra near the neck of the bladder; secretes fluid for transport and activation of spermatozoa in mammal.

PROVENTRICULUS – the anterior division of the bird's stomach, the more posterior being the gizzard; derived from the foregut anterior to the liver diverticulum; the proventriculus becomes glandular as early as the sixth day.

PROXIMAL – nearer the point of reference, toward the main body mass.

PUPIL – the opening into the secondary optic vesicle, occluded in part by the lens, and regulated in diameter by the ciliary muscles of the iris.

PYGOSTYLE – the fused extreme posterior (terminal) vertebrae of the bird.

RAMUS-COMMUNICANS – the connection between the sympathetic ganglion and the spinal nerve, as numerous as the ganglia in any vertebrate; probably originating from the crest cells. Ramus means branch.

RATHKE'S POCKET – the tubular ectodermal rudiment of the hypophysis of most vertebrates (see HYPOPHYSIS).

RAUBER'S CELLS – the remnant of the mammalian trophoblast at the point of junction with the embryonic knob, in rabbit.

RECAPITULATION THEORY – theory that embryonic development reviews the major steps in evolutionary history (see qualifications under BIOGENETIC LAW).

RECTUM – posterior portion of the hindgut, lined with thickened endodermal epithelium, which opens directly into the cloaca.

REDUCTIONAL MATURATION DIVISION – one of the two important divisions in the maturation of gametes which results in the separation of allelomorphic (homologous) pairs of chromosomes so that the resulting cells are invariably haploid. Syn., meiotic division, disjunctional division. Opposed to equational division.

REGENERATION – repair or replacement of lost part or parts by growth and differentiation after the phase of primordial development. The vast organizing potencies of the different regions of the early embryo are lost after the completion of development, and there remain only certain regions of the body which are said to be capable of regeneration. Regenerative powers are more extensive among embryos and adults of phyletically low forms.

REGIONS, PRESUMPTIVE – regions of the blastula which, by previous experimentation, have been demonstrated to develop in certain specific directions under normal ontogenetic influences.

REGULATION – a reorganization toward the normal pattern of the whole; the power of pregastrula embryos to utilize materials remaining, after partial excision, to bring about normal conditions; more flexible power than regeneration.

RENAL CORPUSCLES – derivatives of the intermediate cell mass (glomerulus and Bowman's capsule) located adjacent to the median face of Wolffian body in the chick embryo. Syn., Malpighian body.

RENAL PORTAL SYSTEM – the venous system which carries blood to the mesonephric kidneys, involving the lateral portions of the caval veins, parts of the posterior cardinals, the iliacs, and the dorsolumbars. Found in adult amphibia, the most striking evidences of recapitulation.

RETE CORDS – strands of epithelial cells, containing many primordial germ cells which connect with the seminiferous tubules and later become the vasa efferentia in the bird. Syn., rete testis.

RETINAL ZONE – ectodermal derivatives of the optic cup consisting of the internal limiting membrane, retinal and lenticular layers, and outer pigmented layer. The retinal layers include portions from internal limiting membrane to the rods and cones.

RHINENCEPHALON – most primitive part of the telencephalon, concerned with the olfactory sense and including the olfactory bulb, olfactory tract, and pyriform lobe.
RHOMBENCEPHALON – (see HINDBRAIN).
RHOMBOIDAL SINUS – (see SINUS RHOMBOIDALIS).
RUTTING PERIOD – period of heightened sexual desire on the part of some male mammals, which generally coincides with sexual receptivity on the part of the female.
RUTTING SEASON – brief period of sexual activity occurring in some males; term used by animal breeders.

SACCULE – the outer and ventral portion of the inner ear from which is derived the cochlea. Associated with the auditory nerve (VIII). Syn., sacculus.
SACCUS ENDOLYMPHATICUS – the original endolymphatic duct, closed off from the exterior, which (in the 20-mm frog tadpole) grows up over the myelencephalon to join the other sac and form a vascular covering of the brain. In the chick, the saccus on the sixth day lies between the utricle and the hindbrain and by the eighth day becomes folded, vascular, and glandular, lying above the brain.
SACCUS VAGINALIS – scrotal pouch, peritoneal pocket in scrotal pouch.
SACHS' LAW – all cells tend to divide into equal parts and each new plane of division tends to intersect the preceding one at right angles.
SAGITTAL – a mesial plane, or any plane parallel to it, dividing the right parts of the body from the left. Right angles to both the frontal and transverse planes.
SCLEROTIC COAT – a tough mesenchymatous and sometimes cartilaginous coat outside of the choroid coat of the vertebrate eye. Syn., sclera.
SCLEROTOME – loose mesenchymal cells proliferated from the inner and ventral edges of the somites (5-mm frog) which contribute to the formation of the axial skeleton. In the chick (96 hours), three parts are distinguished: the narrow, undivided perichordal part, the dense aggregations of caudolateral cells, and the cephalic portion, all of which contribute to the axial skeleton.
SCROTUM or SCROTAL SAC – a single or subdivided chamber, external to the body proper, within which the testes of mammals (except elephant) are retained, the internal body temperature being too high for survival of the spermatozoa.
SECONDARY OÖCYTE – the stage in oögenesis between primary oöcyte and ovum, may be either haploid or diploid, depending upon species considered and which maturation division occurs first.
SECONDARY SPERMATOCYTE – the stage in spermatogenesis whose next division results in haploid spermatids, these spermatocytes being either haploid or diploid depending upon species considered (see POSTREDUCTION and PREREDUCTION).
SECRETORY TUBULE – the portion of the kidney tubule actually involved in the excretory process.
SECTION – generally a slice of an embryo, often of microscopic dimensions, taken in any one of the various planes such as frontal, transverse, or sagittal (see SERIAL SECTIONS).
SEESSEL'S POCKET – the remnant of the preoral gut of the chick embryo, after rupture of the oral plate, which gradually flattens out and disappears (see PREORAL GUT).
SEGMENTAL PLATE – (see VERTEBRAL PLATE).
SEGMENTATION – repetition of structural pattern; used as synonym for cleavage as well as for metamerism.
SEGMENTATION CAVITY – the cavity of the blastula; in the chick, it appears as result of cleavage parallel to the surface of the blastoderm, but the cavity never extends beneath the marginal cells of the periblast. Syn., subgerminal cavity, blastocoel.
SEMEN – mixture of secretions from the bulbourethral glands (Cowper's glands), prostate gland, seminal vesicles, and the suspended spermatozoa.
SEMICIRCULAR CANALS – tubular derivatives of the utricle lined with ectoderm from the otocyst which constitute the major balancing mechanisms of vertebrates.
SEMILUNAR VALVES – cuplike pockets within the aortic and pulmonary divisions of the bulbus of the heart, appearing during the fifth day of the chick embryo, which prevent the backflow of blood.

SEMINAL VESICLE – glandular dilatation of the distal end of the ductus deferens where spermatozoa collect prior to ejaculation.

SEMINATION – the act of fertilizing by the discharge of spermatozoa. The deposit of seminal fluid within the vagina. Syn., insemination.

SEMINIFEROUS TUBULE – tubular divisions of the testis derived from sexual (rete) cords, covered by a connective tissue theca and containing supporting (Sertoli) cells and (all) stages of spermatogenesis.

SEMIPLACENTA – type of placenta in which the uterine mucosa and the chorion do not actually grow together so that there is no tearing at birth. Syn., contact placenta.

SENSE PLATE – narrow band of elevated ectodermal tissue which passes transversely across the anterior end of the amphibian embryo ventral to the level of the fused neural folds, with the ends of the band bending dorsally to merge with the neural folds. Lower margins represent the mandibular arch, the plate giving rise to the mucous glands (oral suckers) of amphibia and to parts of the olfactory organs, lens of the eye, and possibly to part of the inner ear.

SEPTUM – a partition.

SEPTUM SPURIUM – an embryonic partition or ridge which soon undergoes retrogression, but which is a prolongation of the dorsal wall of the right atrium from one of the valvulae venosae, that effectively guards the sinus orifice against the backflow of blood from the heart. Syn., false septum.

SEPTUM TRANSVERSUM – a partition which separates the peritoneal and pericardial cavities of the chick embryo, appearing on the third day, and composed of three parts: (1) a median mass made up of the liver, sinus, and ductus venosus and the dorsal and ventral ligaments, (2) the lateral mesocardia, and (3) the lateral closing fold which extends from the mesocardia to the ventrolateral body wall. Beginning of diaphragm of the pig.

SERIAL SECTIONS – thin (often of microscopic dimensions) sections of embryos which are mounted on slides in the order of their removal from the embryo, so that a study in sequence will provide an understanding of all organ systems from one region of the embryo to the other.

SEROAMNIOTIC CONNECTION – the final point of fusion of the various amniotic folds as they converge above the chick embryo, the knot of tissue remaining throughout the entire embryonic period of the chicks which modifies the formation and disposition of the embryonic membranes. Original ectodermal knot becomes displaced by mesoderm which later develops a lumen allowing the connection between the albumen sac and the amniotic cavity (eleventh day). Syn., seroamniotic raphe.

SEROSA – syn., chorion.

SERTOLI CELL – derivative of the sexual cords of the testis, found within the seminiferous tubule and functionally similar to the follicle cell in the ovary in that it is the nutritive, supporting, or nurse cell of the maturing spermatozoa. The heads of adult spermatozoa may be seen embedded in the cytoplasm of Sertoli cells.

SEX-CELL CORD – division of the sex-cell ridge or gonad primordium, not to be confused with sexual (rete) cords.

SEX DETERMINATION – generally means either the conditions of fertilization which determine the ultimate sex of the embryo or the predetermination of sex by experimental means.

SEXUAL CORDS – derivatives of the germinal epithelium from which they become separated and give rise to the bulk of the gonads of both sexes.

SEXUAL CORDS OF THE OVARY – sex cords of the originally indifferent gonad primordium which form only the cords of the medulla of the ovary, the functional follicles coming from the germinal epithelium (of the chick).

SEXUAL CORDS OF THE TESTIS – sex cords of the originally indifferent gonad primordium which give rise to the seminiferous tubules of the testis, forming a rather solid mesenchymatous reticulum to the twentieth day in the chick embryo when cavities lined with spermatogonia (from primordial germ cells) and Sertoli (from peritoneal cells) begin to appear, the whole constituting the seminiferous tubules.

SEXUAL CYCLE, of the uterus – periodic sequence of changes in the ovary and uterine mucosa of the female placenta in mammals. Apart from pregnancy, the cycle in nonprimates includes uterine changes as follows:

Diestrus – period of quiescence, if short.
Anestrus – period of quiescence, if long.
Proestrus – period of construction, destruction, and some repair.
Estrus – period of repair.

The cycle is related to ovulation and to endocrine cycle.

SHEATH, MYELIN – myelin covering of axones in the so-called white matter of the spinal cord and peripheral nerves.

SHELL – a calcareous covering deposited on the chick (bird) egg during the 12-16 hour period it remains in the uterus, made up of carbonates and phosphates of calcium and magnesium; structurally consisting of an:
inner mammillary layer (see MAMMILLAE) of calcareous particles,
intermediate spongy layer of matted calcareous strands,
outer, surface cuticle, which is porous, but otherwise structureless.

SHELL GLAND – the glandular portion of the bird uterus which provides the materials of the shell.

SHELL MEMBRANE – double membrane, consisting of a thick outer and thin inner layer of matted organic fibers crossing each other in all directions, the two membranes separated only by the air space at the blunt end of the egg.

SINUS RHOMBOIDALIS – the region of the receding primitive streak around which the posterior ends of the neural folds are diverted in the chick embryo.

SINUS TERMINALIS – an encircling blood vessel which appears at the outer margin of the area vasculosa of the chick embryo before the end of the first day; found also in eggs of marsupials although no yolk is present. Syn., vena terminalis.

SINUS VENOSUS – the point of fusion of vitelline veins of the amphibian embryo or the most posterior of the original four chambers of the (chick) heart at 33 hours; bilaterally symmetrical and related to the ducts of Cuvier and the ductus venosus to 45 hours; right horn elongated and left reduced with shifting of sinoauricular aperture (60-70 hours) and at 96 hours has the shape of horseshoe between atrium and the septum transversum. Involved in formation of right auricle.

SITUS INVERSUS – an inversion or mirror transposition of the usual relations, sometimes used to describe an abnormal position of the gut.

SKELETOGENOUS SHEATH – sclerotomal cells which first form a continuous layer around both the notochord and nerve cord.

SKIN – (see DERMIS and EPIDERMIS). Syn., integument.

SOMATIC – relating to body in contrast with germinal cells; or relating to the outer body in contrast to inner splanchnic mesoderm.

SOMATIC UMBILICUS – a short, thick, hollow stalk composed of ectoderm and somatic mesoderm continuous with the amnion, which connects the chick embryo with the underlying yolk sac and the extraembryonic membranes; umbilicus closed just before the time of hatching after the remnants of the yolk have passed into the gut.

SOMATOBLAST – blastomeres with specific germ layer predisposition, i.e., ectodermal somatoblasts.

SOMATOPLEURE – the layer of somatic mesoderm closely associated ectoderm, the extension of which (from the body wall) gives rise to both the amnion and chorion.

SOMITE – blocks of paraxial mesoblast metamerically separated by transverse clefts, derived from enterocoelic or gastral mesoderm and giving rise to the dermatome, myotome, and sclerotome.

SPAWNING – the act of expelling eggs from the uteri of anamniota, e.g., amphibia.

SPERM – the germ cell characteristically produced by the male. Syn., spermatozoon, sperm cell, male gamete.

SPERMATID – a product of the second maturation division in spermatogenesis, the spermatids having certain cytological characteristics and being invariably haploid; cells which go through a metamorphosis into functionally mature spermatozoa.

SPERMATOCYTE – stages in spermatogenesis between the time the primordial germ cell (spermatogonium) begins to grow, without division, until after the division which results in spermatids (see PRIMARY and SECONDARY SPERMATOCYTES).

SPERMATOGENESIS – the entire process which results in the maturation of the spermatozoon.

SPERMATOGONIUM – the primordial germ cell of the male gonad, indistinguishable from somatic cells, which are also diploid; stage prior to maturation when the presumptive spermatozoon undergoes rapid multiplication by mitosis.

SPERMATOPHORE – sperm-bearing bundle, such as are shed by male urodeles, the bundles later to be picked up by clocal lips of the female.

SPERMATOSPHERE – (see IDIOSOME).

SPERMATOZOON – the functionally mature male gamete. Syn., sperm.

SPINA BIFIDA – a split involving the spine in a developing embryo caused by a variety of environmental conditions, most of which act through interference with normal gastrulation and neurulation.

SPINAL CORD – that portion of the central nervous system, excluding the brain, which is derived from the epithelium and neural plate, consisting of ependyma, glia, neuroblasts, and their derivatives and connecting cells.

SPINDLE – a group of fibers between the centrosomes during mitosis to which the chromosomes are attached and by means of which (mantle fiber portion) the chromosomes are drawn to their respective poles.

SPINOUS PROCESS – prolongation of spinal processes fused dorsally to the spinal canal. Syn., dorsal spine of vertebra.

SPIRACLE – short funnel between the body wall and the operculum on the left side of the head of the frog tadpole which is the only exit for water passing through the gill chambers to the exterior.

SPIREME – a continuous chromatin thread once believed to be characteristic of the so-called resting cell nucleus.

SPLANCHNIC – refers to the viscera, opposed to somatic or body.

SPLANCHNIC MESODERM – the visceral mesoderm, or that nearest the embryo in the lateral plate.

SPLANCHNOCOEL – that portion of the enterocoel (of Amphioxus) which lies between the somatic and splanchnic mesoderm within the body. Syn., coelom.

SPLANCHNOCRANIUM – that portion of the skull which is preformed in the cartilage of the visceral arches. Opposed to neurocranium.

SPLANCHOPLEURE – the layer of endoderm and inner mesoderm (splanchnic) within which develop the numerous blood vessels of the area vasculosa and later the yolk sac septa; the layers within the body which give rise to the lining and to the musculature of the alimentary canal.

SPLEEN – this organ, seen first at 96 hours, arises in the chick as a proliferation from the peritoneum covering on the left side of the dorsal mesentery just anterior to the pancreas.

SPONGIOBLASTS – cells of the mantle layer of the developing spinal cord destined to form merely supporting tissue.

SPONGIOSA – the glandular layer of the uterus adjacent to the muscularis to which the trophoderm is attached, the other portion of the uterine mucosa being the compacta.

STEREOBLASTULA – solid blastula as found in Crepidula.

STIGMA – (see CICATRIX).

STOMODEUM – ectodermal invagination (pit) which fuses with the pharyngeal endoderm to form the oral plate, which later ruptures to form the margins of the mouth cavity. The stomodeal portion of the mouth lining is therefore ectodermal.

STRATUM GRANULOSUM – layer of follicle cells surrounding the mammalian ovum.

STROMA – the mesodermally derived, medullary, supporting tissues of an organ.

SUBCARDINAL VEINS – embryonic veins ventral to the nephric tissue the posterior portions of which fuse to contribute to the formation of the inferior vena cava in the chick.

SUBCLAVIAN VEINS – these arise primitively as branches of the posterior cardinals and receive blood (in the chick) from the wings and walls of the thorax.

SUBGERMINAL CAVITY – (see BLASTOCOEL or SEGMENTATION CAVITY).

SUBNOTOCHORDAL ROD or BAR – the hypochordal rod in the amphibian embryo, or the point of fusion of the caudal sclerotomal (cephalic vertebral) components beneath the notochord of the chick embryo.

SUBSTANTIA PROPRIA – mesenchyme of the cornea continuous with the sclera.

SUBZONAL LAYER – the layer of cells within which lies the morula of the mammalian blastocyst, later called the trophoblast.

SUCKER – (see ORAL SUCKERS).

SULCUS LIMITANS – longitudinal groove between the dorsal alar plate and the ventral basal plate best seen at the level of the myelencephalon; mark for locating the nuclei and fiber tracts.

SUPERNUMERARY NUCLEI – nuclei of excess spermatozoa which invade the chick egg at fertilization, many causing accessory cleavages and then degenerating.

SUSTENTACULAR CELL – a cell which provides nourishment for another, such as the Sertoli or follicle cells of the gonads.

SYLVIUS, AQUEDUCT OF – (see AQUEDUCT OF SYLVIUS).

SYMPATHETIC SYSTEM – originating from ectodermal cells of the neural crests, organizing as a chain of ganglia near the dorsal aorta, and controlling the involuntary (visceral) musculature.

SYNAPSIS – the lateral (parasynapsis) or terminal (telosynapsis) union of embryos; or pairing of homologous chromosomes.

SYNAPTENE STAGE – the stage in maturation between the leptotene and the synezesis (contraction) stage wherein the chromatin is in the form of long threads intertwined in homologous pairs. Syn., zygotene or amphitene.

SYNCYTIAL TROPHODERM – layer of trophodermal cells of the human embryo outside of but probably derived from the cell layer of Langhans.

SYNCYTIUM – nuclei with cytoplasm without cellular boundaries; multinucleate protoplasm without cell boundaries.

SYNENCEPHALON – the third or most posterior of the three primary divisions of the forebrain which gives rise to the posterior portion of the diencephalon. (The other divisions are parencephalon and telencephalon.)

SYNEZESIS – the stage in maturation between synaptene and pachytene when the chromatin threads are short and thick and the ends away from the centrosome are tangled.

SYNGAMY – specifically the fusion of the gamete pronuclei but also the union of the gametes at fertilization. Syn., fertilization.

TAILBUD – concurrent with the development of the tail fold a mesodermal tailbud mass may be distinguished, at 48 hours in the chick embryo.

TAIL FOLD – a depression begins to develop beneath the posterior end of the chick embryo at about 48 hours, giving rise to a tail fold similar to the head fold except that from the beginning it is made up of ectoderm and somatic mesoderm. In mammals, the tail fold often appears prior to the head fold and is longer. Syn., amniotic tail fold.

TELA CHORIOIDEA – Syn., thin roof of third and fourth brain ventricles.

TELENCEPHALON – the portion of the forebrain anterior to a plane which includes the posterior side of the choroid plexus and the anterior side of the optic recess of the 5-mm frog embryo or the portion of the forebrain ventral to a plane passing from the posterior wall ventral to the optic recess to the anterior wall in the center of the velum transversum in the chick. Gives rise to cerebral hemispheres, corpora striata, paraphysis, anterior choroid plexus, olfactory lobes, lateral ventricles, and the foramina of Monro.

TELOBIOSIS – fusion of embryos end-to-end (see PARABIOSIS).

TELOCOEL – cavity or ventricle of the telencephalon.

TELOLECITHAL – (see EGG, TELOLECITHAL).

TELOPHASE – last phase in mitosis when the respective chromosome groups have reached their respective astral centers and are beginning to reform a resting cell nucleus; the stage often accompanied by the beginning of cytoplasmic division.

TELOSYNAPSIS – end-to-end fusion of chromosomes (see PARASYNAPSIS).

TERATOLOGY – the study of anomalies or monster formation.

TERTIARY MEMBRANES – the soft membraneous and hard calcareous coverings of the chick egg, deposited on the egg within the uterus.

TETRADS – paired (homologous) chromosomes which have become duplicated longitudinally in anticipation of the meiotic (reductional) division. When viewed from one end will appear as a group of four chromosomes, hence a tetrad.

THALAMUS – lateral wall of the diencephalon which becomes thickened by masses of central gray matter and by the development of fibers passing to the cord from the more anterior parts of the cerebral hemispheres.

THECA – connective tissue covering, generally refers to coverings of ovarian follicle.

THECA EXTERNA – the outermost of the coverings of the ovarian follicle, rather loose connective tissue with abundant blood supply. Continuous with ovarian stroma.

THECA FOLLICULI – refers to membranous and cellular coverings of the ovum within the ovary.

THECA INTERNA – the layer of connective tissue consisting of closely packed fibers, possibly some of smooth muscle, immediately external to the ovarian follicle of birds and mammals. Less vascular and more compact than theca externa.

THYMUS – derivatives of first pair of branchial pouches of the frog (12 mm) embryo which separate from the pouches and migrate to a position posterior to the auditory capsules near the surface of the head; in the chick this gland is derived from the intermediate part of the third pair of visceral pouches (first branchial) and a small portion of the fourth pair (second branchial) which form a voluminous tract of lobulated material extending the entire length of the neck of the later embryo and finally all but atrophying at sexual maturity. Endocrine function.

THYROGLOSSAL DUCT – a temporary tubular connection between the thyroid primordium and the pharynx near the base of the tongue.

THYROID BODY or GLAND – originates as an endodermal thickening in the floor of the pharynx between the second pair of visceral arches (48-hour chick); evaginates to form a vesicle temporarily connected with the gut by the thyroglossal duct; separates from gut by the fourth day; becomes divided by the seventh day; and migrates to junction of subclavian and common carotid arteries. Somewhat similar history in all vertebrate embryos. Endocrine function.

TISSUE CULTURE – *in vitro* culturing of isolated tissues; the excision of tissues or organs and their maintenance in an artificial medium, generally consisting in part of embryonic extracts or blood plasma.

TONGUE – solid mesodermal mass covered with endoderm, derived by cell proliferation from the floor of the pharynx beginning in the 9-mm frog tadpole; from two primordia in the chick, one anterior (tuberculum impar) and the other (pars copularis) posterior to the thyroid diverticulum, including parts of the bases of the first three visceral arches; the parts fusing after closure of the thyroglossal duct. Parts of the chick tongue become cornified.

TONSILS – lymphatic structures derived from the endoderm and mesoderm of the second pair of visceral pouches.

TORSION – the twisting of the (chick) embryo so that it lies on its left side, caused in part by the passive resistance of the underlying inert yolk plus the development of the various cephalic flexures.

TORUS TRANSVERSUS – thickening in the median ventroanterior wall of the lamina terminalis of the telencephalon, just exterior to the optic recess, representing the rudiment of the anterior commissure.

TOTIPOTENCY – related to the theory that the isolated blastomere is capable of producing a complete embryo.

TRACHEA – that portion of the respiratory tract between the larynx and the lung buds, lined with endoderm, probably derived from the posterior portion of the original laryngotracheal groove. Tube lengthens rapidly in chick by 6 days.

TRACHEAL GROOVE – (see LARYNGOTRACHEAL GROOVE).

TRANSPLANT – an embryonic area (cell, tissue, or organ) removed to a different environment.

TRANSVERSE – a plane (or section) which divides the anterior-posterior axis at right angles, separating the more anterior from the more posterior. Syn., cross (section), but this synonym is not generally satisfactory.

TRANSVERSE NEURAL FOLD – the continuation of the lateral neural folds (ridge) of the early frog embryo around the anterior end (i.e., region of face), the region of the temporary anterior neuropore. Syn., transverse medullary fold or ridge.

TRIGEMINAL GANGLIA – cranial (V) ganglia which consist of motor and sensory portions and arise from the most anterior crest segments in conjunction with cells from the inner (ganglionic) portion of the corresponding placode. They give rise to ophthalmic, mandibular, and maxillary branches; associated with the myelencephalon at the level of the greatest width of the fourth ventricle.

TROCHLEARIS NERVES – cranial (IV) motor nerves emerge from the dorsal surface of the brain near the isthmus, coming from neuroblasts in floor of midbrain and innervating the superior oblique muscles of the eye.

TROPHECTODERM – region of continuity of ectoderm and outer layer of mammalian trophoblast; extraembryonic ectoderm following germ layer differentiation.

TROPHOBLAST – thin layer of cells which constitute the wall of the mammalian blastocyst; outer layer of blastocyst prior to differentiation of the primary germ layers.

TROPHOBLASTIC VILLI – fingerlike projections of chorionic ectoderm comprising the trophoderm of mammalian embryos.

TROPHODERM – trophectoderm reinforced by a layer of somatic mesoderm. Syn., extra-embryonic somatopleure, serosa.

TRUE PLACENTA – a placenta in which the chorion and uterine mucosa are intimately associated, in contrast with the primitive contact type. Syn., burrowing placenta.

TRUNCUS ARTERIOSUS – anterior continuation of the bulbus arteriosus beneath the foregut, becomes divided by a spiral septum which is continuous through the bulbus to the ventricle; gives off the aortic arches which join the dorsal aorta. Syn., ventral aorta.

TUBAL FISSURE – longitudinal slit in the roof of the pharynx which connects the median chamber of the tubo-tympanic cavity with the oral cavity.

TUBAL RIDGES – the primordia of the Müllerian ducts or oviducts arising on the fourth day in the chick embryo lateral to each mesonephros and adjacent to the respective Wolffian ducts, the anterior ends of which (in the female) become the left ostium abdominalia tuba (only a left oviduct and ovary).

TUBERCULUM IMPAR – median thickening in the floor of the mouth associated with the formation of the tongue, covered by endoderm and derived largely from the hyoid arch.

TUBERCULUM POSTERIUS – a thickening in the floor of the brain at the region of the anterior end of the notochord, representing the posterior margin of the diencephalon.

TUBO-TYMPANIC CAVITY – remnants of the dorsal parts of the first pair of visceral (hyomandibular) pouches and the lateral walls of the pharynx, connecting the pharynx and the middle ear, represented by the Eustachian tube of the adult bird or mammal.

TUBULES – (see under specific names such as COLLECTING, MESONEPHRIC, PRONEPHRIC, SEMINIFEROUS).

TUNICA ALBUGINEA – (see ALBUGINEA OF TESTIS).

TYMPANIC CAVITY – cavity of the middle ear, a vestige of the hyomandibular pouch (see TUBO-TYMPANIC CAVITY).

TYMPANIC MEMBRANE – membrane made up of ectoderm, mesenchyme, and endoderm which separates the tympanic cavity from the exterior. Syn., ear drum.

ULTIMOBRANCHIAL BODY – (see POSTBRANCHIAL BODIES).

UMBILICAL ARTERIES – branches of the sciatic arteries of the chick embryo which appear on the fourth day, the right member being the smaller; carry blood to the allantois.

UMBILICAL VEINS – at first paired veins in the lateral body wall of the chick embryo (fourth day) which bring blood from the allantois and join the ducts of Cuvier, right vein disappearing and the left changing its connection to join the anterior half of the ductus venosus. At hatching, only the proximal portion persists as a vein in the ventral body wall.

UMBILICUS – the stalklike connection between the embryo and all extraembryonic structures, including the somatic stalk, allantoic stalk plus its arteries and veins, and the yolk stalk with its arteries and veins.

UMBILICUS, SOMATIC – (see SOMATIC UMBILICUS).

UMBILICUS, YOLK SAC – (see YOLK-SAC UMBILICUS).

URACHUS – the canal which connects the allantois and the urinary bladder in mammalian embryos.

URETER – diverticulum from the posterior end of the Wolffian (mesonephric) duct appearing paired at the end of the fourth day in the chick embryo; or the mesonephric ducts of the amphibian embryo; functioning as excretory ducts of the adult.

URETHRA – single duct of the mammal which discharges urine from the bladder and also through which semen is liberated from the male genital tract into the genital tract of the female during coitus; mesodermal.

URINARY BLADDER – an endodermally lined vesicle derived from the hindgut, homologous to the allantois of the chick. Connected with the mesonephric (excretory) ducts of the frog only through the cloaca; connected with ureters and urethra of the mammal.

URINIFEROUS TUBULE – functional kidney tubule of both mesonephros and metanephros.

URINOGENITAL DUCT – ducts which open into the cloaca of the male amphibia and convey both excretory and genital products, derived from the mesonephric (Wolffian) ducts.

URINOGENITAL RIDGES – ridges on the lateral surface of the Wolffian bodies (chick) which consist of three parts.

(1) anterior or sexual division, representing about half of the Wolffian body and containing the gonad rudiments;
(2) nonsexual region of the Wolffian body, posterior to the gonad rudiments;
(3) posterior to the Wolffian body as the Wolffian and Müllerian ducts.

URODEUM – intermediate portion of the embryonic cloaca, that portion which persists (in the chick) as the adult cloaca (see PROCTODEUM and COPRODEUM).

URODELE – tailed amphibia (e.g., salamanders). Syn., Caudata.

UROGENITAL SYSTEM – the entire excretory and reproductive systems, some embryonic parts of which degenerate before hatching or birth. Shows various degrees of common origin and ultimate function (see specific excretory and reproductive components).

UROSTYLE – the fused skeletogenous elements of the last two somites in the frog embryo which surround the end of the notochord as cartilage and finally ossify.

UTERINE GLANDS – glands situated within the uterine mucosa of mammals which secrete "uterine milk" a combination of fat, proteids, and glycogen which is absorbed by the trophoblast of the blastocyst.

UTERINE MILK – a viscid fluid secreted by the uterine mucosa consisting of fats, proteids, and glycogen, a source of embryonic nutrition in mammals.

UTERINE MUCOSA – mucosal lining of the uterus which shows cyclic changes in the mammal, associated with reproductive activity. Syn., endometrium.

UTRICLE – a vesicle, generally referring to the superior portion of the otocyst which gives rise to the three semicircular canals of the ear and into which these canals open. Lined with ectoderm.

VAGINA – short, thin-walled cavity which opens from the uterus into the cloaca, of the bird; cavity within which the hen's egg is retained about 4 p.m. until the following morning but the walls of which no contributions are received; the cavity of the female mammal possesing, at its external boundaries, a homologue of the male penis called the clitoris.

VALVES, SEMILUNAR – (see SEMILUNAR VALVES).

VASA DEFERENTIA – mesonephric or Wolffian ducts of the frog, which persist as the male gonoducts of the bird and mammal, connecting with the testes through the vasa efferentia and epididymis and functioning as sperm ducts after the degeneration of the embryonic mesonephros and the development of the gonads. Sing., vas deferens.

VASA EFFERENTIA – the ducts which convey (frog) sperm from the collecting tubules through the mesorchium to the Malpighian corpuscles of the mesonephric kidney; derived (in the chick) from rete cords and connected with the mesonephric tubules of the anterior (sexual) half of the mesonephric or Wolffian body which become the epididymis. Sing., vas efferens.

VEGETAL POLE – the pole of a telolecithal egg where there is the greatest concentration of yolk, generally opposite the animal pole and the location of the germinal vesicle. Syn., vegetal or vegetative hemisphere; abapical or antipolar hemisphere (see ANIMAL POLE).

VEIN – (see under specific names).

VELAR PLATE – folds or flaps developing anterior and posterior to the branchial regions of the frog (anuran) embryo, derived from the pharyngeal wall and serving as a gross sifting organ between the pharynx and the gill (branchial) chamber.

VELUM TRANSVERSUM – depressed roof of the telencephalon just anterior to the lamina terminalis, which later becomes much folded (chick) and vascular as the anterior roof of the third ventricle. The division point between the telencephalon and diencephalon.

VENA CAVA ANTERIOR – junction of inferior jugular (anterior cardinal) and (in the chick) the subclavian and vertebral veins which empty into the ductus Cuvieri and later the right auricle. Syn., superior vena cava, or superior caval veins.

VENA CAVA POSTERIOR – the single median ventral vein which represents the remnant of the right posterior cardinal; subsequently, it receives the hepatic vein prior to joining the sinus venosus and still later is connected directly to the right auricle. Syn., inferior vena cava.

VENA PORTA SINISTRA – some of the veins from the gizzard and proventriculus of the chick embryo which enter the left lobe of the liver.

VENTRAL – belly surface. Ventrad means toward the belly surface.

VENTRAL LIGAMENT OF LIVER – (see FALCIFORM LIGAMENT).

VENTRAL MESENTERY – double layer of splanchnic mesoderm which connects the alimentary canal with the ventral body wall and extraembryonic splanchnopleure in the embryo; in the region of the hindgut includes both somatic and splanchnic mesoderm as a thick mass of mesoblast which binds the hindgut to the somatopleure. In the region of the foregut and midgut of the later chick embryo, this includes the ductus venosus and the liver material.

VENTRICLE III – main cavity (diocoel) of the forebrain, related to paired lateral ventricles or telocoels, by way of the foramina of Monro.

VENTRICLE IV – main cavity of the hindbrain (rhombencephalon) connected anteriorly with the aqueduct of Sylvius and posteriorly with the neural canal and extending through both the metencephalon and myelencephalon (chick and mammal), having as a roof the vascular posterior choroid plexus.

VENTRICLE, LATERAL – (see LATERAL VENTRICLES OF THE BRAIN).

VENTRICLE OF THE HEART – singular (in frog) or double (bird and mammal) and very muscular chamber of the heart developing from the myocardium, subdivided by septa and provided with valves; connected with bulbus arteriosus anteriorly.

VERTEBRAE – derivatives of the sclerotome which surround the nerve cord and notochord and finally incorporate the notochord by chondrification and ossification (centrum).

VERTEBRAL ARCH – (see NEURAL ARCH).

VERTEBRAL PLATE – (see AXIAL MESODERM). Syn., segmental plate.

VESICLE – (see under specific names).

VILLUS – a fingerlike or hairlike projection, such as the chorionic villus which is a projection of the mammalian chorion into folds of the uterine mucosa.

VISCERAL – pertaining to the viscera.

VISCERAL ARCHES – generally, six pairs of mesodermal masses between the visceral pouches and lateral to the pharynx of all vertebrate embryos, including the mandibular, hyoid, and four branchial arches. Each arch is bounded by the endoderm on the pharyngeal side and ectoderm on the outside. The chick has only rudimentary fifth and no sixth arches, the maximum development being on the fourth day. (See derivatives in connection with specific arches.) Syn., visceral arches IV to VI are also called branchial arches I to IV respectively; pharyngeal arch.

VISCERAL CLEFTS – slitlike openings between the pharynx and the outside, found in vertebrate embryos on either side of visceral arches II to V or less, consisting of peripheral lining of ectoderm and mesial lining of endoderm. Syn., pharyngeal, and some may be called gill or branchial clefts.

VISCERAL FURROWS – ectodermal invaginations which may meet endodermal pharyngeal evaginations to form visceral clefts. Syn., visceral groove.

VISCERAL GROOVE – (see VISCERAL FURROWS).

VISCERAL MESODERM – (see SPLANCHNIC MESODERM and SPLANCHNOPLEURE).

VISCERAL PLEXUS – a network of sympathetic neurones which control the viscera, having grown posteriorly from the tenth (vagus) cranial ganglia.

VISCERAL POUCH – endodermal evagination of the pharynx which, if it meets the corresponding visceral furrow, often breaks through to form the visceral cleft. Syn., pharyngeal pouch.
VITAL STAIN – localized staining of living embryonic areas with vital, nontoxic, dyes.
VITELLINE – pertains to yolk, e.g., vitelline vein brings blood from the yolk, vitelline membrane is that which covers the yolked egg.
VITELLINE ARTERIES – paired omphalomesenteric vessels which, in the chick, later fuse (as the dorsal mesentery forms) to go out through the umbilical stalk to the yolk as the vitelline (yolk) arteries, originating from the dorsal aorta.
VITELLINE MEMBRANE – delicate, outer, nonliving egg covering derived while the egg is still within the ovary, probably by joint action of the egg and its follicle cells; probably the same membrane that is elevated as the fertilization membrane after successful insemination (see FERTILIZATION MEMBRANE).
VITELLINE SUBSTANCE – yolk.
VITELLINE VEIN – (see OMPHALOMESENTERIC VEINS).
VITREOUS HUMOR – the rather viscous fluid of the eye chamber posterior to the lens, formed by cells budded from the retinal wall and from the inner side of the lens, hence ectodermal and probably also mesenchymal in origin (see AQUEOUS HUMOR.)
VIVIPAROUS – animals which bring forth young in advanced state of development, more advanced than eggs, e.g., humans.

WING BUD – ectoderm and mesenchyme, probably of sclerotomal origin, which forms paired buds at the level of the fourteenth to sixteenth somites by the fourth day of chick development; the primordia of the wings and possibly parts of the pectoral girdle.
WOLFFIAN BODY – (see MESONEPHROS).
WOLFFIAN DUCT – (see MESONEPHRIC DUCT, URINOGENITAL DUCT, and VASA DEFERENTIA).

YOLK – highly nutritious food (metaplasm) consisting of nonnucleated spheres and globules of fatty material found in all except alecithal eggs. Syn., lecithin.
YOLK NUCLEI – darkly staining chromatinlike substances within the cytoplasm of the young (immature) eggs around which the yolk is accumulated during the growth phase of oögenesis. May be derived from nucleoli which escape from the nucleus.
YOLK PLUG – a plug formed by large yolk cells which are too large to be immediately incorporated in the floor of the archenteron of the amphibia embryo, hence are found protruding slightly from the blastopore. Size of the plug is often used to determine the approximate stage of gastrulation.
YOLK SAC – extraembryonic splanchnopleure which grows around and finally all but encircles the yolk mass, remaining connected with the midgut by the yolk stalk; finally taken into the midgut through the stalk just before hatching. Function is to provide extensive surface area for absorption of nutriment.
YOLK-SAC ENDODERM – the endoderm of the splanchnopleure surrounding the yolk itself, which develops glandular structures within the septa for digestion of the yolk at points of contact.
YOLK-SAC PLACENTA – type of marsupial placenta in which the blood vessels are derived from the area vasculosa of the yolk sac, e.g., Dasyurus.
YOLK-SAC SEPTA – deep folds of splanchnopleure which surrounds the yolk, the folds projecting into the yolk to accomplish digestion by means of glands which secrete enzymes.
YOLK-SAC SPLANCHNOPLEURE – that portion of the splanchnopleure which is in direct contact with the yolk (chick).
YOLK-SAC UMBILICUS – the region where the yolk of the chick egg is not surrounded by the yolk sac until the development of the albumen sac is complete.
YOLK STALK – thick walled stalk which connects the midgut and the yolk sac of the developing chick embryo, obvious at 5 days.

ZONA PELLUCIDA – transparent, noncellular, secreted layer immediately surrounding the mammalian ovum, corresponding to the vitelline membrane of lower forms.
ZONA RADIATA – zona pellucida which exhibits radial striations, not to be confused with the corona radiata. Syn., zona pellucida of the mammalian egg.
ZONARY PLACENTA – the placenta of carnivores in which the villi are arranged in bands or zones.
ZONES, LENTICULAR – (see LENTICULAR ZONE).
ZONES OF JUNCTION – portion of marginal periblast of chick embryo which is not separated from the germ wall beneath by any cytoplasmic cleavages, i.e., region of syncytium, peripheral to which is the zone of overgrowth. Syn., region of the germ wall.
ZONE OF OVERGROWTH – narrow superficial rim of marginal periblast cells of the early chick embryo beyond the zone of junction which represents epibolic growth out over the unsegmented yolk. Syn., margin of overgrowth.
ZYGOTE – the diploid cell formed by the union of two haploid gametes. Syn., fertilized egg.

A supplementary list of some 350 specialized terms may be found in Rugh *Experimental Embryology.* See also all-inclusive glossary on embryology in Rugh *Vertebrate Embryology* pp. 528-73.

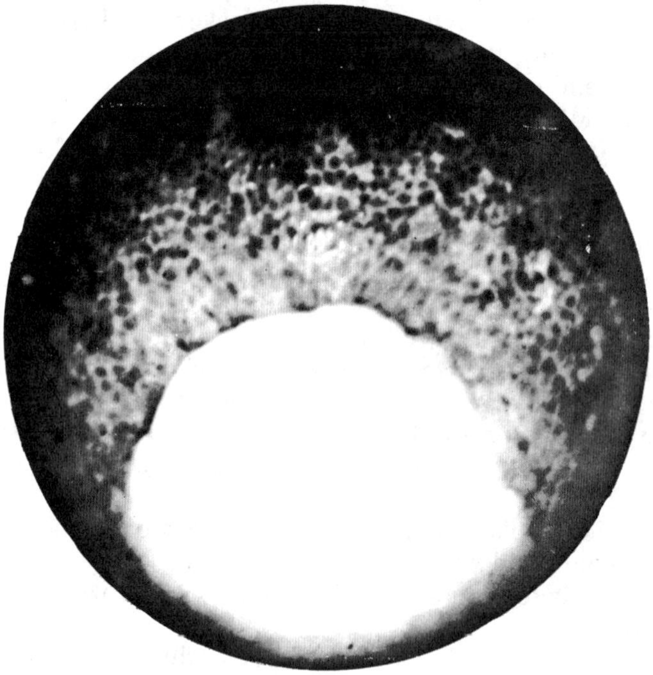

"OMNE VIVUM EX OVO"
PASTEUR

"OMNIS CELLULA E CELLULA"
VIRCHOW

If we have unwittingly omitted acknowledgement of ownership or have been unable to obtain permissions for usage of any material in this *Guide*, we sincerely apologize. We will be most happy, however, upon being presented with proof of ownership, to pay an appropriate fee as if we had been able to obtain permission prior to publication.

ILLUSTRATED PLATES

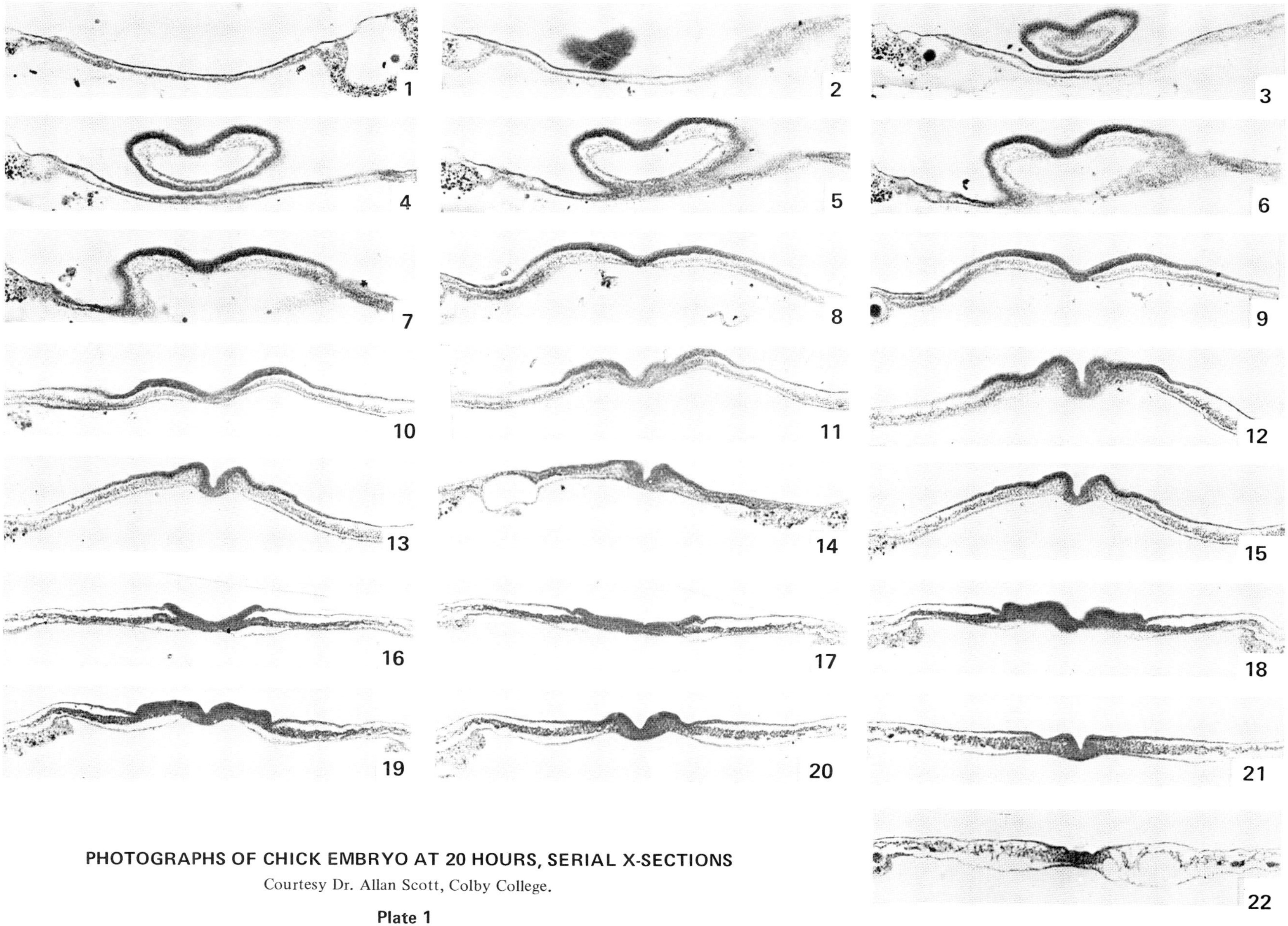

PHOTOGRAPHS OF CHICK EMBRYO AT 20 HOURS, SERIAL X-SECTIONS

Courtesy Dr. Allan Scott, Colby College.

Plate 1

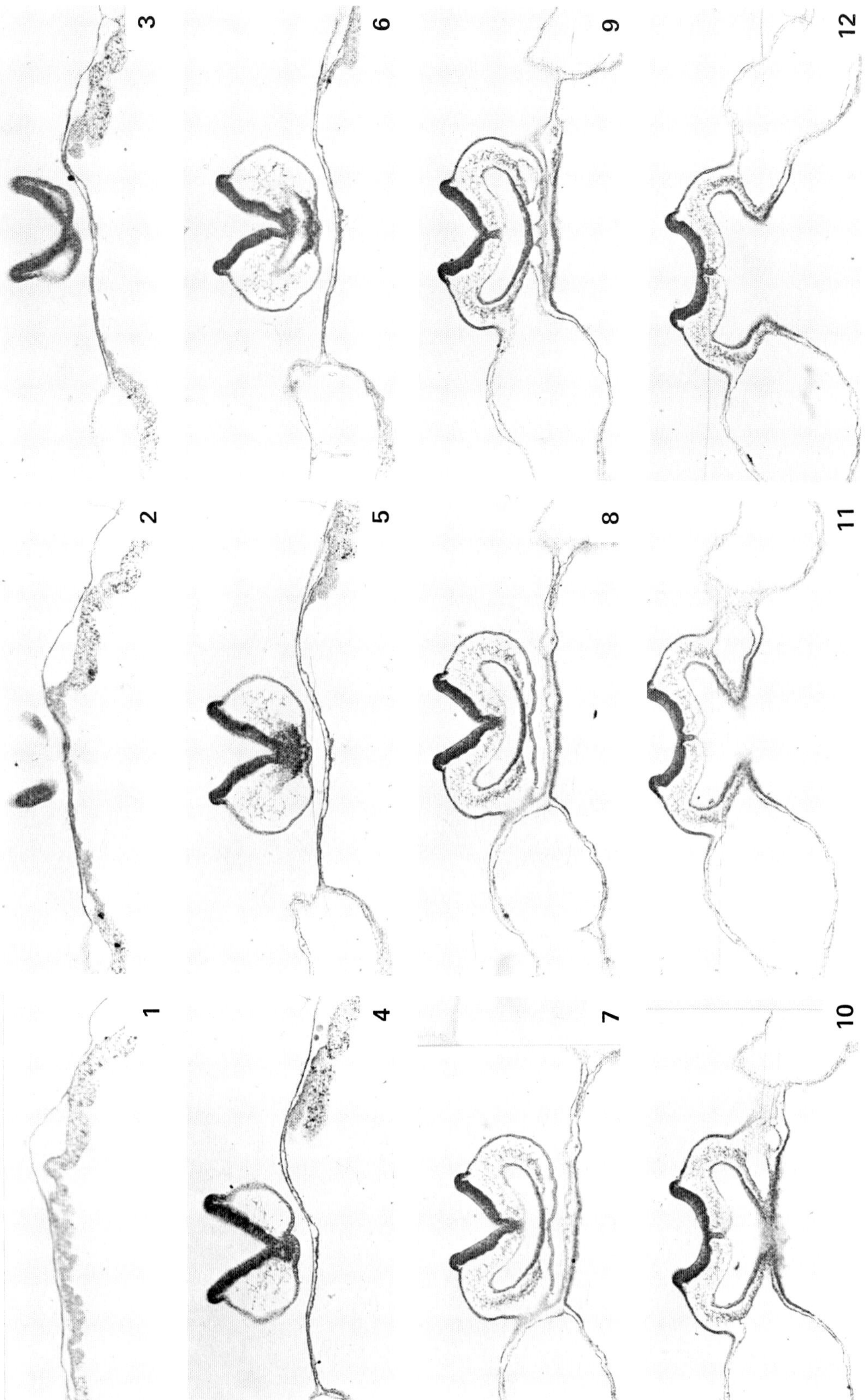
1
2
3
4
5
6
7
8
9
10
11
12

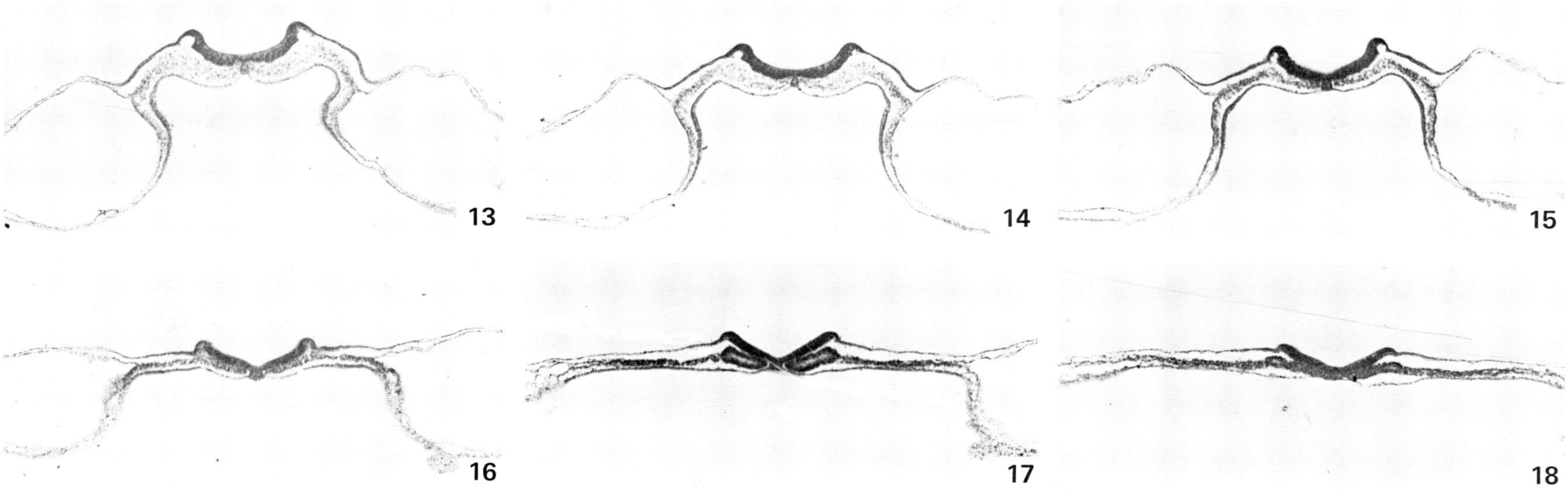

PHOTOGRAPHS OF CHICK EMBRYO AT 24 HOURS, SERIAL X-SECTIONS

Courtesy Dr. Allan Scott, Colby College.

Plate 2

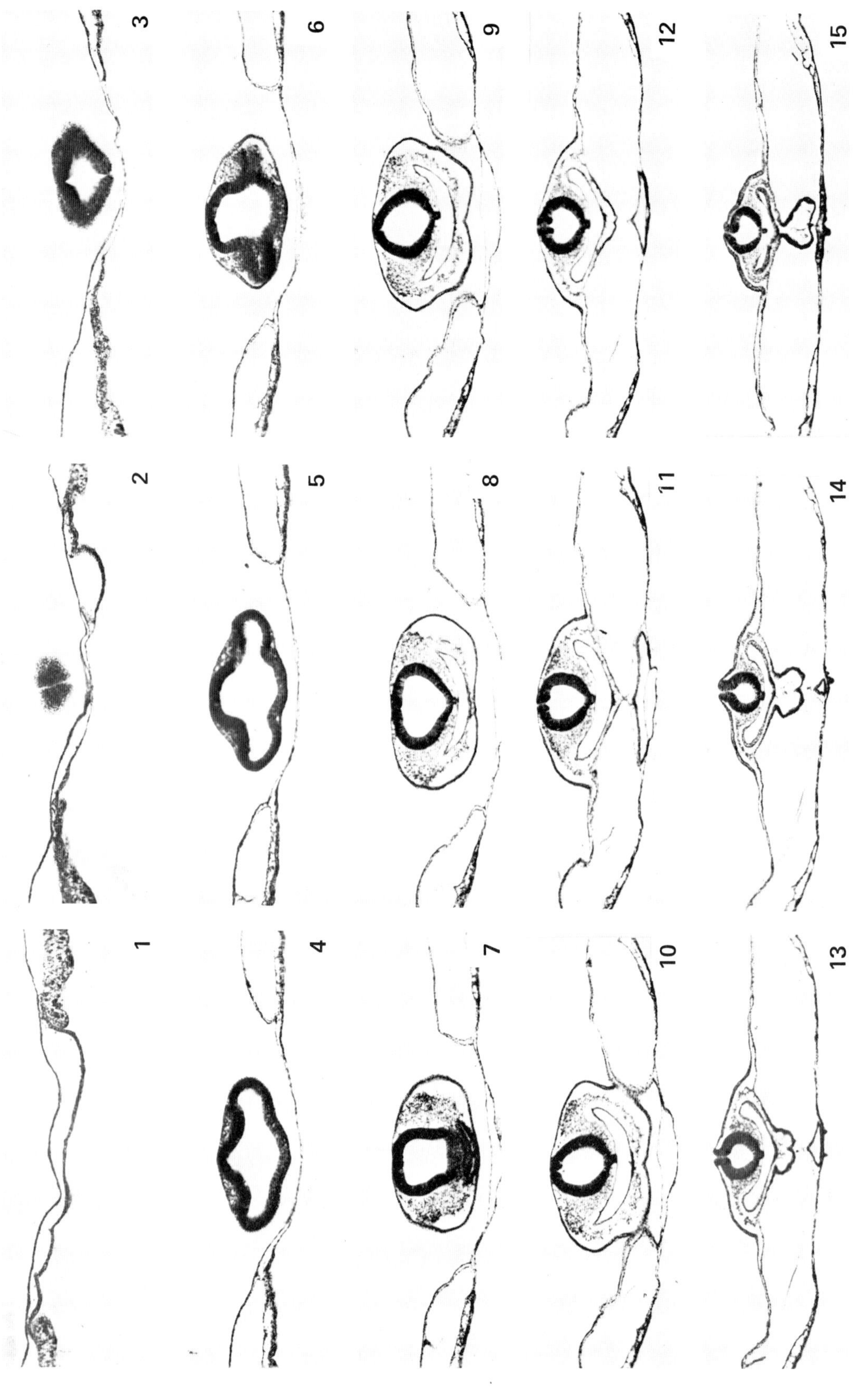
1
2
3
4
5
6
7
8
9
10
11
12
13
14
15

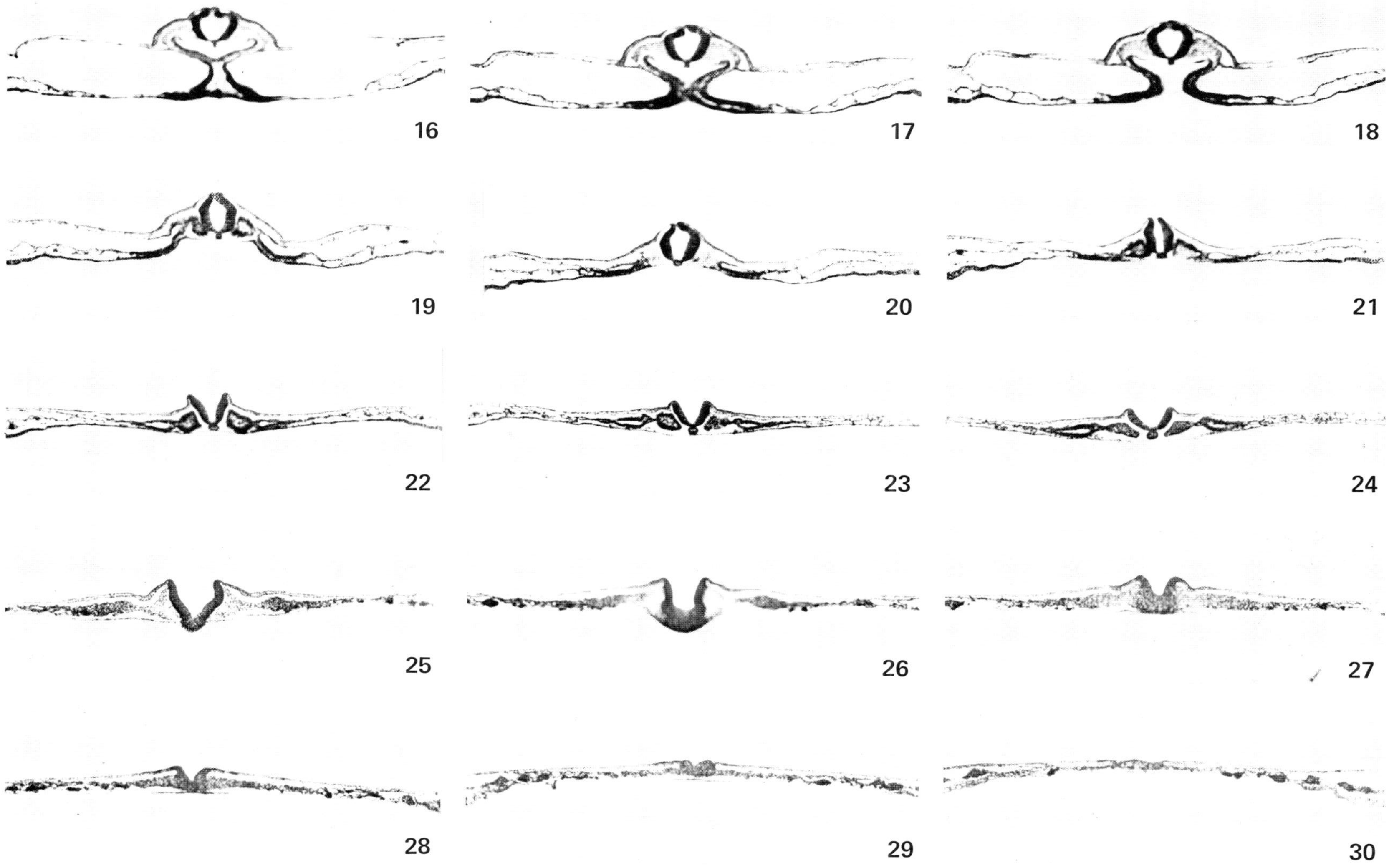

PHOTOGRAPHS OF CHICK EMBRYO AT 33 HOURS, SERIAL X-SECTIONS
Courtesy Dr. A. C. Scott, Colby College.

Plate 3

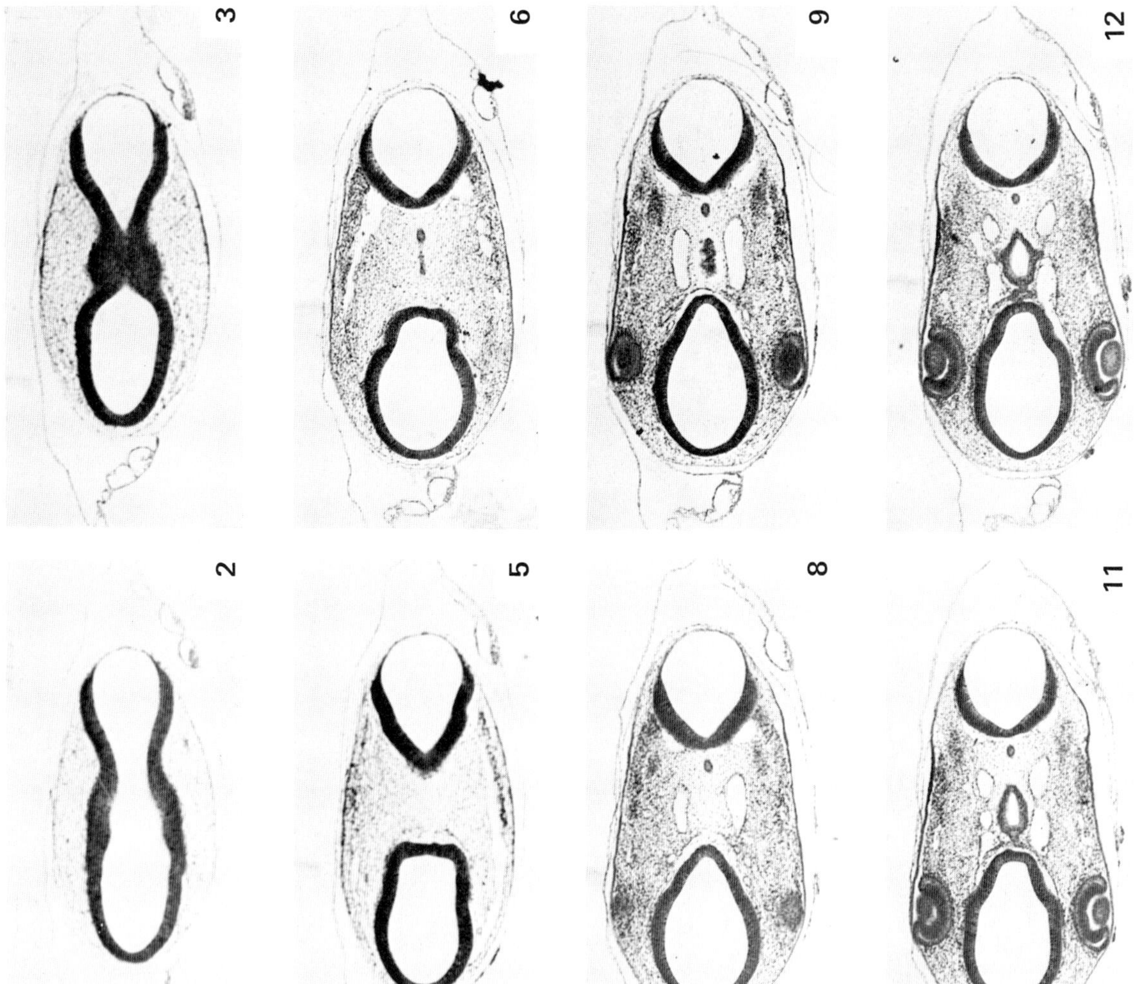
1
2
3
4
5
6
7
8
9
10
11
12

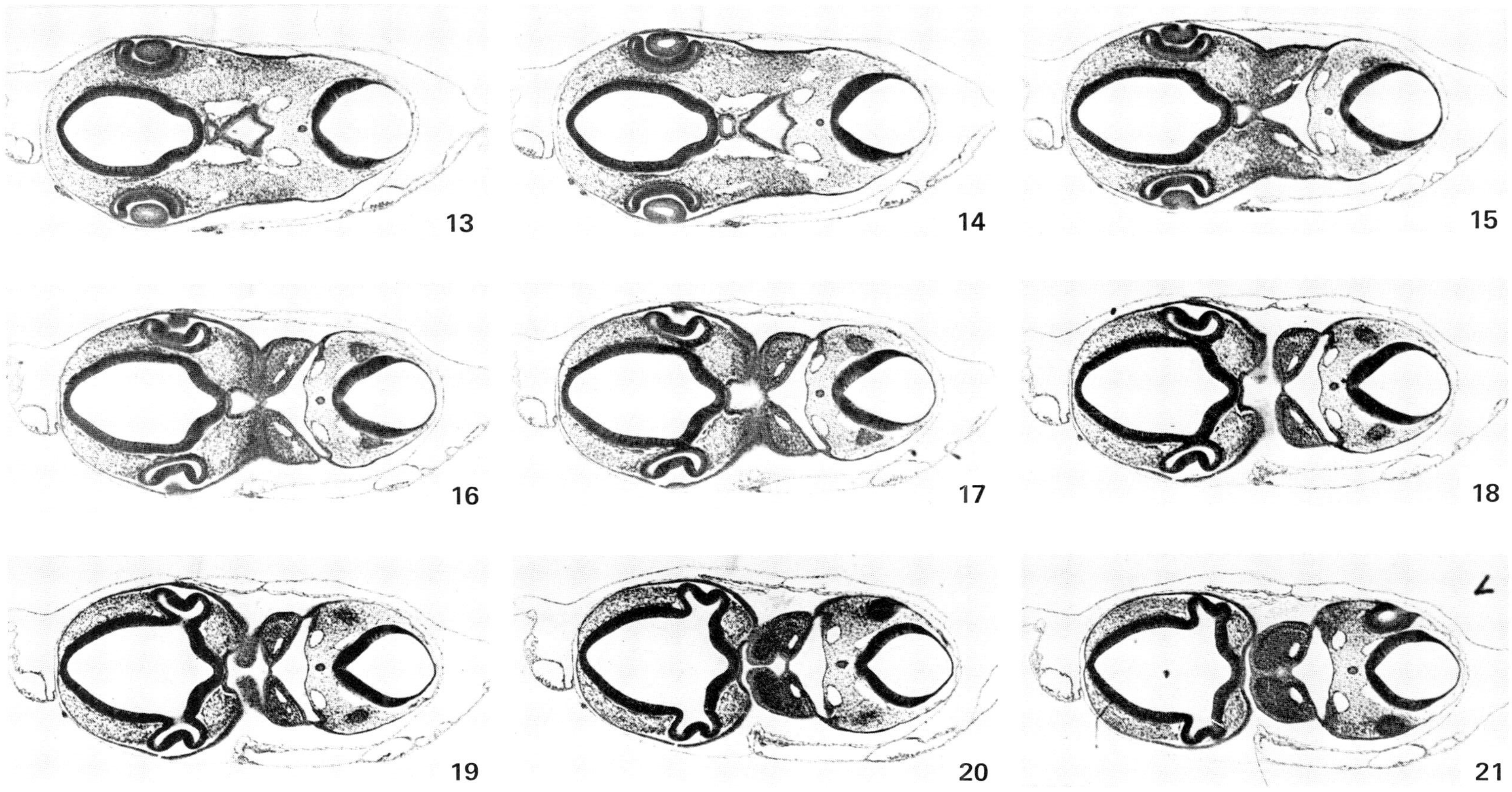

PHOTOGRAPHS OF CHICK EMBRYO AT 48 HOURS, SERIAL X-SECTIONS

Courtesy Dr. A. C. Scott, Colby College.

Plate 4

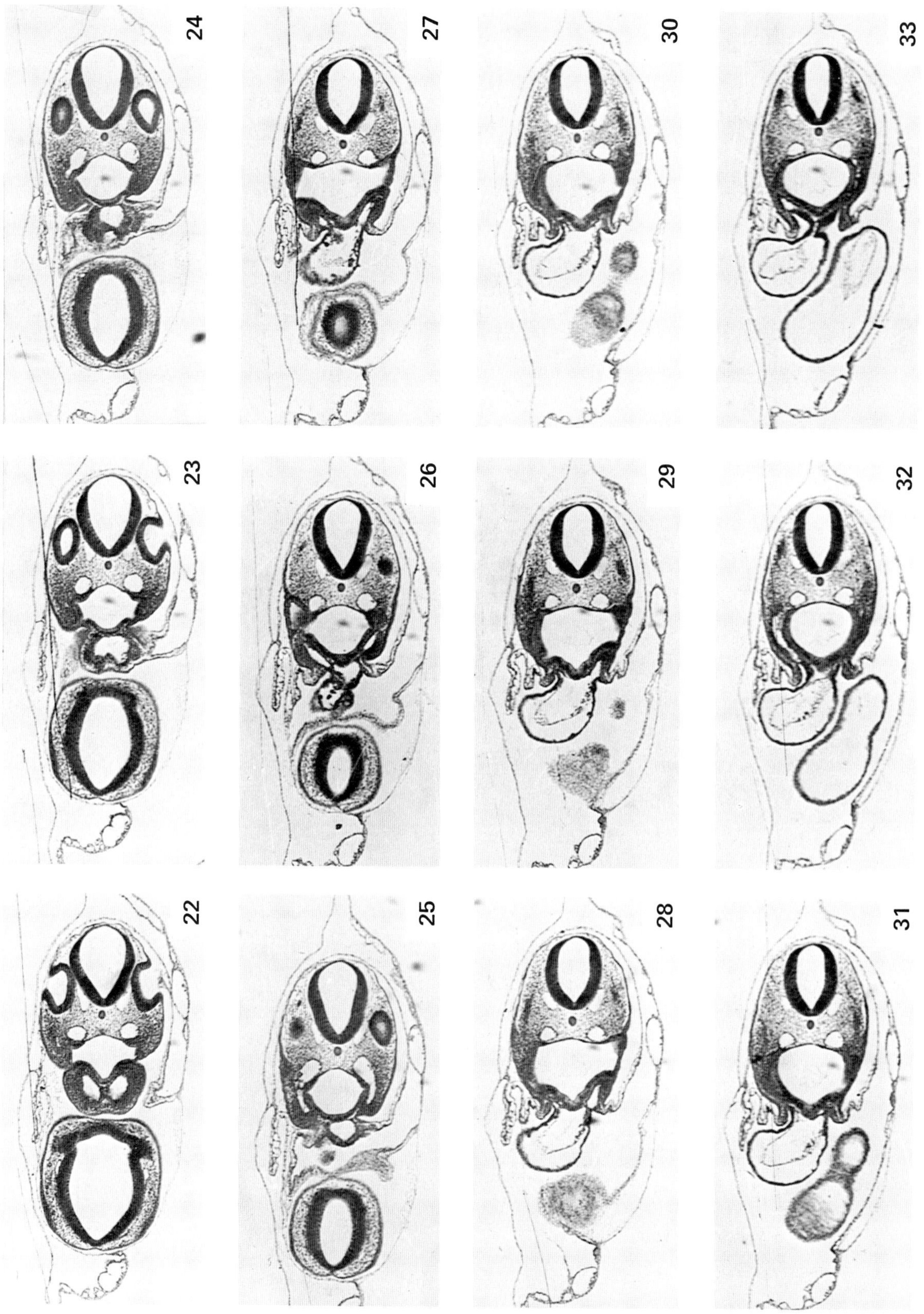
22
23
24
25
26
27
28
29
30
31
32
33

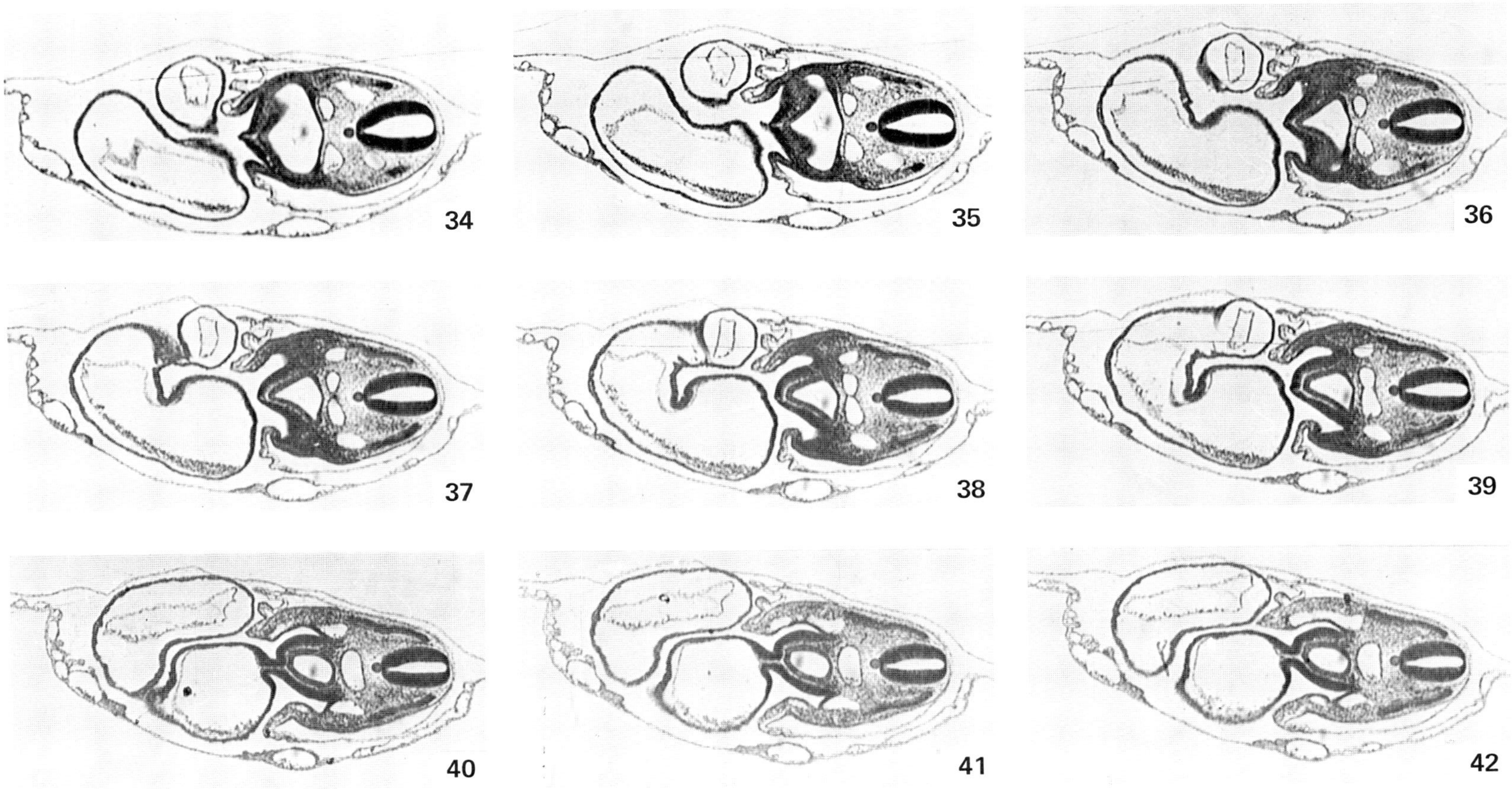

PHOTOGRAPHS OF CHICK EMBRYO AT 48 HOURS, SERIAL X-SECTIONS

Courtesy Dr. A. C. Scott, Colby College.

Plate 4

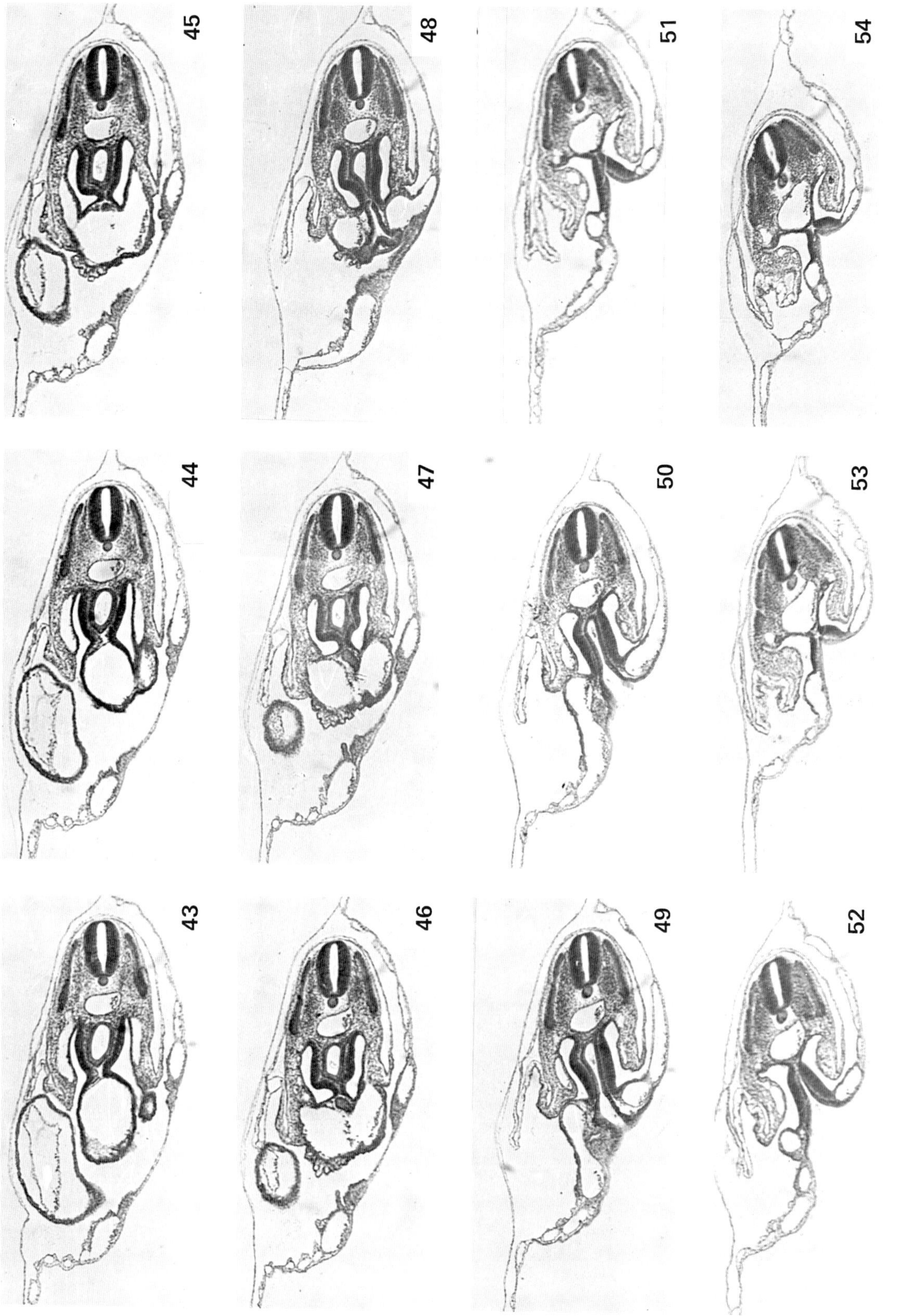
45
48
51
54
44
47
50
53
43
46
49
52

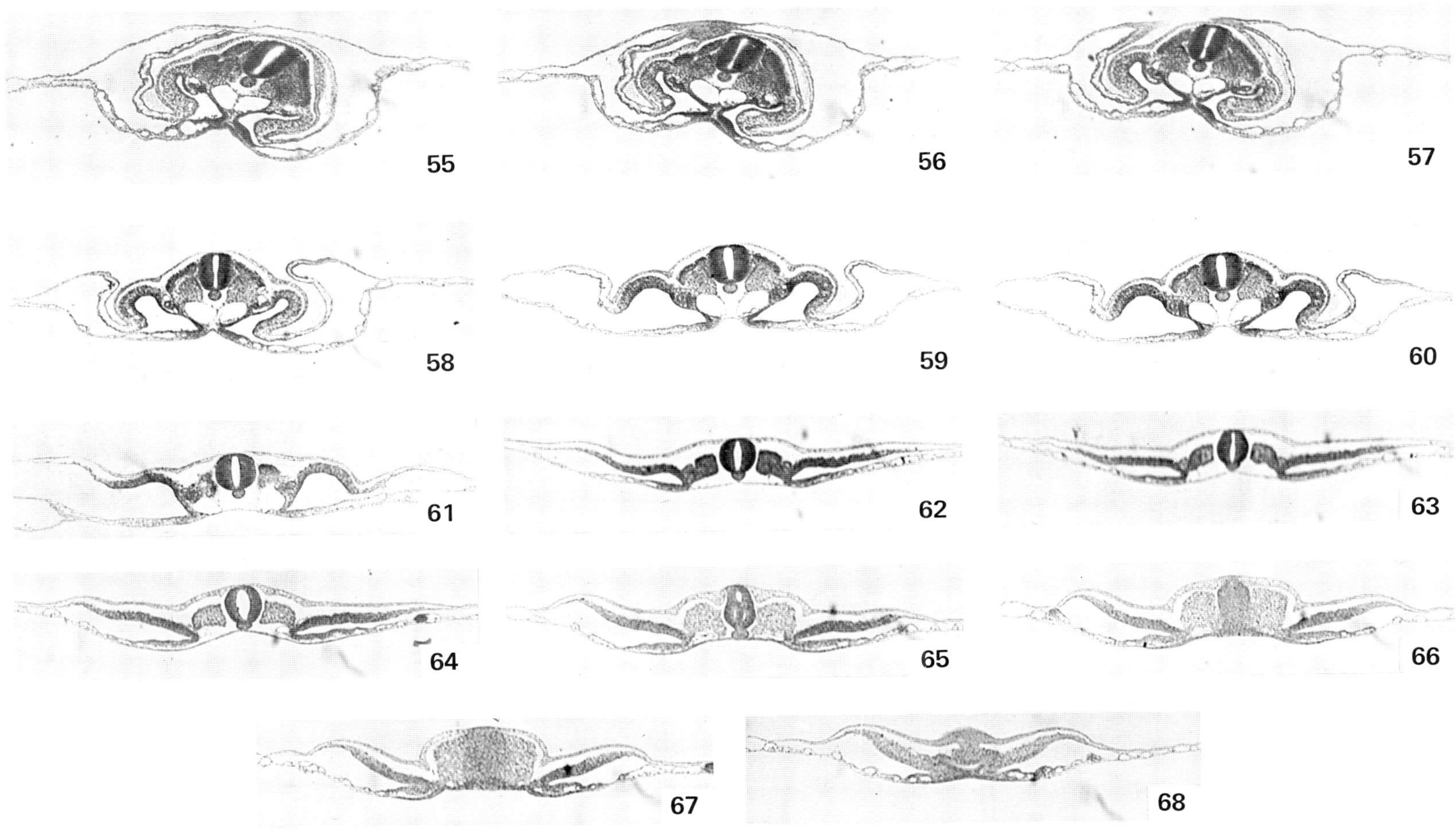

PHOTOGRAPHS OF CHICK EMBRYO AT 48 HOURS, SERIAL X-SECTIONS

Courtesy Dr. A. C. Scott, Colby College.

Plate 4

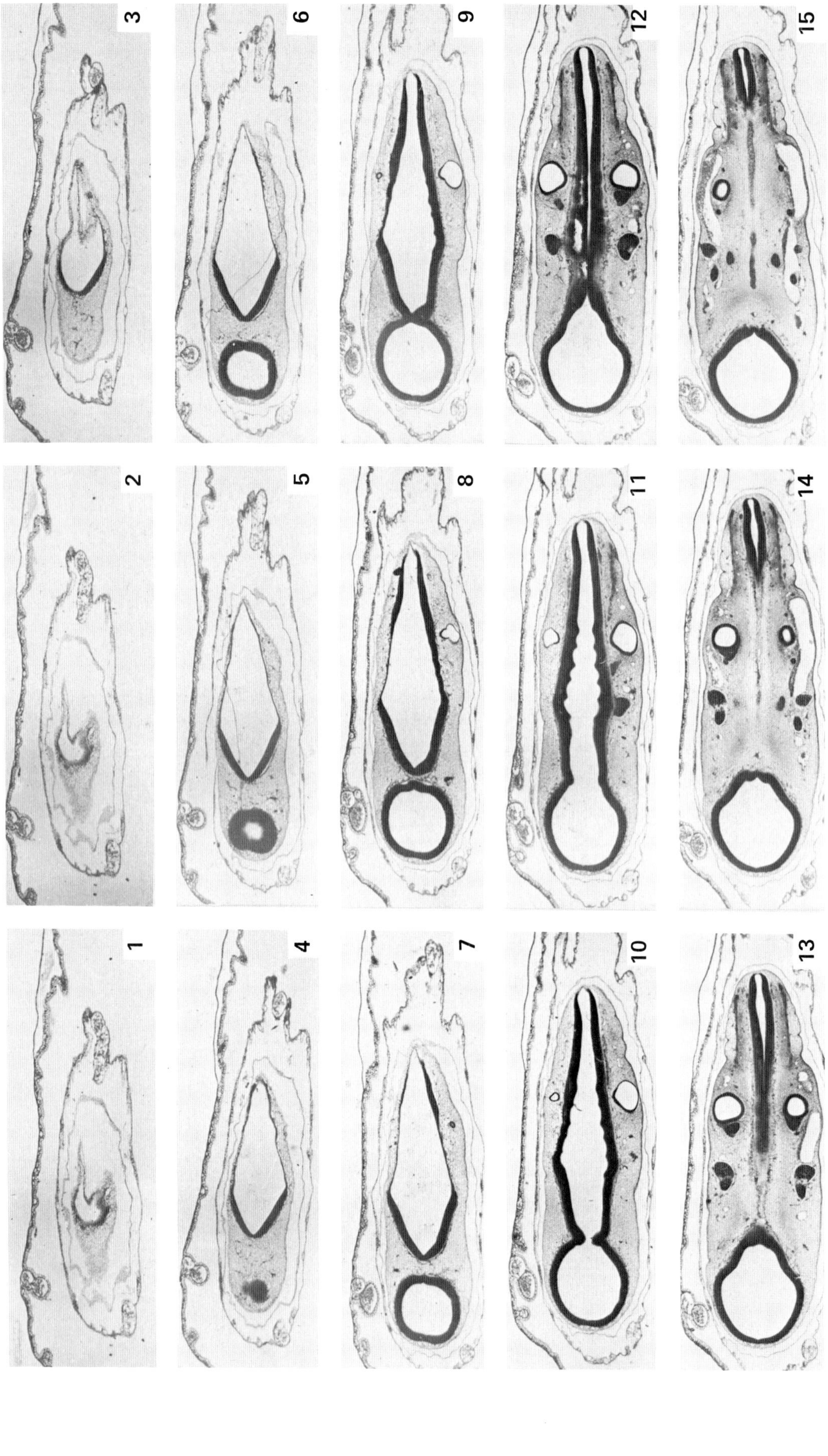
1
2
3
4
5
6
7
8
9
10
11
12
13
14
15

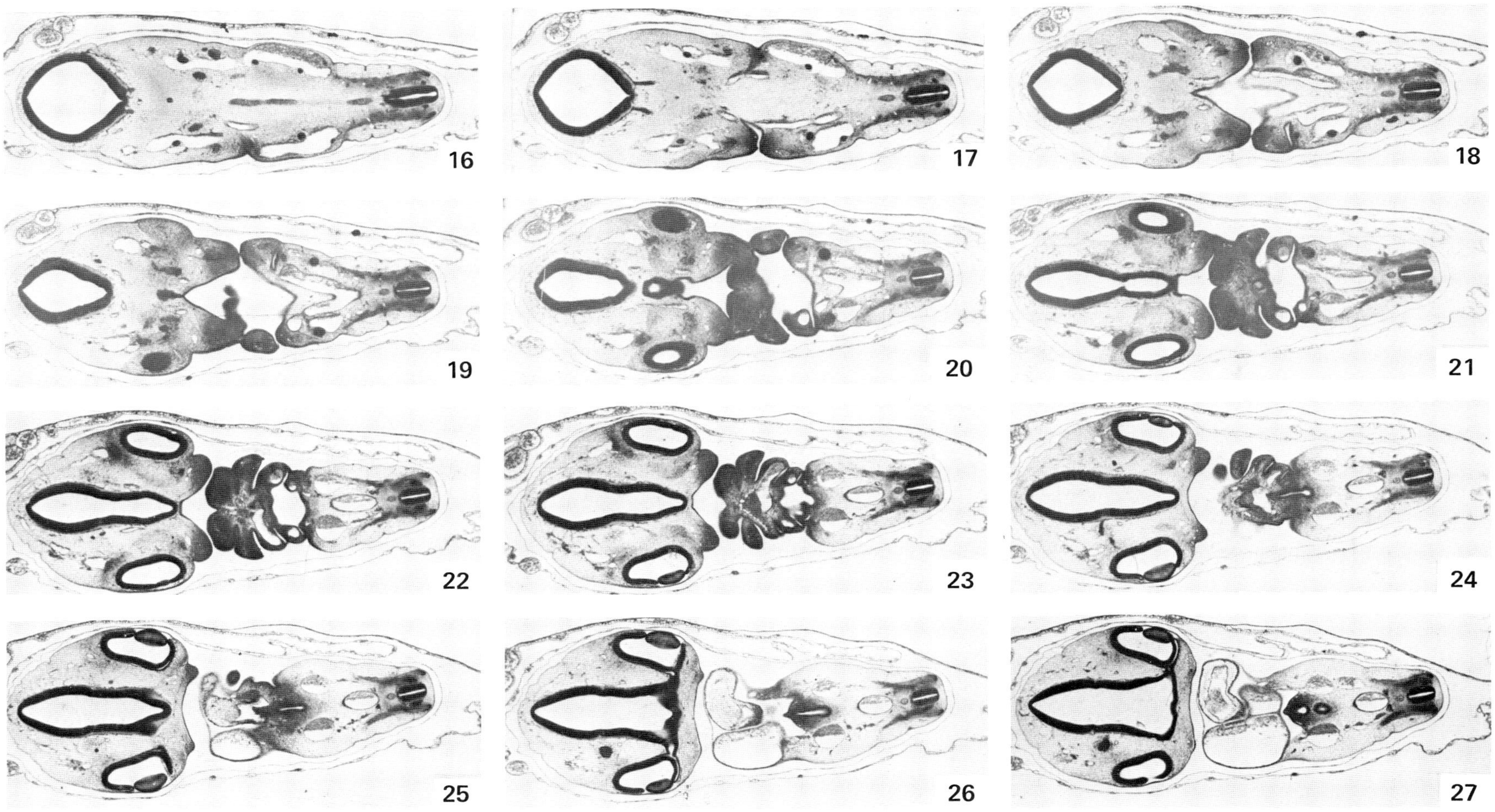

PHOTOGRAPHS OF CHICK EMBRYO AT 86 HOURS, SERIAL X-SECTIONS

Courtesy Dr. Allan Scott, Colby College.

Plate 5

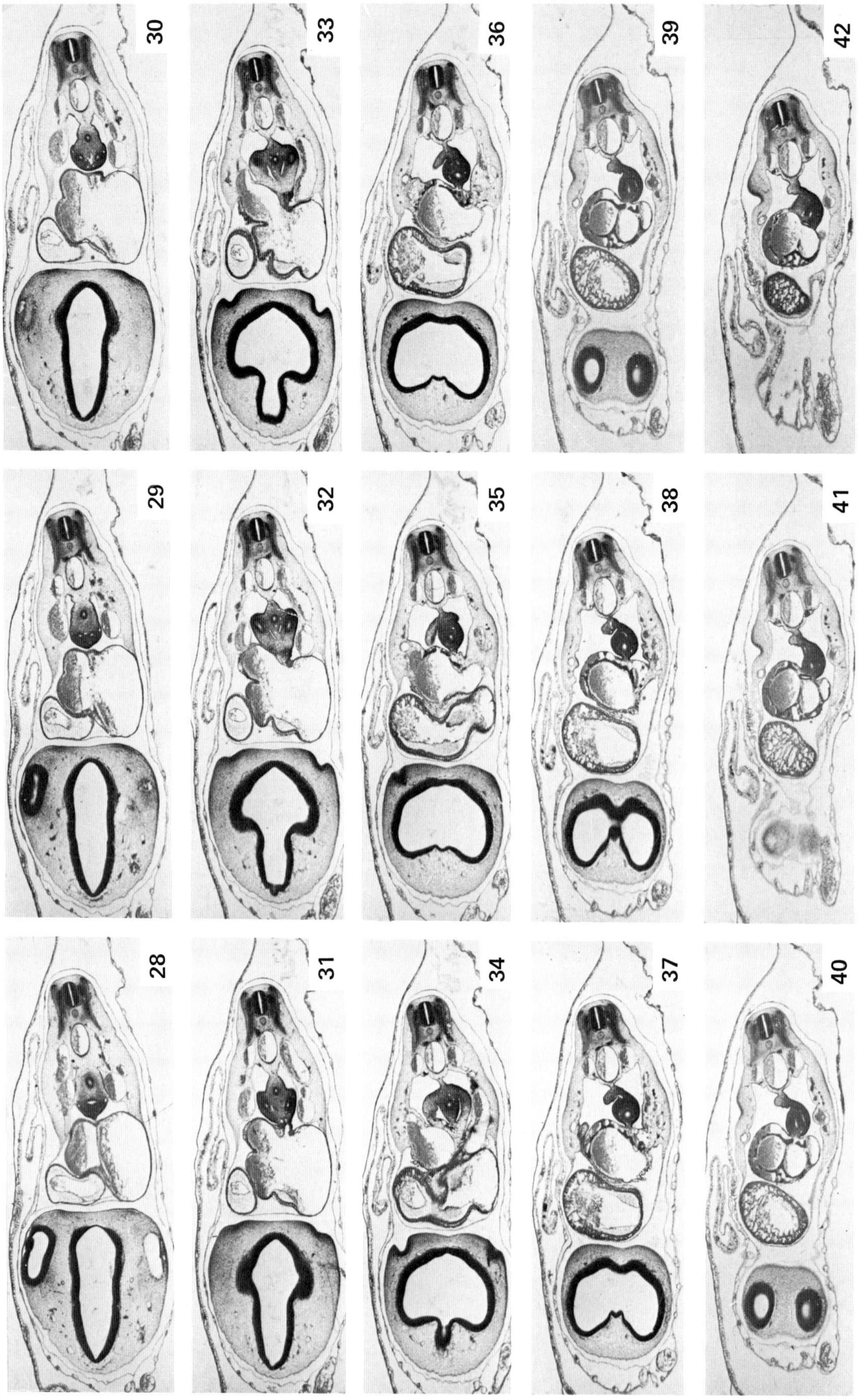
28
29
30
31
32
33
34
35
36
37
38
39
40
41
42

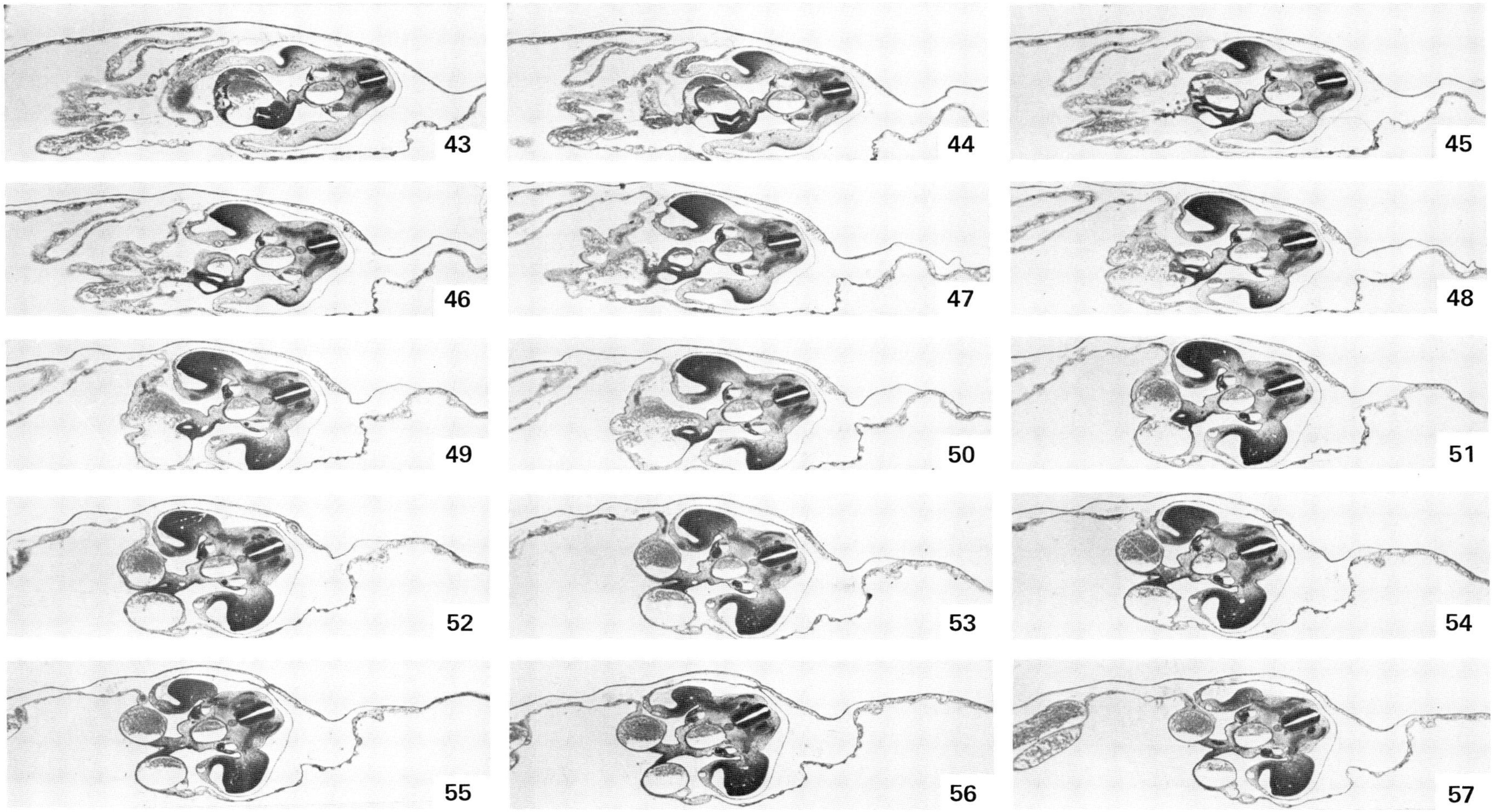

PHOTOGRAPHS OF CHICK EMBRYO AT 86 HOURS, SERIAL X-SECTIONS

Courtesy Dr. Allan Scott, Colby College.

Plate 5

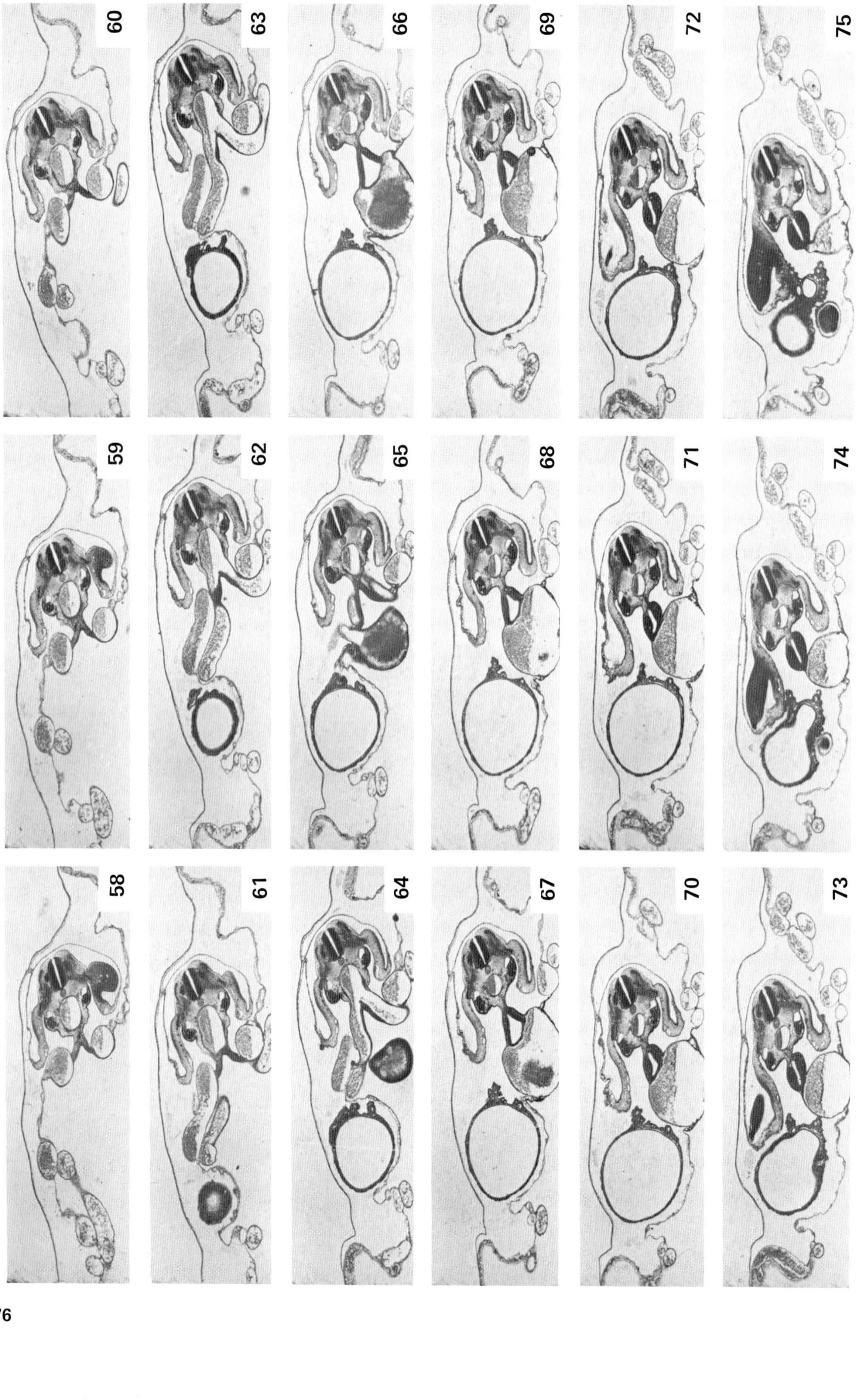
58
59
60
61
62
63
64
65
66
67
68
69
70
71
72
73
74
75

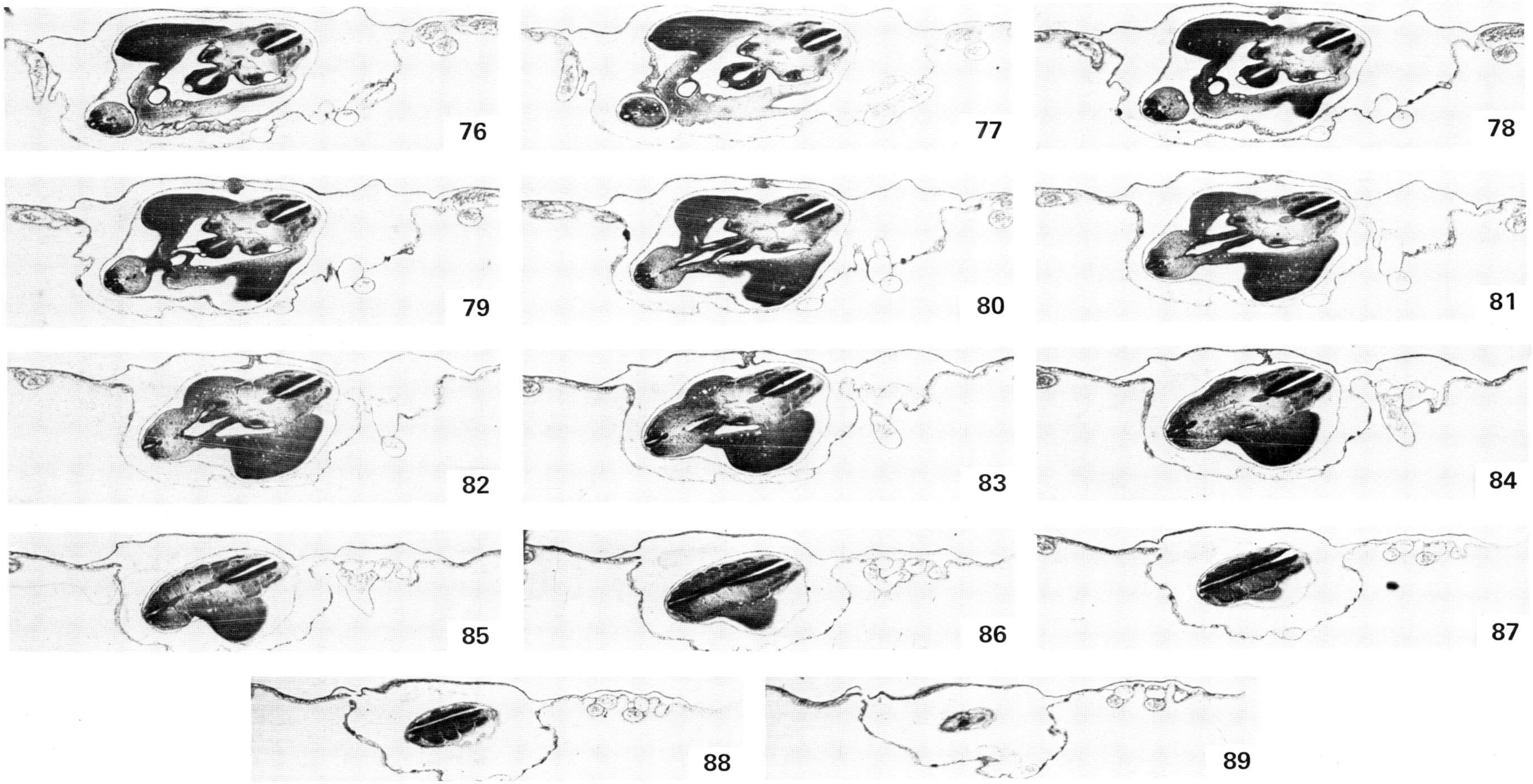

PHOTOGRAPHS OF CHICK EMBRYO AT 86 HOURS, SERIAL X-SECTIONS

Courtesy Dr. Allan Scott, Colby College.

Plate 5

PIG EMBRYO — 6 MM

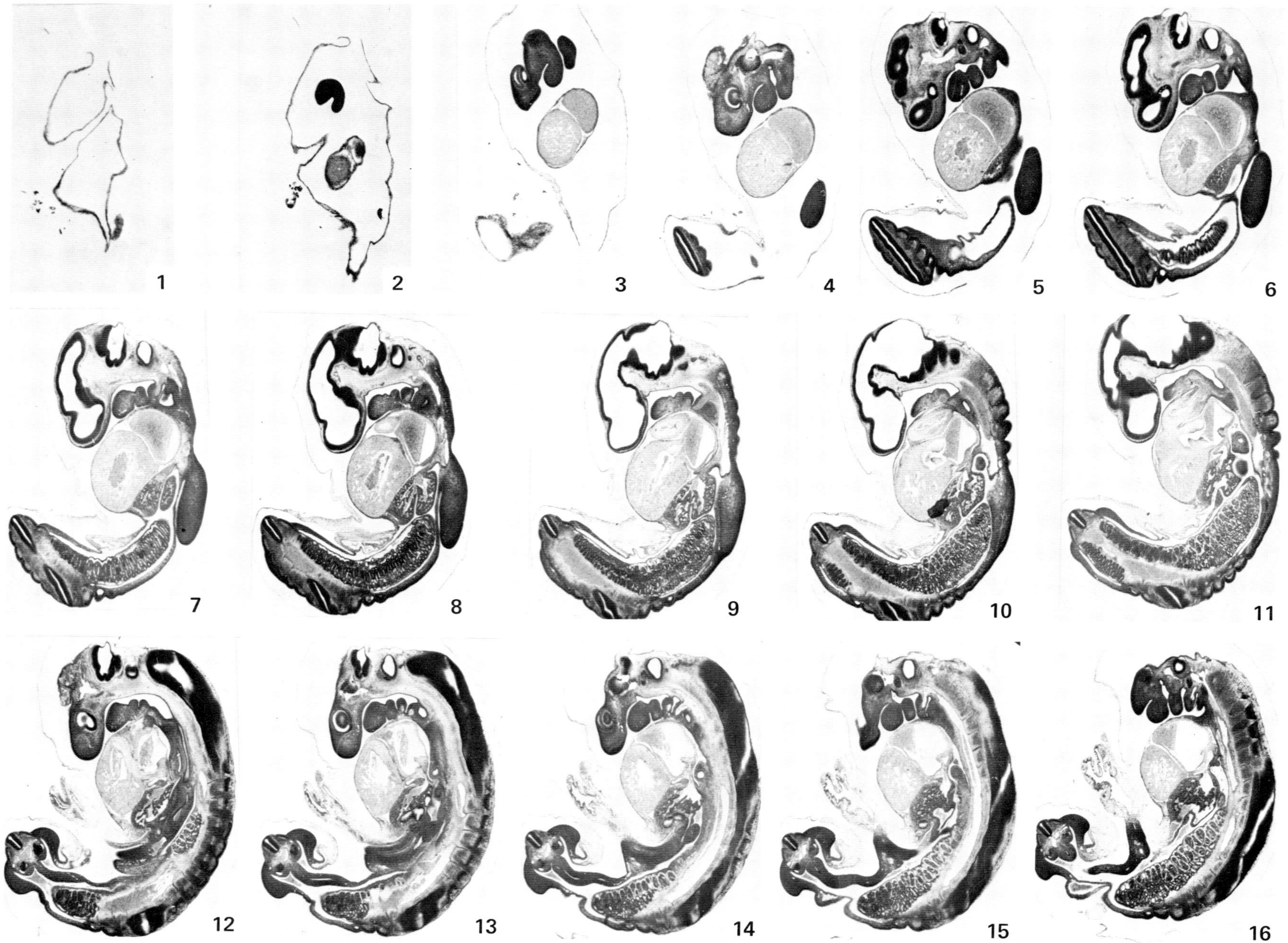

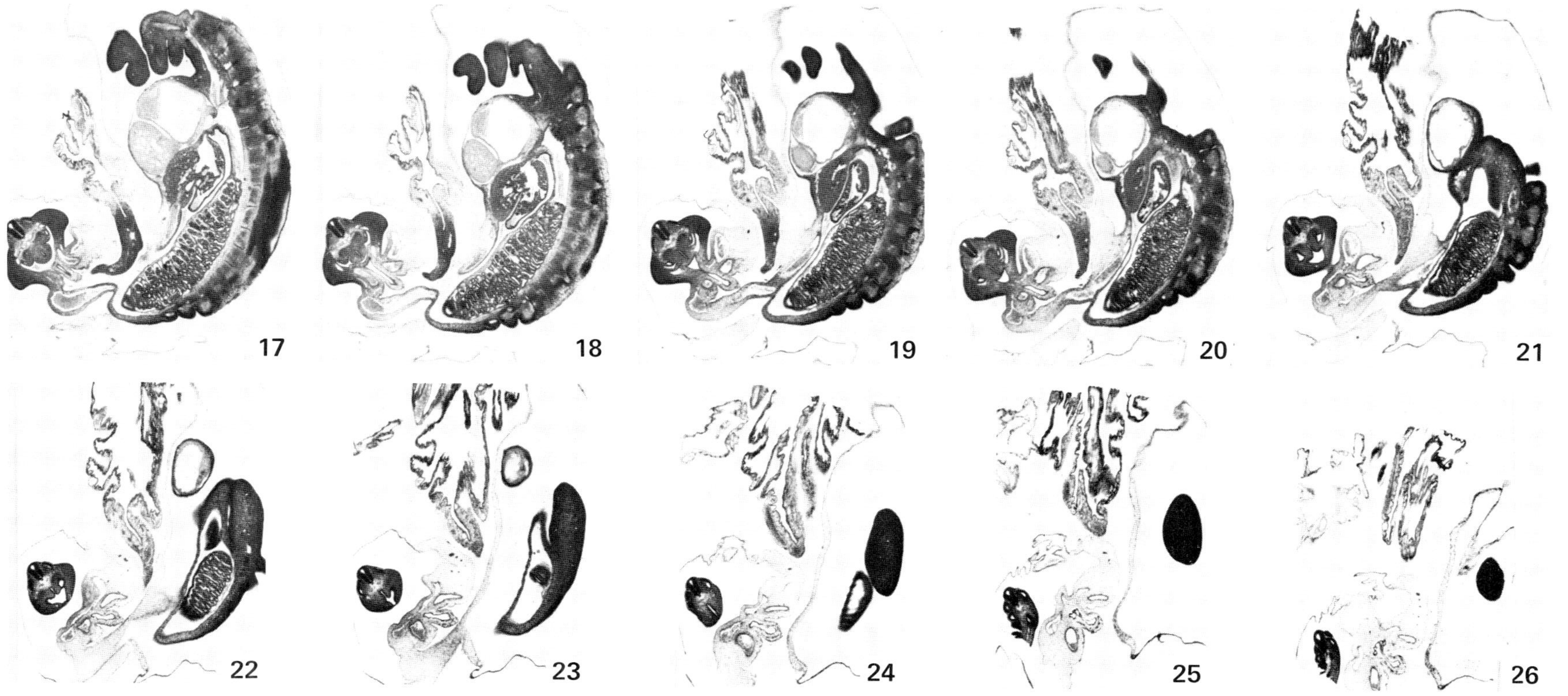

PHOTOGRAPHS OF PIG EMBRYO AT 6-mm, SERIAL SAGITTAL SECTIONS

Courtesy Dr. Allan Scott, Colby College.

Plate 6

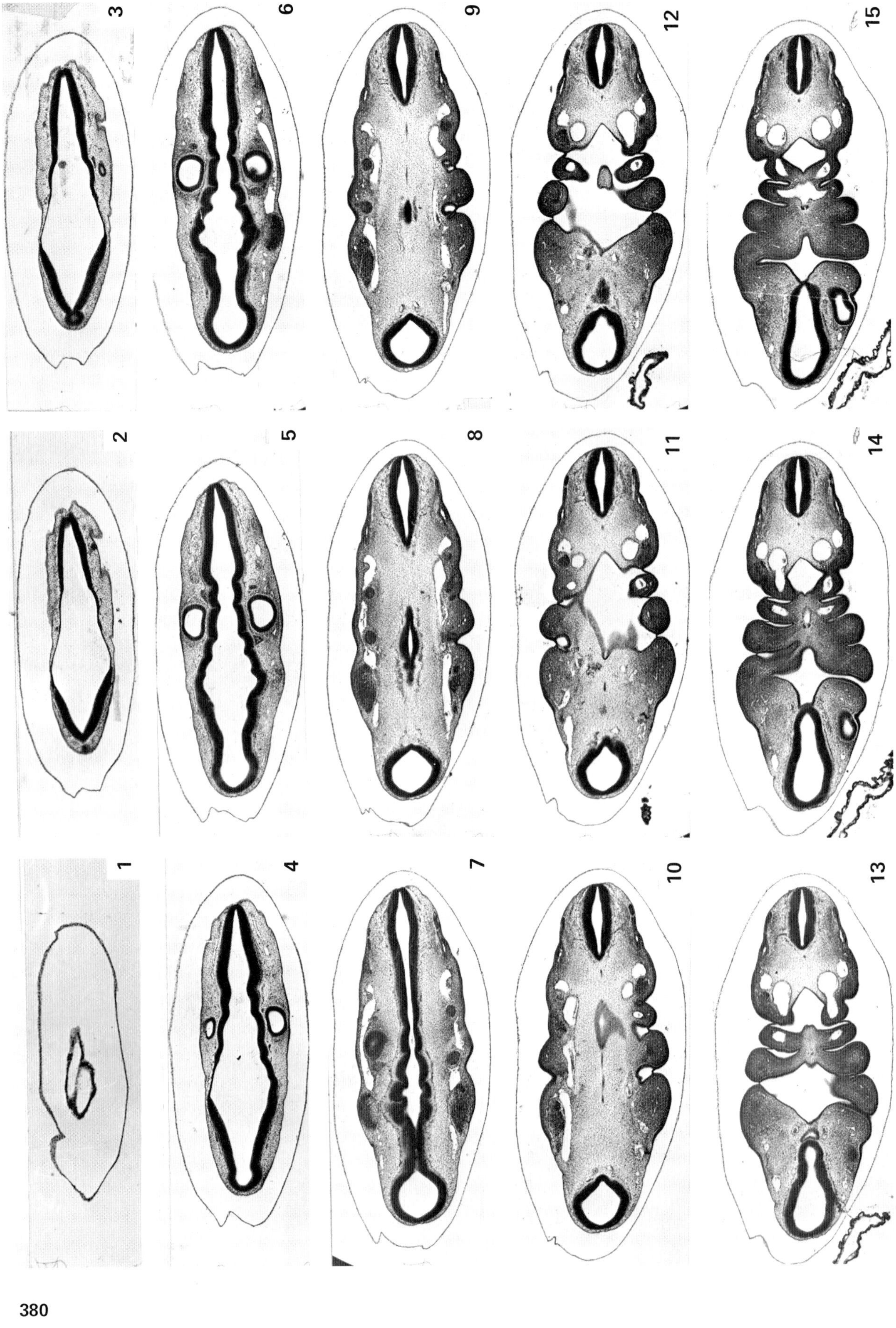
1
2
3
4
5
6
7
8
9
10
11
12
13
14
15

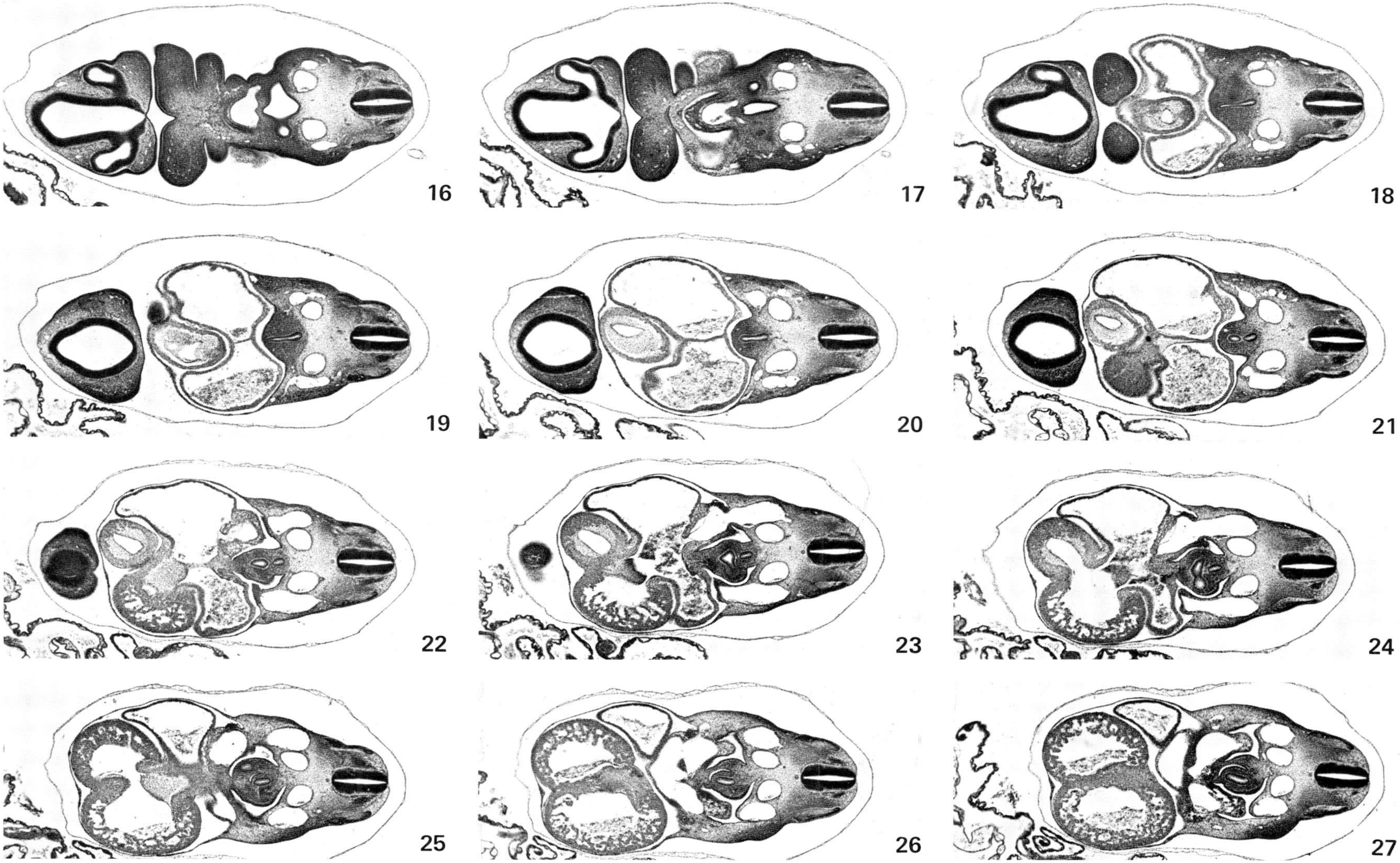

PHOTOGRAPHS OF PIG EMBRYO
AT 6-mm, SERIAL X-SECTIONS
Courtesy Dr. Allan Scott, Colby College.

Plate 7

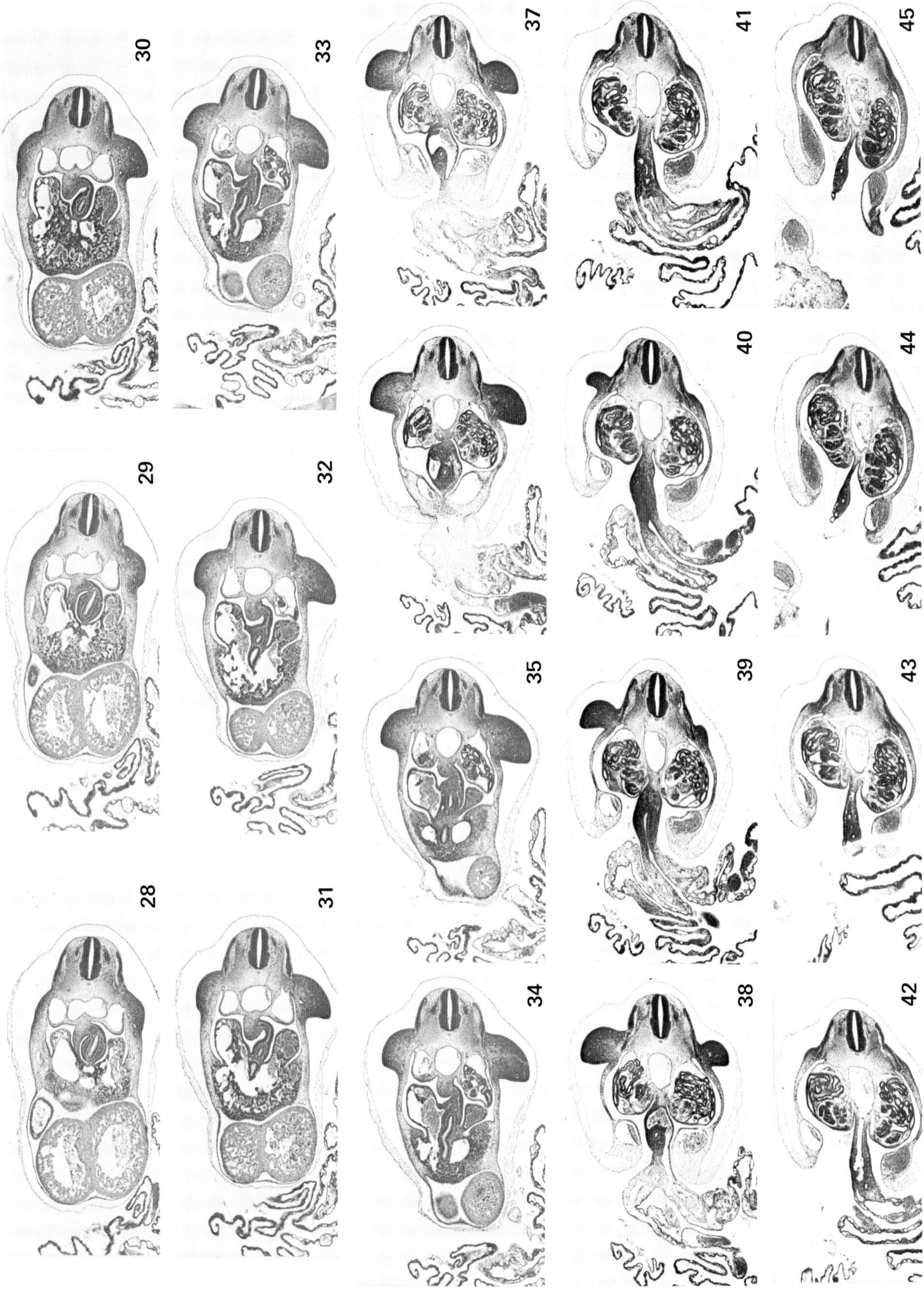
28
29
30
31
32
33
34
35
37
38
39
40
41
42
43
44
45

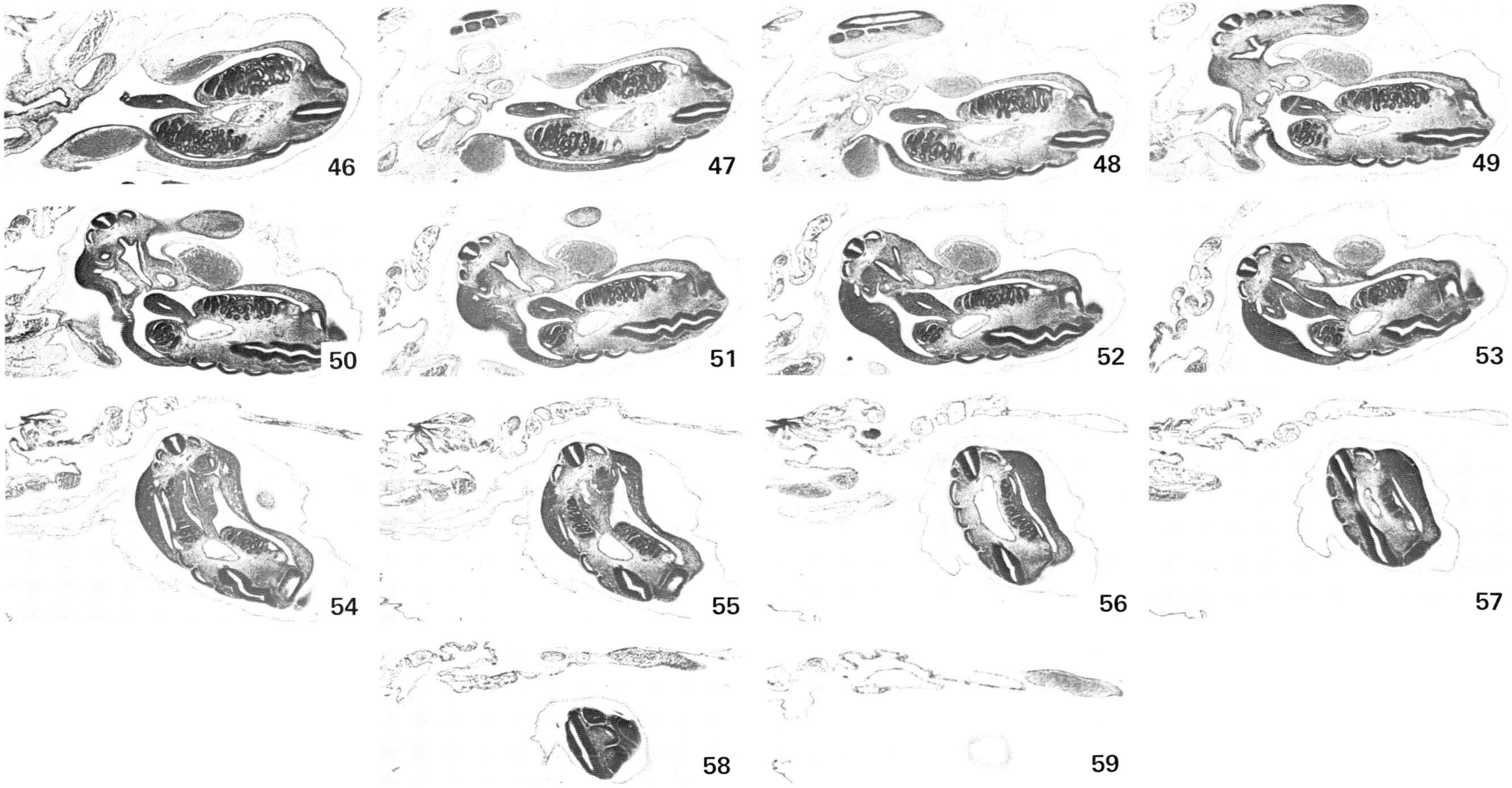

PHOTOGRAPHS OF PIG EMBRYO
AT 6-mm, SERIAL X-SECTIONS

Courtesy Dr. Allan Scott, Colby College.

Plate 7

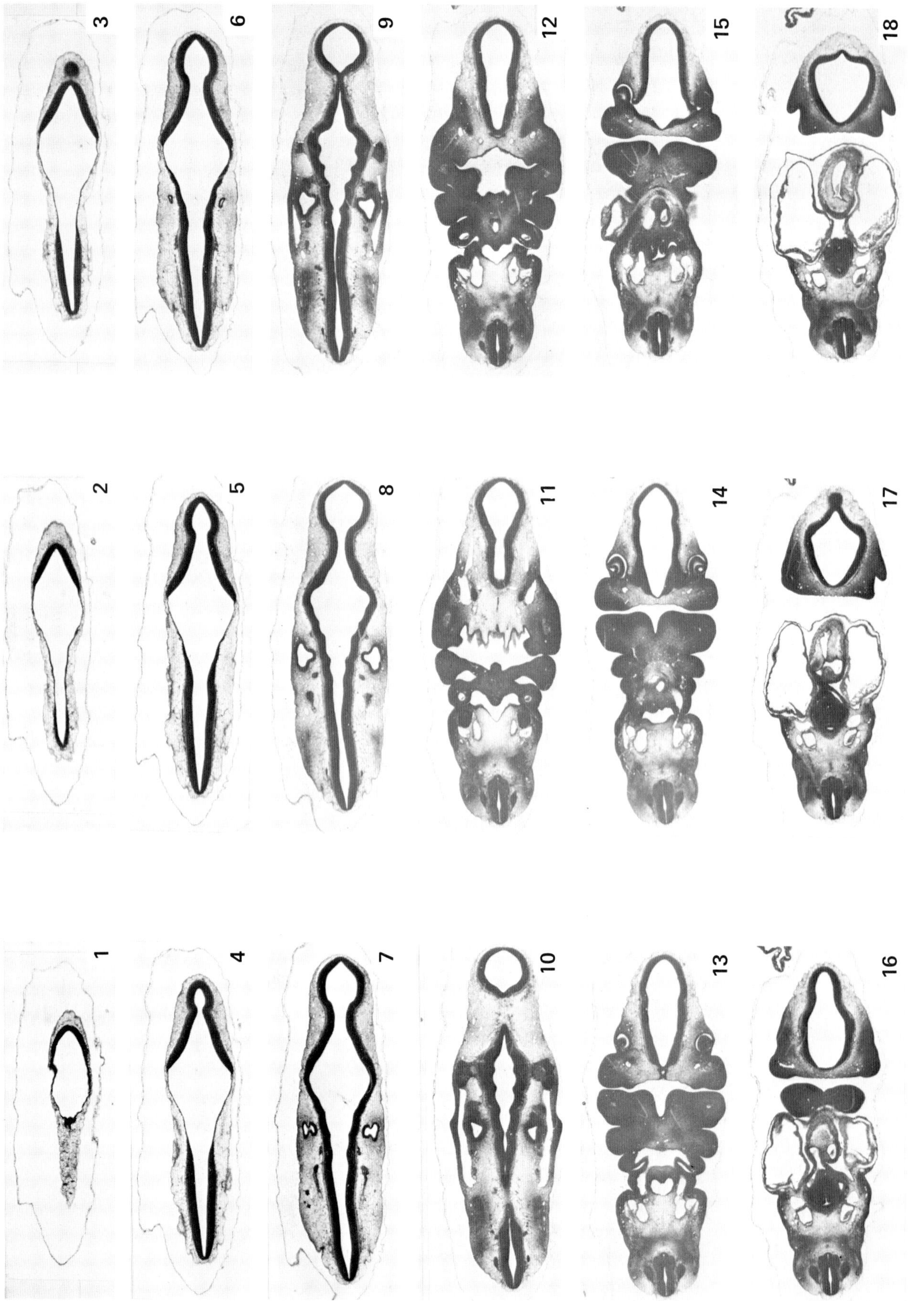
1
2
3
4
5
6
7
8
9
10
11
12
13
14
15
16
17
18

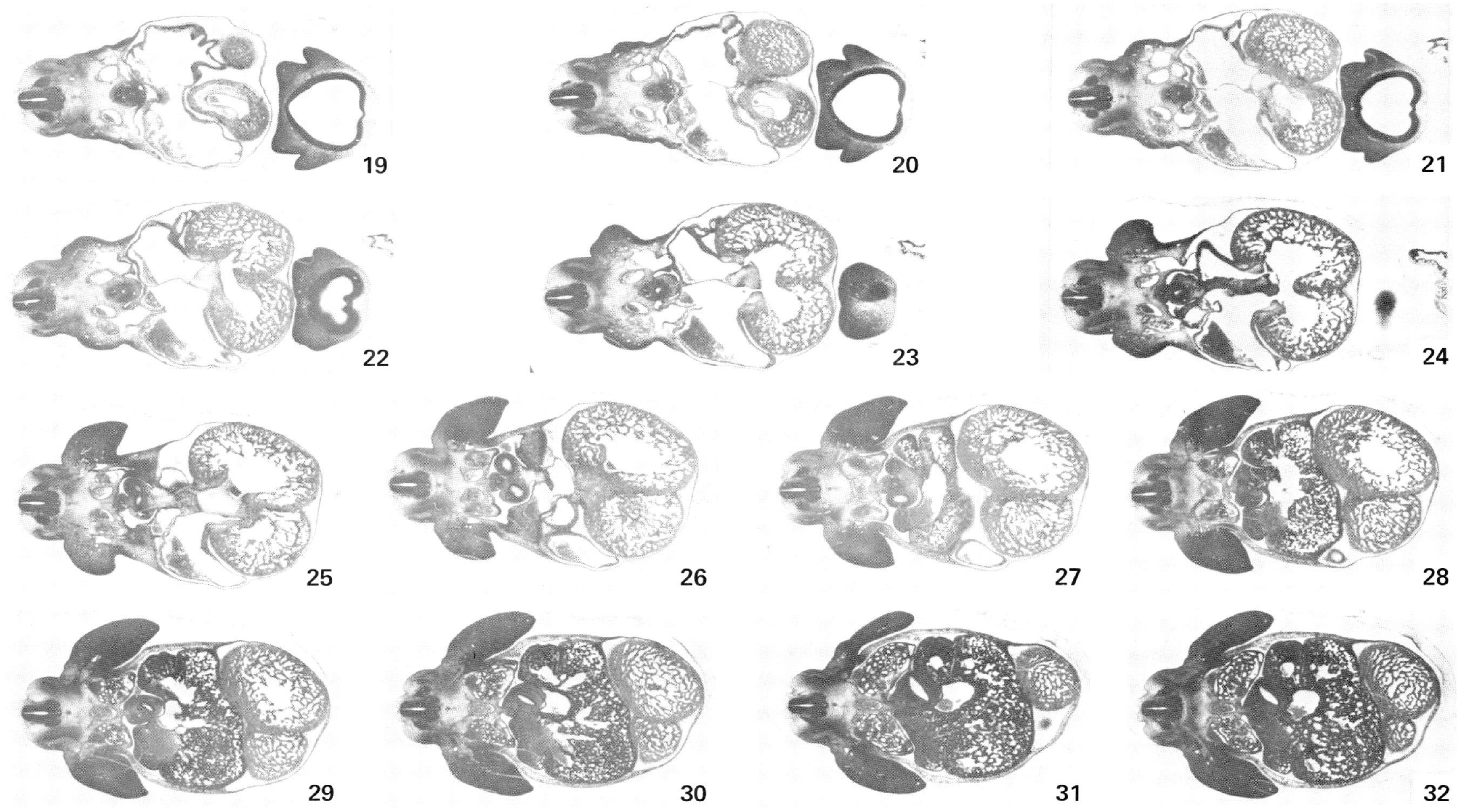

PHOTOGRAPHS OF PIG EMBRYO AT 10-mm, SERIAL TRANSVERSE SECTIONS

Photographs by Mark Lederman, 1966, Colby College.

Plate 8

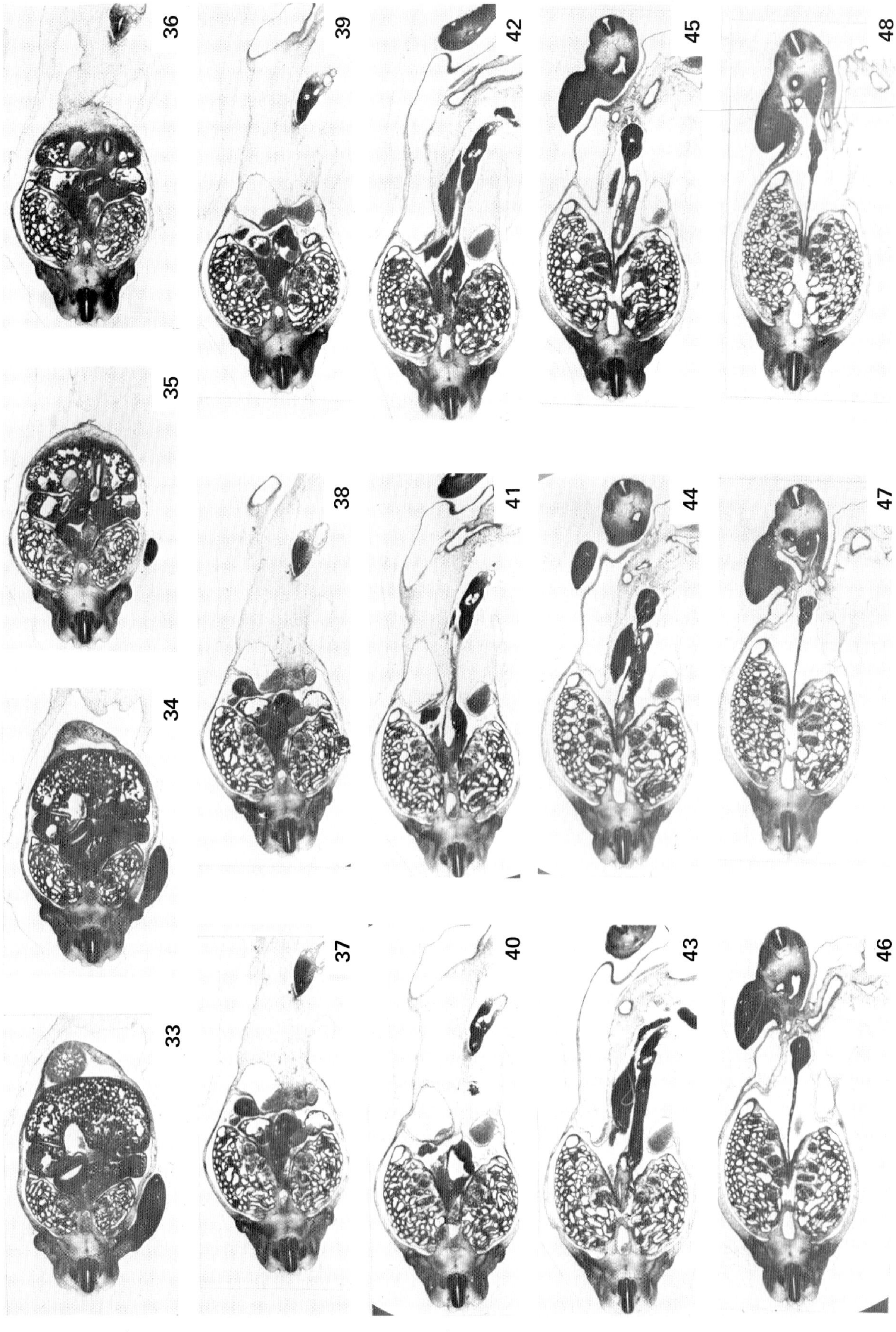
33
34
35
36
37
38
39
40
41
42
43
44
45
46
47
48

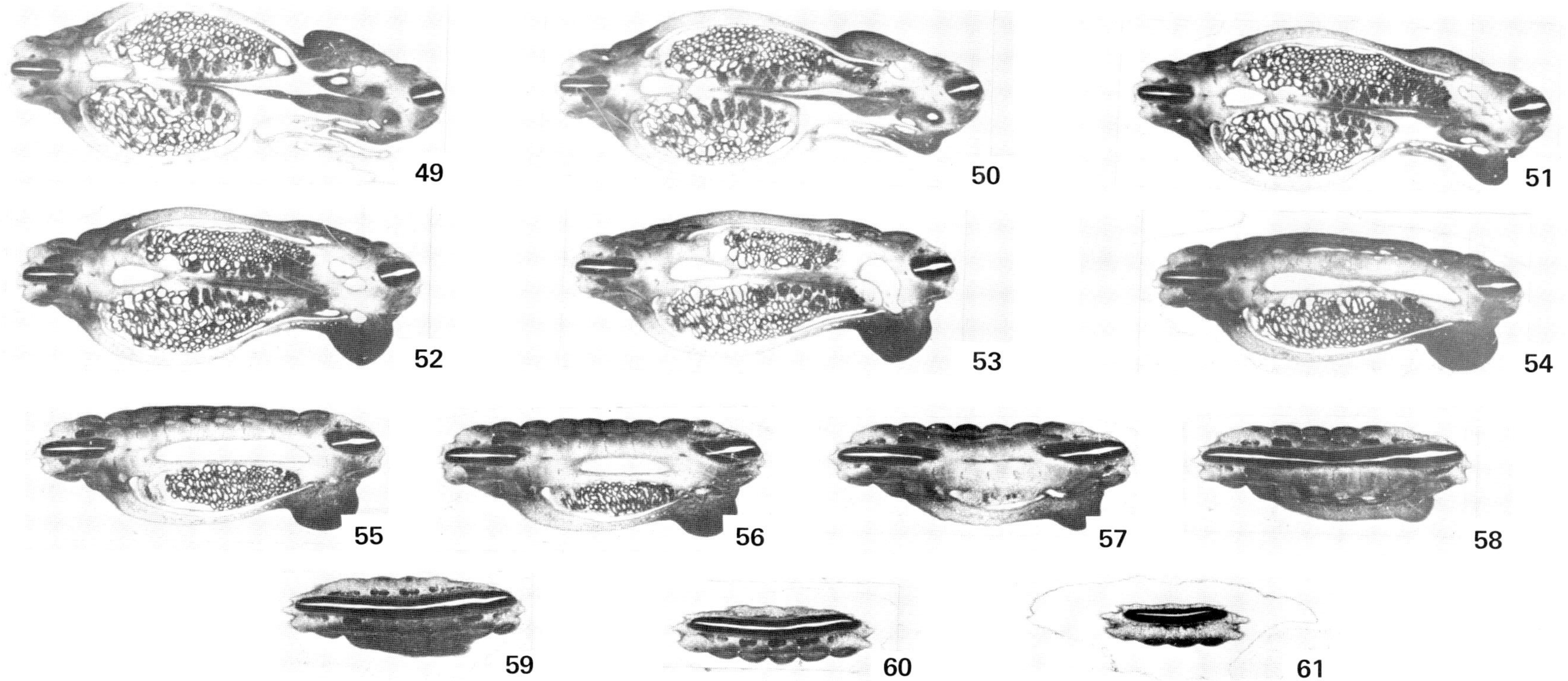

PHOTOGRAPHS OF PIG EMBRYO
AT 10-mm, SERIAL TRANSVERSE
SECTIONS

Photographs by Mark Lederman, 1966, Colby College.

Plate 8

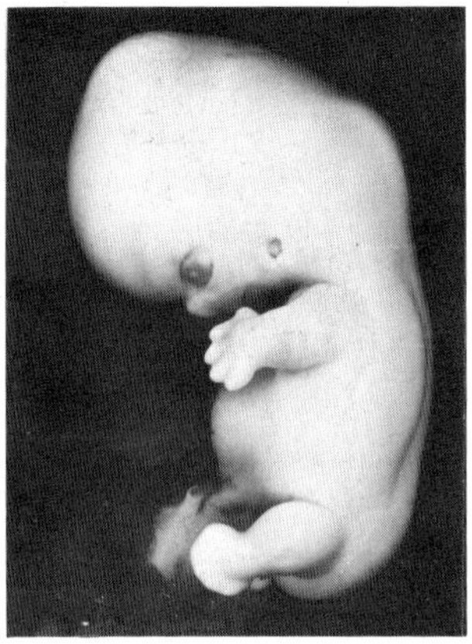

FETAL AGE: 7 WEEKS, $^{11}/_{16}$ INCH CROWN RUMP LENGTH, 0.035 OUNCE.

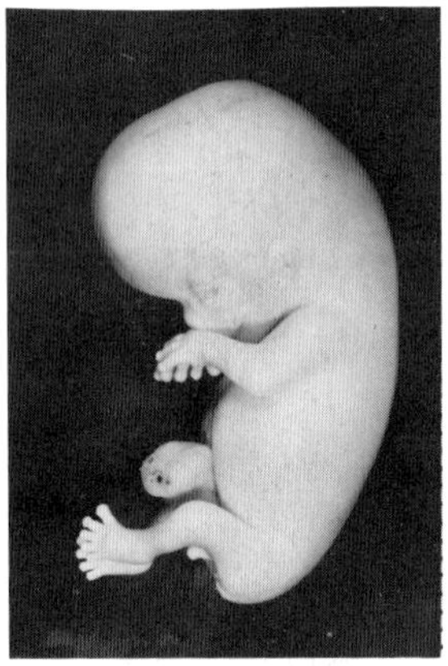

FETAL AGE: 8 WEEKS, $1^{1}/_{8}$ INCH C. R. LENGTH, 0.05 OUNCE.

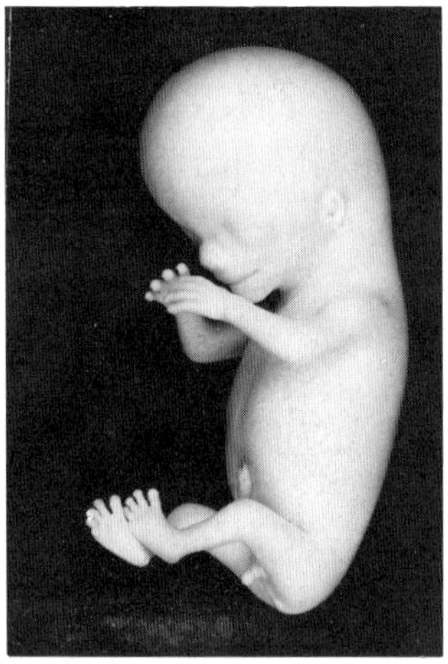

FETAL AGE: 9 WEEKS, 1½ INCH C. R. LENGTH, 0.1 OUNCE.

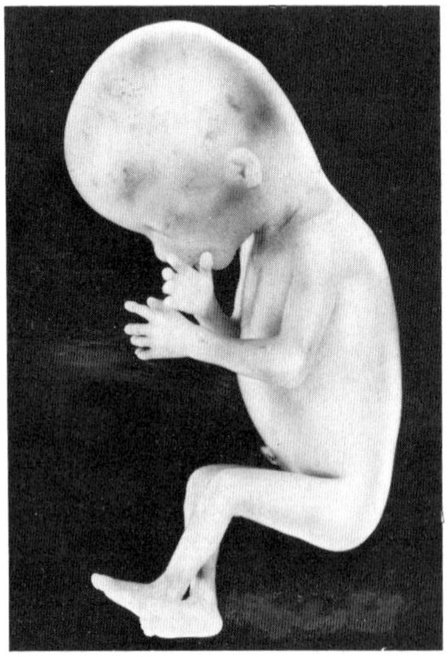

FETAL AGE: 12 WEEKS, 3.0 INCHES C. R. LENGTH, 0.7 OUNCE.

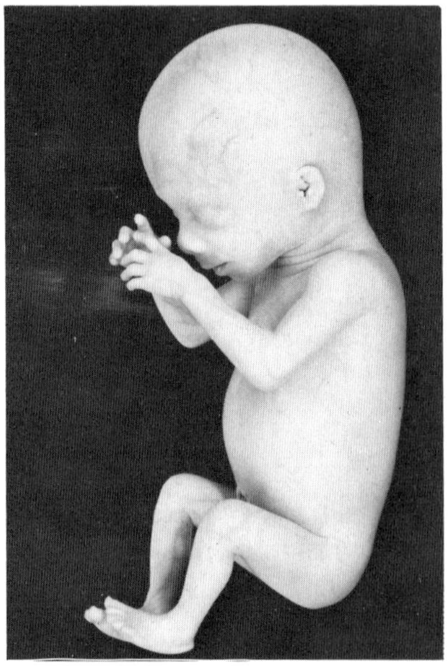

FETAL AGE: 15 WEEKS, 4¾ INCHES C. R. LENGTH, 1.5 OUNCES.

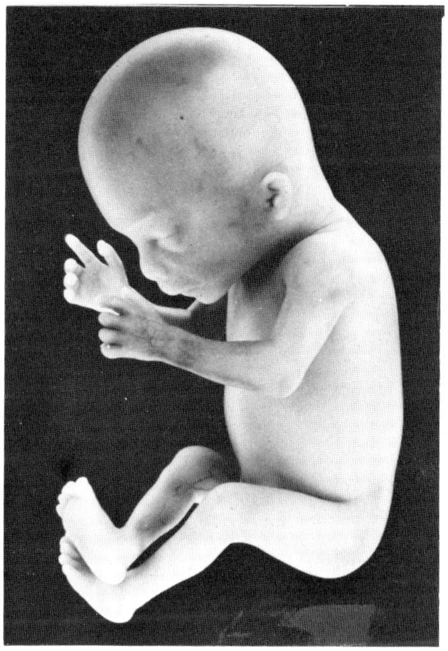

FETAL AGE: 16 WEEKS, 5¼ INCHES
C. R. LENGTH, 3.0 OUNCES.

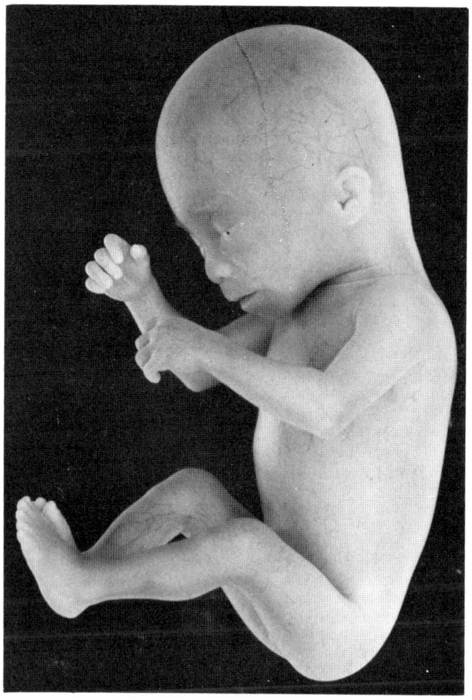

FETAL AGE: 18 WEEKS, 6¼ INCHES
C. R. LENGTH, 4.0 OUNCES.

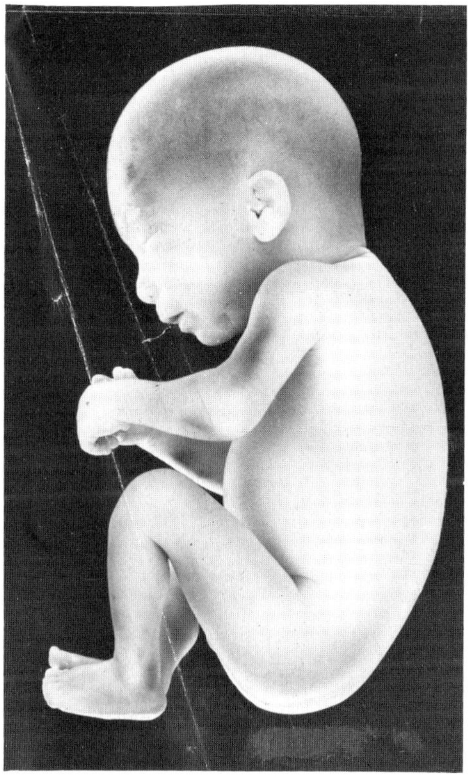

FETAL AGE: 21 WEEKS, 7¾ INCHES
C. R. LENGTH, 8.0 OUNCES.

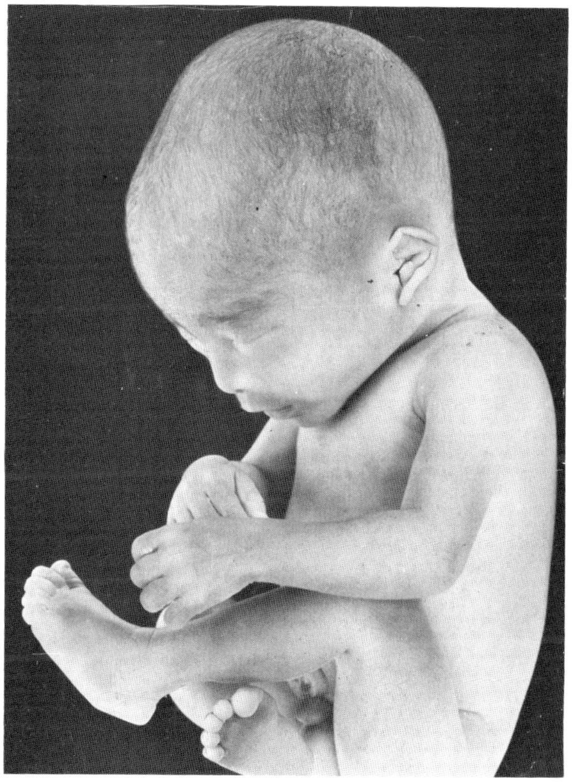

FETAL AGE: 23 WEEKS, 8¼ INCHES
C. R. LENGTH, 1 POUND, 2 OUNCES.

SERIES OF HUMAN FETUSES: These are Japanese fetuses photographed in the same aspect and position but at different ages. All are perfectly normal specimens. These photographs are slightly misleading, however, because they are not enlarged so as to represent accurately the relative growth from 7 to 23 weeks of gestational age. Courtesy Dr. H. Nishimura.

PLATE 9

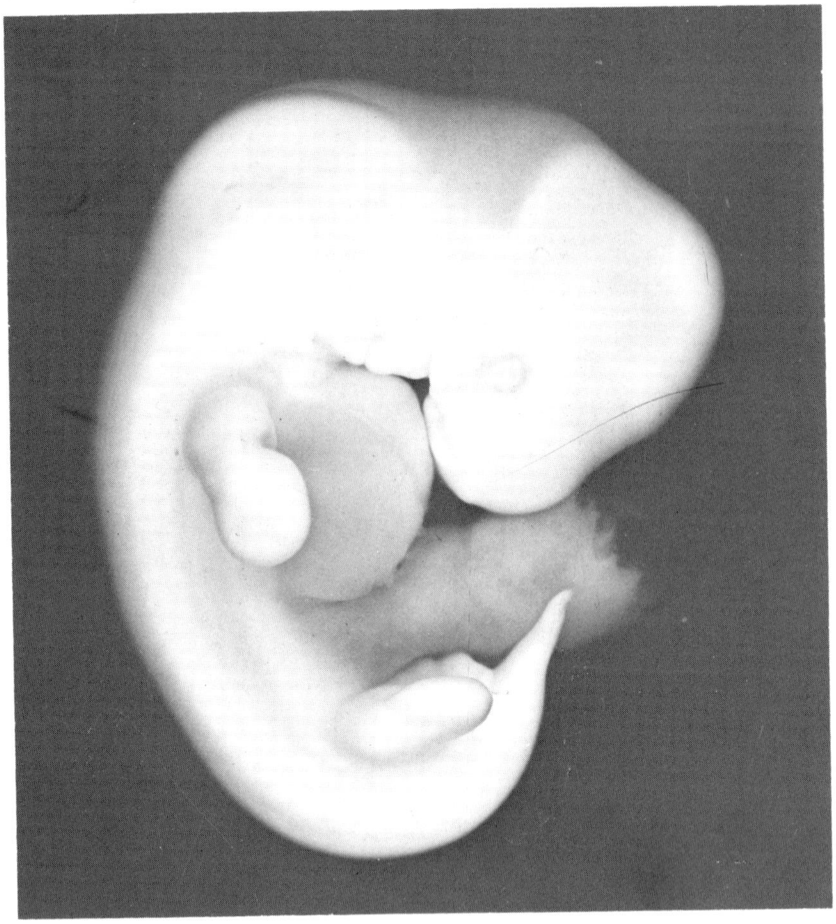

HUMAN EMBRYO AT 37 DAYS

CROWN RUMP LENGTH 10.2MM

Courtesy Dr. H. Nishimura.

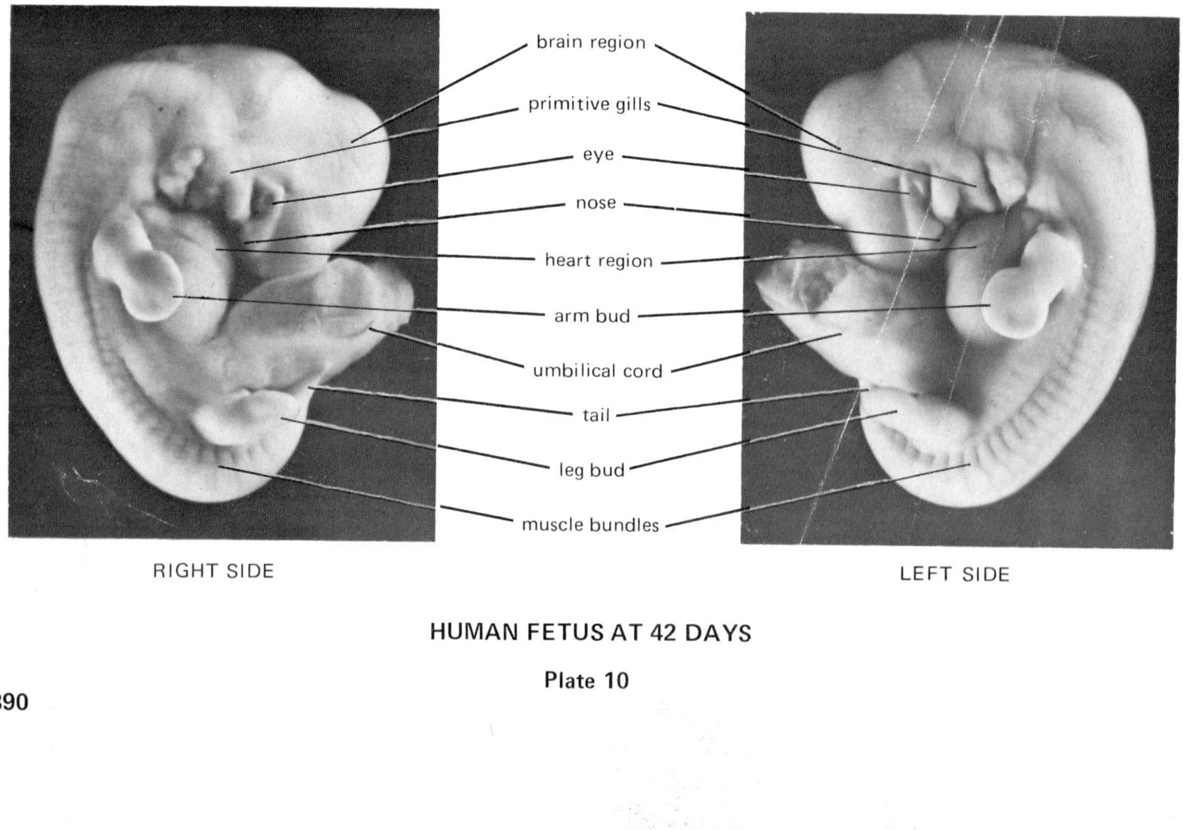

RIGHT SIDE

LEFT SIDE

HUMAN FETUS AT 42 DAYS

Plate 10